Life Histories of North American Flycatchers, Larks, Swallows, and Their Allies

PLATE 1

Tecpan, Guatemala, June 16, 1933.   Jalapa, Mexico.

Nest indicated by arrow.

A. F. Skutch.   Courtesy American Museum of Natural History.
          F. M. Chapman.

NESTING SITE AND NEST OF ROSE-THROATED BECARD.

Life Histories of North American Flycatchers, Larks, Swallows, and Their Allies

by Arthur Cleveland Bent

Dover Publications, Inc., New York

75744

Published in Canada by General Publishing Company, Ltd., 30 Lesmill Road, Don Mills, Toronto, Ontario.

Published in the United Kingdom by Constable and Company, Ltd., 10 Orange Street, London WC 2.

This Dover edition, first published in 1963, is an unabridged and unaltered republication of the work first published in 1942 by the United States Government Printing Office, as Smithsonian Institution United States National Museum *Bulletin 179*.

International Standard Book Number: 0-486-21090-1

Manufactured in the United States of America
Dover Publications, Inc.
180 Varick Street
New York, N. Y. 10014

ADVERTISEMENT

The scientific publications of the National Museum include two series, known, respectively, as *Proceedings* and *Bulletin.*

The *Proceedings* series, begun in 1878, is intended primarily as a medium for the publication of original papers, based on the collections of the National Museum, that set forth newly acquired facts in biology, anthropology, and geology, with descriptions of new forms and revisions of limited groups. Copies of each paper, in pamphlet form, are distributed as published to libraries and scientific organizations and to specialists and others interested in the different subjects. The dates at which these separate papers are published are recorded in the table of contents of each of the volumes.

The series of *Bulletins*, the first of which was issued in 1875, contains separate publications comprising monographs of large zoological groups and other general systematic treatises (occasionally in several volumes), faunal works, reports of expeditions, catalogs of type specimens, special collections, and other material of similar nature. The majority of the volumes are octavo in size, but a quarto size has been adopted in a few instances in which large plates were regarded as indispensable. In the *Bulletin* series appear volumes under the heading *Contributions from the United States National Herbarium,* in octavo form, published by the National Museum since 1902, which contain papers relating to the botanical collections of the Museum.

The present work forms No. 179 of the *Bulletin* series.

ALEXANDER WETMORE,
Assistant Secretary, Smithsonian Institution.

WASHINGTON, D. C., *January 20, 1942.*

CONTENTS

PLATE 2

Grand Traverse, Mich., May 27, 1928.   W. E. Hastings.

Duval County, Fla.   S. A. Grimes.

NESTS OF EASTERN KINGBIRD.

INTRODUCTION

This is the fourteenth in a series of bulletins of the United States National Museum on the life histories of North American birds. Previous numbers have been issued as follows:

107. Life Histories of North American Diving Birds, August 1, 1919.
113. Life Histories of North American Gulls and Terns, August 27, 1921.
121. Life Histories of North American Petrels and Pelicans and their Allies, October 19, 1922.
126. Life Histories of North American Wild Fowl (part), May 25, 1923.
130. Life Histories of North American Wild Fowl (part), June 27, 1925.
135. Life Histories of North American Marsh Birds, March 11, 1927.
142. Life Histories of North American Shore Birds (pt. 1), December 31, 1927.
146. Life Histories of North American Shore Birds (pt. 2), March 24, 1929.
162. Life Histories of North American Gallinaceous Birds, May 25, 1932.
167. Life Histories of North American Birds of Prey (pt. 1), May 3, 1937.
170. Life Histories of North American Birds of Prey (pt. 2), August 8, 1938.
174. Life Histories of North American Woodpeckers, May 23, 1939.
176. Life Histories of North American Cuckoos, Goatsuckers, Hummingbirds, and Their Allies, July 20, 1940.

The same general plan has been followed, as explained in previous bulletins, and the same sources of information have been utilized. The nomenclature of the 1931 check-list of the American Ornithologists' Union has been followed.

An attempt has been made to give as full a life history as possible of the best-known subspecies of each species and to avoid duplication by writing briefly of the others and giving only the characters of the subspecies, its range, and any habits peculiar to it. In many cases certain habits, probably common to the species as a whole, have been recorded for only one subspecies; such habits are mentioned under the subspecies on which the observations were made. The distribution gives the range of the species as a whole, with only rough outlines of the ranges of the subspecies, which in many cases cannot be accurately defined.

The egg dates are the condensed results of a mass of records taken from the data in a large number of the best egg collections in the country, as well as from contributed field notes and from a few published sources. They indicate the dates on which eggs have been actually found in various parts of the country, showing the earliest and latest dates and the limits between which half the dates fall, indicating the height of the season.

The plumages are described in only enough detail to enable the reader to trace the sequence of molts and plumages from birth to maturity and to recognize the birds in the different stages and at the different seasons.

No attempt has been made to describe fully the adult plumages; this has been well done already in the many manuals and State books. Partial or complete albinism is liable to occur in almost any species; for this reason, and because it is practically impossible to locate all such cases, it has seemed best not to attempt to treat this subject at all. The names of colors, when in quotation marks, are taken from Ridgway's Color Standards and Nomenclature (1912). In the measurements of eggs, the four extremes are printed in bold-face type.

Many who have contributed material for previous volumes have continued to cooperate. Receipt of material from over 475 contributors has been acknowledged previously. In addition to these, our thanks are due to the following new contributors: W. L. Bailey, H. Brackbill, F. G. Brandenburg, I. McT. Cowan, W. V. Crich, J. R. Cruttenden, L. I. Davis, J. D. Daynes, D. Grice, E. N. Harrison, H. L. Heaton, T. A. Imhof, C. Kinzel, Mrs. F. C. Laskey, Miss Katherine Merry, E. F. Porter, P. Steib, W. E. Unglish, and University of Colorado Libraries. If any contributor fails to find his or her name in this or some previous bulletin, the author would be glad to be advised. As the demand for these volumes is much greater than the supply, the names of those who have not contributed to the work during the previous 10 years will be dropped from the author's mailing list.

Dr. Winsor M. Tyler rendered valuable assistance by reading and indexing, for these groups, a large part of the literature on North American birds, and contributed four complete life histories. Dr. Alfred O. Gross and Alexander Sprunt, Jr., contributed two each; and Bayard H. Christy, Edward von S. Dingle, Rev. F. C. R. Jourdain, Dr. Gayle Pickwell, and Robert S. Woods contributed one each.

Egg measurements were furnished especially for this volume by American Museum of Natural History (Dean Amadon), Griffing Bancroft, Colorado Museum of Natural History (F. G. Brandenburg), California Academy of Sciences (James Moffitt), F. R. Decker, C. E. Doe, J. B. Dixon, Field Museum of Natural History (R. M. Barnes collection), J. H. Gillin, W. C. Hanna, H. L. Harllee, T. E. McMullen, Museum of Comparative Zoology (J. C. Greenway), Museum of Vertebrate Zoology (Margaret W. Wythe), J. S. Rowley, G. H. Stuart, 3d, and the United States National Museum (J. H. Riley).

Our thanks are due also to F. Seymour Hersey for adding and figuring the egg measurements, and to W. George F. Harris for sorting over and arranging the egg dates. Through the courtesy of the Biological Survey, the services of Frederick C. Lincoln were again obtained to compile the distribution and migration paragraphs. The author claims no credit and assumes no responsibility for this part of the work.

The manuscript for this bulletin was completed in July 1940. Contributions received since then will be acknowledged later. Only information of great importance could be added. The reader is reminded again that this is a cooperative work; if he fails to find in these volumes anything that he knows about the birds, he can blame himself for not having sent the information to—

THE AUTHOR.

PLATE 3

Avery Island, La.

A. M. Bailey.
Courtesy Colorado Museum of Natural History.

EASTERN KINGBIRD.

LIFE HISTORIES OF NORTH AMERICAN FLYCATCH-
ERS, LARKS, SWALLOWS, AND THEIR ALLIES

ORDER PASSERIFORMES (FAMILIES COTINGIDAE, TYRANNIDAE, ALAUDIDAE, AND HIRUNDINIDAE)

By Arthur Cleveland Bent

Taunton, Mass.

Order PASSERIFORMES

Family COTINGIDAE: Cotingas

PLATYPSARIS AGLAIAE ALBIVENTRIS (Lawrence)

XANTUS'S BECARD

HABITS

Xantus's becard belongs to the western Mexican race of the rose-throated becards, which are widely distributed in Mexico and Central America. The species has been split up into some seven or eight subspecies, and there are several closely related species in Jamaica and South America. Our bird (*P. a. albiventris*) is much paler than the type race, both above and below, the under parts being largely pure white or nearly white. The type race (*P. a. aglaiae*), one of the dark races, extends its range into the valley of the lower Rio Grande, in Tamaulipas, Mexico, and may some day be taken on our side of the river in southern Texas. (Pl. 1.)

Xantus's becard has been taken only once within the limits of the United States and must be regarded as a very rare straggler in the mountains of southern Arizona, where W. W. Price (1888) established our only record for the species, of which he writes:

On June 20, 1888, I secured an adult male, in breeding plumage, of this species in the pine forests of the Huachuca Mountains, at an elevation of about 7500 feet, and seven miles north of the Mexican boundary. I am certain there were a pair of these birds, as I heard their very peculiar notes in different places at the same time, but the locality being so extremely rough and broken I only secured the one above recorded. Several times while collecting at high altitudes I have heard bird notes that I thought were these, but they were al-

1

ways on almost inaccessible mountain sides. * * * From observing the actions of the bird I killed, I am sure its mate was in the vicinity, and probably nesting, although I have since carefully searched the place without success. This species will doubtless be found breeding in Arizona.

My attention has recently been called to several sight records of what was apparently this subspecies, made by L. Irby Davis near Harlingen, Tex., in the lower Rio Grande Valley. On numerous occasions the bird was examined carefully through powerful binoculars, often at short range, by Mr. Davis and several other observers, so that there seems to be little doubt about the identification of the subspecies, as it was clearly seen to have a *white* breast. The typical rose-throated becard (*Platypsaris aglaiae aglaiae*), the form we should naturally expect to see there, has a *gray* breast. Mr. Davis has sent me a very full account of his various experiences with it, which is most convincing. He first saw, on October 18, 1937, an adult female and a young male of this species, which at first he was at loss to identify. The female soon disappeared and was never seen again. The young male was found again in March 1938 in the same locality, and in another locality, heavy hackberry woods about a quarter of a mile away, on April 27, 1938. Here he saw it, sometimes within 10 feet and at all angles, on May 1, 8, 15, and 28, June 6, November 19, and December 4, 1938.

In explanation of this strange occurrence, he writes to me: "My theory of the becard record is that an adult female, accompanied by a very young male, came into this area in the fall of 1937 from the west. This is a regular migration route for far western birds, as a number of mountain species regularly winter here—for example, the Sierra hermit thrush. I believe that the adult moved on to the south shortly, and the juvenile who was left here took up permanent residence. Since no females came back here in the spring, no breeding records have been made." This seems to be a plausible theory, as the eastern boundary of Nuevo León, where *albiventris* has been known to occur, is only 65 miles west of Harlingen; a wandering migrant might easily cover that short distance. I believe Mr. Davis has not seen the bird since.

William Beebe (1905) thus describes his meeting with this species in the lowlands of Colima, Mexico:

One day while walking quietly through a dense part of the jungle, where tall, thick-leaved trees shut out the light and hence caused an absence of thick undergrowth, I saw a bird fly from a perch, catch an insect in mid-air and dart back. I had not found any flycatchers heretofore in this thickly wooded section, and, though my heart sank when I saw its back and wings of the usual indefinite flycatcher-hues of light gray, and knew that exact identification without a gun would be next to impossible, I approached the bird. It again flew into the air and again returned to its favourite twig, this time facing me, when one glance removed all doubt as to its identity; for its breast was stained a rich pink, which burned out brightly amid the dark shadows. It was the Xantus Becard, the second member of the family Cotingidae we had met. From time to time it uttered a low, indefinable lisp, and soon flew away. Three other individuals were seen after that, all solitary, all flycatching, all in such deep woods as our Wood Pewee would love.

Dickey and van Rossem (1938) say of the haunts of the gray becard (*P. a. latirostris*) in El Salvador: "Becards are ordinarily rather quiet, sedentary birds, usually to be found in pairs in thin, second growth and about the edges of clearings and open places such as trails and roads. Their habit of sitting motionless for minutes at a time is one which may cause them to be easily overlooked. Though the normal habitat is gallery forest, they are by no means averse to brushy, cut-over land and were quite common in the taller mimosa growth about Davisadero. A few were found in the heavy, swamp forest at Puerto del Triunfo, where they were observed in the thin foliage between the ground and the thick forest crown."

Courtship.—Very little seems to be known about the habits of our race of the rose-throated becard, and so little has been published about this and other habits of the species that it seems desirable to include here some of the observations of Alexander F. Skutch on one of the Central American races, Sumichrast's becard (*Platypsaris aglaiae sumichrasti* Nelson), which probably does not differ materially in its habits from our subspecies. Much that follows is quoted from his unpublished manuscript on the birds of the Caribbean lowlands, which he very kindly lent me for this purpose.

While he was watching, in Guatemala, a pair of these birds building the nest referred to below, he noticed a display by the male that was probably a part of the courtship performance, of which he says: "Sometimes, as he approached the female, he spread and displayed his white epaulets, which appeared very fluffy and conspicuous, standing above his shoulders and contrasting with his dark gray back. These downy white feathers on the shoulders seemed intended for her alone; except when in her presence he wore them laid flat, and so completely covered over with the dark gray plumage of his back that one would never have suspected their presence."

Nesting.—Mr. Skutch's observations on the nesting of Sumichrast's becard follow: "About a mile from the house the road passed through a clearing in the woods, where the subterranean waters welled up diffusely through the surface and gave rise to an open, sedgy marsh, through the center of which flowed a little rill. Beside the rivulet, in the middle of the marsh, four alder trees grew in a clump. Hanging from the extremity of one of the finer twigs of an alder, 50 feet above ground and quite unapproachable, was a large, globular bird's nest, nearly a foot in diameter.   When I first noticed it, in February, it appeared old and weathered, and seemed to have been constructed the previous year.

"On the morning of April 20, as I passed by the marshy opening, something falling from the old nest caught my eye, and looking upward I beheld the male becard clinging to the structure and attempting to pull a fragment from it.   He flew off with some shreds of material in his bill and carried them to a new nest, only recently begun, a sprawling weft of varied constituents attached to several of the fine twiglets at the tip of a slender branch, a few yards from the old structure and very slightly lower.   Working with him was another bird of the same size, which was without doubt his mate, although her plumage was strikingly different from his.

"A week later I returned to watch the becards continue their task. The new nest was now nearly of the size of the old one when first I found it, a roughly globular structure, higher than wide, provided with two entrances, one facing the east, the other the south.   Meanwhile the becards had pulled at the old nest until every trace had fallen or had been incorporated in the new.   In the remains on the ground I found fibrous plant stems, much gray lichen, spider cocoons, thistledown, and sheep wool.

"After another week (the second since the structure was begun) I found the becards putting the finishing touches on their commodious nest.   They had closed up the aperture on the southern side, leaving only that which looked upon the rising sun as their permanent entrance.   Black bird and brown bird continued to bring material as from the first, but their manner of disposing of it was not very different.   In an hour and a quarter the female came 24 times with material for the nest, among which were pine needles, long fibers, and downy substances.   Twenty times she went directly into the nest with her burden, flying skillfully through the entrance without alighting first on the exterior.   Thrice she deposited long fibers on the roof, and once she worked them into the side of the nest.

"The male brought material only 13 times, and everything he carried, whether fibrous or downy, he added to the roof of the nest.   I did not see him enter even once.   His desire to be close to his mate was far stronger than his instinct to aid her in her work, and he accompanied her trips to and from the nest oftener than he brought anything in his bill."

After the birds had been working on the nest for over 2 weeks, and before the nest was quite finished, the tree was cut down and Mr. Skutch had a chance to examine the fallen nest.   "It measured a foot in height and 9 inches in transverse diameter.   The most conspicuous constitutent throughout was a kind of long, slender, much-branched gray lichen, which accounted for three-quarters of the bulk of the material. There were many pieces of fibrous bark of various kinds, dry and partially disintegrated; a wiry length of orange-colored dodder vine; many long, dead pine needles; many yellow spider cocoons and tufts of spiders' silk; many tufts of sheep wool, a few downy feathers, some thistledown, a few pieces of green moss, some slender, dry vine stems, a coiled tendril, a piece of the fabric of a bushtit's nest, probably of last year. The thickness of the walls varied from 1½ to 2½ inches, and the interior cavity, as large as my fist, was lined chiefly with thistledown and fibrous bark.

"Three days later I found that the becards, in no way discouraged by their disaster, had begun a new nest in another tree in the same clump, only 20 feet from the site of their first ill-fated attempt. But alas poor becards. Their second nest followed the first in disaster. Their inexorable enemy, if it was the same, had climbed the tree and chopped down the supporting branch.

"But the becards were as dauntless and as full of hope as Nature herself. They set about at once to build a third nest in the very tree where the second had met with disaster, directly below the position of the last. In a week it reached its full size. This time they had better luck and succeeded in completing the nest and laying the eggs. The construction of the great globular nest of the becards was a very large undertaking for birds no bigger than a sparrow, and as usual in such cases they continued to be preoccupied with it until the eggs hatched. Almost every time that the female returned to her eggs after a brief recess she carried back some bit of material to add to her already bulky structure. The entrance to the completed nest was at the bottom, a little to one side of the center. Just how the aperture communicated with the interior, and what arrangement there was to prevent the eggs rolling out when the bough swayed in the wind, it was impossible to determine without taking down the structure.

"The brown becard's periods on the eggs, as well as her recesses, were of variable duration, but usually brief. I watched her in all kinds of weather, but mostly bad, and found her one of the most restless sitters I have ever known. Her periods in the nest, during a day and a half, ranged from 3 to 38 minutes, with an average of 12. Half of her sessions lasted 8 minutes or less. Her recesses fluctuated from 2 to 19 minutes, with an average of 8½. At the end of the incubation period she remained no more constantly in the nest than at the beginning.

"To enter the downward-facing entrance, with no perch or point of support below it, was not an easy matter, but the bird accomplished the feat with admirable skill. Sometimes she started from a perch below and to one side of the nest, inclined her course sharply upward until it was vertical as she neared her goal, hit with an audible slap the alder leaves draped below the doorway, and disappeared into the interior. At other times she would take off from a branch above the level of the nest, fall almost vertically downward, turn sharply upward in midair and rise directly to the entrance, describing a narrow **U**. Whichever mode of approach she chose, her course was so well calculated from the start that it followed a perfectly smooth and even curve, without kink or angle."

He tells me of another nest that he saw under construction at Colomba, Guatemala, on July 18, 1935; it was 50 feet above the ground at the end of a slender drooping twig of a shade tree in a coffee plantation. This was at an elevation of about 2,600 feet, whereas the nests described above were at an elevation of about 8,500 feet, which is about the altitudinal limit of the species.

A. J. van Rossem (Dickey and van Rossem, 1938) says of the nests of Sumichrast's becard, as found by him in El Salvador: "The nests are very large structures built of grass and other loose, pliable material, resembling in type nests of *Todirostrum cinereum finitimum*. They are, of course, very much larger, some of them a foot long and eight inches in diameter, not including the long streamers of grass hanging from the lower part. The nest cavity is reached from a hole in the side, the entrance of which is protected by overhanging strands from the sharply sloping roof, and the cup is well padded and felted with the softest possible material. The usual site is the spray of foliage at the end of a long, drooping branch, twenty or thirty feet above the ground and, more often than not, entirely inaccessible."

There are two sets of eggs of this species in the Thayer collection in Cambridge, collected by W. Leon Dawson on June 16 and 20, 1925, near Tepic, Nayarit, Mexico; these are apparently referable to *Platypsaris aglaiae albiventris*, as the range of this form is given in the 1931 Check-list.

In a letter to Colonel Thayer, that came with the eggs, Mr. Dawson has this to say about one of the nests: "Taken June 20, 1925, at a point about 3 miles below Tepic, and half a mile from the bridge. This nest was found on the 10th, at which time it was empty. It was placed 15 feet up and 8 or 10 out in a very slender sapling. The limb from which it depended was so slender and limber that I succeeded on both occasions in bending it in for examination without cutting or breaking. The nest itself was the usual 'bushel basket' of vegetable miscellany, closely compacted, evidently with a view to moisture resistance. It had an unusually flat top, but the entrance hole was well below the middle, and so well concealed that I did not bother to look for it till after I had hauled the nest down, it being always easier and safer to dig a new entrance into the nesting cavity and pull the sides together again after the eggs are removed. [The data slip gives the outside diameter of this nest as 14 inches and the outside depth as 12 inches, making a nearly globular structure.]

"The nesting cavity proper is always placed in the lower half of the ball, and is then of very modest dimensions—about the size of the doubled fists placed together. The lining invariably contains broken fragments of soft, dry husks, which, in ensemble, act as a cushion in which the eggs are more or less imbedded. The nests, since they depend from the very tips of the branches, are sometimes subjected to violent thrashing by the wind, yet I have never found a broken egg or lost any through rough handling.

"The Xantus becard almost invariably forms one member of a colony whose nests are under the protection of a pair of champion kingbirds (*Tyrannus crassirostris*), usually in the very tree occupied by the kingbirds. In this instance, however, the sapling containing the becard's nest stood at one side in the shadow of a spreading *higuera* tree, which contained, besides the nest of the kingbird, those of two others of its wards, the Giraud flycatcher (*Myiozetetes texensis texensis*) and the scarlet-headed oriole (*Icterus pustulatus*)."

Mr. van Rossem (Dickey and van Rossem, 1938) does not mention the kingbird association but says: "It seems to be the invariable custom of this species to swing its nests close to the nests of the three common, breeding species of *Icterus*, namely, *gularis*, *sclateri*, and *pectoralis*."

About the time that the 1931 Check-list was going to press, or at least too late for the committee to consider it, Mr. van Rossem (1930) described and named a new northern race of this species, which he called *Platypsaris aglaiae richmondi*. He says that the adult males are "slightly paler and very much grayer with no buffy or brownish tones"; and in the adult females, the under parts are "*very much* paler than in *albiventris*." The range extends from Sonora, and perhaps Chihuahua, northward into southern Arizona. The record specimen from Arizona, now in the Museum of Comparative Zoology, compares favorably with other specimens from Sonora and Chihuahua but is only very slightly paler than what few specimens of *albiventris* we have from more southern States in Mexico. So, if this new subspecies proves to be recognizable in nomenclature, we have no nesting records for the race that belongs on our list.

Dr. Frank M. Chapman (1898) says that this species and its home are well known at Jalapa, Mexico, and that its nest "is some fifteen inches long and about eight in width, with an entrance at one side near the middle. It is a remarkable structure, composed largely of coarse weed-stalks and grasses, in part covered with fresh green mosses, the walls of the cavity being lined with mud. These nests are attached to the end of a limb of one of the taller trees, and sometimes overhang a public road." See plate 1.

Eggs.—The two sets of eggs in the Thayer collection consist of six and four, respectively. One egg is rather long-ovate, and the others are all typically ovate; they are only slightly glossy. In the set of six, the ground color varies from dull white to creamy white; they are thickly marked about the larger end with flecks, spots, and small blotches of dull browns, "wood brown," "buffy brown," or "olive-brown" and are sparingly dotted elsewhere with the same colors and with a few spots of pale gray. In the set of four, the ground color varies from "pale pinkish cinnamon" to "tilleul buff"; and the eggs are somewhat irregularly and more or less generally scrawled with long marks and dotted with minute specks of the same shades of brown. This is a rather heavily marked set of pretty eggs.

Mr. van Rossem (Dickey and van Rossem, 1938) says of the eggs of the closely related Sumichrast's becard:

Two sets of eggs were taken by the simple expedient of shooting off the slender branches to which the nests were attached and catching the nests as they fell. One of these, taken at Lake Guija on May 23, 1927, held five eggs, two of which were broken in the thirty-foot drop. The remaining three measure 23.9×17.5; 23.3×16.9; and 22.4×7.[?] In ground color the larger two are between "vinaceous buff" and "avellaneous" with shell markings of "bone brown," "natal brown" and "army brown," thinly scattered as small streaks and irregular spots over the whole surface, but coalesced into a wreath of heavy blotches about the larger end. The third egg of this set is very different, having a ground color of very light "pale olive gray" with scattered spots and small irregular markings of "mouse gray," "quaker drab," and various shades of pale brown. The markings are more numerous, but do not form a wreath, about the larger end. It may have been laid by a bird other than the parent of the larger two. The second nest, taken at Lake Guija, on May 24, 1927, contained a single egg nearly ready to hatch. It is very different from any of the three eggs in the first set both in size and color. It measures 25.0×16.3. In color it is immaculate "pale ochraceous salmon" with a solid cap of minute, coffee-colored spots at the larger end.

The measurements of 22 eggs of the species average 23.2 by 16.9 millimeters; the eggs showing the four extremes measure **26.0** by 17.9, 24.0 by **18.1, 20.3** by 15.5, and 21.0 by **14.9** millimeters.

Young.—Mr. Skutch learned by close observation that the male takes no·part in the incubation of the eggs; he never saw him enter the nest during the course of incubation. But he does his share in the care and feeding of the young. Of this he writes (MS.): "There came a day when the becards no longer brought leaves and lengths of vine to the nest, but appeared to approach it with empty bills. By looking carefully through the binoculars, I could now and then discover a portion of some small insect projecting from the mandibles. Doubtless at other times they brought insects so small that they could be carried completely inside the mouth, and therefore passed unseen. Now at last the male began to enter the nest.

"As the nestlings became older, their parents brought them portions that were larger and more easily discerned. Small green larvae were the things that I most often recognized, and there were a number of small butterflies and moths. The male and female were equally assiduous in feeding the nestlings, but only the female brooded them. The becards hunted in the manner characteristic of their family; that is, they remained quietly perched until they sighted their prey, then made a rapid dart to snatch it from the air, or from the foliage upon which the creature was crawling, without themselves alighting there."

When the nestlings were 10 days old Mr. Skutch began to hear their weak little calls, and soon they could utter the typical calls of the adults, but in a weaker voice; eventually they became rather noisy nestlings. When the parents were last seen carrying food to the young, the latter had been in the nest 18 or 19 days. Sometime during the next 2 days, the nest was broken open, after which nothing was seen of either parents or young; thus the story of these dauntless becards ended.

Plumages.—I have seen no specimens of *albiventris* or *richmondi* in either nestling or juvenal plumages, but, as the plumage changes are doubtless similar to those of the other races, with due allowance made for subspecific differences, the following remarks by Mr. van Rossem (Dickey and van Rossem, 1938) are significant:

Sequence of plumage in the males has been worked out with the combined series of *latirostris* and *sumichrasti*, since the two are identical in this respect. Two juvenal males are identical in coloration with the black-headed phase common to both juvenal and adult females. After the postjuvenal molt the young males resemble the black-headed females, except that the upperparts are darker and more grayish brown, the underparts are paler and more grayish buffy, and the throats are frequently tinged with pale salmon-pink. The primaries, secondaries, and rectrices are not replaced at this molt, but are worn until the first annual molt the subsequent fall. During the first winter and spring, occasional feathers are added to the body plumage (there seem at this stage always to be a few pin feathers about the head) and in the spring the innermost tertials are renewed, but there is no definite spring molt which results in a change of the type of plumage worn. Males, at least, breed in this immature plumage. Eight examples of this one-year-old-stage were taken. They were collected in September, October, January, April, May, July, and August; those of the last two months are in the first annual molt.

The second-year plumage, which is attained at the first annual molt, is very similar to that of the fully adult male, but the underparts and rump are strongly tinged with brownish or olive, the rectrices show terminal edgings or mottling of cinnamon and the throat averages less extensively pink. The abbreviated ninth primary is acquired at this time. Of these second-year males there are six specimens taken in January, February, May, and September, besides two critical examples which show the transition from first-year to second-year plumage.

Our record specimen, taken June 20, 1888, in the Huachuca Mountains, Ariz., by W. W. Price, is a young male in second-year plumage; it is largely in adult plumage, but the rose of the throat is more restricted and paler, "light jasper red" instead of "rose red," and the wings and tail are of the immature type, pale and worn. It is labeled *albiventris* but is apparently referable to *richmondi*.

Both adults and immature birds evidently have the complete annual molt in September and October.

Food.—Mr. van Rossem (1938) records the stomach contents in 10 specimens of *latirostris;* he found berry seeds and fruit pulp in six stomachs; berry seeds, pulp, and insects in three; and insects exclusively in only one stomach. Mr. Davis says (MS.) of the bird he watched: "It ate only insects, as far as I could observe. On one occasion, I watched it swallow a large insect that required considerable effort to get down. This appeared to be a dragonfly, and the wings stuck out of the bird's mouth for some time, as the insect's body was slowly choked down."

Behavior.—By its general behavior Mr. Davis, at first, thought that the strange bird might be an immature Derby flycatcher or a freak crested flycatcher. He noted that the Derby flycatchers and green jays showed considerable hostility toward it, but the becard made very little attempt to defend its territory against them; it would give its call repeatedly when approached by them, but would quickly fly away if they became quarrelsome. "It tended to keep hidden in the thick foliage of large trees and never perched out in the open, as do the flycatchers."

Voice.—Mr. Price (1888) says: "Their note reminds one of the song of Stephens's Vireo (*Vireo huttoni stephensi*), but is not so long continued, and is harsher." The bird observed by Dr. Beebe (1905) uttered from time to time "a low, indefinable lisp."

Mr. Davis writes to me as follows regarding his impressions of the call of the becard: "It is quite a lot softer than the whistle of the crested flycatcher and is of entirely different quality and is more drawn out. While very soft, the call carries well and can be heard for some distance. From what I have observed, it seems that the adults almost always use some preliminary chattering notes, which are still softer than the main call or seem to be so because they are of lower pitch. Possibly the whole call could be stated as *chu-chu-chu-chu, tee-oooooo,* or *chatter, chatter, chee-oooooo.* The preliminary notes are low and rapid, and the latter part starts high, though rather thin as compared to the crested flycatcher, and rapidly drops down and trails off to nothing in the drawn out *oooooo*."

Range.—Western Mexico, accidental in southern Arizona; probably only slightly migratory.

The range of Xantus's becard extends **north** to northern Sonora (Saric, San Rafael, and Guirocoha); and southern Chihuahua (San Rafael). From this region it is found **south** to Morelos (Puente de Ixtla) and Guerrero (Chilpancingo, Acapulco, and Coyuca). It appears to be resident north to southern Sonora (Alamos and Tesia) and southern Chihuahua (San Rafael).

Other races of this species are found in eastern Mexico on the Tres Marias Islands, and in other Central American countries.

Casual records.—A single specimen has been collected in the United States, a bird obtained on June 20, 1888, in the pine forests of the Huachuca Mountains, Ariz. According to Ridgway (1907) a specimen of this race was collected in March at Cerro de la Silla, Nuevo Leon, Mexico, which is east of the usual range. One bird was seen by L. Irby Davis near Harlingen, Tex., on numerous occasions in 1937 and 1938 (see account of it above).

Egg dates.—Mexico: 6 records, May 8 to June 20.

Family TYRANNIDAE: Tyrant Flycatchers

TYRANNUS TYRANNUS (Linnaeus)

EASTERN KINGBIRD

PLATES 2–4

HABITS

CONTRIBUTED BY WINSOR MARRETT TYLER

When we think of the kingbird, even if it be winter here in the north, and he is for the time thousands of miles away in the Tropics, we picture him as we see him in summer, perched on the topmost limb of an apple tree, erect in his full-dress suit—white tie, shirt-front, and waistcoat—upright, head thrown back, his eye roaming over his domain, on the watch for intruders. We see him sail out into the air, moving slowly, although his wings are quivering fast, then gaining speed and mounting higher as he comes near his enemy—a crow, a hawk, any bird that has stirred his resentment. We hear his high, sibilant, jerky voice ring out a challenge; we watch him dive at the big bird, striking for his back, and drive him off, and then come slithering back to his watchtower, proclaiming victory with an explosion of stuttering notes.

Spring.—Unlike most of our migrant birds, the kingbird arrives in New England unobtrusively—about the tenth of May in the latitude of Boston—and for a few days remains quiet, both in voice and demeanor. We are apt to see our first kingbird of the season sitting silent and alone on a fencepost or a wire or making a short flight out from a tree and back again. It appears listless, as if not interested in its surroundings, as if it were tired. There is none of the exuberance of the Baltimore oriole, in full song when he returns to his breeding ground, or of the showy arrival of the bronzed grackles as they come pouring into New England in vast clattering hordes. It is not long, however, before the kingbird throws off his lethargy and appears in his true colors—the tyrant of tyrants.

Alexander F. Skutch has sent Mr. Bent an excellent account of the kingbird's northward passage through Central America, where, during the early stages of their long journey, the birds are concentrated in large numbers. He says: "Although only a bird of passage through the great isthmus that stretches from Tehuantepec to Darien, the kingbird, because of its large size, active habits, and its custom of migrating by day in flocks, is the most conspicuous of the flycatchers that visit Central America from the north. The birds appear to enter Central America from their winter home in South America about April 1, and the last do not leave the region until nearly the middle of May.

"Kingbirds travel chiefly in the early morning and the latter half of the afternoon. At these times I have on numerous occasions watched them fly overhead in loose-straggling flocks of irregular formation, sometimes containing, according to a rough estimate, more than a hundred individuals. Thus, soon after dawn on April 28, 1935, as I was paddling along the shore of Barro Colorado Island in the Canal Zone, a large flock of kingbirds flew across Gatun Lake from east to west, or from the South American to the North American side. They came to rest in the tops of some small trees from which a few birds made sallies into the air to snatch up insects, but after a pause of a minute or so they continued on toward the west.

"During both years of my residence at Rivas, in the deep, narrow, north-and-south valley of the Rio Buena Vista in southern Costa Rica, I witnessed numerous northward flights of kingbirds in April, always in the afternoon. On April 18, 1936, about half-past four in the afternoon, I beheld several multitudinous flocks of small birds come up the valley from the south, a few minutes apart, flying high and straight as if they were journeying. There were barn swallows, rough-winged swallows, kingbirds, and small black swifts. The kingbirds were the first to drop out of the flock. They settled in some low scattered trees to rest. There were scores of them, and they made a substantial addition to the large company of kingbirds already in the valley.

"On both their northward and southward passage through Central America the kingbirds may break their journey and delay for considerable periods in some locality which pleases them. Although it is possible that the kingbirds one sees during the course of several weeks in the same vicinity may represent a population whose members change from day to day, the fact that they roost every night in the same spot is to me rather convincing evidence that the same individuals linger for more than one night's lodging.

"During the spring migration of 1936 the kingbirds roosted nightly for nearly a month, from April 16 to May 11, on a small islet covered with low trees, behind my cabin. I frequently watched them congregate for the night or begin their day's activities. On April 17, late in the afternoon, a large, straggling flock settled in the riverwood trees on the brink of the stream and from these sallied in their spectacular manner into the open space above the channel, or high into the air, to capture flying insects. Long before dark they began to congregate on the little island. They did not immediately settle to rest but wove gracefully among the branches and the long leaves of the wild cane and skimmed above the foliage to snatch up some insect that blundered temptingly close to them. Finally, as dusk deepened, they became quiet among the inner recesses of the foliage where they were so well concealed that I could not, even with my glasses, pick out a single one. While in Central America they rarely utter a sound."

Courtship.—Ralph Hoffmann (1927) says that the kingbird's "mating performance consists in flying upward, and then tumbling suddenly in the air, repeating the manoeuvre again and again, all the time uttering its shrill cry." Dr. Charles W. Townsend (1920a) says of it: "The Kingbird executes a series of zig-zag and erratic flights, emitting at the same time a harsh double scream. This is a true courtship flight song."

These flights take place at no great height from the ground—15 or 20 feet, perhaps, above the top of an apple tree. The dives are usually short, quick dips, accompanied by accented notes, and in between them the wings flutter jerkily as the bird rises again or progresses a short distance on a level. Occasionally, however, the dip is much deeper—a long, slow dive. I find in my notes of July 28, 1909, that I observed their curious flight evolutions many times. They flew out from a treetop, half flying and half hovering, then, with wings almost still, but just quivering, they slowly dropped almost to the ground, the while jerking out in a high, squeaky, tremulous voice their *ki-ki-ki*, etc.

A. Dawes DuBois wrote to Mr. Bent of a pair of kingbirds courting on May 21, 1910. "One of them," he says, "went through some very remarkable antics in the air, turning backward somersaults while flying."

I once watched two kingbirds not 20 feet away whose behavior strongly suggested a courtship of milder form than the wild display in the air. Both were adult birds. One, a male I thought, was perched not far from the other with feathers puffed out and head erect and drawn back a little way. He twitched his tail sharply downward over and over again, at the same time fanning it out. These actions were plainly addressed to the other bird. Twice he flew toward her(?), and she(?) retreated. Both birds were silent except when once or twice one gave some sibilant notes. These notes were not uttered while the bird was posturing; they were not uttered with any emphasis; and they did not suggest the kingbird's song at all.

Although the actions of this bird might well be suitable to courtship—it is to be observed that the two ornamented parts of the plumage (the crown patch and the tail) were displayed—the date (July 25, 1917) is too late to expect courtship with breeding intent. I do not doubt, however, that the performance represented some form of nuptial display.

Nesting.—Like many birds whose breeding range extends over a widely diverse country, such as the mourning dove, the kingbird chooses a variety of nesting sites. Here in eastern Massachusetts where a large part of the country consists of farmland, orchards, acres of scattered trees, and woodland of small, thin growth, a typically situated kingbird's nest is built well up in an apple tree, often on a horizontal limb, generally well out from the trunk, almost always shaded by branches higher up. It is a rather large nest for the size of the bird, and a little bulky. The outside is rough and unkempt, a heap of twigs, straw, and twine, not finished off like the nest of the wood pewee. Another favorite location here is in trees or low shrubs growing along a river, often on branches overhanging the water. In the West, however, in regions where there are few trees, the kingbird may place its nest in the open, on a fencepost or a stump, in a situation without concealment or shade.

Even in the East, the bird occasionally resorts to this practice. Fred H. Kennard (1898) reports such a case of nesting, in Bedford, Mass., a farming district, on a fencepost "within 35 feet of the railroad, and immediately· beside a road, over which men are travelling back and forth all day long. * * * This post was made of an abandoned railroad tie, whose end had been somewhat hollowed by decay. * * * The top of the post was only about four feet above the ground."

The kingbird often shows its fondness for water by nesting on stumps or snags on submerged land. For example, Ralph Works Chaney (1910), writing of Michigan, says: "This species might be considered almost aquatic in its nesting habits, as the nests were invariably placed in stumps projecting out of the water, often at a considerable distance from the shore," and William Brewster (1937) speaks of kingbirds breeding "along stub-lined shores bordering on northerly reaches of the Lake [Umbagog] or on shallow lagoons in the heavily-timbered bottom-lands of the Lower Megalloway. Those frequenting all such localities nested mostly within hollow tree-trunks. * * * Of the nests thus placed some were sunk eight or ten inches below the upper rim of the cavity and hence invisible save from above, others so near it that the sitting bird, and perhaps also a small portion of the nest, could be seen by any one passing beneath."

Of eccentric situations where kingbirds have nested, we may note two instances in which a nest was built on the reflector of an electric street light—A. C. Gardner (1921) and Rolf D. Rohwer (1933)—and a very remarkable report of its nesting in a rain gauge, Lincoln Ellison (1936). Stranger still, perhaps, are two cases of kingbirds appropriating oriole's nests for their own use. Henry Mousley (1916) tells of a pair of kingbirds that "took possession of an old Baltimore Oriole's nest in the top of a maple tree in front of my house, in which strange home they laid a third set of eggs and brought up a brood," and Clarence Cottam (1938) cites "the successful occupancy by an Eastern Kingbird * * * of a deserted hanging nest of a Bullock Oriole. * * * The nest * * * was attached about 12 feet above the ground to some terminal and partly drooping branches of a cottonwood tree." Edward R. Ford (MS.) writes that he saw a kingbird sitting on a nest which had been built and used by cedar waxwings the year before.

The height above the ground of the kingbird's nest varies considerably: J. K. Jensen (1918) gives the extremes as "from two to sixty feet."

Of "a typical nest" taken in Minnesota, Bendire (1895) says: It "measures about 5½ inches in outer diameter by 3¼ inches in depth; its inner diameter is 3 inches by 1¾ inches deep. Its exterior is constructed of small twigs and dry weed stems, mixed with cottonwood down, pieces of twine, and a little hair. The inner cup is lined with fine dry grass, a few rootlets, and a small quantity of horsehair." Continuing, he says: "Mr. E. A. McIlhenny tells me that in the willow swamps in southern Louisiana these birds construct their nests entirely out of willow catkins, without any sticks whatever, and that the nests can be squeezed together in the hand like a ball." Here in New England, where the birds breed in orchards and dooryards and near farm buildings, they often pick up bits of cloth, straw, feathers and pieces of string and add them to their nest. J. J. Murray (MS.) says that in Lexington, Va., a favorite material is sheep wool and that the birds often nest in trees along the edge of pastures where wool is easily obtained.

Kingbirds appear to have a very strong attachment to the nesting site they have chosen and return year after year to its immediate vicinity. Roy Latham (1924) gives a striking illustration of this tendency at his home on Long Island, N. Y. In spite of "development" that changed the face of the country, the kingbirds did not desert it. He says: "The wild cherries are gone, the old line-fences are gone, and the Bob-whites are gone. But year after year a pair of Kingbirds return each May and carefully select a nesting-tree. Every tree on the homestead has been used—some thrice over. In all those thirty-five summers the Kingbirds have not failed once to bring off a full brood. Yet in the entire period there has never been a second pair breeding on the premises, or to my knowledge, making any attempt to nest within the limits of the yard."

Kingbirds are averse to having another pair of kingbirds nest near them, but they do not object to nesting near other species of birds. As an extreme illustration of this habit, Charles M. Morse (1931) published a photograph of two occupied nests 14 inches apart, one a kingbird's, the other a robin's. He says: "The two families lived in perfect harmony."

S. F. Rathbun wrote to Mr. Bent: "Once, when I was in eastern Washington, I ran across a nest of this kingbird in a small tree at the edge of a stream. Within 300 feet a pair of Arkansas flycatchers had a nearly completed nest in another tree, and not far away a pair of ash-throated flycatchers were nesting in a box placed under the eaves of a dwelling. To me it was of interest to see these three species of flycatchers nesting so near each other."

M. G. Vaiden, of Rosedale, Miss., wrote to Mr. Bent: "I found located in a large native pecan tree a nest of the kingbird, wood pewee, red-eyed vireo, two English sparrow nests, and the nest of a Baltimore oriole. All seemed more or less in perfect accord except the wood pewee, whose nesting territory had been crowded by the home of one of the sparrows. The wood pewee seemed to do most of the fighting, with little if any attention paid by the sparrow. Each probably had a vertical and horizontal area that they defended, should the occasion arise."

According to Bendire (1895), "the male assists in the construction of the nest, and to some extent in the duties of incubation. He relieves the female from time to time to allow her to feed, guards the nesting site, and is usually perched on a limb close by, where he has a good view of the surroundings."

A most unusual nesting site for the eastern kingbird is reported, in a letter to Mr. Bent, by Capt. H. L. Harllee, of Florence, S. C. This pair of birds built a nest and laid a set of eggs in a gourd that was suspended from a pole at the edge of a yard in Beaufort County. The gourd, such as are commonly used by purple martins in the South, happened to have large openings on two opposite sides, which gave the birds convenient entrance and exit, as well as some visibility while on the nest.

Eggs.—[AUTHOR's NOTE: The eastern kingbird lays three to five eggs to a set; three is the commonest number and five decidedly uncommon or rare. The eggs are commonly ovate, with variations toward short-ovate or elongate-ovate or, rarely, elliptical-ovate. They are only slightly glossy. The ground color is pure white, creamy white (most commonly), or pinkish white, and very rarely decidedly pink. As a rule they are quite heavily and irregularly marked with large and small spots, or small blotches, but some are quite evenly sprinkled with fine dots. The markings are in various shades of brown, "chestnut-brown," "chocolate," "liver brown," "claret brown," or "cinnamon," with underlying spots and blotches of different shades of "Quaker drab," "brownish drab," "heliotrope gray," or "lavender." Very rarely an egg is nearly immaculate. The measurements of 50 eggs average 24.2 by 17.7 millimeters; the eggs showing the four extremes measure 27.9 by 18.3, 23.6 by 19.8, 22.1 by 18.5, and 23.9 by 16.2 millimeters.]

Young.—Hatched from what some oologists consider the most beautiful of eggs, the young kingbirds remain in the nest for about 2 weeks. Gilbert H. Trafton (1908) gives the time as 10 to 11 days, and A. D. Whedon (1906) as about 18 days. Most writers, however, agree on 13 to 14 days as the average time.

Francis Hobart Herrick (1905) studied the nest life of the kingbird in detail from a blind placed close to a nest which, with the limb supporting it, he had moved a short distance to facilitate observation. The nest contained four young birds, two of them transferred from another nest. Writing of the day when the nestlings were 10 days old, he says:

In the space of four hours * * * the parents made one hundred and eight visits to the nest and fed their brood ninety-one times. In this task the female bore the larger share, bringing food more than fifty times, although the male made a good showing, having a record of thirty-seven visits to his credit.

* * * During the first hour the young were fed on an average of once in one and a half minutes. * * * The mother brooded eighteen times, and altogether for the space of one hour and twenty minutes. The nest was cleaned seven times, and the nest and young were constantly inspected and picked all over by both birds, although the female was the more scrupulous in her attentions. * * * One of the birds while perched near by was seen to disgorge the indigestible parts of its insect food, a common practice with flycatchers, both old and young. * * *

The last [young bird] to leave [the nest] flew easily two hundred feet down the hillside on the thirteenth of July [i. e., when 18 days old].

Raymond S. Deck (1934) speaks of the effect of sunlight on the behavior of the young birds. He says: "The nestlings appeared to respond to the sun in a quite sunflowerly way. Early in the morning they lay in the nest facing the rising sun. As the morning wore on and the sun moved south, the birds shifted their position to face constantly toward it. During the hottest part of the day they lay facing north-east, directly away from the sun, but when evening came the birds were lying with their faces toward the sunset. On every subsequent day when I visited the nest the young birds were facing east in the morning and they always went to sleep at night facing west."

Early in July, here in New England, fledgling kingbirds are full-grown, although their tails may be rather short. We may see a brood of them perched not far apart on a wire, or on an exposed branch of a tree, waiting for their parents to bring them food. They keep up a frequently repeated, high, short, emphatic note, *tzee*, snapping their bills open and shut as they utter it, showing the bright orange color of their throats, and when they see the old bird approaching, they lean eagerly forward, and their voices become rough and harsh. At times they fly out and meet the parent bird in the air, where, to judge from their actions, food is transferred to them with a good deal of chippering and fluttering.

Burns (1915) gives the incubation period as 12 to 13 days, and several other observers agree with him closely.

Plumages.—[AUTHOR'S NOTE: The natal down that soon appears on the otherwise naked nestling is "mouse gray." The young bird in juvenal plumage is much like the adult, but there is no orange crown patch; the nape and rump are faintly edged with "cinnamon"; the wing coverts are edged with pale buff, and the other paler edgings of the wing feathers are pale buffy or yellowish white; the white tips of the tail feathers are tinged with brownish, especially the outer ones; there is a grayish band, tinged with buff, across the upper breast; and the two outer primaries are not attenuated as in the adult.

A postjuvenal molt begins before the birds migrate, but the birds go south before even the body molt is complete. Dr. Dwight (1900) says: "Birds taken in Central America, unfortunately without dates, show that the species reaches the tropics without any moult of the flight feathers or of the wing coverts and often in full juvenal plumage. It is an interesting problem whether the wings and tail are renewed at the end of the postjuvenal moult or at a prenuptial moult, the former conclusion being most probable. A bird from South America taken March 31 (which may possibly be an adult) shows a recently completed moult the sheaths still adhering to the new primaries."

That young birds have a complete postjuvenal molt during fall, winter, and early spring is shown by the fact that they arrive in spring in fresh plumage, including the two outer emarginate primaries (in the male), the new white-tipped tail, and the orange crown patch. Young birds, which were alike in juvenal plumage, now show the sex differences of the adults.

The molts of the adults apparently follow the same sequence as in the young birds. Adult males have the two outer primaries attenuated, or emarginated, and the adult females only one, as a rule. There is not enough winter material available to work out the molts with certainty.]

Food.—F. E. L. Beal (1897) summarizes the results of his analysis of the kingbird's food thus: "Three points seem to be clearly established in regard to the food of the kingbird—(1) that about 90 per cent consists of insects, mostly injurious species; (2) that the alleged habit of preying upon honeybees is much less prevalent than has been supposed, and probably does not result in any great damage; and (3) that the vegetable food consists almost entirely of wild fruits which have no economic value. These facts, taken in connection with its well-known enmity for hawks and crows, entitle the kingbird to a place among the most desirable birds of the orchard or garden."

In regard to the eating of bees, Beal (1897) states: "The Biological Survey has made an examination of 281 stomachs [of kingbirds] collected in various parts of the country, but found only 14 containing remains of honeybees. In these 14 stomachs there were in all 50 honeybees, of which 40 were drones, 4 were certainly workers, and the remaining 6 were too badly broken to be identified as to sex."

In a later paper Beal (1912) lists over 200 kinds of insects found in kingbirds' stomachs, and the fruit or seeds of 40 species of plants.

To itemize the kingbird's diet more in detail, we may mention the following:

Hairy caterpillars are reported by Mary Mann Miller (1899), who says: "The Flycatchers darted upon the caterpillars as they swung suspended by their webs or fed on pendant leaves."

H. H. Kopman (1915) states that "in the piney sections of southeastern Louisiana and southern Mississippi, the Kingbird feeds extensively in the fall on the ripened seeds of the two common native magnolias (*M. foetida* and *M. virginiana*)."

William L. Bailey (1915), speaking of the feeding of nestling kingbirds, says: "To my amazement a large green dragon-fly with great head and eyes, measuring across the wings at least four inches, was jammed wings and all, into the mouth of one of the little ones. After a few minutes, as if for dessert, a large red cherry fully one-half inch in diameter was rammed home in the same manner."

Robert T. Morris (1912) relates the following: "There is a sassafras tree * * * at my country-place at Stamford, Connecticut, which bears a heavy crop of fruit every year, and about the last of August the Kingbirds gather in numbers, spending the entire day in the tree, and strip it entirely of its fruit. * * * At the time when they are gorging themselves with sassafras berries, they seem to devote little time to catching insects."

Dr. Harry C. Oberholser (1938) includes "small fishes" as an item in the kingbird's diet.

The kingbird captures most of its food by pursuing a flying insect and catching it in the air. Rarely, it snaps up a larva suspended by a thread; and Dr. Charles W. Townsend (1920b) reports: "I have seen a Kingbird swoop down and pick up an insect from the calm surface of a pond without wetting a feather. I have also seen one flying and picking off berries from a shad-bush without alighting."

Of "terrestrial feeding kingbirds" William Youngworth (1937) says:

On June 3, 4, and 5, 1935, the Waubay Lakes region in northeastern South Dakota was swept by high winds from the north and the temperature during the night dropped to near the freezing point. Heavy frost was visible on two mornings and it was such weather that caught the last migrating wave of kingbirds and orioles. It was a common sight to find hundreds of Common Kingbirds, Arkansas Kingbirds, and Baltimore Orioles in the lee of every small patch of trees or brush. The dust-filled air was not only extremely cold, but apparently was void of insect life. Thus the birds resorted to ground feeding, and here they hopped around picking up numbed insects. Usually the birds just hopped in a rather awkward manner from one catch to the next. However, occasionally the kingbirds would flutter and hop while picking up an insect.

Behavior.—Dr. Harry C. Oberholser (1938) exactly describes the habitat of the kingbird when he writes that it "lives in the more open country, and is not fond of the deep forests. Cultivated lands, such as orchards and the borders of fields, highways, brushy pastures, or even open woodlands, are frequented also. It is not usually found in any considerable flocks, but during migration sometimes many are found within a relatively small area."

If we were limited to one adjective to suggest the kingbird's character as impressed on us by his behavior, I think most of us would use the word "defiant"; if we were allowed one more, perhaps we should add "fearless." In contrast to most birds, whose concern is restricted to the immediate vicinity of their nest, the kingbird's attention reaches far out. His perch always commands a good view of the surrounding country; he is always on the watch for the enemy. He reminds us of those delightful young men in *Romeo and Juliet* who, let a Capulet appear, flash out their swords and rush into a fight.

The kingbird seems to consider any big bird his enemy; he does not wait for one to come near but, assuming the offensive, dashes out at crow, vulture, or a big hawk—size seems to make no difference to him—and practically always wins.

A. D. DuBois (MS.) testifies to the genuineness of the kingbird's attack thus: "The kingbird can be more than a mere annoyance to its traditional enemy. I saw a pair attack a crow which was flying near their nest. They made him croak, and one of them perched on his back and pulled out a lot of his feathers, which came floating down."

Gilbert H. Trafton (1908) also speaks of a fierce attack upon himself at a nest he was watching. He says: "Whenever I approached near enough the nest to set up the camera, the Kingbirds flew at me furiously, poising themselves above me and then darting quickly at my head, now coming near enough to strike me with their bill. In no case was blood drawn, but, as they usually struck about the same spot each time, I was glad of an excuse to cover my head with a cloth while focusing the camera. * * * They never attacked me unless both birds were present, and even then only one came near enough to strike me."

Frederick C. Lincoln (1925), writing of North Dakota, says: "On July 20 I watched a Kingbird attack a Hawk and saw it alight on the back of the larger bird, to be carried 40 to 50 yards before again taking flight." J. J. Murray (MS.) reports a similar observation: "Near Lexington, Virginia, I saw a kingbird chase an American egret for a hundred yards or more, practically riding on its back."

Florence Merriam Bailey (1918) speaks thus of a kingbird attacking so swift a bird as a black tern: "I saw [the tern] beating over the open slough close by when suddenly chased after by a Kingbird, chased so closely and persistently and rancorously that if he were not pecked on the back, a deep dent was made in his gray matter, for he fled precipitately through the sky, going out into its grayness."

John R. Williams (1935) tells of a kingbird which repeatedly attacked a low-flying airplane. He says: "The courage and audacity of this bird in attacking a noisy and relatively huge airplane was certainly extraordinary."

Isaac E. Hess (1910) states: "I have seen the Kingbird victor in every battle except one. In this dispute 'Tyrannus' beat a hasty retreat from the onslaughts of an angry Yellow Warbler."

William Brewster (1937) relates another instance of the defeat of *Tyrannus:*

Despite his notorious daring in attacking hawks and crows, the Kingbird sometimes turns tail and flees ignominiously, like many another bully, when boldly faced by birds no larger or better fitted for combat than himself. An instance of this happened to-day [August 10, 1907] when I saw a Sapsucker pursue and overtake a Kingbird in a cove of the Lake [Umbagog]. * * * As the two were passing me within ten yards I could see the Sapsucker deal oft-repeated blows with his sharp bill at the back of the Kingbird who was doubling and twisting all the while, with shrill and incessant outcry. * * * After the birds had separated the Sapsucker alighted very near me on a stub, when I was surprised to note that it was a young one, apparently of female sex.

The kingbird's flight varies considerably both in form and tempo. In his quiet hours he may flutter calmly and steadily along, neither rising nor falling, his long axis parallel to the ground, moving slowly and evenly, his wings quivering in short, quick vibrations—as Francis Beach White (1937) says, "hovering all the way just over the top of the tall grass." At other times, in his wilder moments, to quote Ned Dearborn (1903), "the bird becomes a veritable fury, and dashes upward toward the clouds, crying fiercely, and ever and anon reaching a frenzied climax, when its cry is prolonged into a kind of shriek, and its flight a zigzag of blind rage. These exhibitions are frequently given in the teeth of the premonitory gust before a thunder storm, as if in defiance of the very elements."

I find an entry in my notes that shows how seldom kingbirds move from place to place except by the use of their wings: "June 1910. A pair of kingbirds spent much of their time one afternoon feeding in a newly cultivated field of about an acre in extent. They stationed themselves on small lumps of earth, sometimes near together and sometimes in different parts of the field, and watched for insects. When they saw one they flew to capture it and then returned to the same little elevation, or to another one. The wind was blowing hard, and invariably they alighted facing it, turning just before perching. I did not, during half an hour or so, see either bird take a step or make a hop. They always flew, even to a point less than a foot away."

Francis H. Allen (MS.) states: "Kingbirds sometimes hover, facing into the wind as they feed, taking insects from the air. Sometimes a strong breeze will blow them back, so that they seem to be flying backward."

See also a note on flight under "Fall."

It is the custom of the kingbird to bathe by dashing down over and over onto the surface of water as he flies along, as swifts and swallows do. Dr. Charles W. Townsend (1920b) remarks: "It is not uncommon to see a Kingbird plunge several times into the water from a post or tree, evidently for a bath, and afterward preen itself. I have also seen this method of bathing in a small shallow birds' bath."

Of the brilliant feathers on the kingbird's crown, made visible only by the parting of the surrounding feathers, J. A. Spurrell (1919) says: "I have never seen the red crest on a living kingbird except when displayed by a victorious male after defeating a rival." Bayard H. Christy (1932), however, gives a vivid picture of a kingbird using his crest for intimidation:

On the river side of the [golf] course, at a clump of young pines, a Kingbird was hovering and screaming, and, as I came near, I easily discovered the nest, about twenty feet up, on a bough of one of the trees. As I stood at the base of the tree, at the edge of the circle of the lower branches, the Kingbird came plunging from above, directly toward my upturned face, and as it did so it flashed out broadly its brilliant vermilion crown-patch. The effect was astonishing: it gave the impression of a gaping mouth, venomous and menacing, and, in spite of myself, I bowed my head before the attack. The bird did not indeed strike, but passing me narrowly it rose to repeat the manoeuvre. This was a sudden demonstration of an unsuspected value of this splendid but ordinarily concealed item of decoration. Is it decoration? It seemed to me that a wandering squirrel or snake, potent for mischief, might well by such a display be driven off, before ever it had found the prize.

Voice.—The voice of the kingbird is shrill, not overloud, with only moderate carrying power and without a wide range of pitch. The letters *tzi* suggest the simplest of his notes, although perhaps Bendire's (1895) "*pthee*" is as good a rendering.

This note is delivered as a single, short, sharp exclamation, and when lengthened or modulated in pitch forms the basis of several more complicated utterances. It is often given alone, repeated slowly over and over with a short pause between each note, or repeated rapidly as a high, squeaky chatter, and it is frequently combined with its lengthened form *tzeee*, preceding or following the longer note, which is strongly accented. *Tzi, tzee* is a common form.

Such phrases are characteristic of the bird when in a quiet mood, but when he is aroused to belligerency we hear him utter another note as he flies out to battle, a double note with falling inflection (often rendered *kipper*) cried out in long series which alternate with emphatic shrieks. This battle cry is somewhat similar to the courtship song mentioned under "Courtship." Rev. J. Hibbert Langille (1884) indicates very well the mode of the kingbird's enunciation when he says: "His sharp screeping note [is] coughed out and accompanied by a jerk of the tail."

The formal song of the kingbird is prettily described by Olive Thorne Miller (1892), who was the first to publish an account of it. She heard it, "a sweet though simple strain," early in the morning when, as she says, "it was so still that the flit of a wing was almost startling." She continues: "It began with a low king-bird 'K-r-r-r' (or rolling sound impossible to express by letters), without which I should not have identified it at first, and it ended with a very sweet call of two notes, five tones apart, the lower first, after a manner suggestive of the phoebe—something like this: 'Kr-r-r-r-r-ree-bé'!"

I remember the first time I heard the kingbird's song. It was on July 8, 1908. I was walking home early in the morning from a professional engagement. It was almost dark; an hour before sunrise; about 3 o'clock. Soon the robin chorus began feebly; the east was becoming pale now, and, after a little, a song sparrow and a catbird woke in the dim bushes beside the road. I took a short cut across a meadow, and as I was feeling my way along, I heard a new bird note break out of the darkness in front of me. The bird was beyond the meadow on a rise of ground where I knew there were shade trees, and farther on was an orchard.

I suspected the singer at once, but I was not sure. The voice was high and sharp, with the squeaky quality characteristic of *Tyrannus*, but the arrangement of the notes was wholly strange. They formed a short musical theme of three syllables repeated again and again with a long pause after each one. As I came nearer, however, I found that a part of each pause was filled in by a series of high, short, stuttering notes, given in a hesitating fashion. These notes led up to and immediately preceded the clearly enunciated, emphatic theme. I wrote down the whole song as *i-i-i-i-i, ee, tweea*, with both double e's strongly accented. It was all on one even tone, or nearly so, except at the very end where the pitch either dropped a little (suggesting the song of *Sayornis phoebe*), or rose still higher.

I sat down on a wall near the invisible singer and waited. Again and again the song came from overhead; the bird was singing virtually in black night, shouting out a sharp song, which, in spite of its high, squeaky pitch, was in tune with that peaceful, shadowy hour before the morning twilight.

Gradually dawn brightened the east; green spread over the dark gray meadow. I looked up and saw a kingbird, quietly perched on a branch above my head.

A few days later Walter Faxon, after listening to the song, remarked to me, "He is trying to pronounce the word 'explicit,' but he is making a miserable, stuttering failure of it."

Although heard oftenest in the morning before dawn, the song is occasionally given in the daytime. I have heard it several times on misty summer afternoons—gray, almost colorless days—and once, August 12, 1909, at noon, under a blue sky.

Dr. Leon Augustus Hausman (1925) has made a careful study of "The Utterances of the Kingbird" to which readers are referred. To quote from his summary: "The various cries and calls of the Kingbird, as well as the flight song, are all built up from the simple call notes, which are best represented by the syllables *kitter* and *kit*, and differ from one another in grouping, length and intensity. The flight song may be regarded as a true song, and is given only during the mating season. The mating song is seldom heard; is more musical in character than the flight song; possesses a definite song-rhythm and two new, true song-notes."

Albert R. Brand (1938), who has recorded on film the songs and calls of almost 100 species of birds, summarizes the results of his investigation thus: "I believe that these studies are sufficiently comprehensive to warrant the conclusion that passerine song averages above 4,000 vibrations per second or around the highest note of the piano keyboard." He records the kingbird's voice as 6,225 vibrations per second (approximate mean), very close in pitch to the song of the redstart (6,200).

Field marks.—The eastern kingbird is a large flycatcher with a broad white line across the tip of its black tail, two very inconspicuous wing bars, and no yellow in its plumage. Of the two flycatchers that resemble the eastern kingbird in general appearance, the gray kingbird and the Arkansas kingbird, the former has no white in its gray tail, and the latter has the tail margined with white and has a yellowish breast. Ralph Hoffmann (1904) says: "The *black* tail, broadly *tipped with white*, and the *white under parts* make the Kingbird an easy bird to identify, even from a car window."

Enemies.—The kingbird has few enemies. A hawk may occasionally catch him off guard, and once in a while a misguided apiarist or proprietor of a cranberry bog may turn against him.

Formerly man was the bird's deadly enemy. Both Wilson and Audubon deplored the wholesale slaughter of kingbirds in their day by farmers for fancied depredations on their bees. Nowadays, however, the kingbird is protected as a song bird.

Dr. Herbert Friedmann (1929) says that "the Kingbird is a very uncommon victim of the Cowbird, there being only a very few actual cases on record, although several writers have listed it, probably all based on the same published instances."

Fall.—Kingbirds keep mostly in family units until well into August; when migration time is near, these small groups coalesce and form flocks of a dozen birds or more. Now, nearly silent, they sit about on wires, fences, and trees, or in open country on the ground, loosely associated, showing little tendency to move in unison, although individual birds take short flights from time to time. Occasionally, however, they become more active and restless. For example, on August 15, 1936, I saw a gathering of 15 or 20, flying about over a meadow just before sunset. They were not noisy but gave frequently a subdued *z-z-z-z-zee*. Sometimes they flew out in groups of three or four, making swoops at each other; sometimes they perched for a moment, a few together, in the top of a tree, their feathers drawn in close, and their necks stretched out, posturing as cedar waxwings often do. In making long flights the wings were carried backward in full, free strokes—almost as far as a robin's. When they flew thus, as they did most of the time, they moved through the air very rapidly and lost all resemblance to kingbirds. Occasionally they flew for short distances with the characteristic mincing fluttering.

P. A. Taverner and B. H. Swales (1907) describe an impressive flight at Point Pelee, Ontario, Canada. They say: "In 1907, when we arrived August 24, Kingbirds were very common and distributed all over the Point and the adjoining mainland. Each day brought more, until by the 27th there were a greater number of Kingbirds present than any of us had ever seen at one time before. Most of them were in the waste clearings near the end of the Point, where at times we saw flocks numbering hundreds of individuals. The dead trees scattered about the edges of these clearings were at all times more or less filled with them and it was no uncommon sight to see from fifteen to twenty in one small tree."

To quote again from A. F. Skutch's notes: "The southward migration of kingbirds passes through Central America during September and the first half of October. In 1930 I saw more kingbirds during the autumn at Tela, on the northern coast of Honduras, than I have seen in any other locality. Here I kept watch over a roost of kingbirds during the southward migration. The site they selected as their sleeping place was a patch of tall elephant grass, higher than a man's head and very dense, which already was the nightly shelter of myriads of small seed-eaters of four species, of the resident Lesson's orioles and of the flocks of orchard orioles that had arrived somewhat earlier. It was a surprise to find the kingbirds, those creatures of high and open spaces, consorting in slumber with the humble seed-eaters, yet all got along most amicably together. The new arrivals were silent among all that chattering throng. At dusk I would see them hovering on beating wings, or moving slowly between the tall grass stalks, often circling and turning, more rarely making a short dart into the open space above, picking up a few final morsels before they settled down in sleep. Because of their active habits and indifference to concealment, the kingbirds were, during their sojourn in the valley, one of the most conspicuous members of its avian community."

DISTRIBUTION

Range.—North and South America.

Breeding range.—The breeding range of the eastern kingbird extends **north** to southern British Columbia (Courtenay, Westminster, and Swan Lake); central Alberta (Edmonton, Belvedere, and Lac la Biche); southern Saskatchewan (Wiseton and Quill Lake); central Manitoba (Chimawawin and Grand River); southern Ontario (Gargantua, Cobalt, and Ottawa); southern Quebec (Montreal, Quebec, and Kamouraska); New Brunswick (Chatham); Prince Edward Island (Tignish); and the Magdalen Islands. The **eastern** limits of the range reach the Atlantic coast from the Magdalen Islands, Quebec, south to southern Florida (Royal Palm Hammock). To the **south,** the Gulf coast is reached from Florida (Royal Palm Hammock, St. Petersburg, and Pensacola); **west** to Texas (Houston and Refugio), thence in the interior to northern New Mexico (Ribera and Santa Cruz); northern Utah (Salt Lake County); and Oregon (Malheur Lake, Burns, and Wasco). **West** to western Oregon (Wasco and Maupin); western Washington (Nesqually Plains, Seattle, Dungeness, and Bellingham); and southwestern British Columbia (Courtenay).

During the summer season kingbirds also have been recorded at many points well outside their normal breeding range, as in the north to central British Columbia (Hazelton); Mackenzie (Fort Simpson, Fort Resolution, and Fort Rae); and Labrador (Cape Mokkovik and Killinik Island). There are a number of summer records for California, and the species has also been recorded at this season in Arizona (Kayenta) and Nevada (Alamo, Lovelock, and Big Creek Ranch).

Winter range.—The winter range extends **north** to Costa Rica (Villa Quesada and Volcan de Trazu); eastern Panama (Gatun); northern Colombia (Trojas de Catoca and Bonda); and British Guiana (Abary River and Blairmount). From the latter region the range extends southward, probably through western Brazil, to Bolivia (Santa Cruz de la Sierra and Caiza). **South** to southern Bolivia (Caiza); and Peru (Lima). **West** to Peru (Lima); Ecuador (Zamora and Gualea); and Costa Rica (Villa Quesada).

Spring migration.—Early dates of arrival in the United States are: Florida—Basinger, March 14; Kissimmee, March 20. Georgia—Beachton, March 25. South Carolina—Mount Pleasant, March 25. North Carolina—Raleigh, April 13. Virginia—Variety Mills, April 17. District of Columbia—Washington, April 18. New Jersey—Caldwell, April 28. New York—Ballston Spa, May 1. Connecticut

—Hadlyme, April 26. Massachusetts—Boston, April 30. Vermont—St. Johnsbury, May 5. New Hampshire—Hanover, May 3. Maine—Presque Isle, May 5. Quebec—Sherbrooke, May 15. New Brunswick—Chatham, May 12. Nova Scotia—Pictou, May 16. Prince Edward Island—May 19. Louisiana—New Orleans, March 19. Tennessee—Sewee, April 17. Kentucky—Eubank, April 12. Missouri—St. Louis, April 15. Illinois—Odin, April 16. Indiana—Brookville, April 18. Ohio—Oberlin, April 22. Michigan—Petersburg, April 23. Ontario—Ottawa, May 3. Iowa—Keokuk, April 23. Wisconsin—Milwaukee, April 20. Minnesota—Lanesboro, April 24. Manitoba—Aweme, May 10. Texas—Kerrville, April 22. Kansas—Onaga, April 19. Nebraska—Syracuse, April 25. South Dakota—Rapid City, May 8. North Dakota—Larimore, May 10. Saskatchewan—Indian Head, May 14. Colorado—Denver, May 7. Wyoming—Cheyenne, May 9. Montana—Terry, May 13. Alberta—Edmonton, May 21. British Columbia—Edgewood, April 24.

Fall migration.—Late dates of fall departure are: British Columbia—Okanagan Landing, September 13. Montana—Columbia Falls, September 11. Wyoming—Yellowstone Park, September 30. Colorado—Fort Morgan, September 15. Saskatchewan—Eastend, September 9. North Dakota—Fargo, September 18. South Dakota—Rapid City, September 24. Nebraska—Lincoln, September 22. Kansas—Clearwater, October 8. Oklahoma—Copan, September 23. Texas—Brownsville, October 1. Minnesota—St. Paul, September 23. Iowa—Des Moines, September 30. Ontario—Toronto, September 24. Michigan—Blaney, September 27. Ohio—Wauseon, September 28. Illinois—Chicago, September 25. Missouri—Columbia, September 23. Kentucky—Danville, September 29. Mississippi—Biloxi, October 20. Prince Edward Island—September 4. New Brunswick—Scotch Lake, September 16. Maine—Portland, September 12. New Hampshire—Durham, September 11. Massachusetts—Hudson, September 20. New Jersey—Milltown, September 17. District of Columbia—Washington, September 23. North Carolina—Raleigh, September 18. South Carolina—Charleston, October 9. Georgia—Atlanta, September 19. Florida—Pensacola, October 6; Orlando, October 12.

Casual records.—Aside from the summer occurrences immediately north of the regular breeding grounds, there are not many cases where this species has been detected outside of its normal range. It was observed along the Humber River, Newfoundland, during the summer of 1911, and ultimately it may be found to breed on that island. The kingbird has been credited also to Greenland, but the evidence upon which the record was based is not now known. On June 17, 1931, an adult female was captured by an Eskimo at Point Barrow, Alaska.

Egg dates.—British Columbia: 10 records, July 6 to July 7.

Colorado: 13 records, June 7 to 25.

Florida: 5 records, May 3 to June 11.

Illinois: 20 records, May 2 to July 27; 10 records, June 11 to 21, indicating the height of the season.

Massachusetts: 34 records, May 30 to June 30; 18 records, June 5 to 20.

Ontario: 19 records, May 30 to July 21; 10 records, June 12 to 25.

Pennsylvania: 21 records, May 23 to July 14; 11 records, June 1 to 11.

TYRANNUS DOMINICENSIS DOMINICENSIS (Gmelin)

GRAY KINGBIRD

PLATE 5

HABITS

CONTRIBUTED BY ALEXANDER SPRUNT, JR.

Every field ornithologist can call to mind certain observations that stand out indelibly in memory, not so much because of the rarity of the species involved but because of the general combination of circumstances surrounding it. Indeed, such a recollection might deal with a locally abundant bird, or one that the observer has seen many, many times, but the particular situation and conditions are such as to frame it permanently in memory. I have such a one in mind in connection with the gray kingbird.

In company with a northern ornithologist, I was, one May afternoon, at the Pan American Airport in Coconut Grove, Miami, Fla., watching one of the great "clippers" being hauled up the ramp by a puffing little tractor. The ship was placed on level ground and the tractor departed elsewhere. As we stood there, marveling at the intricate fabric before us, a quick shadow fell across us and a small gray bird swept overhead, chattered once or twice, and came to rest on the tail fin of the plane. My companion promptly lost interest in the latter and with an exclamation, focused a pair of glasses on the newcomer and stared at it intently. It was the first gray kingbird ever seen through those glasses; a new bird for the "life list," another of many "hoped for" species on that trip.

It is always a satisfaction to show a companion a new bird, and I enjoyed it from that angle, but there was something else about that sight that was tremendously appealing and eminently fitting. That trim gray bird sitting there on the huge gray plane inevitably started a train of thought. Two travelers of the sky they were; the one a tiny, fragile mechanism of flesh, bone, and feather, marvelously efficient and imbued with life. The other a gigantic combination of metal, wood, glass, and rubber; man-conceived, man-made, man-governed, but now silent, inert, lifeless. Without a guiding brain it could never be anything else than that. Both were birds of the air, but the one created by an alchemy beyond the ken of even the master minds that built the other.

Both had just recently come from the Tropics to land upon Florida's shore; one of them silently, surely, guided by something of unerring accuracy, which shaped its living course without chart or compass; the other, boring northward with four great motors roaring steadily, its sharp hull cleaving the upper air like some fabulous juggernaut. And now, both had reached the goal and the tiny one rested upon its huge imitator, to give a glimpse in comparison to two bird students, and in the memory of one of them, at least, that brief tableau will remain forever etched.

Spring.—There is a rather irregular and varied series of dates for the spring arrival of the gray kingbird in Florida. Arthur H. Howell (1932) states that Atkins mentions it as arriving at Key West on April 11. This checks closely with what observations I have been able to gather in that area. Edward M. Moore, the Audubon Society's representative in Key West, was instructed to pay particular attention to this, and in 1938 the bird was first seen on April 12. In 1939 it appeared on April 10. And yet it has been reported in Fort Lauderdale on March 25, 1918; New Smyrna, April 3, 1924; and Chokoloskee, April 5, 1928 (Howell). The earliest record for the spring arrival was at the Dry Tortugas, March 16, 1923, and Scott took two birds there on March 23, 1890 (Howell).

It is my experience that the bird can be seen anywhere in the Florida Keys after April 15, but anyone visiting south Florida prior to April 10 might well be disappointed in not seeing it. In extreme northwestern Florida Francis M. Weston says that "the gray Kingbird is a late migrant into the Pensacola region, the westernmost limit of its range, and the earliest arrival dates in my journal are April 26, 1935, and April 27, 1936."

In South Carolina the few occasions of its arrival have been early in May. Nests with one and two eggs were secured on May 28 and 30, which means that the birds probably arrived early that month.

On May 17, 1927, Arthur T. Wayne and I saw a gray kingbird on Oakland Plantation, Charleston County, only a few hundred yards from Mr. Wayne's house. The bird was perched on a plow handle standing in a field. Mr. Wayne was sure that it had a mate and was settled for the summer, so we did not take it. Search of the vicinity, however, failed to reveal another, and nothing more was seen of the bird. Wayne (1927) recorded the observation and ends his note with this comment: "This makes the fifth gray kingbird I have seen in South Carolina since 1885. * * * These birds have longer wings, culmen and middle toe than specimens from the Bahamas, Florida, Greater Antilles and Caribbean Sea showing that the birds that breed on the coast of South Carolina have a much longer distance to travel and hence possess longer wings."

Since Georgia intervenes between South Carolina and Florida, one would expect rather more instances of the occurrence of *dominicensis* there than in South Carolina, but in reality they are about the same, though not so well known. Indeed, present-day workers in Georgia ornithology seem under the impression that there are fewer records than actually exist! For instance, Ivan R. Tomkins, of Savannah, says (1934) that "Rossignol [Gilbert H.] who is thoroughly familiar with the species in its normal range, saw and heard a single bird near Quarantine Station June 8, 1933. Quarantine is near the river mouth [Savannah] fourteen miles east of the city. [The Savannah River is the South Carolina–Georgia line.] There are no other Georgia records."

Mr. Tomkins was in error in making this last statement, and it seems important to clear the record. There were two records previous to Rossignol's, and two have appeared since, making a total of five. Of these, two are nesting records, which will be mentioned under the heading "Nesting," while the others consist of a specimen taken by Arnow at St. Marys, on August 1, 1905, and recorded by Troup D. Perry (1911), and a sight record by Hoxie, which was reported by Fargo (1934, p. 190). These South Carolina–Georgia records constitute all that is known of the spring movements of the gray kingbird outside of Florida. Extralimital records of fall occurrence will be discussed later.

Courtship.—There seems little to add to Audubon's account (1840) on this phase of the gray kingbird's history. He puts it so well and covers the ground so accurately that observations of others since have only been corroboration of it. He says: "During the love season, the male and the female are seen rising from a dry twig together, either perpendicularly, or in a spiral manner, crossing each other as they ascend, twittering loudly, and conducting themselves in a manner much resembling that of the Tyrant Flycatcher." Baird, Brewer, and Ridgway (1905) quote Richard Hill, of Spanish Town, Jamaica, as saying practically the same thing as Audubon, but more recent observers are reticent, probably because there is nothing much to add. I have noted no variation of the above proceedings in my observations of the mating flight except that on a few occasions there was some pronounced snapping of the bill as the birds ascended in the spiral manner. The courtship is usually performed in open situations and is therefore conspicuous.

Nesting.—There is little time between the arrival of this species from the Tropics and the commencement of its domestic duties. This fact is quite apparent to any one familiar with the bird, and I quote pertinent notes sent me by Francis M. Weston, of Pensacola, Fla.: "The gray kingbird apparently starts nesting almost as soon as it arrives. In 1932, a completed nest was found on May 15, only a few days after the normal date of arrival, and the first egg was laid on May 29. This was destined to be an unusual nest for by June 2 it contained four eggs instead of the almost universal three. Another early nesting, in 1933, resulted in the young birds being on the wing by June 21, which means, if we assume that periods of incubation and nest life are the same as for the eastern kingbird, that the first of three eggs was laid not later than May 29. Occupied nests can be found from early in June until early in August, though the late nests may be the result of repeated attempts to raise a single brood rather than true second nestings."

During the breeding season, as at other times except migration, this species is essentially a seaside lover. It displays little fear of man, and the nest may be approached closely, with the bird remaining either on it or in the immediate vicinity. One nest that I examined on Pavilion Key, amid the Ten Thousand Islands, Fla., was very jealously guarded, and the birds remained in close proximity during the time it was being photographed. This strictly littoral habit seems general throughout its range, and at the western extremity (Pensacola, Fla.) F. M. Weston (MS.) has described it as follows: "During the breeding season it is confined almost exclusively to the dense jungle of saw palmetto, vines, scrubby live oak, and stunted magnolia that clothe the landward side of the high range of sand dunes that front the Gulf beach on this part of the coast. The only departure from this habit that has come to my attention was the location of a pair for three consecutive summers among the trees and gardens of the officers' quarters at the Naval Air Station, still on the waterfront but more than a mile from the outer beach."

Of the only two nesting sites of this species recorded from South Carolina, one was in the city of Charleston itself, and since the latter occupies a peninsula, salt water is nowhere far off. As it happens, this urban site was about midway between the two rivers bounding the strip of land that the city occupies, about three-quarters of a mile from either. The nest itself must strike anyone who has seen it as decidedly unsubstantial. The term "flimsy" has been employed by both A. H. Howell and F. M. Weston in describing it. All the nests seen by me have been of that character, and in several the contours of the eggs were visible from beneath, through the nest material. Thus, it differs materially from the nest of the eastern kingbird, which is distinctly well made, compact, and bulky. Rather coarse twigs form the foundation of the nest, and the lining may be of various grasses, sometimes salt-marsh grass if near a locality where this plant is found.

As to the choice of a nesting tree, the mangrove, usually the red mangrove (*Rhiziphora mangle*), is almost invariably chosen throughout the bird's range in Florida. Where the species penetrates and this tree does not, an oak is usually taken, the stiff twigs of which form excellent support for the frail nest. The mangrove occurs as far up the east coast of Florida as New Smyrna and Daytona, and it is here that one usually encounters the first gray kingbirds on the way south. On my frequent trips to Florida, I never think about watching the wires along the road for this species until I reach Daytona, and the birds are fairly common in that town itself.

Low altitudes are to be expected in view of the favored growths. While the mangrove reaches considerable height along the southwest coast of Florida, the birds seem consistently to prefer the normal types and build no higher than 10 to 12 feet from the ground or, as is often the case, over the water. Many nests are no more than 3 or 4 feet up. Variations will, of course, occur. The two greatest recorded elevations I can discover are those of a nest seen by Dr. Wetmore (Wetmore and Swales, 1931) near Constanza, Haiti, which was 40 feet from the ground on the limb of a lofty pine. The other was a nest found by J. H. Riley (1905) on Abaco Island, Bahamas, which was "about 50 feet high in a pine."

Strong attachment is exhibited by the gray kingbird for a nesting site. Weston has noted that "apparently a pair of birds returns to the same tree or the same clump of trees year after year, for several nests in progressive stages of decay are usually found within an area of a few square yards." In the account of a nesting many years ago in South Carolina, Dr. John Bachman stated that the birds returned year after year to the same clump of trees in Charleston to build.

In his full notes on this species, Weston (MS.) mentions another possible habit as follows: "Egg destruction may be imputed against this species from a single circumstantial instance noted on July 25, 1928. A nest containing three eggs was found on July 9. As it turned out later, another pair of birds had built less than 30 feet away on the far side of a small dune. On July 25, the known nest was found to have been deserted and all three eggs had been punctured as by the beak of a bird, while in the second nest (just then discovered) were well-grown young birds. The inference is that the established pair had destroyed the eggs of the intruders."

The only authentic nesting records of the gray kingbird outside of Florida have been for South Carolina and Georgia. So far as South Carolina is concerned, the known nesting dates are many years apart, but close together geographically. Since I have a "home town" pride in them, plus the fact that they are ornithological history, they are quoted herewith in full, and they constitute the most northerly nesting of *dominicensis*. Audubon (1840) records the first instance, though he himself did not observe the birds or the nest. He writes:

After I had arrived at Charleston in South Carolina, on returning from my expedition to the Floridas, a son of Paul Lee, Esq., a friend of the Rev. John Bachman, called upon us, asserting that he had observed a pair of Flycatchers in the College [Charleston] yard, differing from all others with which he was acquainted. We listened, but paid little regard to the information, and deferred our visit to the trees in the College yard. A week after, young Lee returned to the charge, urging us to go to the place, and see both the birds and the nest. To please this amiable youth, Mr. Bachman and I soon reached the spot; but before we arrived the nest had been destroyed by some boys. The birds were not to be seen, but a common King Bird happening to fly over us, we jeered our young observer, and returned home. Soon after the Flycatchers formed another nest, in which they reared a brood, when young Lee gave intimation to Mr. Bachman, who, on visiting the place, recognised them as of the species described in this article. Of this I was apprised by letter after I had left Charleston. * * * The circumstance enforced upon me the propriety of never suffering an opportunity of acquiring knowledge to pass, and of never imagining for a moment that another may not know something that has escaped your attention.

Since that time, three years have elapsed. The birds have regularly returned every Spring to the College yard, and have there reared, in peace, two broods each season, having been admired and respected by the collegians, after they were apprised that the species had not previously been found in the State.

Young Paul Lee deserves more credit than he ever received, for had it not been for his persistent visits to Dr. Bachman the knowledge might well never have gone farther than his own conviction that he had seen something unusual.

The College of Charleston occupies a city block bounded by St. Philip, George, Green, and College Streets. There are many very large oaks in the "yard," and the nests must necessarily have been at considerably greater altitudes than those usually utilized by this species. This illustrates the tendency of the gray kingbird to return to the same tree "or clump of trees," as Weston points out. Since Bachman's day, no further instance of the nesting has occurred there; at least none has been noted.

Many years elapsed before the gray kingbird was seen again in coastal South Carolina, this time by William Brewster and Arthur T. Wayne. Wayne (1910) prefaces his account of this record as follows: "Since Audubon wrote, I have been the next observer who has seen and taken this rare species in the State." It so happens that Wayne was in error here, though the record previous to his and Mr. Brewster's was not made known publicly until many years afterward. Herbert Ravenel Sass, of Charleston, who was once connected with the newspaper business, and who conducted a nature column in the

"News and Courier," wrote, under date of March 17, 1921, that "Mr. W. B. Gadsden of Summerville [S. C.] supplies the interesting information that his father, the late Prof. John Gadsden, obtained one of these birds [Gray Kingbird] on the grounds of Porter Military Academy in Charleston, somewhere between 1881 and 1885. Mr. Gadsden says that his father was struck by the unusual appearance of some birds in the Porter elms, and shot one of them with a small rifle. He then took the bird to the College of Charleston Museum [now the Charleston Museum] where it was definitely identified as the Gray Kingbird."

Unfortunately, the exact date cannot be ascertained now, nor can the supposition that this was another city nesting record be substantiated. However, this information fills a gap between the Lee-Audubon-Bachman instance and the Brewster-Wayne record, though much closer to the latter in date. As far as cities are concerned, Charleston certainly seems more in the limelight in regard to the gray kingbird than any other in the country!

The remaining nesting records of South Carolina occurred on Sullivans Island, which was made famous by the Battle of Fort Moultrie in the Revolution and which lies just across the harbor from the City of Charleston. The first of these is described by Arthur T. Wayne (1894):

In the early part of May, 1885, Mr. William Brewster and myself saw a pair of Gray Kingbirds at Fort Moultrie, Sullivan's Island, S. C. I determined to secure these birds with their nest and eggs, and after several visits to the Island I located their range, and on May 28, I found their nest which contained one egg and shot the female bird. The nest was built in a silver-leaf poplar, in a gentleman's yard [Maj. W. J. Gayer] only a few feet from his dwelling house. The nest, as I remember it, was very frail.

As might be supposed by anyone who knew Mr. Wayne, this discovery fired him with intense enthusiasm, and he had the species in mind every spring. He was always greatly attracted to the barrier islands of the coast and visited them monthly throughout the year with one bird or another in view. He ever connected Sullivans Island with the gray kingbird, but it was not until 1893 that he was again successful in finding the species there. His graphic account (1894) of that experience follows:

On May 30 of this year [1893], I determined to search Sullivan's Island carefully for this rare visitor, and accordingly I arrived there early in the morning of the above date. After walking the entire length of the Island near the front beach [about 5 miles], and having failed to discover this species, I leisurely searched the back beach. At twelve o'clock—mid-day—a bird I saw flying about three hundred yards away I took to be this species. I followed the direction of its flight until it was lost to view—over half a mile away. I at once hastened to the spot, and to my delight found a veritable Gray Kingbird perched on the top of a flag pole about fifty feet high in a private yard. The law on the Island

forbids shooting, under penalty of $10.00 fine.   My only chance was for the bird
to light on the Government property—Fort Moultrie grounds—six yards away,
where I could not be molested.   I did not have long to wait before the male
which was perched on the flag pole flew into the Government lands where I at
once shot it.   Upon my shooting the bird its mate flew directly over me, and I
soon had it stowed carefully away in my collecting basket.   The nest which was
found in the private yard, close to the flag pole, was built in the top of a small live
oak tree about twenty feet high.   It is a very frail structure, and is composed
of sticks, jesamine vines, and lined apparently with oleander rootlets.   One article
in its composition which is quite curious is a long piece of fishing cord.   The
nest contained two eggs, and upon dissecting the female I found one more egg
which would have been laid the following day.   It will be seen that all the specimens of the Gray Kingbird which have been actually taken in South Carolina were
from this famous Island—a favorite summer resort for the people of Charleston.

As noted above, Mr. Wayne was in error in this last statement, for
the specimen seen by Professor Gadsden in the grounds of the Porter
Military Academy, between 1881 and 1885, was taken.   Indeed, particular significance attaches to the P. M. A. specimen, for, since it was
not stated by Audubon that Bachman took any of the College of
Charleston campus birds, Prof. Gadsden's specimen was the first one
secured in South Carolina.   The species has not been observed, or at
least there is no record of it, since the May 1927 bird seen by Mr.
Wayne and myself.

The history of the gray kingbird in Georgia is similar.   One specimen has been taken, two have been seen, and there are two definite nesting records.   The first instance is that of a specimen secured by Isaac
Arnow at St. Marys (near the Florida line) on August 1, 1905, and
recorded by T. D. Perry (1911).   Sight records made by Walter Hoxie
on the Savannah River (South Carolina State line) are vague and
without dates.   The other was an observation of Gilbert R. Rossignol,
at the Quarantine Station, mouth of the Savannah River, June 8, 1933,
this having been already mentioned here.   The two sight records, viz.,
those of Hoxie and Rossignol, were within an ace of being South Carolina records, as well as Georgia observations.

H. B. Bailey (1883) gives some notes on the nesting in Georgia but
mentions only one instance specifically.   This concerns a set of eggs
collected by Dr. S. W. Wilson, "between the years 1853 and 1865."   That
more than one breeding record was concerned is evidenced by his statement that the gray kingbird nests "chiefly on St. Simon's Island and
in Wayne and McIntosh Counties."   This leaves a good deal to be
desired in the way of precise information, but one can be certain of the
nest found by Dr. Wilson if nothing else.   Probably the other records
will always be shrouded in obscurity, which is unfortunate.   Mr.
Bailey states that the species "nests on the horizontal branches of oak
trees, near the top, and loosely constructed of twigs 'with little or no
lining'; eggs always three."   The most recent instance of the breeding

of this species in Georgia was the discovery of a nest and three eggs on July 3, 1938, by Don Eyles and Ivan R. Tomkins, on the grounds of the Quarantine Station, at the mouth of the Savannah River. It is recorded by Eyles (1938). He was under the impression that there were no former records of the nesting of this species in the State, for he states this in his account. He was also either unaware of or ignores the Hoxie sight records. The young of the above nest successfully hatched, and photographs of them were taken. The Quarantine Station is on Cockspur Island and virtually on the South Carolina line.

Eggs.—[AUTHOR'S NOTE: The gray kingbird lays usually three or four, perhaps rarely five, very handsome eggs. They vary from ovate to elliptical-ovate, less often elongate-ovate, and they are only slightly glossy. The ground color varies from "seashell pink" to pale "salmon-buff." The eggs are irregularly but rather profusely spotted and blotched with dark, rich browns, "chocolate," "burnt umber," "claret brown" or "cinnamon-brown," and with shades of "Quaker drab," "brownish drab," or "lavender." The measurements of 50 eggs average 25.1 by 18.2 millimeters; the eggs showing the four extremes measure 27.5 by 18.8, 25.7 by 19.4, 22.6 by 17.5, and 22.9 by 17.0 millimeters.]

Plumages.—[AUTHOR'S NOTE: In four nestlings that I have examined, the natal down still adheres to the growing juvenal plumage; the down varies in color from "cream-buff" to "cartridge buff"; the crown is "hair brown," and the back is "deep brownish drab"; the wing coverts are tipped with "cinnamon"; and the under parts are white, tinged with buff on the sides and flanks.

Ridgway (1907) describes the young, in full juvenal plumage, as "essentially like adults, but without orange on crown; gray of upper parts browner; upper tail-coverts broadly margined with rusty brown or chestnut, rectrices edged and terminally margined with cinnamon, lesser wing-coverts margined with cinnamon or cinnamon-buff, and other paler wing-markings more or less tinged with cinnamon."

I have seen birds in this juvenal plumage as late as October 22, which would indicate that the postjuvenal molt is prolonged through the winter, with perhaps a partial body molt after the birds have migrated, and a molt of the wings and tail later in the winter, when the attenuated three or four outer primaries are acquired. I have seen one young male, taken February 21, that was molting its tail and in which the wings were much worn; it also had very little yellow in the crown. By the time the young birds come north they have probably acquired a plumage that is nearly or quite adult.

Adults apparently molt mainly while they are in their winter quarters, and we have not enough winter specimens to determine just

how or when this is done. Probably there is a postnuptial molt of
the body plumage late in fall and a molt of the flight feathers late
in winter. I have seen an adult female, taken March 20, that was
molting about the throat and apparently molting the tail, the wings
apparently having been renewed.]

Food.—With a species as restricted in range in this country as the
gray kingbird is, one cannot find a great deal in regard to the specific
character of its food. Most of the information available relates to
the range in the Tropics. However, being what it is, a typical flycatcher, the general nature of the diet is obvious. It is inevitable
that insects predominate, but exact information is limited. As I have
had no experience in the stomach analysis and have not collected
the species, I depend entirely upon the findings of others.

Arthur H. Howell (1932) gives results on two stomachs only. One
"from St. George Island contained only insect remains, of which
Hymenoptera (bees, wasps, etc.) composed 61 percent, a large wood-
boring beetle 31 percent, and bugs 5 percent." Another was "a specimen taken at Cape Sable" which had eaten "3 large dragon flies, 1
bee, and 10 berries of the gumbo-limbo, or West Indian birch
(*Elaphrium simaruba*)."

The berry-eating habit seems to be characteristic, as it has been
noted in other parts of the range, foreign as well as domestic. Relative to the former, Richard Hill, of Spanish Town, Jamaica, is
quoted in Baird, Brewer, and Ridgway (1905) as saying that the gray
kingbird eats "wild sweet berries, especially those of the pimento."

South Florida is replete with many tropical plants, and since the
gray kingbird is so abundant in that part of the State it is natural
that its berry-eating propensities would include several species unknown in other parts of the country. The gumbo-limbo, mentioned
by Howell, is a very characteristic tree of extreme southern Florida,
a remarkable growth in many respects, commonly found in nearly
every "hammock" and a favorite tree of the beautiful *Liguus*, or
arboreal snail.   The soapberry (*Sapindus saponaria*) is also very
common, particularly about the Sable Capes, where I have found it
in several hammocks. Though not specifically mentioned in the food
of the gray kingbird, its great bunches of berries could hardly fail
to attract this species, and examination of stomachs in that locality
might well reveal evidences of this growth. Wetmore (1916) states
that berries of the royal palm (*Roystonea borinquena*) are freely
eaten in Puerto Rico, and since this magnificent tree is still found
in isolated spots in south Florida, it undoubtedly figures in the diet
of the species also.

In cultivated districts the positively beneficial results of the gray
kingbird's food habits are illustrated by the reputation imparted to
it in the Virgin Islands, where it destroys cotton pests. Charles E.

Wilson (1923) states that these birds "eat large numbers of the larvae and adults and are of great value in controlling the cotton-worm" (*Alabama argillacea*), which is the second most important pest of cotton in the Islands. In that locality it also feeds on the fall army-worm (*Laphygma fragiperda*), southern green stink bug (*Nezara viridula*), and bollworm (*Heliothis obsoleta*). All these are injurious to cotton.

Junius Henderson (1927) gives a table listing as complete a summary of the gray kingbird's food as I have found. It is an account of an analysis of 89 stomachs studied by Dr. Alexander Wetmore (1916) in Puerto Rico.

A digest of the full information given by Dr. Wetmore follows: Vegetable matter comprised some 22.44 percent and animal matter 77.60 percent. Of the former, the items are broken down as follows:

Seeds and fruits comprise 22.06 percent, while vegetable rubbish amounts to only 0.38 percent. The berries borne by the royal palm (*Roystonea borinquena*) and other species of the same family are favorites, as are those of the espino (*Xanthoxylum* spp.), a fruit with little pulp and a peculiarly reticulated seed. Seeds of various euphorbias and of plants of the nightshade family are also sought greedily, and one third had eaten seeds of the Santa Maria (*Lantana* sp.), a pernicious weed not of major importance in Puerto Rico though very troublesome under similar conditions in Hawaii. The moral (*Cordia* spp.) was perhaps the favorite, being found in 12 stomachs. The fruits eaten were all wild and none are of commercial importance.

The animal matter (77.56 percent) is listed as follows:

	Percent		Percent
Mole crickets	2.36	Coleoptera	1.3
Other Orthoptera	0.95	Honeybees (workers)	2.21
Earwigs	4.72	Bees, mostly wild	15.28
Homoptera (largely cicadas)	1.97	Other Hymenoptera (mostly	
Stink bugs	2.12	wasps)	11.33
Other bugs	0.68	Moths and caterpillars	4.75
Cane-root weevils	17.19	Miscellaneous invertebrates	2.42
Stalk-boring weevils	5.30	Lizards	3.64
Miscellaneous weevils	1.34		

Richard Hill, of Spanish Town, Jamaica, already quoted in regard to the berry-eating habit, imparts a rather remarkable predaceous practice of the gray kingbird in "seizing hummingbirds, as they hover over blossoms * * * killing the prey by repeated blows struck on a branch and then devouring them." Strange tactics for a flycatcher certainly, and probably confined to the tropical portion of its range. It will be recalled that some of the goatsuckers include small birds in their diet, whether intentionally or not being yet somewhat obscure, but such action on the part of the gray kingbird appears to be an utterly deliberate and unique habit. The case of egg destruction implied by F. M. Weston cannot be laid to dietary motive, but rather to punitive measures taken by a bird whose territory was invaded by another of the same species.

Wetmore (1916) sums up the food habits as far as Puerto Rico is concerned thus: "Detailed study of the food of the gray kingbird shows it to be beneficial almost without exception. A few honeybees are eaten, but they are more than made up for by the large bulk of injurious weevils, mole crickets, and Hemiptera destroyed. Though not so great an enemy of the changa (*Scapteriscus didactylus*) as has been commonly believed, it accomplishes practically as much good in consuming cane root- and stalk-boring weevils and coffee leaf-weevils."

Though it is unfortunate that more is not known about the diet of this species in its United States range, there is no reason to believe that its beneficial tendencies in the Tropics do not extend to Florida and elsewhere in this country. It is often common about citrus groves and must perform distinct service in reducing the insect pests in such places, as well as about cultivated fields of vegetables. The swarming insect life of Florida must offer unlimited food in such line, and anyone who has had to work afield in many parts of the gray kingbird's range in that State will have wished for more reducing agencies to curb the incredible abundance of these creatures. The species should be encouraged in every possible way.

Behavior.—There is nothing really outstanding in the behavior of the gray kingbird differing from that of the other Tyrannidae; it is very characteristically a kingbird! It is fond of an exposed perch in order that detection and pursuit of insect prey can be accomplished with despatch. Any sort of perch may be utilized, but wherever telephone wires are available these are preferred to anything else.

One of the best localities for observation in the whole of the bird's range is the Florida Keys—certainly a very appropriate place, for here it was made known to science by Audubon. Coming in by Indian Key, he probably saw his first ones on the Matecumbes, or perhaps Lignum-vitae Key. Today the visitor there can see them as easily as he did, and under similar conditions, for the species is very abundant there, perhaps more so than any part of the State. Driving along the Overseas Highway (Florida 4–A), one encounters long stretches where the wires are close to the road, and at times gray kingbirds seem to be everywhere. Of course, numbers vary, and on some days fewer are seen than on others. I have made 16 trips over this highway, from the mainland to Key West, and in some parts of it have counted as many as five individuals in 10 miles, or one every 2 miles. They may be even commoner that this, or much scarcer. The characteristic contour of the bird can be recognized at considerable distances, and the dashing manner of the feeding habits is quite spectacular. Forays after insects are carried out with much verve and considerable speed.

As mentioned previously, they are essentially a littoral species, and in the Keys one is never far from water. Thus, one is impressed by the kingbird population in that locality, and even some of the smallest of the keys in the Bay of Florida have their quota of the birds.

In the habit of tolerating no intrusion upon its domain or territory, this kingbird is like others of the family. It does not hesitate to attack any other bird and sometimes mammals. In the Keys, the two species more often assailed seem to be the turkey vulture and the insular red-shouldered hawk. I have frequently seen both these birds of prey dodging and twisting about in the air with a tiny, dancing speck above, and upon them, which resolves itself into a gray kingbird. Herons are sometimes attacked, even the majestic great white, and the contrast exhibited under such conditions is hardly more spectacular than ludicrous.

John R. Williams (1935) gives an instance of the eastern kingbird (*T. tyrannus*) attacking an airplane. I do considerable flying in the Florida Keys and have often wondered whether *dominicensis* would make such an attack, but as yet no case has been noted. As Mr. Williams points out, "A case of this sort could scarcely have occurred except where a slow, low-flying plane was involved." While the planes used by me are not particularly slow, they are certainly often low-flying, as one of the primary objects of these flights is to check numbers and nests of the great white heron, and often altitudes of under 100 feet, and sometimes much less than that, are flown. It would not surprise me should such an encounter take place, and indeed, it may well have already occurred and escaped notice.

I have never observed the gray kingbird eating berries so cannot describe it, though it is unlikely that there is anything unusual about the habit. I do recall seeing one of these birds near Cuthbert Lake one day (Cape Sable area), hovering in a rather peculiar manner about the outer branches of a large gumbo-limbo tree and moving rather jerkily about. At the time I thought it was catching insects, but it might have been picking berries. Probably this action is usually indulged in, however, while the bird is perched, and not on the wing.

Insect prey is often taken close to the surface of water and also actually from it. In open situations the bird will make long swoops from a mangrove, snap up the prey a foot or so above the water, describe a curve, and swing back to the perch. Or, if necessary, it will hover momentarily, pick up the insect from the surface, and return. In more restricted quarters, the tactics vary. In the canal, for instance, which parallels the highway on Key Largo from the Card-Barnes Sound Bridge, I have watched kingbirds virtually dive from a mangrove perch to the water. The recovery and ascent appear very loosely accomplished, as if the wings were almost completely pivoted. This must be what Richard Hill noted in Jamaica and so aptly described as having the "appearance of tumbling, and, in rising again, ascends with a singular motion of the wings, as if hurled into the air and endeavoring to recover itself." Though most of the kingbirds flutter their wings rather rapidly in flight, this species seems to exhibit the habit to a greater extent, the vibrations being very noticeable.

The tameness of the gray kingbird is marked. It allows close approach and appears indifferent to observers, though excitable enough when the nest area is invaded. It is very noisy then and indulges the habit of snapping the beak. Some pairs of birds are more truculent than others and defend the nest strenuously, even against human beings. Without actual contact being made, one sometimes dodges involuntarily as the excited, chattering birds swoop and dive about one's head. J. H. Riley (1905) speaks of a nest found on Eleuthera Island, Bahamas, where the owner or owners darted almost into his face as he was investigating it. I had a somewhat similar experience with a nesting pair on Pavilion Key in the Ten Thousand Islands of Florida.

While the gray kingbird frequents the remotest wilderness areas, such as the Cape Sable district of south Florida, it occurs freely about cities and towns as well. It seems to make no difference whether absolutely primeval conditions prevail or the roar of city traffic resounds on all sides. I have watched the bird from a Miami hotel window as easily as I have studied it among the mangroves of Cuthbert Lake. Many of the east and west coast Florida towns have sizeable populations of the species, and it abounds in Key West. This inclination makes it easy to know, and the observer who visits its range will have no difficulty whatever in finding and studying the bird at will.

Voice.—The gray kingbird is a noisy species. Its voice is certainly one of its outstanding characteristics and draws mention by all who have written about it. Most of these observers are in universal agreement and unite in voting it vociferous.

The eastern kingbird too is a noisy creature, but it is outdone by its larger southern cousin. As C. J. Maynard (1896) has expressed it, "The northern species are noisy birds but in this respect they are excelled by the Gray King Birds which are constantly chattering." While not literally true, this statement covers the ground very well. Perhaps it would be a little more accurate to say that the gray kingbird is seldom quiet! Whether the bird is sitting on the watch for prey or actually engaging in the chase, whether on guard at the nest or "making a passage" somewhere, the shrill, chattering notes are more likely to be uttered than not. Descriptions of bird notes by means of words are often wide of the mark. It is a difficult matter except in some few striking cases to render avian vocal sounds into English or other words. However, it seems to be the only way to give any idea as to what they are. In the case of *dominicensis* there are several interpretations, but all convey much the same idea.

Richard Hill says that it utters "a ceaseless shriek," being a repetition of three notes like the word "pe-cheer-y." This is the commonest term used to describe the notes. They are, definitely, three-syllabled, with the accent on the second. F. M. Weston's comment on the voice follows:

"The gray kingbird is even noisier than the eastern kingbird. The usual written description of its notes, although poor, is adequate, and, in my own experience, gave me instant recognition of the first gray kingbird I ever heard, even before I saw it. However, it was the quality of the sound that indicated a kingbird of some sort, and my recollection of the accenting of the written words that led to the recognition of the species. On one occasion, I heard an eastern kingbird give an exact imitation of the notes of its larger relative."

The last sentence is commended to readers living in the range of the gray kingbird. As characteristic as are the notes of this bird, it will be noted that *T. tyrannus* is capable of mimicry, and records of birds heard but not seen should be verified. Doubtless this is unusual; indeed, it is the sole instance of the kind that has come to my notice, but if one common kingbird can imitate notes, others can. It is an extremely interesting and important bit of information.

Frank M. Chapman (1912) describes the gray kingbird's notes as "pitirri[e]," a term applied in Cuba for the common name of the species. Indeed, the majority of vernacular names are derived from the voice, as might be expected. James Bond (1936) lists other names as follows: Petchary (Jamaica); fighter, christomarie, pick-peter (Bahamas); pewitler (Barbados); pipiri (Lesser Antilles); chicheri (Virgin Islands); pitirre (Cuba, Puerto Rico); titirre (Dominica); pipirite (Haiti). It will be recalled that Audubon wrote of the bird under the name of pipiry flycatcher. The phonetic similarity between all these names is striking, particularly the last five, which are all but identical. The name "fighter" (Bahamas) is obviously drawn from other characteristics, and in view of J. H. Riley's experience with nesting birds there, it might be inferred that pugnacity is more developed in Bahaman birds than elsewhere!

B. S. Bowdish (1903) states that the gray kingbird sometimes utters a note "at times quite similar to some the phoebe occasionally utters." I have been impressed with this also and have heard it several times in the Florida Keys.

Field marks.—The gray kingbird is a very distinctive and individual species. I can see no reason for confusion or doubt about identification arising from any similarity to others of its genus, even on first sight. True, it is similar to the common kingbird in contour and general appearance at a distance, but in any position, or almost any distance within reason, it is instantly to be recognized as a kingbird. Audubon (1840) though of course recognizing the first ones he saw as something different, seemed much impressed by the similarity to *tyrannus*. He says that "its whole demeanour so much resembles that of the Tyrant Flycatcher, that, were it not for its greater sizer, and the difference in its notes, it might be mistaken for that bird, as I think it has been on former occasions by travellers less intent than I on distinguishing species."

Curiously enough, he makes no allusion whatever to what I consider the bird's most striking character as compared to other kingbirds. That is its color! This has certainly appealed to nearly everyone who has seen the species and can hardly fail to do so. There is no mistake in this bird's name, for the gray kingbird is eminently and strikingly gray. The first one I ever saw vividly brought to mind the loggerhead shrike (*Lanius ludoviciana*), not in shape, size, or anything save the color. Somehow ever since the shrike has come to my mind when I have seen a gray kingbird.

Referring again to F. M. Weston's excellent notes, we find that he characterizes the shade of gray as an "almost ghostly paleness." That it impressed him markedly is evidenced in the following comment: "On first acquaintance with the gray kingbird, an observer familiar only with the eastern kingbird is struck by the much larger size, particularly the larger beak, of the gray bird. The lack of the white terminal band on the tail is immediately noticed. The most striking feature though is the almost ghostly paleness of the whole bird when seen in its chosen habitat of pale vegetation and glaring sunlight. Among the sparse grayish-green foliage and pale-gray bushy twigs of the scrubby live oaks, the bird enjoys almost perfect color protection."

Among mangroves this blending of bird and background is not to be noted. The mangrove does not occur in the Pensacola region, however, and Weston has not seen the bird in far southern Florida.

Against the dark, glossy green of the mangroves, the gray kingbird stands out sharply, but in such vegetation as described above the protective coloration would undoubtedly be impressive.

Weston considers the difference in size between *dominicensis* and *tyrannus* to be considerable, viz, "much larger." E. H. Forbush (1927), in his "Field Marks" of the species, says that it is "similar in size and shape to the kingbird, but somewhat larger etc." Mr. Forbush admitted that he was not at all sure he had seen the species himself. Had he done so, he would have undoubtedly been impressed with the distinctly larger size. The beak of the gray kingbird is so much larger than that of *tyrannus* that it seems over-sized and top-heavy, being very broad and flattened at the base. So, between size, color, larger beak, and absence of the white-tipped tail, the gray kingbird stands alone and can hardly be mistaken for any other bird.

Fall.—While the gray kingbird has a very limited range in this country—virtually one State, Florida—it does wander at times, as so many birds will. Extralimital records, however, are exceedingly uncommon. Other than the few breeding records for South Carolina and Georgia, occurrence of *dominicensis* in the United States is confined to but four Eastern States and one western Canadian Province, the latter the most remarkable of all. Most of these wanderings appear to be indulged in during the autumnal migration, and some of the distances reached and the dearth of records in intervening territory are noteworthy.

In Florida the fall migration is rather early. Arthur H. Howell (1932) notes departures from "St. Marks [Gulf coast] September 26, 1917, and New Smyrna [east coast] September 18, 1924." A specimen seen by Dr. H. C. Burgess in the lower Everglades on December 26, 1917, will be commented on later. F. M. Weston says of the Pensacola area: "Fall migration is early, often by the last week in August. The latest dates in my journal are September 6, 1931, and September 18, 1927." He has this to add on the behavior of the adults and young in late summer: "After the nesting season, the family group remains together until time for departure. At this time they sometimes wander a bit from the beach habitat and have been seen in clearings in the pine woods several miles from the Gulf. The Naval Air Station birds come out into the bare industrial section of the station and catch their prey of flying insects from the vantage point of a concrete coping or a steel tower."

In the Florida Keys I have seen numbers of the birds late in September; at Tavernier, Key Largo, on October 2; and Key West on October 4. They no doubt remain in the Keys until well after these dates. Usually I do not visit Key West in November but start my winter trips to that point early in December. No birds were in evidence then in any of the past five years in any part of the Keys, Upper or Lower, nor at Cape Sable. It is my belief that late October or early November sees the last of the migrants depart.

Outside of Florida, South Carolina, and Georgia (for which last two States there are no fall records) occurrence of the gray kingbird is accidental, but the few instances are worth mention. Progressing northward from the normal range, the Middle Atlantic States exhibit no records, and New Jersey is the first northern State to show one. Its presence in that State is recorded at Cape May by Julian K. Potter (1923). A remarkable thing about this note is that no date is mentioned in connection with it!

While a party of members of the Delaware Valley Ornithological Club were exploring the meadows and dunes at Cape May Point, at the mouth of Delaware Bay, we were attracted by a bird which flew out from a growth of wind swept and half dead Red Cedars in Pond Creek Meadow. It dashed out into the air seized an insect and returned to its perch. It had all the actions of a Kingbird and such we supposed it to be. But when a dozen glasses were levelled at it we saw to our surprise that the bird lacked the characteristic white tip to its tail; the upper surface was found to be gray and in addition a dark line extended through the eye like that of a shrike though broader and not so distinct. In actions and general appearance the bird was like our ordinary Kingbird. He made no sound of any kind while under observation. We were trying to place the bird when someone produced one of those ever ready bird identifiers, Reed's 'Pocket Bird Guide,' and turned to the Kingbird and there on the opposite page was the Gray Kingbird. The bird in the tree was compared with the picture in the book and was found to be identical in every detail. For further confirmation a description was written and sent to Dr. Witmer Stone with the question attached 'What is it?' The answer came back next morning over the wire 'Gray Kingbird.' Several of those who saw the bird examined skins the next day at the Academy of Natural Sciences in Philadelphia and further confirmed the identification. This is the first record of the species for New Jersey and, we believe, with one exception, the first record north of South Carolina.

I wrote to Mr. Potter and asked him the date of this observation, and he answered by stating that it was May 30, 1923. (It would seem from this date that the above observation should have been included in the division "Spring," but since the occurrence is purely accidental, it was thought best to mention it in this discussion of extralimital records.)

Still farther north, the gray kingbird has been taken once in New York State on Long Island. This is said by Ludlow Griscom (1923) to have been "at Setauket, Long Island, about 1874," but no details are given. Strangely, all remaining gray kingbird records for the United States come from New England: Two of them are from Massachusetts, many years apart. The first is mentioned by Forbush, that of an immature bird, taken at Lynn on October 23, 1869, by C. F. Goodale. Mr. Forbush stated that "it has been reported more than once in Massachusetts, but the record given above appears to be the only one substantiated by a specimen taken in the state." The disposition of this bird was not mentioned by Mr. Forbush but is cleared up in the quotation that appears below.

The second Massachusetts occurrence of the gray kingbird was noted on November 22, 1931, and was recorded in graphic style by F. H. Allen and Ludlow Griscom (1932):

On November 22, 1931, a party of observers was working in West Newbury, Essex County * * * when Allen spied a bird on the telegraph wires, which Griscom thought was a large flycatcher * * * We were properly astounded to recognize a Gray Kingbird * * * We all had a perfect study of the kingbird, easily noting all the diagnostic characters.

It seemed highly desirable to collect the specimen, but the party was weaponless. Griscom accordingly walked to the nearest house to borrow a shotgun * * * obtained the gun * * * but the available ammunition consisted of two No. 2 shells. Griscom was devoid of experience in collecting small land birds with No. 2 shot * * * but the bird was secured comparatively undamaged.

The specimen proved to be an adult female and has been presented to the Peabody Museum at Salem * * * The date is the height of the fall migration of the species from the West Indies to South America. A low pressure area which blanketed New England in rain, fog, and mist for five days during the preceding week may have been attended by strong winds farther south. The only other record for New England is based on a bird shot in 1869, also in Essex Co., Mass., and preserved in the Boston Society of Natural History, the latter fact unrecorded by Forbush.

Since this note was published, the gray kingbird has again appeared in New England, this time in Maine, even farther north than Massachusetts. The following notes concerning this occurrence were furnished by my friend Martin Curtler and are considered absolutely authentic. Having been afield with Mr. Curtler on western and southern trips, I have had his care and accuracy in field identification impressively demonstrated. This record, representing the farthest north occurrence of the gray kingbird in the United States, has never before been published and is quoted as Mr. Curtler sent it:

"My gray kingbirds were seen at the southwest end of Deer Isle (1½ miles from Stonington) [Maine], on September 14, 1938. I observed them at close quarters for upwards of an hour; and they were particularly tame, allowing a close approach. They can have been nothing else. At first I thought the single one I saw first was a shrike, being pale gray, with a dark line through the eye. But I soon saw I was mistaken, i. e., on getting my glasses on him. And then, the behavior of the pair later was just like a couple of common kingbirds, sitting on vantage-points, swooping down and out into the air after insects and returning to where they had been before. I caught a glimpse of pale yellow under the wings . . . they were quite silent. It was brilliantly sunny, about 12 noon."

While Curtler does not stress the grayness of these birds in so many words, he was certainly impressed by it because he thought at first he was seeing a shrike! He is the only correspondent who likens the species to that bird, an impression I have always entertained since seeing my first specimen! It will be noted too that these three New England records have each occurred in a different fall month; Maine in September, Massachusetts in October and November.

The most extraordinary extralimital occurrence of the gray kingbird is the sole Canadian record, of which P. A. Taverner (1934) has very aptly said: "An accidental straggler that may never occur again within our borders." The bare facts of this amazing record are that on September 29, 1889, a specimen was secured by a Miss Cox at Cape Beale, on the west coast of Vancouver Island, and was presented to and still exists in the Provincial Museum, Victoria. What it was doing there will forever remain a mystery. So far as I am aware, the possibility that it was a cagebird was not raised.

It is extremely doubtful whether the gray kingbird ever remains in south Florida in winter. Everything points to the fact that it leaves this country completely. There is but a single record of its occurrence in winter in the United States, this being quoted by Arthur H. Howell (1932). He states that "Dr. H. C. Burgess saw one at Royal Palm Hammock, December 26 to 28, 1917, which occurrence seems to indicate that a few pass the winter in extreme southern Florida." This seems to be a very tenuous thread on which to base an assumption of this sort. It is tremendously more likely that the Burgess specimen was a belated migrant. December 26 is not an excessively late date, and the apparently complete absence of January and February records would be much more indicative that the species does not pass the winter in south Florida.

To those thoroughly familiar with south Florida, Howell's nomenclature of the locality of the Burgess specimen tends toward confusion, for he consistently confuses Royal Palm Hammock with Paradise Key, considering the two to be synonymous terms for the same locality (Howell, 1932, p. 61). As a matter of fact, the two are entirely different places, removed from each other by a hundred miles of road. Paradise Key is Royal Palm *State Park*, situated about 12 miles southwest of Homestead, Dade County, in the lower Everglades. (This is the location of the Burgess observation.) Royal Palm Hammock is about 14 miles west of Carnestown, Collier County, on the Tamiami Trail, between Everglades City and Naples. It even appears on most road maps.

In connection with the sanctuary work of the Audubon Society, I spend about two weeks out of every month in south Florida, from October through May, except November, every year. This results in at least seven trips during the fall and winter, and has been the case since 1935. The time is spent entirely in the field, and in all 35 trips have been made to date. I have never seen the gray kingbird in winter in south Florida, either on the mainland or in the Keys. Paradise Key is often visited and so is the Cape Sable region, and all the area between Northwest Cape and Madeira Bay, east and west of Flamingo. Key West is always the terminus of these trips, several days being spent there each time. The Ten Thousand Islands and the many rivers of the southwest coast are frequently investigated, as is the whole of Florida Bay.

Therefore, I can state definitely that, as far as five years of intensive field work are concerned, the gray kingbird is absent from this region in winter, and questions asked of residents who say they know the bird have resulted in similar conclusions. It is by no means impossible that it may appear in future, but the probability is that any bird seen after December 1 is a belated migrant and not a wintering specimen.

DISTRIBUTION

Range.—Southeastern United States, the West Indies, and northern South America; casual on the coast of Central America; accidental in New Jersey, New York, Massachusetts, and British Columbia.

Breeding range.—The gray kingbird breeds **north** to northern Florida (Pensacola, Santa Rosa Island, and St. Marks); and rarely South Carolina (Sullivans Island). **East** to rarely South Carolina (Sullivans Islands); probably rarely Georgia (Blackbeard Island and St. Marys); eastern Florida (St. Augustine and New Smyrna); the Bahama Islands (Elbow Cay, Eleuthera, and Watling Island); Haiti (Tortue Island); the Dominican Republic (Sanchez); Puerto Rico (Aguadilla and Yabucoa); the Lesser Antilles (Anegada, St. Bartholomew, Dominica, Barbados, and Grenada); and eastern Venezuela (Rio Uracoa). **South** to Venezuela (Rio Uracoa, La Pedrita, and Ciudad Bolivar); and Colombia (Noarama). **West** to western Colombia (Noarama, Blanco, Varrud, and Santa Marta); Jamaica (Port Henderson); western Cuba (Isle of Pines, Pilotes, and Habana); and western Florida (Dry Tortugas, Sevenoaks, and Pensacola).

Winter range.—The species is apparently resident in the South American portion of its range and in most of the West Indies, **north** at least to Puerto Rico (Vieques Island and Cartagena Lagoon); the Dominican Republic (Monte Cristi); Haiti (Port-au-Prince); and the Isle of Pines.

The range as outlined is for the species, which has been separated into two subspecies. The typical gray kingbird (*T. d. dominicensis*) is found over the entire range except the southern islands of the Lesser Antilles and the island of Trinidad, which are occupied by the large-billed kingbird (*T. d. vorax*).

Spring migration.—Early dates of spring arrivals are: Cuba— Trinidad, March 19. Bahamas—Watling Island, March 27. Florida—Fort Myers, March 12; New Smyrna, April 3; St. Marks, April 14.

Fall migration.—Late dates of fall departure from Florida are: St. Marks, September 29; St. Augustine, October 27; Royal Palm Hammock, December 28.

Casual records.—The species has been recorded either by sight or the collection of specimens from Quintana Roo (Cozumel Island), Honduras (between La Ceiba and Puerto Castilla), Nicaragua (Greytown), Costa Rica, and Panama (Gatun, Colon, Permé, and Obaldia), but it can be considered as of only casual occurrence in this region. It was recorded from Cayenne, French Guiana, on October 16, 1902, and November 18, 1902.

One was seen at Cape May Point, N. J., on May 30, 1933, and one was taken at Setauket, Long Island, N. Y., about the middle of July 1874. A specimen was collected at Lynn, Mass., early in October 1869, and another was taken at West Newbury on November 22, 1931. A remarkable record is that of one taken at Cape Beale, British Columbia, on September 29, 1889.

Egg dates.—Florida: 58 records, April 5 to July 30; 30 records, May 20 to 29, indicating the height of the season.

South Carolina: 1 record, May 30.

West Indies: 9 records, March 23 to July 8.

TYRANNUS MELANCHOLICUS COUCHI Baird

COUCH'S KINGBIRD

HABITS

Couch's kingbird is the name of one of the northern races of a widely distributed species that ranges through Central and South America. Its breeding range extends from the valley of the lower Rio Grande in southern Texas southward through northeastern Mexico to Veracruz and Puebla.

It is a large, pale race of the species; in comparing it with Lichtenstein's kingbird, the more southern race, Ridgway (1907) says: "Similar to the lighter colored examples of *T. m. satrapa* but decidedly larger, grayish brown of tail and wings paler, chin and upper throat more purely white, color of chest more yellowish, and 'mantle' more uniformly yellowish olive-green."

At the time that Baird described this bird (Baird, Cassin, and Lawrence, 1860), it had not been recorded north of the Mexican border, though it was supposed to range north to the valley of the Rio Grande in Mexico. To George B. Sennett (1878) belongs the honor of adding it to our fauna; he writes: "On May 8th, I saw a number of this species at Lomita Ranch, on the ebony-trees. Three were shot, but only one secured, the others being lost in the tall grass and thickets. At this point is the finest grove of ebonies I saw on the river. On the hillside, back of the buildings, they overlook the large resaca, then filled with tasseled corn. It was the tops of these grand old trees that these Flycatchers loved, and so persistent were they in staying there that I thought they were going to settle in the neighborhood for the season. There was a company of some six or eight scattered about."

When I visited southern Texas, in 1923, we found Couch's kingbird fairly common during May in Cameron and Hidalgo Counties, where it was breeding. It was one of the characteristic birds of the chaparral, where it was often seen, and oftener heard, in that pigmy forest of mesquite, ebony, retama, granjena, persimmon, madrona, and shittim wood, with an undergrowth of various thorny bushes, such as the fragrant cat's-claw, round-flowered devil's-claw, and that thorniest of all thorny bushes, the *Corona christi*. A fully fledged young, evidently recently from the nest, was discovered on May 23; its noisy parents were making a great demonstration of anxiety over it. But we did not discover its nest.

Nesting.—Mr. Sennett's collector, Mr. Bourbois, took what was probably the first set of eggs of this kingbird to be taken north of the Mexican boundary. It was taken, with the parent birds, at Lomita Ranch, on the Rio Grande, Texas, in 1881. Mr. Sennett (1884) describes the nest as follows: "The nest was situated some twenty feet from the ground, on a small lateral branch of a large elm, in a fine grove not far from the houses of the ranch. It is composed of small elm twigs, with a little Spanish moss and a few branchlets and leaves of the growing elm intermixed. The sides of the nest are lined with fine rootlets, the bottom with the black hair-like heart of the Spanish moss. The outside diameter is 6 inches, and the depth 2 inches. The inside diameter is 3 inches, and the depth 1.25 inches."

A set of five eggs in my collection was taken in Tamaulipas, Mexico, on May 6, 1895, by or for Frank B. Armstrong; the nest was said to be made of Spanish moss, strips of bark, and plant down; it was placed near the end of a limb of a tree in open woods and only 8 feet from the ground.

Eggs.—Couch's kingbird evidently lays three to five eggs, oftener three or four. While showing the usual kingbird characteristics, they are usually distinctive. Mr. Sennett (1884), in describing his first set, says that they "are quite distinct in form, size, and ground-color from any others I have seen. The blotches, too, are more numerous and smaller. The large end is very round, and the small end quite pointed. * * * The ground-color is a rich buff. The general color of the blotches is similar to that of the Kingbird's eggs, and their distribution irregular over the entire egg, but massed about the greater diameter. If this set proves to be typical I should have no trouble in selecting the eggs of this species from any number of eggs of other species of the genus."

Based on a study of 13 eggs in the United States National Museum collection, Major Bendire (1895) says: "The ground color of the eggs is a delicate creamy pink, and they are moderately well blotched and spotted with chocolate, claret brown, heliotrope purple, and lavender. These markings are, in some instances, scattered pretty evenly over the entire surface of the egg; in others they are mainly confined to the larger end. They are readily distinguishable from the eggs of the balance of our Kingbirds by their peculiar ground color, while their markings are very similar to those found on the eggs of the other species of this family. The shell is close-grained and rather strong, and in shape the eggs are generally ovate or elongate ovate."

The measurements of 43 eggs average 24.8 by 18.4 millimeters; the eggs showing the four extremes measure 27.2 by 19.7, 22.6 by 19.0, and 23.4 by 17.0 millimeters.

Plumages.—As the plumages of this race seem to correspond very closely to those of Lichtenstein's kingbird, the reader is referred to the remarks on this subject under that race, on which we have more information.

Food and behavior.—I can find nothing recorded on the food of this subspecies, which probably does not differ materially from that of the other tyrant flycatchers of this genus. Mr. Sennett (1878) found Couch's kingbirds associated with common eastern kingbirds in the tops of the large ebony trees, where they were doubtless in pursuit of flying insects. He "did not find them shy, for after our firing they would almost immediately return to the same trees." The birds that we saw in Texas were very noisy and apparently quite aggressive.

Field marks.—Couch's kingbird bears a superficial resemblance to Cassin's, and to a less degree to the Arkansas kingbird, though the white outer webs of the outer tail feathers of the latter, which the former two do not have, should eliminate any confusion. Couch's is much like Cassin's on the upper parts, but the tail is browner in Couch's and blacker in Cassin's. In Cassin's only a small part of the chin is white, the throat and chest being extensively gray, "light neutral gray" to "pale neutral gray," and the rest of the under parts are paler yellow, "lemon yellow" to "pinard yellow"; whereas in Couch's the chin and whole throat are extensively white, and the under parts are a deep, rich "lemon chrome" or "empire yellow," slightly tinged with olive or olive-gray on the chest. These differences in color patterns should enable the observer to distinguish any of the subspecies of *melancholicus* from *vociferans*.

DISTRIBUTION

Range.—Lower Rio Grande Valley in Texas and northeastern Mexico.

While there are several races of this species in South and Central America and in the West Indies, this is the only form that is a regular visitor to the United States. Its range extends north to the Rio Grande Valley in Texas (Lomita and Brownsville); and from this district south through eastern Mexico; Nuevo Leon (Ceralvo, Monterey, and Rio San Juan); Tamaulipas (Matamoros, Sierra Madre, Aldama, Altamira, and Tampico); to Veracruz (Papantla, Jalapa, and Orizaba).

In winter it appears to withdraw entirely from the Texas area, but at this season it is found in northern Neuvo Leon (Ceralvo) and Tamaulipas (Matamoros). Early dates of spring arrival in Texas are: Brownsville, March 12, and Hidalgo, March 21. No dates of fall departure are available.

Casual records.—A specimen *of T. m. couchi* taken at Kerrville, Tex., on September 11, 1908, is the most northerly record for this race. The west Mexican kingbird (*T. m. occidentalis*) has been recorded twice in the United States: A specimen taken at Fort Lowell, Ariz., on May 12, 1905, and one collected in Jefferson County, Wash., on November 18, 1916. Another race of this species, known as Lichtenstein's kingbird (*T. m. chloronotus*), also has been recorded on two occasions at widely separated points. One was collected at Scarborough, Maine, on October 31, 1915, and another was taken at French's Beach, British Columbia, in February 1923. Reexamination of the latter specimen would probably prove it to be the west Mexican race, *occidentalis*.

Egg dates.—Arizona: 2 records, May 11 and June 13.

Mexico: 16 records, April 6 to July 28; 8 records, June 6 to 14, indicating the height of the season.

Texas: 5 records, May 7 to 21.

TYRANNUS MELANCHOLICUS CHLORONOTUS Berlepsch

LICHTENSTEIN'S KINGBIRD

HABITS

The normal range of this subspecies is from southern Veracruz, Mexico, southward to Colombia, Venezuela, and the lower Amazon Valley in Brazil. But, strangely enough, there seem to be no records of its occurrence in any part of the Southern United States; and still more strangely, its inclusion in our Check-list is based on two widely separated records of occurrence in Maine and on Vancouver Island, far remote from its normal habitat. Arthur H. Norton (1916) reported: "On October 31, 1915, Mr. George Oliver observed this stranger near his house in Scarborough, and secured it for the collection of the Portland Society of Natural History. Mr. Oliver said that it was seen the day before it was taken, and was thought to have been a shrike. Upon reaching the identification given, it was sent to the United States National Museum, where it was confirmed by Mr. H. C. Oberholser, and Mr. Robert Ridgway. The bird was a young male, in very good condition. * * * It should be recalled in connection with this waif that two very intense tropical cyclones visited the United States, one in August, the other in September, 1915."

The second specimen "was collected at French's Beach, Renfrew District, Vancouver Island, in February 1923 by J. G. French." This bird was identified by Maj. Allan Brooks, through the interest of J. A. Munro; and it was suggested that "it may have strayed so far north through the medium of a steamer" (Kermode, 1928).

A. J. van Rossem (Dickey and van Rossem, 1938) says that, in El Salvador, it is a "common resident of open or semiwooded country in the Arid Lower Tropical Zone. The species is most numerous on the coastal plain and in the lower foothills and only rarely straggles to an elevation of 4,500 feet. * * * Lichtenstein's kingbirds are generally distributed over open country everywhere in the lower levels and may, locally, be very common indeed. Such places as Colima and Divisadero, where much of the terrain is tree-dotted agricultural land, are eminently suited to their needs, and they were very numerous in both localities. They are much less common in wooded areas such as Lake Olomega and Puerto del Triunfo, where their spheres of activity are necessarily limited to clearings or waterfronts."

Nesting.—The same observer says that "the nests differ greatly from the bulky, padded structures of the northern species. One found at Zapotitán on June 12, 1912, was placed six feet from the ground in the foliage of a horizontal branch of a small mimosa tree. It was so thin and so poorly constructed that the three eggs could easily be seen from below. The body was of small twigs, and the nest cup was lined with fine round grasses. Another in an almost exactly similar situation, found at Lake Guija May 28, 1927, was somewhat better built, for its contents could not be seen from below. Like many other native species this one often takes advantage of wasps' nests by placing its own home close by."

Eggs.—There is a set of four eggs in the Thayer collection in Cambridge. These are ovate and show a slight gloss. The ground color is characteristic of the species, varying from cream-white to pale "seashell pink"; they are marked much like other kingbirds' eggs with the same colors, and the spots are mainly grouped about the larger end. The measurements of 9 eggs average 24.52 by 18.15 millimeters; the eggs showing the four extremes measure 25.4 by 18.6, 24.3 by 18.9, 23.8 by 18.0, and 24.1 by 17.6 millimeters.

Plumages.—Ridgway (1907) describes the young, evidently in juvenal plumage, as "essentially like adults, but without orange on crown, gray of head browner (smoke gray or drab gray), back, etc., duller olive, yellow of under parts usually paler, and wing-coverts and rectrices conspicuously margined with pale cinnamon or buffy." Mr. van Rossem (Dickey and van Rossem, 1938) writes:

The plumage sequences parallel those of *Tyrannus verticalis* and *Tyrannus vociferans.* At the postjuvenal body molt a body plumage like that of the adults is acquired. The juvenal wing feathers and rectrices are retained, sometimes until the annual molt of the following fall, but are usually replaced either in part or entirely during the first winter and spring. The concealed colored feathers of the crown also are delayed in their appearance until the spring molt. The annual molt commences in some birds as early as the middle of July, and in one specimen is as yet unfinished at so late a date as November 12. About August 1 to October 1 is probably the average molting period. The spring molt is extensive and includes a varying number of rectrices. It occurs in February, March, or April.

The degree of rapidity with which the dorsal plumage fades from olive-green to gray is astonishing. Just after the annual and postjuvenal molts the back is uniformly a solid, bright olive-green, but within a few weeks becomes duller and by midwinter is definitely gray. New feathers coming through at any time of the year are bright olive-green and this, contrasted with the older, gray ones, gives a mottled appearance.

Food.—The only information I can find on this subject is the report made by W. L. McAtee to Mr. Norton (1916) on the contents of the stomach of the bird taken in Maine, which were probably not typical of its normal food except in a general way. These contents were "remains of at least 16 *Muscidae*, part of them *Pallenia rudis*, and part of a metallic kind, probably *Phormia*, 96%; 1 *Scatophaga furcata* and 1 *Syrphus* sp. 4%; bits of unidentified vegetable matter tr."

Behavior.—Mr. van Rossem (Dickey and van Rossem, 1938) says that "in general, these kingbirds resemble in habits their congeners of the north. The most noticeable differences are their comparatively placid and less pugnacious natures, and the very different character of the call-notes. Instead of the sharp, raucous clatter of sounds so characteristic of the northern species, the voice of *chloronotus* is subdued and at times almost musical."

TYRANNUS MELANCHOLICUS OCCIDENTALIS Hartert and Goodson

WEST MEXICAN KINGBIRD

HABITS

The 1931 Check-list includes this Mexican race as a straggler, based on a specimen in the Dickey collection in the California Institute of Technology, collected by Carl Lien in Jefferson County, Wash., on November 18, 1916. In recording this specimen, A. J. van Rossem (1929) says:

This specimen was purchased by Mr. Dickey from Paul Trapier as part of a general collection of Washington birds mostly taken by Mr. Lien. It was labelled by the original collector as "Ash-throated Flycatcher."

The specimen here recorded is somewhat soot-stained, but is clearly of the west-Mexican race which differs from the Central American in having paler, less intensely yellow underparts and slightly larger bill. Except for the darker tinge caused by soot-stain, it is very similar to two birds from Escuinapa, Sinaloa. * * *

In view of the subspecific status of the Washington bird, it would appear that a re-examination of Mr. Kermode's specimen is desirable. Logically, it should be of the north-west Mexican race rather than the Central American race."

Since the 1931 Check-list was published, James Lee Peters (1936) has discovered a specimen of the west Mexican kingbird in the Thayer collection in Cambridge; it was taken by H. H. Kimball at Fort Lowell, near Tucson, Ariz., on May 12, 1905.

And now we have strong evidence to indicate that this kingbird may be a regular summer resident and breeder in southern Arizona. Allan R. Phillips (1940) reports that he has collected both adults and young birds, the latter evidently hatched in the vicinity, near Tucson, Ariz.; and during the summers of 1938 and 1939 he and his companions saw other families of these birds on several occasions, "making a total of possibly four pairs present in the area covered in 1939." Mr. Phillips writes on the subject:

It seems evident that the West Mexican Kingbird is a regular summer resident at the present time near Tucson, from May 12 to September 3 at least. The numbers present are not great. The birds have been seen by a few other observers, also, including Dr. A. A. Allen, A. H. Anderson, Dr. Wm. L. Holt, F. W. Loetscher, Jr., and Gale Monson.

The call of this kingbird is strikingly different from those of the three northern kingbirds, being of a metallic rather than a throaty quality. It consists of a rapid series of short, staccato notes in an ascending, high-pitched series, and might be rendered as *pit—it–it–it–it–it–it-it-it*. In form the call somewhat resembles that of the Vermilion Flycatcher, but it is much louder, sharper, and higher-pitched. Besides the call, the heavy bill, whitish throat, bright yellow belly, and brownish, emarginate tail all help distinguish it in the field, and the tail characters are obvious in flight even at some distance. In spite of these several easy distinctions, it seems probable that the birds have been allowed to pass for Arkansas Kingbirds by the few ornithologists who have entered their restricted ranges in the summer months.

We seem to have no evidence that the nesting habits and the eggs of this kingbird are in any way different from those of the other races of the species. The measurements of 45 eggs average 23.7 by 17.2 millimeters; the eggs showing the four extremes measure 25.2 by 17.1, 20.1 by 17.9, and 24.9 by 16.2 millimeters.

TYRANNUS VERTICALIS Say

ARKANSAS KINGBIRD

PLATE 6

HABITS

We formerly regarded the Arkansas kingbird as a western bird, when it was known by the appropriate name of western kingbird. It was then merely a straggler east of the Mississippi River and was more or less rare as a wanderer or as a migrant in the States immediately west of that river. And, as a straggler, it wandered as far east as the Atlantic States during the latter half of the nineteenth century. For example, there is an early record for Eliot, Maine, in October 1864 (Haven, 1926), and one for Riverdale, N. Y., on October 19, 1875 (Bicknell, 1879). Since what is apparently the first Massachusetts record, Chatham, October 20, 1912 (Kennard, 1913), there have been so many New England records that this bird might almost be considered a frequent visitor. And since the beginning of the twentieth century there have been numerous records from other Atlantic Coast States—New Jersey, Delaware, Virginia, and Florida.

There is abundant evidence, also, that this kingbird has been extending its wanderings, and even its breeding range, eastward in the interior during the past 40 years. Without devoting too much space to the subject, it seems worth while to cite a few examples. Referring to Manitoba, P. A. Taverner (1927) says: "This species is another recent arrival in Manitoba. E. T. Seton does not mention it in his 1891 list of 'Birds of Manitoba' and gives only adjoining records in his 'Fauna of Manitoba,' 1909. The first record for the province appears to have been a specimen taken at Oak Lake, August 10, 1907." Other records followed until now he calls it "rather common throughout southwestern Manitoba."

For Minnesota, Dr. Thomas S. Roberts (1932) writes: "Three pairs were found in the Big Stone-Traverse lakes region in June, 1879 by Franklin Benner and the writer, the first record for the state; in 1889 it was found by Cantwell in Lac qui Parle County; in 1893 it was present in Pipestone (Roberts) and Otter Tail (Gault) counties and by 1898 had become a common nesting bird all over southwestern Minnesota as far east as Redwood, Cottonwood, and Jackson counties." Later on, it extended its range farther north until it is "now an abundant summer resident throughout the western prairie portion of the state"; and "it has increased rapidly and spread eastward until in recent years it has reached the eastern part of the state south of the evergreen forests."

Arkansas kingbirds have been seen in summer and collected in breeding condition in Michigan since 1925 (Van Tyne, 1933), indicating an extension of range eastward from Minnesota. And now comes a breeding record for Ohio. Louis W. Campbell (1934), referring to a sight record in Lucas County in 1931, reports: "On July 29, 1933, some three miles east of the location of the sight record mentioned above I found a family of Arkansas kingbirds consisting of one adult and three young. * * * Two of the young, a male and a female, were collected. Although both of these birds were well able to fly, all the tail feathers and all but two or three of the primaries were still more than one-fourth sheathed. The condition of these feathers and general lack of development pointed to the conclusion that the birds had been out of the nest only a short time."

A similar eastward advance has been noted farther south around the turn of the century in Nebraska, Kansas, and Oklahoma. Swenk and Dawson (1921) tell the story for Nebraska; and Mrs. Nice (1931a) gives the records for Oklahoma. If this kingbird continues to advance, it may yet reach Arkansas and its name may be justified.

The Arkansas kingbird is a bird of the open country, associated in my mind with the prairie regions of the Middle West, the ranches and tree claims and the timber belts along the streams. In southwestern Saskatchewan, in 1905 and 1906, we found this kingbird breeding commonly in such situations, but it was not so common in the timber belts as the eastern kingbird and was more likely to be seen about the ranch buildings and railroad stations.

As to its haunts in Minnesota, Dr. Roberts (1932) says: "On the prairies this bird avoids the natural groves of timber, seeking the vicinity of habitations. Where there are no buildings on a tree-claim or the farm is deserted it is rarely found, probably because there is less insect food where there are no cattle. The bird invades the towns everywhere and builds commonly in the shade trees even on business streets. At times it wanders far from home out over the prairies, and may frequently be seen perched on a wire or fence at a considerable distance from woodlands."

In Cochise County, Ariz., we found the Arkansas kingbird abundant at the lower levels, on the soapweed plains, about the ranches, and along the dry washes, where there were trees, but not venturing far up into the timbered canyons. Evidently the birds prefer the open country to the more restricted canyons.

Courtship.—I have never seen the courtship performance of this bird, but E. S. Cameron (1907) makes the following brief statement in regard to it: "The male indulges in a curious display when courting the female. He makes successive darts in the air, fluttering, vibrating his quills, and trilling as he shoots forward. Propelling himself thus for several hundred yards, he looks like a bird gone mad."

Nesting.—The Arkansas kingbird builds its nest in a great variety of situations, almost anywhere except on the ground. It apparently prefers to build in trees, where suitable trees are available, which very often is not the case. If in a tree, the nest may be placed against the trunk, in a crotch, or, more often, out on a horizontal branch, or occasionally on a dead branch. It may be placed at any height from 8 to 40 feet above the ground, but oftener 15 to 30 feet. Nests may also be placed on bushes as low as 5 feet from the ground. Sometimes several pairs build their nests close together in a grove or small group of trees; and two or more nests have been found in a single tree on rare occasions. The trees most often chosen are cottonwoods, oaks, sycamores, and willows, perhaps because they happen to be the commonest trees available. Nests have also been recorded in elms, eucalyptus, junipers, apple trees and other orchard trees, locusts, aspens, alders, and rarely pines; probably a number of others might be added to the list, for these birds do not seem to be at all particular in their choice. Claude T. Barnes writes to me that he has many times noted the fondness of these birds for locust trees, and thinks this may be due to their height and comparative openness of foliage. These kingbirds seem to prefer an open situation where they can command a clear outlook.

Most of the nests that I have seen have been in trees, but in Alberta near Many Island Lake, on June 16, 1906, we stopped to examine a Swainson's hawk's nest that was situated in a little patch of large brush on a steep hillside; the hawk flew from the nest as we approached; and within a few yards of the hawk's nest we found an Arkansas kingbird's nest with three fresh eggs in it; it was placed only 5 feet from the ground in a "stony berry bush." On the soapweed plains in Cochise County, Ariz., on June 1, 1922, while we were hunting for nests of Scott's oriole, we found a nest of this kingbird built in the upright, dead flower stalk of a soapweed yucca 14 feet from the ground.

Where no suitable trees are available, the Arkansas kingbird is most adaptable and versatile in the selection of a nesting site, being satisfied to utilize almost any form of human structure that will hold its nest, preferably such stable structures as telephone or telegraph poles, fence posts, stationary towers, parts of buildings, or boxes set up for that purpose; but in many cases the chosen site is far from stationary or secure, with resulting disaster. The following quotations will illustrate its versatility.

Major Bendire (1895) writes:

Mr. William G. Smith informs me that in Colorado they nest occasionally on ledges. Dr. C. T. Cocke writes me that a pair of these birds nested in the summer of 1891 in a church steeple in Salem, Oregon, and Mr. Elmer T. Judd, of Cando, North Dakota, informs me that he found a nest on a beam of a railroad windmill pump, about 6 feet from the ground, where trains passed close by the nest constantly; another was found by him on a grainbinder which was standing within a couple of rods of a public schoolhouse.

I have examined many of their nests in various parts of the West. * * * One nest was placed in the top of a hollow cottonwood stump, the rim of the nest being flush with the top; another pair made use of an old nest of the Western Robin; and still another built on the sill of one of the attic windows of my quarters at Fort Lapwai, Idaho. They probably would not have succeeded in keeping this nest in place had I not nailed a piece of board along the outside to prevent the wind from blowing the materials away as fast as the birds could bring them. They were persistent, however, and not easily discouraged, working hard for a couple of days in trying to secure a firm foundation before I came to their assistance. Both birds were equally diligent in the construction of their home until it was nearly finished, when the female did most of the arranging of the inner lining, and many a consultation was evidently indulged in between the pair before the nest was finally ready for occupation, a low twittering being kept up almost constantly. It took just a week to build it.

John G. Tyler (1913), writing of the Fresno district, Calif., says:

Formerly they resorted to the framework of flumes, windmills, outbuildings, and even the tops of fence posts; but of recent years the rural telephone lines that have thrown their network of wires and poles all over the valley have provided nesting sites galore, and of a kind seemingly exactly suited to the requirements of these birds. * * * Where the lines cross entrances to farmhouses or intersecting roads, * * * the wires are raised several feet. * * * This additional height is attained by nailing two two-inch pieces to the original pole on opposite sides, thus leaving a four inch platform protected on two sides, in which a nest just fits sungly. A drive through the country during the summer months now reveals a pair of kingbirds tenanted in nearly every such pole.

Lee Raymond Dice (1918) mentions a nest near Wallula, Wash., on a hay derrick; the bird remained on the nest even while the derrick was in use; a nest was observed in a barn, and another on a fence post, in the bunchgrass hills; he also saw a nest on a rocky cliff. J. A. Munro (1919) says that in the Okanagan Valley "for two seasons, a pair built in the eaves-trough of" his "house, directly over the vent. Both years the eggs were destroyed by rain storms and washed into the rain barrel. * * * The residents along some of the country roads nail up small soap or starch boxes on their gate-posts for the reception of milk bottles, etc.; these are frequently used as nesting sites. I have known them to build on a ledge above the kitchen door of a farm house, which was opened and shut fifty times during the day. Frequently they used abandoned Flicker holes, or the roughened, decayed top of a fence post."

Clarence Hamilton Kennedy (1915) has published an interesting paper on the adaptability of this kingbird, illustrated with drawings of nesting sites, including one in an old nest of Bullock's oriole.

The Arkansas kingbird is almost as versatile in its selection of nesting materials as it is adaptable in its choice of a site. Bendire (1895) gives this general description of the nests: "Generally they are compactly built structures, the foundation and outer walls being composed of weed stems, fine twigs, plant fibers, and rootlets, intermixed with wool, cocoons, hair, feathers, bits of string, cottonwood, milkweed, and thistle down, or pieces of paper, and lined with finer materials of the same kind." He says that a typical nest "measures 6 inches in outer diameter by 3 in depth; the inner cup is 3 inches wide by 1¾ deep."

Lt. C. A. H. McCauley (1877) describes a pretty nest, made by a pair of these birds in a cottonwood tree in Texas, as follows:

Using large quantities of the fibrous, coarse cotton of the tree, they had matted this well for fully an inch above the limbs; through which, well interwoven, ran bits of sage-brush, coarse grasses, and fine twigs, with a few dried leaves. Above this part came finer grasses and small fibrous roots, whilst for the interior they had apparently carefully selected choice bits of cotton, at times arranged in strata, over which was buffalo-wool liberally placed and neatly fastened and bound with finest threads. The lining of the interior was unusually soft, being padded in a peculiar manner with the wool itself within. In the homes of the two preceding species [swallow-tailed flycatcher and eastern kingbird], the eggs rest upon the slender roots and threads which thickly cover and bind together the underlying wool; but this Flycatcher is more select; he fastens the tufts of wool below in such a way that the eggs have a resting-place almost as soft as down, and the intertwining threads are scarcely visible.

The nests that I have seen have been made of many of the materials listed above by Bendire and constructed in much the same manner. In all cases the interior has been thickly and warmly lined with firmly matted, or felted, cow's hair, sheep's wool, cotton, or cottonwood down. Cow's hair is easily obtainable about the cattle ranches and is freely used. On the sheep ranges, the barbed-wire fences are well decorated with wool, offering a bountiful supply of this desirable material. About the poultry yards and farms the kingbirds make free use of hen's feathers both in the body of the nest and decoration around the rim, where the upstanding feathers not only add to the beauty of the nest, but help to conceal the sitting bird; one especially pretty nest was profusely decorated with brown hen's feathers around the rim and was lined with white hen's feathers and thistledown, making a pleasing contrast.

R. C. Tate (1925) says that in Oklahoma these birds use "seven and eight inch pieces of the thin inner bark from dead cottonwood trees, rags, string, sandbur rootlets, horse-hair, wool from old sheep carcasses, and cotton from discarded quilts. Pieces of dried snakeskin are frequently made use of also."

Eggs.—The Arkansas kingbird lays three to five eggs, usually four, and occasionally six and even seven. They are almost exactly like those of the eastern kingbird, though averaging slightly smaller. The shape is ovate or short-ovate as a rule, but sometimes elongated-ovate or even elliptical-ovate. The ground color varies from pure white to pinkish white, or creamy white, and the eggs are more or less heavily marked with small spots or blotches of various browns, "chestnut-brown," "chocolate," "liver brown," or "claret brown," with underlying spots and blotches of different shades of "Quaker drab," "heliotrope gray," or "lavender." The markings are sometimes grouped about the larger end. The measurements of 50 eggs averaged 23.5 by 17.7 millimeters; the eggs showing the four extremes measure 26.4 by 18.8, 25.4 by 19.1, and 19.8 by 15.7 millimeters.

Young.—The period of incubation, as given by different observers, is said to be 12, 13, or 14 days. Major Bendire (1895) says:

This duty is mostly performed by the female, but I have also seen the male on the nest, and he can generally be observed close by, on the look out for danger. Both parents are exceedingly courageous in the defense of their nest and young, and every bird of this species in the neighborhood will quickly come to the rescue and help to drive intruders off as soon as one gives the alarm. The young grow rapidly and are able to leave the nest in about two weeks. They consume an immense amount of food, certainly fully their own weight in a day. I have often watched the family previously referred to, raised on the sill of my attic window, and also fed them with the bodies of the large black crickets while one of the parents was looking on, and apparently approvingly, within a few feet of me. I have stuffed them until it seemed impossible for them to hold any more, but there was no satisfying them; it certainly keeps the parents busy from early morning till late at night to supply their always hungry family. They are readily tamed when taken young, and are very intelligent, making interesting pets [see Ridgway, 1869, and Pinckney, 1938]. I believe that only one brood, as a rule, is raised in a season, excepting possibly in the extreme southern portions of their range, in southern Arizona and California, as I found fresh eggs on Rillito Creek, near Tucson, as late as July 20, in a locality where these birds had not been previously disturbed, which seems to indicate that they occasionally may rear a second brood.

Mrs. Irene G. Wheelock (1904) says of the young:

At first they are fed by regurgitation, but after the third day large insects are torn apart and given fresh. Fourteen crickets in ten minutes was the record of one busy forager. * * *

In two weeks the babies have grown so that they overflow the nest, and one balances himself outside. And now his lessons begin. As soon as he has learned to use his wings he is taught to catch his food in the same way in which he must obtain it all his life. I have seen the parent bring a dragon-fly or other insect, alight with it opposite above the young bird, and call his attention to it in a peculiar low twitter. Then, when quite ready, he releases the prey, which half falls, half flutters, downward. Nearly always the nestling is out after it and back with it in his beak before you can realize how it is done. Many times we have watched them, and the lesson is always given in this way, and always repeated until there can be no fear of missing. Then the young are taken to the meadows and taught to dart down after butterflies or grasshoppers. In some way they learn that the worker bees have stings and must not be caught, but that the drones are delicious morsels.

Plumages.—I have not seen the natal down, which is probably gray, as in the eastern kingbird. The juvenal plumage is largely acquired before the young bird leaves the nest, except that the wings and tail are not fully grown. In this plumage the sexes are alike, and both lack the orange crown patch; the crown is "pale smoke gray" with pale edgings, the back "ecru-olive," the rump and upper tail coverts pale "clove brown"; the tail is dull black, tipped with pale brownish, and the outer webs of the outer rectrices are white; the wings are pale "clove brown," with yellowish white edgings; the throat and chest are "pale smoke gray," and the abdomen is "empire yellow"; the first primary is not attenuated.

This plumage is worn through the summer and most of the fall; I have seen birds in this plumage that do not show much sign of molt even in November; one, taken November 19, still wears the juvenal plumage, but the wings and tail are somewhat worn, as if there was to be a complete postjuvenal molt. Usually the post-juvenal molt is accomplished mainly in October and November, and the first winter plumage is assumed, in which old and young birds become indistinguishable; the orange crown patch is acquired and the three outer primaries are attenuated at their tips. There is, ap-parently, a partial molt of the body plumage in March and April, a few new feathers showing among the older, faded ones of the winter plumage. There is a complete postnuptial molt in August and September.

Food.—Professor Beal (1912) in his study of the food of the Arkansas kingbird, examined the contents of 109 stomachs and reports that—

The food is found to consist of 90.61 per cent of animal matter to 9.39 per cent of vegetable. Of the animal portion, Hymenoptera (bees and wasps),

Coleoptera (beetles), and Orthoptera (grasshoppers) constitute over three-fourths.

Beetles of all kinds amount to 17.02 per cent of the food, and include 5.47 per cent of useful species, mostly Carabidae and Cicindelidae. For a flycatcher this is a large record of these useful beetles, as they are largely ground-inhabiting species and not so often on the wing as most others. The remainder, 11.55 per cent, are either harmful or neutral. No special pests were found among them. Hymenoptera (bees and wasps) are the largest item of animal food and amount to 31.38 per cent. * * *

Orthoptera (grasshoppers, crickets, etc.) stand next to Hymenoptera in the diet of the Arkansas kingbird. * * * The total for the year is 27.76 per cent. It is a singular fact that several western species of flycatchers eat more grasshoppers than do the meadowlark and blackbirds, which obtain their food almost wholly upon the ground. The reverse is the case with the corresponding eastern birds. The orthopterous food consists almost entirely of grasshoppers with very few crickets.

The remainder of the animal food consists of bugs, flies, "and a few other miscellaneous insects, millepeds and spiders, the bones of tree frogs in 3 stomachs, and egg shells, apparently those of domestic fowl." He lists in the insect food 8 species of Hymenoptera, 24 of Coleoptera, 1 of Diptera, and 7 of Hemiptera (bugs).

Of the vegetable food, he says: "Although vegetable food amounts to 9.39 per cent, it presents but little variety. A few weed seeds occurred in one stomach. Seed and skins of elderberries (*Sambucus*) were found in 11 stomachs, woodbine (*Psedera*) in 2, hawthorn berries (*Crataegus*) in 1, an olive in 1, and skin and pulp of fruit not further identified in 2."

As to the destruction of bees, of which this kingbird has been accused, Beal writes elsewhere (1910): "It is said that the birds linger about the hives and snap up the bees as they return home laden with honey. Remains of honey bees were searched for with special care, and were found to constitute 5 per cent of the food. Thirty-one individuals were discovered in 5 stomachs. Of these, 29 were drones, or males, and 2 were workers. In 3 stomachs containing males there was no other food, and when it is borne in mind that there are thousands of worker bees to one drone, it appears that the latter must be carefully selected. As a rule, the destruction of drones is not an injury to the colony, and often is a positive benefit."

In the stomachs of California birds, he found that "miscellaneous insects, consisting of caterpillars and moths, a few bugs, flies, and a dragonfly, constituted 10 percent. Several stomachs contained a number of moths, and one was entirely filled with them."

Like other flycatchers, the Arkansas kingbird captures most of its prey in the air, sallying out after flying insects, often from some low perch, such as a fence post or wire, a low tree or bush, or even the top of some tall weedstalk. Francis H. Allen saw one feeding near the seashore in Massachusetts, of which he says (MS.): "Its flight, as I saw it, was swallowlike, darting this way and that, but this may have been due to the abundance of the flies upon which it was feeding, which made it unnecessary to return to the perch after each capture. While it was perched on a point of rock near the beach, where the flies were particularly thick, it often simply reached out after a fly and picked it out of the air."

The fact that so much of its food consists of grasshoppers and ground-inhabiting beetles indicates that much of its prey must be secured on or near the ground. Its fondness for grasshoppers is shown by the fact that Mr. Ridgway's (1877) captive bird was fed 120 grasshoppers in one day.

Behavior.—I cannot do better than to quote the following well-chosen words of Mrs. Bailey (1902b) on the behavior of this spirited bird:

The Arkansas kingbird is a masterful, positive character, and when you come into his neighborhood you are very likely to know it, for he seems to be always screaming and scrimmaging. If he is not overhead twisting and turning with wings open and square tail spread so wide that it shows the white lines that border it, he is climbing up the air claw to claw with a rival, falling to ground clinched with him, or dashing after a hawk, screaming in thin falsetto like a scissor-tail flycatcher. A passing enemy is allowed no time to loiter but driven from the field with impetuous onslaught and clang of trumpets. Be he crow, hawk, or owl, he is escorted to a safe distance, sometimes actually ridden by the angry kingbird, who, like the scissor-tail, enforces his screams with sharp pecks on the back.

While the above described behavior is doubtless characteristic, this bird is not always as hostile toward hawks as Mrs. Bailey's remarks indicate. I found a pair of kingbirds occupying a nest within a few yards of an occupied nest of a Swainson's hawk. And Major Bendire (1895) says: "They are undoubtedly more social than the common Kingbird, as I have seen two pairs nesting in the same tree, apparently living in perfect harmony with each other. While they are by no means devoid of courage, they appear to me to be much less quarrelsome on the whole than the former, and they are far more tolerant toward some of the larger Raptores. For instance, in the vicinity of Camp Harney, Oregon, I found a pair of these birds nesting in the same tree (a medium-sized pine) with Bullock's Oriole and Swainson's Hawk, and, as far as I could see, all were on excellent terms."

Claude T. Barnes tells me that, in Utah, some of these kingbirds "were living in perfect harmony with Bullock orioles, mourning doves, yellow warblers and domestic sparrows." But sometimes the kingbirds have to fight to protect their nests; M. French Gilman (1915) tells of such a case, where a Bendire's thrasher tried to take possession of the kingbirds' nest: "The Thrasher would bring some nesting material, and settle down in the nest. Then the Kingbirds would appear, scolding and trying to drive her away. As long as they kept flying at her she stayed on the nest, but if one came close and alighted she would fluff out her feathers and make a vicious dive at him, or her, as it might be. Had her mate been as much on the job the Kingbirds would have lost out, but he sang and did nothing else, so she finally gave it up, and the Kingbirds raised three young."

Most observers agree that the Arkansas kingbird is intolerant toward intruders on its domain, just as the eastern kingbird is, and there is plenty of evidence to show that it will attack any hawk from a large redtail down to the little sparrow hawk, or any other large bird that comes too near its nest; and often, perhaps, it attacks them without any such good excuse. Mr. Swarth (1904) writes: "During the breeding season the large numbers of White-necked Ravens and Swainson Hawks found in the vicinity afford the Kingbirds exceptional opportunities for exhibiting their peculiar talents, and during the summer months these wretched birds' lives are made a burden to them through the incessant persecution they receive. The hawks usually leave as soon as possible on being attacked; but the ravens, though beating a hasty retreat often try to fight back, twisting from side to side in vain endeavor to reach their diminutive assailant; cawing a vigorous protest, meanwhile, at being treated in such a disrespectful fashion."

James B. Dixon tells me that he "noted one that was nesting in a sycamore where an eagle had a nest and was raising young. Every morning the kingbird spent a good part of its time heckling the young eagle."

Voice.—Major Bendire (1895) gives a very good general idea of this, as follows: "This Flycatcher is, if anything, more noisy than our common eastern Kingbird, and utters also a greater variety of notes; some of these resemble the squeaking sounds of our Grackles; others are indifferent efforts at song—a low, warbling kind of twitter—while occasionally it gives utterance to shrill, metallic-sounding notes with more force to them than those of the Kingbird. During the mating season they are especially noisy, and begin their love songs, if they may be called such, at the earliest dawn, and keep up their concerts with but slight intermission during the greater part of the day; but after they are mated and nidification commences they are more quiet."

He quotes R. H. Lawrence as follows: "On the night of July 30, 1893, I frequently heard a queer cry; sometimes only a single note, and again this was repeated three or four times, followed by a crying or wailing sound, as if made by a very young kitten. I heard these notes on successive nights. On August 2, about 4:30 a. m., I succeeded in shooting the performer out of a pepper tree standing close to the house, and it proved to be an Arkansas Flycatcher."

Dr. O. A. Stevens writes to me: "These birds are among the early risers. During midsummer they are about the first to be heard in the morning, before the first streaks of daylight have appeared. Dead limbs of trees and telephone wires are favorite perches. They seem to have essentially one note, and while they certainly do not sing, they make as much noise as any of our common birds. While sitting quietly they repeat at frequent intervals a single *kip*. Occasionally this is varied with a two-syllable *quer-ich*, and frequently a four-syllable, rapidly delivered combination, the first two notes being the most accented. Then there is the familiar clatter of still more rapid notes, delivered especially while two of the birds hover in midair, apparently about to engage in mortal combat. The voice has a very different quality from that of the eastern kingbird, lacking the very shrill, high-pitched character of the latter."

Claude T. Barnes tells me that they make "while awing or while resting, harsh, discordant notes like *ker-er-ip-ker-er-ip*, the number of repetitions apparently depending upon the whim of the moment."

W. Leon Dawson (1923) makes the amusing suggestion that "the love song is, curiously, a sneeze. For the early notes are ridiculously like the frantic protests of a prospective victim of cachination, followed by an emphatic and triumphant relief:

$$añ \; a \qquad\qquad\qquad añ \; a$$
$$añ \qquad\qquad\qquad\qquad añ$$
$$añ \qquad kucheź \; iwick, \quad añ \qquad kucheź \; iwick!"$$

Enemies.—These kingbirds and their eggs and young are doubtless preyed upon by the ordinary mammalian and avian predators that attack other small birds, but they are valiant and often successful in driving their enemies away from their homes. Dr. Roberts (1932) tells of the vigilance of a pair of these birds in defending their nestlings. The watchful parent "paid no apparent attention to other birds that might enter the tree, but when a red squirrel appeared a hundred feet away it made directly for it, attacking it vigorously and forcing it to leave the neighborhood."

Lt. McCauley (1877) tells an interesting story of how a "tree-mouse" camped in the lining of a nest and devoured the contents of the eggs; he says:

"The rascally little mouse had made himself completely at home. Burrowing in the buffalo-wool, he had as warm and cosy a retreat as mouse ever dreamt of or wished for. When hungry, he quietly reached up, and his meal was ready and warm. It was purely a case of 'free lunch' in nature. He had eaten all the eggs but two, his retreat being full of fine pieces of egg-shells. Of those remaining, he had sucked out nearly all the contents of one, and upon the other he had also begun; a hole had been gnawed in the side of it, and the embryo, which had been well advanced, was lifeless."

The Arkansas kingbird is a "a very rare victim" of the cowbirds (Friedmann, 1929).

Field marks.—Although the behavior of the Arkansas kingbird is similar to that of the eastern kingbird, and although their ranges somewhat overlap, there is no excuse for confusing the gray and yellow western with the black and white eastern species. The Arkansas and Cassin's kingbirds are much alike in general appearance, but the head of the former is lighter gray, the gray of the chest is paler, less extensive, and less strongly contrasted with the white chin of the latter, and the tail of the former is blacker than that of Cassin's. But the most reliable field mark of the Arkansas kingbird, which will distinguish it with certainty from any other kingbird, is the conspicuously white outer web of the outer tail feather.

Winter.—The Arkansas kingbird spends the winter in Central America, from western Mexico to Nicaragua. Dickey and van Rossem (1938) record it as a "fairly common migrant and winter visitant in the foothills and mountains [of El Salvador], arriving late in the fall and remaining until well along in the early summer. * * * The limits in altitude were 800 to 7,200 feet."

DISTRIBUTION

Range.—Western North America, north to southern Canada.

Breeding range.—The normal breeding range of the Arkansas kingbird extends **north** to southern British Columbia (Clinton); and the southern part of the Prairie Provinces, Alberta (Morrin); Saskatchewan (Last Mountain Lake); and Manitoba (Oak Point and Portage la Prairie). **East** to southeastern Manitoba (Portage la Prairie); Minnesota (Warren, Duluth, and Minneapolis); Iowa (West Okoboji Lake and Ogden); Missouri (Columbia and Stotesbury); and eastern Oklahoma (Tulsa and Okmulgee). **South** to Oklahoma (Okmulgee and Norman); northern Texas (Wichita Falls and Vernon); northern Chihuahua (Casas Grandes and Whitewater); southern Arizona (Tombstone, Huachuca Mountains, and Baboquivari Mountains); and northern Baja California (La Grulla). **West** along the Pacific coast from northern Baja California (La Grulla and Todos Santos Island); to southwestern British Columbia (Chilliwack, Lillooet, and Clinton). There are scattered casual breeding records as far east as Michigan (Delton) and Ohio (Bono).

Winter range.—During the winter season this species is concentrated chiefly in the western parts of Central America, north to Sonora (Alamos); and south to southern Guatemala (Gualan, Amatitlan, and Patulul); and El Salvador (Colima and Mount Cacaguatique). Records are lacking that would indicate any residence at this season in the eastern part of these countries.

Spring migration.—Early dates of spring arrival are: Texas—Rockport, April 18. Oklahoma—Oklahoma City, April 28. Kansas—Harper, April 23. Nebraska—Red Cloud, April 26. South Dakota—Faulkton, May 3. North Dakota—Fargo, May 6. Manitoba—Aweme, May 17. New Mexico—Fort Webster, March 25. Colorado—Fort Morgan, April 20. Wyoming—Lingle, May 5. Montana—Terry, May 8. Saskatchewan—Old Wives Creek, May 26. California—Buena Park, March 15. Oregon—Coos Bay, April 3. Washington—North Dalles, April 24. British Columbia—Okanagan Landing, April 29.

Fall migration.—Late dates of fall departures are: British Columbia—Okanagan Landing, September 9. Washington—Grand Dalles, September 1. Oregon—Coos Bay, September 10. California—Exeter, October 28. Alberta—Red Deer River, September 1. Montana—Fortine, September 5. Wyoming—Careyhurst, September 6. Colorado—Fort Morgan, September 20. New Mexico—Rio Alamosa, September 28. Saskatchewan—Eastend, August 26. North Dakota—Argusville, September 15. South Dakota—Faulkton, September 15. Nebraska—Red Cloud, September 23. Kansas—Lawrence, October 4. Manitoba—Aweme, September 3. Minnesota—Fosston, September 6.

Casual records.—The Arkansas kingbird has been taken on many occasions at points along the Atlantic seaboard, most of these being in the States of New Jersey, New York, Massachusetts, and Maine. A specimen killed by a cat at Lower Wedgeport, Nova Scotia, on October 26, 1935, appears to be the most northerly record in this region. The species also has been recorded in Maryland, near Washington, D. C., September 30, 1874, and near Denton, September 28, 1931; Virginia, Wallops Island, September 19, 1919; North Carolina, Lake Mattamuskeet, October 1, 1935; South Carolina, Charleston, December 16, 1913, and Bull Island, November 19, 1937; while there are a number of records for Florida from Pensacola and St. Marks, south to Key West, observed or collected from September 22 to January 18. Specimens also have been taken in Illinois, Highland Park, June 6, 1924; Wisconsin, Albion, June 11, 1877, Kenosha, June 2, 1935, near Madison, August 1, 1927, and Roxbury, May 31, 1931. A specimen taken at Lake St. Martin, Manitoba, on June 8, 1932, with others seen in that locality in 1933 and 1936, appears to be the most northerly station.

Egg dates.—Arizona: 18 records, May 14 to July 20; 10 records, May 30 to June 17, indicating the height of the season.

California: 106 records, April 17 to July 9; 53 records, May 10 to June 4.

Colorado: 12 records, June 1 to 17.

South Dakota: 8 records, June 4 to 25.

Washington: 10 records, May 25 to June 20.

TYRANNUS VOCIFERANS Swainson

CASSIN'S KINGBIRD

PLATE 7

HABITS

This is another yellow-breasted kingbird, somewhat resembling the Arkansas kingbird and occupying some of the western, and more especially southwestern, range of the latter. The range of *vociferans* is not nearly so extensive as that of *verticalis*, and its local habitat is often quite different. The two species are often found in the same general habitat, especially during the migration periods, but during the breeding season and to a certain extent at other seasons Cassin's ranges higher in the foothills and the mountains than the Arkansas kingbird. We found Cassin's kingbirds very common, in April and May, in the lower portions of the canyons, among the large syca- mores, in the Catalina and Huachuca Mountains in Arizona. Harry S. Swarth (1904), referring to the Huachucas, says: "I have occa- sionally, but not often seen the birds as high as 7500 feet, and found one nest quite at the mouth of the canyon, 4500 feet; but as a rule, the territories occupied by this species and *verticalis* during the breeding season hardly overlap." The majority of the nests he found were between 5000 and 6000 feet. Referring to its range in the Catalinas, W. E. D. Scott (1887) writes: "At the higher limits of its range in the breeding season—about 9000 feet—it is much more common than *T. verticalis*, though the reverse is true as re- gards the lower limit of its range—about 3500 feet—in the breeding season." Dr. Alexander Wetmore (1920) says that at Lake Burford, N. Mex., "they frequented rocky hillsides where scattered Yellow Pines rising above the low undergrowth made convenient perches from which to watch for insects and look out over the valleys."

Henshaw (1875) found it frequenting open country in Arizona and New Mexico, saying, "I have seen it much on the sage brush plains, though never very far from the vicinity of timber; and the sides of open, brushy ravines seem to suit its nature well."

In California, its distribution is more or less irregular, where it seems to be less of a mountain species than in other places, for W. L. Dawson (1923) says: "Cassin's Kingbird, at the nesting season, barely exceeds the upper limit of the Lower Sonoran faunal zone; and it is not even mentioned in the exhaustive reports on the San Jacinto and San Bernardino mountain regions. It is apparently of very irregular distribution over the two California deserts, and in the lowlands of the San Diego-Santa Barbara region."

Courtship.—The only reference I can find to what might be called a courtship activity is the following observation by Dr. Wetmore (1920) at Lake Burford: "This Kingbird was first observed on May 25 and from then on it was fairly common. * * * Males were seen at intervals in crazy zigzag sky dances made to the accompaniment of harsh calls and odd notes, similar to those of none of our other birds. Toward dusk they called constantly their harsh, stirring notes making a pleasing sound that mingled with the songs of House and Rock Wrens, the scolding of an occasional Mockingbird and the cheerful calls of the Robins."

Nesting.—Major Bendire (1895) writes:

The trees generally selected by this species for nesting sites are pines, oaks, cottonwood, walnut, hackberry, and sycamores, and the nests are almost invariably placed near the end of a horizontal limb, usually 20 to 40 feet from the ground, in positions where they are not easily reached. All of the nests examined by me were placed in large cottonwoods, with long spreading limbs, and were saddled on one of these, well out toward the extremity. The majority could only be reached by placing a pole against the limb and climbing to it. They are fully as demonstrative as the Arkansas Kingbird when their nests are disturbed, and are equally courageous in the defense of their eggs and young. The nests are large, bulky structures, larger than those of the preceding species, but composed of similar materials. An average nest measures 8 inches in outer diameter by 3 inches in depth. The inner cup is 3½ inches wide by 1¾ deep. Sometimes they are pretty well concealed to view from below, but they can usually be readily seen at a distance.

Mr. Dawson (1923) noticed, in California, a close association with the Arkansas kingbird:

[In a] region of scattering oak trees and of stream beds lined with cottonwoods, both birds are exceedingly common. As surely as a pair of oak trees boast some degree of isolation from their fellows, one will be occupied by a pair of Cassin Kingbirds and the other by a pair of Westerns. Or if the trees are only members of a series, next door neighbors will be occupied by these paired doubles; and the group may be separated by an interval of a hundred yards or so from the next quartet. The arrangement is evidently studied, and it must be mutually agreeable, for the two species are on the best of terms, and I have never seen evidence of jealousy or ill-will on the part of either, though I have camped right under their nests.

In the Huachuca and Catalina Mountains, the Cassin's kingbirds showed a preference for the sycamores in the lower canyons, and for the evergreen oaks on the foothills. But near Tombstone, Ariz., my companion, Frank Willard, found a nest 10 feet up in a slender willow and another 8 feet from the ground in a small walnut tree. In California nests have been found in blue-gum trees and in box-elders. Robert B. Rockwell (1908) found nests in Mesa County,

Colo., "in scrub-oak, cottonwoods, quaking-asps, and gate frames, on log fences, and on the top of a large farm gate. The birds are of a sociable disposition, nests being rarely found any great distance from human habitation."

The bulky nests of the Cassin's kingbird are much like those of the Arkansas kingbird, but will average somewhat larger and rather more firmly built. The foundation and walls are built up with small twigs, rootlets, weed stalks, strips of inner bark and other plant fibers, mixed with bits of string, rags, or dry leaves; the sides and rim are often decorated with feathers, dry blossoms of the sage, or the dry flower clusters of other plants; and the inner cavity is lined with finer rootlets, fine grass, and perhaps a few small feathers; some nests are profusely lined with the cottony seeds of the cottonwood.

One shallow nest that I measured, taken from a sycamore in the Huachuca Mountains, measured externally 1¾ to 2 inches in height, and 5 to 6 inches in diameter; the inner cavity was 3½ by 4 inches in diameter and 1½ inches deep.

Eggs.—Three or four eggs are most commonly laid by Cassin's kingbird, though quite often the set consists of five and very rarely as few as two. The eggs resemble those of other kingbirds, though they are as a rule less heavily spotted. They are about ovate and are only slightly glossy. The ground color is white or creamy white. The markings, often grouped about the larger end, are small spots or dots of various browns, "chestnut-brown," "dark vinaceous-brown," or "light brownish drab," often, but not always, mixed with underlying spots of "Quaker drab" or shades of pale lavender. The measurements of 50 eggs average 23.5 by 17.4 millimeters; the eggs showing the four extremes measure 27.9 by 19.3, 21.8 by 17.3, and 22.6 by 16.3 millimeters.

Young.—Bendire (1895) says that incubation "lasts from twelve to fourteen days, and is almost always, if not exclusively, performed by the female. I have never noticed the male on the nest." Both sexes, however, assist in the defense and care of the young. Mrs. Wheelock (1904) says that the young remain in the nest about two weeks. In the southern portion of its range, the Cassin's kingbird is said to raise two broods in a season.

Plumages.—The young kingbirds are hatched naked, but the juvenal plumage soon appears and is well developed, except for the shorter wings and tail, by the time the young bird leaves the nest. The fully developed juvenal plumage is much like that of the adult female, but the plumage is of looser texture and all the colors are duller; the sexes are alike; the head, neck, and breast are a lighter gray, the yellow of the abdomen is paler, and the back is more grayish brown; there is no orange-red crown patch; the paler markings on the wings are tinged with buff, and the outer primaries are not attenuated; the tail feathers are somewhat shorter than in the adult, and are narrowly tipped with brown. This plumage is worn through the summer, until an apparently complete postjuvenal molt, mainly in September, produces a first winter plumage, like that of the adult, but slightly paler and still lacking the crown patch. In this first winter plumage the young male acquires the attenuated outer primaries as in the adult; adult males have four or five of the outer primaries sharply attenuated; in adult females these feathers are only bluntly pointed, if at all so. In the first winter tail the feathers are tipped with yellowish gray, or whitish, and more broadly than in the younger bird. Apparently there is a partial prenuptial molt early in spring, at which the orange-red crown patch is acquired. Adults apparently molt at the same times and in the same manner as the young birds; June and July adults show considerable wear.

Food.—Living in much of the territory occupied by the Arkansas kingbird and foraging in much the same manner, the food of Cassin's kingbird is almost identical with that of the better-known species. Professor Beal (1912) examined only 40 stomachs of Cassin's kingbird, in which the food was "found to be composed of 78.57 per cent of animal matter to 21.43 of vegetable." Of the animal food—

beetles of all kinds amount to 14.91 per cent of the food. Of these, about 1 per cent are of species that are more or less useful (Carabidae). * * * Hymenoptera amount to 21.61 per cent and consist for the most part of wild bees and wasps. No honeybees were found, but several predaceous or parasitic species were identified. * * * Lepidoptera, i. e., moths and caterpillars, amount to 18.21 per cent of the food, which is a high percentage for a flycatcher; for while moths may be caught on the wing, caterpillars must be picked from the surface on which they crawl, unless they let themselves down from a tree by a thread and so hang in mid-air. * * * Orthoptera (grasshoppers and crickets) are apparently eaten rather irregularly, but as nearly every month in which they appeared at all showed a goodly quantity, they would seem to be a favorite food, and it is probable that a greater number of stomachs would give a more regular showing. In January they amount to 47.50 per cent in 4 stomachs, while the 1 stomach taken in February shows none at all. * * * The total for the year is 14.67."

Small percentages of Hemiptera (bugs), Diptera (flies), a few dragonflies, and some spiders make up the balance of the animal food, 9.17 percent.

Of the vegetable food, Professor Beal (1912) says: "Although Cassin's kingbird eats more vegetable food than any other flycatcher, there is very little variety to it. Grapes, apparently of cultivated varieties, were found in 9 stomachs, olives in 2, elderberries in 1, blueberries (*Vaccinium*) in 1, and pulp not further identified in 4. With the exception of some grapes found in 1 of the March stomachs, all the fruit was eaten in the months from September to January, inclusive."

Behavior.—Henshaw (1875) quotes some Colorado notes from C. E. Aiken, which give a good idea of the behavior and appearance of Cassin's kingbird as compared with the Arkansas kingbird; he writes:

Although these two birds resemble each other so closely in the skin, in life there are marked differences in notes and actions that even a novice cannot fail to notice. *Verticalis* is a nervous, fickle creature, seldom remaining long in one place, and flying with a quick fluttering motion of the wings. It is also exceedingly noisy, its notes being a high pitched chatter. *Vociferans*, on the other hand, is a more matter of fact bird, often sitting quietly for a long time in the same place, and its notes are harsher and less frequently uttered. Its appearance, too, when alive conveys the impression of a heavier, stouter built bird. When migrating, and indeed at other times, it appears to be restricted to the parks of the foothills, alighting upon weed stalks and low bushes, from which it sallies forth occasionally to seize some passing insect.

Henshaw (1875) also says: "Though found in the same locality, individuals of the two species never meet without displaying their natural enmity. At Camp Grant, my attention being called by the loud outcries of several of these birds, I found that a female and several young of the Arkansas Flycatcher were the objects of a savage assault by a pair of the present species. The mother bird most gallantly stood up and fought for her offspring, repelling each attack with a brave front, and retaliating to the best of her ability. I watched them until I saw that the assailants, having fairly got worsted, were glad to retire, and leave the family to gather together in peace."

This behavior is quite different from that noted by Mr. Dawson (1923) in California, referred to under nesting; evidently there is considerable individual variation in behavior in this, as in many other species. Some observers state that this kingbird does not persecute hawks and crows so much as the other kingbirds do, but others report to the contrary. For example, E. A. Mearns (1890) says: "On the Mogollon Mountains I saw them attack Crows and Western Red-tailed Hawks and drive them from the neighborhood of their nests after the spirited fashion of the Eastern Kingbird." Doubtless all kingbirds would do this. Florence A. Merriam Bailey (1896) writes: "Mr. Merriam told me that when he was plowing and the Blackbirds were following him, two or three of the 'Bee-birds,' as he called them, would take up positions on stakes overlooking the flock; and when one of the Blackbirds got a worm that he could not gulp right down, a Beebird would dart after him and fight for it, chasing the Blackbird till he got it away. For the time the Flycatchers regularly made their living off the Blackbirds as the Eagles do from the Fish Hawks."

Voice.—Mr. Dawson (1923) writes: "Cassin's Kingbird says, *Che bew'*, in a heavy, grumpy tone, whose last flick nevertheless cuts like a whip-lash—*chebeeú*. This is generically similar, but specifically very different from the evenly accented, and more nearly placid *ber' wick* of the Western Kingbird. The note of greeting or of general alarm in Cassin is a breathless *kuh daý kuh daý kuhdaý;* or, as I heard a female render it, *kiddoó kiddoó kiddoó kiddoó kidduck.* For the rest Cassin is a rather more sober and a much more silent bird than is the volatile *verticalis.*"

Ralph Hoffmann (1927) says that "its common call is a harsh, low-pitched *cherr*, followed by a *ke-déar*, which suggests the Ash-throated Flycatcher. In the breeding season it utters a series of high, petulant notes, *ki-dee-dee-dee.*" Mr. Swarth (1904) writes: "Commencing shortly before daybreak, they keep up a continuous clamor, generally on the wooded hillsides, to such an extent that it seems like an army of birds engaged. They do not seem to be quarreling or fighting at these times, for those I have seen merely sat, screaming, on the top of some tall tree. This racket is kept up until about sunrise, when it drops rather abruptly." About our camp in picturesque Apache Canyon in the Catalina Mountains, Ariz., we found Cassin's king-birds very common; they greeted us every morning with their rather melodious notes, as they flitted about in the tops of the grand old sycamores and cottonwoods; they seemed to say "come here, come here," in rather pleasing tones, as their voices mingled with the rich song of the Arizona cardinal and the attractive notes of the canyon wren, the Arizona hooded oriole, the canyon towhee, and the host of other birds that made our mornings in that beautiful canyon so delightful.

Field marks.—Cassin's kingbird bears a superficial resemblance to both Couch's kingbird and the Arkansas kingbird. Where its range overlaps that of the western races of *melancholicus*, Cassin's can be recognized by having a blacker tail and the conspicuous white space on the chin and throat more restricted and more sharply defined against the gray of the chest; the yellow of the under parts is also paler. From the Arkansas kingbird it can be distinguished by the darker gray of the head, neck, and chest, sharply contrasted with the white throat, by the lack of the white outer web of the outer tail feather, and by the white tips of the tail feathers. Moreover, its voice, its apparently heavier build, and its general behavior, as indicated above, should make it easier to recognize Cassin's kingbird in life than in a museum specimen.

DISTRIBUTION

Range.—Western United States and Central America south to Guatemala.

Breeding range.—Cassin's kingbird breeds **north** to central California (Redwood City and probably Tracy); northern Utah (Salt Lake County); and northern Colorado (probably Ault). **East** to eastern Colorado (probably Ault, Agate, and Beulah); western Oklahoma (Kenton); eastern New Mexico (Montoya and Roswell); western Texas (Fort Davis and Alpine); western Tamaulipas (Miquihuana); and Guerrero (Chilpancingo). **South** to Guerrero (Chilpancingo); Jalisco (Ocotlan); southwestern Durango (Salto); and northern Baja California (Aguaita). **West** to northwestern Baja California (Aguaita and San Quintin); and north along the coast of California to Redwood City.

Winter range.—During the winter months these birds are found **north** to southern California (Santa Barbara and Salton Sea); southern Arizona (Nogales); and Chihuahua (Colonia Garcia and Chihuahua City); and **south** through Mexico (chiefly the western part), to Guatemala (Salama and Duenas).

Spring migration.—Early dates of spring arrival are: New Mexico—State College, April 13. Colorado—Pueblo, April 29. Arizona—Tucson, March 24. Utah—Salt Lake City, May 11. Central California—Paicines, March 7.

Fall migration.—Late dates of fall departure are: Central California—San Francisco, October 22. Arizona—Tombstone, November 1; Tucson, November 10. Colorado—Beulah, September 5. New Mexico—Mesilla, October 12.

Casual records.—In common with many other flycatchers, Cassin's kingbird has been taken or observed on several occasions at points outside its normal range. Records in southeastern Wyoming (Douglas, Laramie Peak, Laramie, Cheyenne, and Albany County) indicate the probability of breeding in that area. In the Bull Mountains, Mont., several were seen and specimens taken on August 5 and 6, 1918; one was reported at Beaverton, Oreg., on May 5, 1885, and a pair were recorded from the Warner Valley in that State on June 10, 1922.

Egg dates.—Arizona: 21 records, May 3 to August 1; 11 records, May 28 to June 22, indicating the height of the season.

California: 72 records, April 22 to June 29; 36 records, May 11 to June 2.

Lower California: 8 records, May 13 to June 18.

New Mexico: 8 records, May 4 to June 29.

MUSCIVORA TYRANNUS (Linnaeus)

FORK-TAILED FLYCATCHER

HABITS

This long-tailed, conspicuously black and white flycatcher, a straggler from its range in Mexico and South America, has been recorded north of our southern border in Mississippi, Kentucky, New Jersey, Massachusetts, and Maine. Audubon (1840) gives the earliest records of its occurrence within the United States in the following words:

In the end of June, 1832, I observed one of these birds a few miles below the city of Camden, New Jersey, flying over a meadow in pursuit of insects, after which it alighted on the top of a small detached tree, where I followed it and succeeded in obtaining it. * * *

Many years ago, while residing at Henderson in Kentucky, I had one of these birds brought to me which had been caught by the hand, and was nearly putrid when I got it. The person who presented it to me had caught it in the Barrens, ten or twelve miles from Henderson, late in October, after a succession of white frosts, and had kept it more than a week. While near the city of Natchez, in the state of Mississippi, in August 1822, I saw two others high in the air, twittering in the manner of the King Bird; but they disappeared to the westward, and I was unable to see them again.

The other records are to be found in the paragraph on distribution following this account.

Alexander F. Skutch writes to me that "this is a bird of the drier and more open parts of Central America. I have seen it only at Balboa, Canal Zone; Buenos Aires de Osa, Costa Rica; and a few in the neighborhood of Cartago, Costa Rica, in open pasture-land at an altitude of 4500 feet, where it is said to breed. Perhaps the most conspicuous bird on the *sabanas* or prairies about Buenos Aires was the fork-tailed flycatcher. These might be found, in small, scattered flocks, wherever low bushes, growing amidst the grass, afforded them perches from a foot to a yard above the ground. On such low perches they rested quietly through the day, ever alert to dart forth and snatch up some insect whose movement attracted their keen sight."

W. H. Hudson (1920) says that in Argentina the fork-tailed flycatcher "is migratory, and arrives, already mated, at Buenos Ayres at the end of September, and takes its departure at the end of February in families—old and young birds together. * * * It prefers open situations with scattered trees and bushes; and is also partial to marshy grounds, where it takes up position on an elevated stalk to watch for insects."

Nesting.—According to Hudson (1920), "the nest is not deep, but is much more elaborately constructed than is usual with the Tyrants. Soft materials are preferred, and in many cases the nests are composed almost exclusively of wool. The inside is cup-shaped, with a flat bottom, and is smooth and hard, the thistle-down with which it is lined being cemented with gum."

Evidently these flycatchers are favorite hosts for the Argentine cowbird, for he writes: "One December I collected ten nests of the Scissor-tail (*Milvulus tyrannus*) from my trees; they contained a total of forty-seven eggs, twelve of the Scissor-tails and thirty-five of the Cow-birds. It is worthy of remark that the *Milvulus* breeds in

October or early in November, rearing only one brood; so that these ten nests found late in December were of birds that had lost their first nests. Probably three-fourths of the lost nests of *Milvulus* are abandoned in consequence of the confusion caused in them by the Cowbirds."

Ten out of nineteen sets of eggs of this flycatcher in the United States National Museum contain eggs of this cowbird. The only nest of this flycatcher that I have seen is in the Thayer collection in Cambridge. It was taken at Punta Gorda, British Honduras, on March 31, 1907, by N. Karslund and contains four eggs. It was built in a small pine, 12 feet from the ground. On a foundation of coarse weed stems and coarse grasses is a layer of stiff, black rootlets; above that is the nest of grasses, coarse vegetable down, weed tops, seed and blossom scales, and vegetable rubbish of various kinds; it is lined with a fine grade of usnea; its outside diameter is 5 by 4¾ inches, and the inner diameter is about 3 inches; the external depth is from 2 to 2¼ inches; and the inner cavity is only three-quarters of an inch deep.

George K. Cherrie (1892), in his list of birds found near San Jose, Costa Rica, says that "a nest with three fresh eggs, taken by Don Anastasio Alfaro at Tambor, Alajuela, May 2, 1889, was placed in a small tree, about ten feet from the ground. The parent bird left the nest only very reluctantly and not until almost within the grasp of the collector. The nest is constructed of a mixture of small dry grass and weed stems and soft dry grass rather compactly woven together, with a lining of a few fine rootlets."

Eggs.—The four eggs in the Thayer collection, the only ones I have seen, are short-ovate and only slightly glossy. They are pure white and rather sparingly marked with spots and small blotches of dark brown, dark "chestnut-brown" or "seal brown" and various shades of "quaker drab" or "brownish drab."

Hudson (1920) says that "the eggs are four, sharply pointed, light cream-colour, and spotted, chiefly at the large end, with chocolate."

Baird, Brewer, and Ridgway (1905) say: "An egg of this species obtained by Dr. Baldamus, from Cayenne, exhibits a strong resemblance to the egg of the common Kingbird. It has a clear white ground, and is spotted with deep and prominent markings of red and red-brown. They are of an oblong-oval shape, are tapering at one end, and measure .90 by .68 of an inch."

Evidently the eggs vary considerably in shape and coloration. The measurements of 50 eggs average 22.1 by 16.2 millimeters; the eggs showing the four extremes measure 25.2 by 17.3, 23.1 by 18.0, 20.0 by 15.3, and 20.7 by 15.0 millimeters.

Plumages.—As I have seen no very young specimens of this species and not enough molting birds to give a clear idea of the plumage changes, I am following Dr. Jonathan Dwight, Jr. (1900) in his excellent treatment of this subject. He says that the sexes are alike in the juvenal plumage, which he describes as follows: "Above, including wings and tail, olive-brown; coverts and wing-quills narrowly edged with pale russet. Below, white. Orbital region dull clove-brown. * * * The tips of the primaries are rounded and there is no yellow crown-patch. The tail is but five inches in length."

A complete postjuvenal molt produces, in September or later, a first winter plumage in which young and old birds become practically indistinguishable. "Males become glossy black with yellow crown-patch; the outer pair of rectrices are fully nine inches in length and blacker than those of the juvenal dress; the three distal primaries are deeply incised at the tips, a peculiar emargination."

There is, apparently, a partial prenuptial molt in both young and old birds in February and March, involving usually the body plumage only. The complete annual molt of adults occurs mainly in August and September, in birds reared north of the Equator. "In later plumages the sexes are very similar, the females usually with less emargination"; and the yellow crown patch is smaller in the female.

Food.—The fork-tailed flycatcher feeds largely, although not wholly, on insects, mostly obtained on or near the ground. Perched on some low shrub or stalk, it watches for passing insects and darts out after them in the air. But Mr. Skutch (MS.) also observed that "most of their prey seemed to be snatched from the grass or the ground; on a burnt hilltop above the village, where the fire had consumed the grass but only charred the low bushes, I could see clearly that they picked most of their food from the charred soil, often at a considerable distance from the point where they had been perching, then alighted upon another low perch to devour it."

He also describes, in his notes made at Balboa on December 30, 1930, another method of feeding: "Walking down the magnificent avenue of royal palms that leads to the Administration Building, I saw my first fork-tailed flycatchers this morning. There were dozens of them, perching in the trees bordering the parkway, and sallying out now and again to pluck a red berry from one of the heavy clusters of ripe fruit that hung from the palm, trees in the central square. It was a delight to watch them fly lightly and airily forward, the two narrow and elongated outer tail feathers, each longer than the body of the bird, forming slender black streamers, which undulated gracefully with every movement. As the bird poised on wing before the brilliant cluster to pluck a berry in its bill, the tail was spread until its posterior margin suggested the crescent moon, with its horns extended into narrow ribbons of black. The berry was usually carried back to a perch in a tree above the sidewalk and there eaten whole. Sometimes, however, the flycatcher clung to the spadix of the palm and gobbled down a few berries while hanging there. The berries are about the size of large peas, with a thin yellow pulp enclosing the single large seed."

Behavior.—He says further (MS): "Feeding with the fork-tailed flycatchers were numerous Lichtenstein's kingbirds, whose persistent twitters were in contrast to the silence of the former. Unimportant clashes between the two species were frequent, and the kingbird was usually the aggressor."

Referring to his observations in Costa Rica, where he saw them feeding on the ground, he writes (MS.): "It seems strange that these birds, which appear so aerial with their long, slender, streaming tail feathers, should hug the ground so closely; by day I saw them only resting low in the open places, or flying over them at no great elevation, never in the bushy lands of the valley, where most of the birds of the neighborhood were to be found, nor perching in trees. They are 'ground-grazers' like some of the Argentine flycatchers described by Hudson. As they rested low among the grasslands, their white breasts, when turned toward me, seemed in the distance to be the great snow-white blossoms of some humble prairie herb.

"Each evening, as the sun began to sink low above the forested crests of the hills beyond the Rio Grande de Térraba, all the fork-tailed flycatchers in the vicinity began to stream in toward the village. With their long tails flowing behind them, they are no less graceful in flight than their relatives the scissor-tailed flycatchers; but since they lack the scarlet and pink sides of the latter, they are far less colorful. Just as the scissor-tail slept in the orange trees behind the *jefatura* of Las Cañas in Guanacaste, so the forktails roosted in the orange trees behind the *casa cural* (priest's house) in Buenos Aires. Since the *padre* spends here only a few days during the course of the year, the house had long been unoccupied when I arrived to lodge there, and the birds had been unmolested. I found it impossible to estimate how many of them gathered nightly in the orange trees; there were scores of them, possibly hundreds. If I appeared at the edge of the porch, or on the ground beneath them, while they were arranging themselves for the night, many would dart rapidly forth in all directions, with their long tails whistling as they rushed through the air. Then, when I had vanished into the house, the birds would return to their orange trees. Once darkness had descended and they had fallen asleep, I could move quietly beneath them without disturbing their slumbers.

"In the morning they would linger in their roosting places after awakening. If I appeared beneath them before they had quit their sleeping places, many would rush forth with a great whistling of tailfeathers; but others bravely held their perches. As the day grew older, they flocked over the *sabanas* and were not again to be seen in the trees about the village until the day waned.

"With the fork-tailed flycatchers on the *sabanas*, I frequently found small flocks of wintering myrtle warblers. The warblers not only foraged about the low bushes in which the flycatchers rested, but the two kinds of birds, so dissimilar in appearance and habits, changed their feeding grounds together. While I sometimes found the myrtle warblers alone, I saw them in the company of the fork-tailed fly-catchers too often for the association to be accidental. The warb-lers seem to have a definite preference for the company of the flycatchers, just as their near relatives, the Audubon's warblers, like to accompany the bluebirds in the highlands of Guatemala."

Hudson (1920) refers to a remarkable habit of these flycatchers as follows: "They are not gregarious, but once every day, just before the sun sets, all the birds living near together rise to the tops of the trees, calling to one another with loud, excited chirps, and then mount upwards like rockets to a great height in the air; then, after whirling about for a few moments, they precipitate themselves down-wards with the greatest violence, opening and shutting their tails during their wild zig-zag flight, and uttering a succession of sharp, grinding notes. After this curious performance they separate in pairs, and perching on the treetops each couple utters together its rattling castanet notes, after which the company breaks up."

Voice.—Some of the notes of the fork-tailed flycatcher, as heard by Hudson, are mentioned above. The bird that Audubon (1840) shot uttered a "sharp squeak, which it repeated, accompanied with smart clicks of its bill." Mr. Skutch says in his note: "The only sound I have heard a fork-tailed flycatcher utter is a low, weak, somewhat croaking monosyllable; but as the birds rested quietly in the tops of their orange trees early in the morning, they would sometimes join their slight voices in a sudden wave of sound, which lasted only a few seconds, then died away almost as abruptly as it had begun. When many voices were united in this fashion, they produced a sound of considerable volume."

Field marks.—The fork-tailed flycatcher could hardly be mistaken for anything else, except that, at a great distance and in unfavorable light, it might resemble in general appearance the swallow-tailed fly-catcher, a closely related species of similar shape and behavior. But, under any ordinary circumstances, its pure-white under parts, gray back, black crown, and long black tail are strikingly distinctive.

DISTRIBUTION

Range.—The fork-tailed flycatcher is a tropical species that ranges from southern Veracruz (Tlacotalpan and Playa Vincente); Quintana Roo (Camp Mengel); south through Central America and South America to Argentina (Rosas, Azul, and Zelaya). Several subspecies in this range are recognized.

Casual records.—While the species has been recorded on several occasions in the United States, some of the records are unsatisfactory and cannot be fully accepted. Audubon reports taking one in June 1832 near Camden, N. J., and seeing two on the wing near Natchez, Miss., in August 1822. Another one was brought to him at Henderson, Ky. A specimen is said to have been taken at Lake Ridge, Lenawee County, Mich., in July 1879, but apparently it has since been destroyed. This also is the fate of one alleged to have been taken at Santa Monica, Calif., late in the summer of 1883. There are indefinite reports of occurrence in March 1847 and April 1849 in Bermuda.

A specimen in the Academy of Natural Sciences of Philadelphia labeled merely "New Jersey" has been identified as *Muscivora t. sanctaemartae;* one is known to have been taken near Bridgeton, N. J., during the first week in December, about 1820; another, taken near Trenton in the fall of 1900, has since been lost; while one was seen near Cape May Point during the first three days of November 1939. A sight record for Massachusetts is one reported from Edgartown on Marthas Vineyard, October 22, 1916. One taken at Marion, Washington County, Maine, on December 1, 1908, has been identified as *M. t. tyrannus*, which also is the determination of a specimen collected at Fox Chase, near Philadelphia, Pa., in the fall of 1873.

Egg dates.—Argentina: 4 records, November 24 to December 13.

Brazil: 4 records, November 9 to 23.

British Honduras: 2 records, March 21 and May 20.

Paraguay: 18 records, November 3 to December 4; 9 records, November 13 to 25, indicating the height of the season.

MUSCIVORA FORFICATA (Gmelin)

SCISSOR-TAILED FLYCATCHER

PLATE 8

HABITS

Its delicate and pleasing color contrasts, white, black, soft gray, salmon-pink, and bright scarlet, its trim and graceful form, and its spectacular behavior as an aerial acrobat, all combine to make the scissor-tailed flycatcher one of the most attractive of North American birds. The traveler in Texas cannot fail to notice this charming bird, where it is universally common, or even abundant, and everywhere conspicuous.

Its breeding range extends from southern Nebraska to southern Texas, with casual breeding records in neighboring States. But it has wandered widely beyond this range, as far north as Hudson Bay, as far east as New Brunswick, Massachusetts, South Carolina, and Florida, and as far west as Colorado.

The scissor-tailed flycatcher is a bird of the open country. We found it generally distributed in the coastal prairie region of Texas from Corpus Christi to Brownsville, in the trees along the country roadsides, on the open prairies dotted with small trees and mottes, on the mesquite prairies, in the open chaparral country, about the ranches, and even in the small towns. It was often seen sitting on the telephone wires, on fence posts, and on wire fences, quite familiar and unafraid.

Spring.—A. J. van Rossem (Dickey and van Rossem, 1938) says that in El Salvador "the northward movement starts about April 1. On April 5, 1926, a flight of about 100 birds, strung out with many yards between individual members, was seen passing along the foothills near Divisadero. From April 15 to 27, 1926, numbers were seen each evening over the city of San Salvador, flying westward low over the housetops and stopping frequently to perch on flagpoles or telephone wires. Fully a hundred were seen each evening, the straggling flocks being often accompanied by other migrating species such as *Tyrannus tyrannus* and *Petrochelidon albifrons.*"

Courtship.—Herbert Brandt (1940) writes:

During his courting days and even until the eggs are hatched, the male engages in one of the most fantastic of feathered sky-dances. Mounting the air to a height of perhaps a hundred feet, he starts his routine by plunging downward for about a fourth of the distance, then turns sharply upward to nearly the previous height; and he repeats this up and down zigzag course several times, emitting meanwhile a rolling, cackling sound like rapid, high-pitched hand-clapping. This he seems to produce by loud snapping of the mandibles, or it may be a vocal effort, or both, though I observed it to be the former. The last upward flight may take him still higher, and his path then becomes a vertical line. When the flycatcher reaches the zenith of this flight, so vivacious is his ardor that over he topples backward, making two or three consecutive reverse somersaults, descending like a Tumbler Pigeon, all the while displaying to his mate the soft, effective, under-wing colors.

This active display is remarkably emphasized by the long, flowing tail that becomes an expressive banner of showmanship, and it is then that one realizes its nuptial significance. That dual appendage, the like of which is possessed by no other North American bird, adds to every movement the smooth, effortless rhythm of superb body grace; and consequently the aerial ballet of the Scissor-tail is incomparable in flowing, graceful action and flirtatious courtship interpretation.

Nesting.—The scissor-tailed flycatcher may place its carelessly built nest almost anywhere. George Finlay Simmons (1925) says that its location may be "7 to 30 feet from the ground, generally on a horizontal limb or fork, less commonly in a crotch, in an isolated, open-foliaged hackberry, mesquite, cedar elm, eastern live oak, or hornbeam retama tree, standing alongside a country road, a fencerow, or a quiet city street; in a truck garden; beside a farm house; in pasture-lands; or on the edges of scattered woodlands. Occasionally on knolls in the open country, particularly in post oaks, or in peach and other orchard trees; in metal windmill towers; on the crossbars of telegraph poles; in the iron skeletons of aerial light towers in Austin; and in the iron framework of river bridges."

While I was waiting one day for my guide to appear at a farm-house, I was amused to see one of these flycatchers attempting to build a nest on the wings of a windmill that was in active use; the location would have been satisfactory if the wings had been stationary, but unfortunately they were never still for any great length of time. Each time that the windmill went into action the nest was destroyed; but the flycatchers were not dismayed, and started each time to rebuild it. This happened several times while I was watching.

Mr. Simmons (1925) says of the nest construction: "Roughly built; base and sides composed largely of Indian tobacco weed and small twigs, with some rootlets, weed stems, and cotton, onto which base is built a nest mass of thistledown, cotton, wool, and Indian tobacco, and sometimes cotton waste, corn husks, rags, and twine. Lined with rootlets, horsehair, cotton, or Indian tobacco. * * * Out-side, diameter 4.50 to 6.25 by 6, height 2.75 to 3. Inside, diameter 2.75 to 3 by 3.50, depth 1.75 to 2.25."

Major Bendire (1895) has written quite fully on the nesting habits of these flycatchers, from which I quote:

They nest by preference in mesquite trees, less frequently in live and post oaks, the thorny hackberry or granjeno (*Celtis pallida*), the huisache (*Acacia farnesiana*), honey locust, mulberry, pecan, and the magnolia, as well as in various small, thorny shrubs. Their nests are placed at various distances of 5 to 40 feet from the ground, but on an average not over 15 feet, and often in very exposed situations, where they can be readily seen. Occasionally, when placed in trees whose limbs are well covered with long streamers of the gray Spanish moss, or in shrubs overgrown with vines, they are rather more difficult to discover. * * *

Nests of this species from different localities vary greatly in size and materials from which constructed. The base and sides of the nest are usually composed of small twigs or rootlets, cotton and weed stems (those of a low floccose, woolly annual *Evax prolifera* and *Evax multicaulis*, the former growing on dry and the latter on low ground, being nearly always present); in some sections the gray Spanish moss forms the bulk of the nest, in others raw cotton, and again sheep's wool; while rags, hair, twine, feathers, bits of paper, dry grass, and even seaweeds may be incorporated in the mass. * * *

A nest taken by Dr. E. A. Mearns, United States Army, on April 29, 1893, from an oak tree situated on the edge of the parade ground at Fort Clarke, Texas, is mainly composed of strong cotton twine, mixed with a few twigs, weed stems, and rags; even the inner lining consists mostly of twine. How the female managed to use this without getting hopelessly entangled is astonish-ing. The previous season's nest still remained in the same tree, and a considerable quantity of twine entered also into its composition.

Eggs.—The scissor-tailed flycatcher lays four to six eggs, most commonly five, one egg being laid each day until the set is complete. They are ovate and only slightly glossy. The ground color is clear white, or creamy white, and rarely slightly pinkish white. They are more or less spotted or blotched with dark browns, dark "chestnut brown," "seal brown," or "claret brown," with underlying markings of "heliotrope gray" or shades of "Quaker drab" or "brownish drab." Some eggs are very lightly marked, and some are nearly immaculate. The measurements of 50 eggs average 22.5 by 17.0 millimeters; the eggs showing the four extremes measure 24.1 by 17.8, 23.1 by 19.5, and 20.3 by 15.5 millimeters.

Young.—Incubation is said to last for 12 or 13 days. Bendire (1895) says: "Incubation lasts about twelve days, and the female appears to perform this duty alone, while the male remains in the vicinity, and promptly chases away every suspicious intruder who may venture too close to the nest. The young are fed exclusively on an insect diet, and are able to leave the nest in about two weeks. Both parents assist in their care. In the late summer they congregate in considerable numbers in the cotton fields and open prairies preparatory to their migration south."

Plumages.—The only nestling that I have seen is fully feathered with the remiges and rectrices about half out of their sheaths, and the whole plumage is soft and fluffy; the soft feathers of the upper parts are dull white, more or less tipped with pale drab on the head and upper back; those of the lower back and rump are tipped for half their length or more with darker drab, darkest and most extensive on the rump and upper tail coverts; the under parts are immaculate, dull white, but are washed with pale drab on the upper breast. This plumage is probably not fully developed; when these feathers are fully grown, perhaps before the bird would leave the nest, their white bases would probably be more fully concealed.

Ridgway (1907) describes a young bird, which is evidently in full juvenal plumage, as "somewhat like the paler or duller colored females, but gray of upper parts decidedly brownish (pale drab-gray), the crown darker, and without trace of concealed spot; sides, flanks, abdomen, and under tail-coverts uniform, very pale cream-buff; no orange axillary patch."

I have seen birds in this plumage, in which the outer primary is entire, not attentuated as in the adult, as late as December and even January. But, apparently, the postjuvenal molt begins in October, or earlier, and may continue well into January. This molt produces the first winter plumage, which is more like that of the adult, with the outer primary sharply attenuated at the tip, but the back is somewhat browner, the pink on the flanks is duller, and there is still no orange crown patch, or only a trace of it. I have seen birds in fresh, first-winter plumage, taken in Central America in October; and have seen others in this plumage, showing various stages of wear, in March, April, and May. The material seems to indicate that the postjuvenal molt is complete, or nearly so, and that a partial prenuptial molt, late in winter or early in spring, produces a body plumage approaching that of the adult, which is worn until the complete annual molt in late summer or fall. The annual molt of adults apparently occurs in August or September, perhaps even later.

Food.—Prof. F. E. L. Beal's (1912) report is based on the examination of 129 stomachs of the swallow-tailed flycatcher, taken in every month from April to October, inclusive. The animal food amounted to 96.12 percent and the vegetable food to 3.88 percent of the contents. "Of the animal food beetles amount to 13.74 percent and form a rather constant article of diet. Less than 1 percent belong to theoretically useful families. The others are practically all of harmful species." Bees and wasps account for 12.81 percent and bugs 10.17 percent of the food. He says that flies "do not seem to appeal to this flycatcher as articles of food. They were found in the stomachs taken in April, May, and September only, and amount to but 3.80 percent." Beal continues:

Orthoptera (grasshoppers and crickets) are evidently the favorite food of the scissortail. They were found in the stomachs of every month, with a good percentage in all except April. The average for the year is 46.07 percent— the highest for any flycatcher. The 1 stomach taken in October contained 86 percent of these insects, but it is probable that the month of maximum consumption is July, when they amount to over 65 percent. As this bird is said to seldom light upon the ground, it follows that these insects must be captured when they take their short flights or jumps. * * *

Caterpillars, with a few moths, constitute a small but rather regular article of diet with the scissortail. They amount to 4.61 percent for the year and were found in the stomachs of every month except October. In several stomachs the cotton leaf worm (*Alabama argillacea*) was identified and the cotton boll-worm (*Heliothis obsoleta*) in another. Both of these are well-known pests of the cotton plant and also feed upon a number of other cultivated plants. The latter is also well known as the corn worm, because it feeds upon the sweet corn of the garden. It also preys upon tomatoes and occasionally upon beans and peas. A few dragon flies and some other miscellaneous insects and spiders make the rest of the animal food, 4.92 percent.

In his long list of insects identified in the food of this flycatcher, he names 4 species of Hymenoptera, 35 of Coleoptera, 2 of Lepidoptera, 10 of Hemiptera, 6 of Orthoptera, and 2 of Neuroptera. He says further that "the vegetable food consists of small fruit, or berries, and a few seeds. The total percentage, 3.88, indicates that this is not the favorite kind of food, but is taken for variety.

"It needs but little study of the food of the scissor-tailed flycatcher to show that where the bird is abundant it is of much economic value. Its food consists almost entirely of insects, including so few useful species that they may be safely disregarded. Its consumption of grasshoppers is alone sufficient to entitle this bird to complete protection."

Fortunately, these birds are enjoying the needed protection and are seldom molested. They seem to be increasing in Texas and are extending their range. Bendire (1895) noted that they were far more common in many parts of Texas at that time than they were 20 years previously. Beal (1912) implied that these flycatchers seldom alighted on the ground, but Samuel N. Rhoads (1892) "observed them for hours gleaning insects in the open pastures and salt flats near Corpus Christi, alighting without hesitation in the short grass to secure or devour their food." Probably many grasshoppers, locusts, and crickets are captured on the ground. This ground feeding may account for the worn appearance of the long tails in summer specimens.

Behavior.—When first seen sitting on a bush or telephone wire, with its long tail tightly closed and hanging straight down, the scissor-tailed flycatcher impresses one as a trim, neat bird of soft, pleasing colors and quiet mien. Or, as one flies in direct flight from one tree to another, with its long streamers trailing out behind, there is no indication of the flight gymnastics of this aerial acrobat. But, sooner or later, the observer will be treated to an exhibition, well worth watching, which has been so well described by Mrs. Florence Merriam Bailey (1902a) as follows:

One of his favorite performances is to fly up and, with rattling wings, execute an aerial seesaw, a line of sharp angled VVVVVVVs, helping himself at the short turns by rapidly opening and shutting his long white scissors. As he goes up and down he utters all the while a penetrating bee-bird scream *ka-queé—ka-queé—ka-queé—ka-queé—ka-queé*, the emphasis being given each time at the top of the ascending line. * * *

The head of a family we saw on the Nueces River one day was guarding his mate at the nest when another scissor-tail invaded his preserves. The angry guardian flew at him in fury, chasing him from the field with a loud noise of wings. At the first sound of combat the brooding bird's head appeared above the nest and hopping up on the rim she watched the chase with craned neck till the intruder with her lord and master close at his heels faded into white specks in the blue.

Another day we saw a scissor-tail in pursuit of an innocent caracara who was accidentally passing through the neighborhood. The slow ungainly caracara was no match for the swift-winged flycatcher and with a dash Milvulus pounced down upon him and actually rode the hawk till they were out of sight.

She writes of seeing a scissortail overtake a lark sparrow, which was pursuing an insect on the wing, and snatch the coveted morsel "from under its bill." She and her husband found these flycatchers really abundant in parts of the mesquite prairies of southern Texas. "Near Corpus Christi we once counted thirteen in sight down the road." But the largest number they ever saw together was in an oak mott between Corpus Christi and Brownsville, where these birds were roosting for the night. "At sundown, when Mr. Bailey shot a rattlesnake at the foot of a big oak in camp the report was followed by a roar and rattle in the top of the tree and a great flock of scissortails arose and dispersed in the darkness. They did not all leave the tree, apparently, even then, although some of them may have returned to it, for when daylight came to my surprise a large number of them straggled out of the tree. How one oak top could hold so many birds seemed a mystery. Before the flycatchers dispersed for the day the sky around the mott was alive with them careering around in their usual acrobatic manner making the air vibrate with shrill screams."

Mrs. Margaret Morse Nice (1931a) witnessed the pretty picture of a flock of these beautiful birds taking their evening bath; she writes: "On a day in mid September a dozen or more of these lovely birds gathered in the little willows growing in a small pond; one by one they swooped down to the water, but came up without quite touching it. Finally one brave bird splashed its breast into the water, whereupon they all followed suit, sometimes singly, sometimes two or three at a time, darting down quickly—a sudden dip into the water and then up again. The colors on their sides and under their wings shone pink and salmon and ruby in the late afternoon light. It was a rarely beautiful sight—the exquisite birds in their fairy-like evolutions."

The scissor-tailed flycatcher is a swift flier; its powerful little wings vibrate so rapidly, almost a blur to the human eye, that its stream-lined body is propelled through the air with speed enough to overtake quickly the slower flying hawk or crow that ventures too near its territory; with vicious attacks from the dynamic little warrior the big intruder is driven from the scene, only too glad to beat a hasty retreat.

Voice.—Mrs. Bailey (1902a) records the notes uttered during its flight maneuvers as an oft-repeated *ka-queé—ka-queé*. Bendire (1895) writes: "In all its movements on the wing it is extremely graceful and pleasing to the eye, especially when fluttering slowly from tree to tree on the rather open prairie, uttering its twittering notes, which sound like the syllables 'psee-psee' frequently repeated, and which resemble those of the Kingbird, but are neither as loud nor as shrill; again, when chasing each other in play or anger, in swift flight from tree to tree, when it utters a harsh note like 'thish-thish'." Mrs. Nice (1931a) writes:

Like the Kingbirds, Crested Flycatcher and Wood Pewee the "Texas Bird of Paradise" has a "twilight song" given before dawn during the nesting season. I have only one record of it, obtained at Cashion June 2, 1929 where a pair of these birds had a nest containing one egg. At 5:01 a. m. (26 minutes before sunrise) the male began to shout *pup-pup-pup-pup-pup-pup-perlép* 16 times a minute for about four minutes. Then for three minutes nothing was heard but a few *pups*. At 5:07 he began with a new note, *pup-pup-pup-peróo*, lower and less loud than the first phrase, the number of *pups* varying from one to three, the most common number being two. A minute later he started to fly about, but kept up a continuous chatter of *pup-peróo* till 5:12.

In further description of the same songs, she says elsewhere (1931b): "He and his mate then flew away, but were back at the nest at 5.18 with loud *pups*. At 5.27 just as the sun was rising over the prairie, the female sat on the barbed wire fence with wings held straight out from her body and her tail spread to its fullest extent. Later the male assumed this same attitude, at the same time saying *peelyer per*. At 5:42 he returned to the nest and gave a last *pup-pup-peróo*. * * * The *pup-pup-pup-pup-pup-pup-perléep* was about one second long; the intervals between beginnings of phrases varied from 3.5 to 4 seconds. The *pups* were uttered rapidly, giving the effect of a stutter; the emphasis was on the *perléep*."

Field marks.—There is no excuse for not recognizing a swallow-tailed flycatcher, with its extremely long tail, its soft gray, salmon-pink, white and black colors, offset with a dash of scarlet under the wing. The black upper parts of the fork-tailed flycatcher, and the absence of pink, will distinguish the only other long-tailed flycatcher with which it might possibly be confused.

Enemies.—The dwarf cowbird sometimes lays its eggs in the nest of this flycatcher (Simmons, 1925).

Fall.—Early in September the old and young birds begin to gather into flocks and wander about in preparation for migration; by the end of that month most of them have gone from the northern portions of their range. The migration seems to be made mainly during the night. Referring to southeastern Texas, Henry Nehrling (1882) says: "In September, after the breeding season, they gather in large flocks, visiting the cotton fields, where multitudes of cotton worms (*Aletia argillacea*) and their moths abound, on which they, with many other small birds, eagerly feed; early in October they depart for the South."

According to Dr. J. C. Merrill (1878), "about the middle of October, 1876, just before sunset, a flock of at least one hundred and fifty of these birds passed over the fort [at Brownsville]; they were flying leisurely southward, constantly pausing to catch passing insects; and in the rays of the setting sun their salmon-colored sides seemed bright crimson."

In El Salvador, Mr. van Rossem (Dickey and van Rossem, 1938) observed that "the first scissor-tailed flycatchers to arrive in the fall were noted at Divisadero on October 10, 1925, when a single adult female was taken in a dead-topped tree in an old cornfield. On the

17th a few more were observed along telephone wires and over the pastures, but the species remained rare until the 20th, when it suddenly became more common. A good-sized flight of several scores was noted flying southeastward by singles and couples at sunset on the 23rd, and after that date scissor-tailed flycatchers were conspicuous objects in all types of more open country."

In Guatemala, "at Ocos the migration was in full swing October fifteenth," according to some notes sent by A. W. Anthony to Ludlow Griscom (1932). "On that date a stiff gale was blowing from the southeast, against which the migrants were forced to fly. At times the flycatchers were just able to hold their own and, again, seemed to be forced backward. Often such birds, in order to find more favorable air currents, dropped to within a few feet of the ground, others veered inland and tried the shelter of the forest. The flight was still in full force at dusk. The Scissor-tails were still abundant the last of the month."

Winter.—The scissor-tailed flycatcher spends the winter from southern Mexico to Panama. In El Salvador, Mr. van Rossem (Dickey and van Rossem, 1938) calls it a "common, locally abundant, fall and spring migrant and less common winter visitant in the Arid Lower Tropical Zone." In Guatemala, Mr. Griscom (1932) refers to it as "a common transient and winter visitant to all open country below 5,000 feet." Alexander F. Skutch has sent me some notes on this species, in which he says: "In Central America the scissor-tailed flycatcher is most abundant as a winter resident on the dry Pacific side of the isthmus from the Gulf of Nicoya northward. It spreads sparingly over the cleared lands of the central plateau of Costa Rica; and in November 1935 I saw a few as high as 7,500 feet above sealevel in the pasturelands on the south side of the Volcano Irazu. I have never found the bird in the Caribbean lowlands or in heavily forested country.

"At Las Canas, Province of Guanacaste, Costa Rica, scores of these lovely, graceful birds slept nightly in some tall orange trees behind the *jefatura*, or town-hall, in the heart of the village, in company with a far smaller number of Lichtenstein's kingbirds (*Tyrannus melancholicus*). At or a little before sunset, they were to be seen flying into the village from all directions, high in the air. Soon they began to settle in the tops of the orange trees; but alarmed by the sudden passage of some person beneath them, or seized by a sudden unrest, they would dart swiftly forth again, and circle about in the air before returning to their sleeping-place. At Nicoya, a far smaller number slept in some fig trees with dense foliage which stood in a row along one side of the plaza of the little village. Here also they roosted with Lichtenstein's kingbirds, which here seemed to outnumber them. Although so gregarious in their roosts, by day they do not flock, but spread out over the surrounding country one by one."

Range.—South-central United States south to Panama; accidental north to Manitoba and New Brunswick and west to California.

Breeding range.—The scissor-tailed flycatcher breeds **north** to northwestern Oklahoma (Kenton); and southern Kansas (Harper, Wichita, and Independence). **East** to southeastern Kansas (Independence); eastern Oklahoma (Copan, Tulsa, and Okmulgee); western Arkansas (Fort Smith); and eastern Texas (Commerce, Waco, and Houston). **South** to southern Texas (Houston, Brownsville, Uvalde, and Pecos). **West** to western Texas (Pecos); southeastern New Mexico (Malaga, Carlsbad, and Lovington); and western Oklahoma (Kenton).

Winter range.—In winter the species is found **north** to extreme southern Texas (Brownsville, rarely Rio Grande, and Port Lavaca); hence **south** through eastern Mexico, Puebla (Huehuetlan); western Chiapas (Tapachula and San Benito); western Guatemala (Lake Atitlan and San Jose); El Salvador (Cotima and Puerto del Triunfo); Costa Rica (Miravallas, Bagaces, Tambor, and San Jose); to Panama (Chiriqui).

Spring migration.—Early dates of spring arrival are: Texas—Victoria, March 10; Houston, March 23. Oklahoma—Norman, April 3. Kansas—Harper, April 5.

Fall migration.—Late dates of fall departure, are: Kansas—Harper, October 24. Oklahoma—Norman, October 23. Texas—Abilene, October 16; Atascosa County, October 27; Pecos, November 20.

Casual records.—This flycatcher has been recorded outside of its normal range on many occasions. Among these are the following: Florida, a specimen was taken at Tulford on December 14, 1924, while three were seen on January 18, 1919, and four on November 18, 1930, at Key West; Georgia, a specimen was collected at Quarantine on June 5, 1933; South Carolina, one was seen on November 6, 1928, on Edisto Island, and a specimen was collected on July 16, 1930, at St. Matthews; Virginia, one was taken at Norfolk, on January 2, 1882, and one was reported from Aylett on August 31, 1895; New Jersey, a specimen was obtained near Trenton on April 15, 1872; Connecticut, one was taken at Wauregan on April 27, 1876; Massachusetts, at West Springfield one was collected on April 29, 1933; Vermont, one was taken at St. Johnsbury apparently in 1884; New Brunswick, an adult female was obtained on Grand Manan on October 26, 1924, and another was taken at Clarendon Station on May 21,

1906; Alabama, one was obtained at Autaugaville in the spring of 1889 or 1890; Louisiana, a flock of 10 was seen at Kenner on October 6, 1889, and three specimens were taken at Grand Isle, on April 6 and 10, 1926, and April 2, 1927, while one was collected at Wisner, on April 10, 1933; Ohio, a specimen was taken at Marietta on May 20, 1894; Nebraska, one was reported as seen near Lincoln in the fall of 1872, another was reported as seen near Greenwood in the spring of 1913, while a third actually built two nests near Lincoln in May and June 1921; Wisconsin, an adult male was collected at Milton on October 1, 1895; Minnesota, one was taken at New London some time prior to 1912, and one was seen in Jackson County on June 5, 1930; Manitoba, at York Factory a specimen was taken in the summer of 1880, and another on October 2, 1924, while one was found dead at Portage la Prairie, on October 20, 1884; Colorado, two females were taken at Campo, Baca County, on May 31 and June 1, 1923; Arizona, one was collected at Kayenta on July 8, 1934, and another was seen at Sahuaro Lake, on the Salt River, Maricopa County, on July 12, 1935; California, a specimen was obtained at Elizabeth Lake, Los Angeles County, on June 26, 1915.

Egg dates.—Oklahoma: 6 records, June 4 to 22.

Texas: 116 records, April 7 to July 26; 58 records, May 9 to June 10, indicating the height of the season.

PITANGUS SULPHURATUS DERBIANUS (Kaup)

DERBY FLYCATCHER

HABITS

This handsome and conspicuous giant flycatcher was first introduced to the fauna of the United States by George B. Sennett (1879), who collected the first specimens to be taken north of our boundary in the valley of the lower Rio Grande in southern Texas, where so many other Mexican species enter our borders. He reports: "On April 23 a male and a female of this species were shot at Lake San José, a few miles from Lomita. Both were shot about four feet up on the trunks of small retama-trees standing in the water, and were clinging to them and working their way down to the water, possibly to drink. They were not particularly shy. On May 3 another female was shot in trees bordering the lake, yet not over the water. One or two more were observed in timber about water-holes."

For some time after that the bird was supposed to be only an uncommon summer visitor in Texas, but it has since been shown to be of regular occurrence and a fairly common breeder in that region. When I visited Brownsville in 1923 we found the magnificent Derby flycatcher rather common in the dense forests along the resacas or stagnant watercourses and the old beds of rivers; these often contained large trees, mesquite, huisache, palms, etc., with a thick undergrowth of many shrubs and small trees, such as granjena, persimmons, coffee bean, bush morning-glory, etc. Here these flycatchers were associated with the other characteristic birds of the region, such as the curious chachalaca, the red-billed pigeon, the showy green jay, Audubon's oriole, and the dainty little Texas kingfisher. We also saw them occasionally in the large trees about the ranches of the Mexicans in other parts of Cameron County, where their huge nests were quite conspicuous.

Recent developments in this valley threaten to extirpate this and many other interesting Mexican species and drive them from their only foothold in the United States. The destruction of the chaparral and the forests by wholesale cutting, clearing, and plowing, to make room for more citrus orchards, truck farms, and other agricultural developments, is rapidly transforming the lower Rio Grande Valley into an agricultural and commercial community and is driving its interesting fauna back into Mexico.

There is some hope, however, that the Derby flycatcher may adapt itself to new conditions there, as it apparently has in El Salvador, where Dickey and van Rossem (1938) tell us that it is an "extremely common resident throughout the Arid Lower Tropical Zone and distributed less numerously, though regularly, to 4,500 feet wherever cultivation has cleared the land." They continue:

The center of abundance is along watercourses and lakes on the coastal plain and up to about 2,500 feet in the foothills. Under very favorable conditions the species may reach an altitude of nearly 7,000 feet. * * * The Derby flycatcher seemingly has but one requirement—that of open or semiopen country. Otherwise it is one of the most versatile of birds, adapting itself to almost any conceivable environment. * * * Typically these flycatchers inhabit much the same type of country as do kingbirds, that is, districts given over to agriculture. In El Salvador most of the hill region from the level of the coastal plain to about 2,500 feet has been cleared of timber and is checkerboarded into countless small fields, divided off by rows of trees and cut in every direction by steep-walled ravines. It is the center of human population and the center of the Derby flycatcher population as well. * * * On the coastal plain they do not occur in deep jungle. However, all cleared land is well populated by them, and along the borders of lowland rivers and lakes and about the mangrove lagoons they are exceedingly common also. Large cities as well as small towns and farms are invaded in numbers, and every plaza in which there are trees of any size is sure to have its pair or more of Derbys.

Nesting.—The same authors give the following account of the nesting of the Derby flycatcher:

Such a wide variety of sites is chosen for the bulky nests that to designate any one as typical would be misleading. "Typical" sites about towns and farms are cocoanut trees, the height at which the nests are placed averaging about twenty feet; but extremes of six and fifty feet were noticed. In the lower country where various thorny trees are common along watercourses, a frequent site is in a maze of thorns ten or twelve feet up, usually within a few feet of, or even placed directly on, a wasp's nest. In marshes or lakes the crotch of a dead tree is used probably more often than any other situation. Vine tangles and clumps of parasitic growth are also occupied.

The nest is a large structure resembling an oversized nest of the cactus wren. It is built chiefly of dead grass and any other soft material at hand, such as rags, plant fiber, and feathers; the cup in the interior is rather shallow and of well-packed and smoothed-down grass stems. A nest collected at San Salvador on March 28, 1912, measured 18 inches long by 10 inches wide by 8 inches high. The cavity was 7 inches long by 5 inches wide and 5 inches high, the shallow nest cup itself taking up the entire floor. The entrance was on the side and pointed slightly downward to prevent rain from beating directly into the nest chamber. Most nests are a little more round (less purse-shaped) than this one, but all are very similar in type.

The nest described above is the largest of which we have any record. One that I brought home from Brownsville measured 14 by 10 inches externally, and one in the Ralph collection in Washington is nearly 13½ inches long. Major Bendire (1895) says, of three sets of eggs in this collection, taken in Cameron County, Tex.: "Two of these nests were located in a thicket of huisache trees (*Acacia farnesiana*), about 10 feet from the ground; the other in a large bunch of Spanish moss, pending from the limb of a large tree, about 14 feet up. The last named is now in the collection. The nest proper is an unusually bulky structure, composed principally of gray Spanish moss, dry weed stems, pieces of vines, and swamp grasses, and lined with finer materials of the same kinds. * * * The other nests were lined with wool, feathers, plant down and Spanish moss."

Dr. Herbert Friedmann (1925) observed two nests in the vicinity of Brownsville, Tex.; one "was high up in a tall tree and all around it were bunches of Spanish moss, but none of the moss was actually on or suspended from the nest." The other "was built on top of an old nest of a Mexican Cormorant in a dead tree standing in shallow water." There are two nests in the Thayer collection in Cambridge, taken in British Honduras; these are great, nearly globular bundles of similar materials to those mentioned above, measuring 12 inches or more in diameter; one was 30 feet from the ground in a pine tree, and the other was 15 feet up on the top of a pine stub. Four sets of eggs in my collection came from similar large globular nests, having the entrance on the side.

Eggs.—Two to five eggs are laid to a set by the Derby flycatcher, but four seems to be the commonest number. Most of the eggs are ovate, but some are short-ovate and some elongate-ovate; the shell is smooth and only slightly glossy. The ground color is pale creamy white, and they are sparingly marked with small spots and minute dots, mostly about the larger end, of very dark browns, or almost black, "seal brown" or "liver brown," and shades of "Quaker drab" or lavender. The measurements of 50 eggs average 28.5 by 21.0 millimeters; the eggs showing the four extremes measure 31.4 by 21.8, 30.0 by 22.6, 25.6 by 22.4, and 30.9 by 18.8 millimeters.

Plumages.—I have not seen any very young specimens, but birds in full juvenal plumage are much like the adults, the sexes being alike; but the young birds have no yellow in the crown, the colors generally are duller, the plumage is softer, and the rufous margins of the wing coverts and secondaries are somewhat wider. This juvenal plumage is worn through the summer and early fall, or until October or even later, when a postjuvenal molt replaces the contour plumage with a first winter plumage, which is practically adult. The juvenal wings and tail are, however, retained until the following spring or later.

A complete postnuptial molt of adults occurs in August and September; but there is apparently only a very limited molt of the contour feathers in the spring.

Food.—Dickey and van Rossem (1938) write: "Stomachs of four birds taken at San Salvador in 1912 contained small beetles, wasps, and small grasshoppers in relative abundance in the order named; and, in addition, a mass of smaller-winged insects. At Puerto del Triunfo many birds were seen perched on mangrove roots over the water, sitting motionless and in their attitudes resembling kingfishers. In striking the water, however, they do not make the clean-cut dive of a kingfisher, but after hovering an instant make a headlong splash. The objects of the dives seemed to be tiny fish. This was certainly so in one case and by inference in others. This species is one of the very few larger flycatchers which appears never to take fruit or berries."

Contradicting the last remark, Dr. Charles W. Richmond told Major Bendire (1895) that he had "seen one specimen that had its mouth and throat full of ripe banana." And William Beebe (1905) says: "It was winter, and insects, while fairly abundant, were apparently too scarce to provide the flycatchers with their usual diet, and we found them feeding freely on berries and seeds."

The fishing habit has been observed by others. Dr. Beebe describes it as follows: "Like the kingfishers, the Derby perched upon a rock and watched the eddies, and then dived with all his might two or three times in succession, each time securing a small fish, or sometimes a tadpole. It seemed impossible for him to immerse himself more than three consecutive times, for his plumage became water-soaked, and he then flew heavily to a sun-lit branch to spread himself to the sun. After drying he was at it again."

And Col. A. J. Grayson (Lawrence, 1874) says: "I have often observed them dart into the water after water insects and minnows that were swimming near the surface, not unlike the kingfisher; but they usually pursue and capture on the wing the larger kinds of Coleoptera and Neuroptera, swallowing its prey entire after first beating it a few times against its perch. They are usually in pairs, but I have also seen as many as twenty about a stagnant pool, watching its turbid water for insects and small fish, for which they seem to have a great partiality."

A. J. van Rossem (1914) noted, in the city of San Salvador, that "on two occasions one (probably the same individual) was seen about an arc-light long after dark. It may have been attracted by the light, but in my own mind there is no doubt that the insects which buzzed around the globe in swarms were the real reason for the bird's presence, as it appeared in no way confused and kept well outside the most brilliant circle of illumination. Owing to this fact, and also because the light was quite high up, I could not actually see the bird catch anything, though its frequent short and erratic flights would indicate that this was the object. Its perch seemed to be directly above the shade."

Behavior.—In flight the "bull-headed flycatcher," as it is sometimes called, somewhat resembles a kingfisher; it has even been referred to as a yellow-bellied kingfisher; the resemblance is even more striking as it sits quietly over some stream or pool, watching for its finny prey, or dives into the water to sieze some minnow or water insect. But, when not thus engaged, it is an active, noisy, nervous, and irritable bird, always ready to pick a quarrel. Mr. van Rossem (1914) writes:

In the city of San Salvador are a great many birds which are without doubt non-breeders (as only two nests were found in the city proper), even though they are mostly in pairs. These individuals, having nothing better to do, contrive to keep things lively by scrapping not only with each other but with anything that happens to attract their attention, such as a stray house-cat or a wandering hawk.

A favorite lookout is a tall flag pole or similar point of vantage, and this is taken possession of to the exclusion of all other birds, most especially of their own kind; in fact, the advent of another pair onto their preserve is the signal for a battle royal which generally ends as it should—in favor of the home team. From dawn till an hour or so after sunrise, and in the cool of the late afternoon and early evening, they are most active and noisy. Their call notes can then be heard in every quarter of the city and the birds themselves are most in evidence, snatching flies over heaps of refuse in the gutters, hawking about the plazas, or 'kingbirding' an unlucky Black Vulture. Activity, though, is by no means confined to these periods. Birds may be found at almost any hour of the day.

Voice.—The Derby flycatcher is a noisy bird, especially during the morning and evening hours. It has a great variety of notes, on which several local names are based; Dickey and van Rossem (1938) state that "some of them are 'Bien-te-veo,' 'Dich-oso-fui' (sometimes interpreted as 'Kiss-ka-dee'), and 'Chio'."

Mr. van Rossem (1914) says elsewhere: "As is the case with many other common and well-known varieties, the native name of 'Chio' is derived from the Derby's call note, which may best be written che-oh, or chee-o, generally given rather slowly, but under stress of excitement or anger losing entirely its deliberate quality and becoming shrill and hurried. At such times, too, and particularly at the nest where the parents become almost frantic, these notes are interspersed and plentifully larded with extremely Kingbird-like expletives."

Field marks.—The large size and heavy build of the Derby flycatcher, its crown striped with black, white, and yellow, its conspicuous white throat, the absence of gray on its chest, and the rufous in the wings and tail should serve to distinguish it from any of the yellow-breasted kingbirds with which it might be confused. Its behavior, posture, and voice are all distinctive.

Enemies.—Dr. Beebe (1905) relates the following story:

The unfortunate end of the piscatory Derby Flycatcher came about in this way. Some of the Raccoons usually made their way directly to the water, and drank and splashed about in the darkness. One evening it happened that the Derby was fishing from a sand-bar on the opposite bank. One of the coons must have stealthily made his way through the underbrush to within a short distance of the preoccupied flycatcher. Suddenly we heard a loud rustle and the poor bird gave utterance to the most piercing screams, which echoed and reëchoed from cliff to cliff. The bedraggled feathers of the bird doubtless rendered it an easy prey. An instant more and a dead silence settled over all. Next morning we found a pile of yellow feathers, and the telltale bear-like footprints of the animal. The Raccoon returned the following night, but the bird, which he found ready slain, was tied to the pedal of a steel trap, and by the law of fate we enjoyed a delicious stew, made from the fattest of coons. The Derby was avenged.

Dr. Friedmann (1929) cites two authorities who claim that the Derby flycatcher has been imposed upon by the red-eyed cowbird, but says: "Probably it is seldom parasitized as it is a large, pugnacious bird eminently able to keep off any unwelcome visitors."

DISTRIBUTION

Range.—South America and Central America, north to southern Texas; nonmigratory.

The range of the Derby flycatcher extends **north** to southern Sonora (Alamos); southern Texas (Devils Lake, Hidalgo, Lomita, and Brownsville); Yucatan (Temax); northern Colombia (Barranquilla, Bonda, and Rio Hacha); northern Venezuela (San Julian and Ciudad Bolivar); and Guiana (Georgetown, Paramaribo, and Cayenne).

East to Guiana (Cayenne); eastern Brazil (Goyaz); Paraguay (Sapucay); and Uruguay (Rio Negro and Minas). **South** to southern Uruguay (Minas, Montevideo, San Jose, and Colonia); and central Argentina (Isla San Martin, Buenos Aires, and Venado Tuerto). **West** to east-central Argentina (Venado Tuerto and Santa Elena); Bolivia (Montes); Ecuador (Rio Suno and Rio Coca); western Colombia (Rio Frio and Bogota); Costa Rica (Puntarenas and Puerto Humo); El Salvador (San Salvador and Santa Ana); Oaxaca (Tapariatepec and Chivela); Nayarit (Tepic and San Blas); Sinaloa (Escuinapa and Mazatlan); and Sonora (Agiabampo and Alamos).

The range as outlined is for the entire species, which has been separated into several races. All but one of these are confined to South and Central America. The form occurring in the United States (*Pitangus sulphuratus derbianus*) ranges southward from the lower Rio Grande Valley in Texas to western Panama.

Casual records.—A specimen was collected at Inglewood, Calif., on September 4, 1926, and one was taken at Chenier au Tigre, La., on May 23, 1930.

Egg dates.—British Honduras: 4 records, May 7 to 24.

Mexico: 23 records, March 28 to July 13; 12 records, May 1 to 17, indicating the height of the season.

Texas: 12 records, March 27 to June 23; 6 records, May 12 to 28.

MYIODYNASTES LUTEIVENTRIS SWARTHI van Rossem

ARIZONA SULPHUR-BELLIED FLYCATCHER

PLATE 9

HABITS

The above name was bestowed on the northwestern race of the sulphur-bellied flycatcher, which breeds in the high mountains of southern Arizona and probably adjacent parts of Mexico, by A. J. van Rossem (1927). He describes it as similar to the typical race of southern Mexico and Central America, "but under parts paler yellow; streaking of under parts less conspicuous and averaging narrower, particularly on flanks; upper parts paler with edgings of feathers grayer (less yellowish) buff; edgings on wing coverts whiter and unusually broader; cinnamon rufous of lower rump, upper tail coverts, and tail paler."

We had been in the Huachuca Mountains, Ariz., for nearly a month before we saw and collected my first pair of sulphur-bellied flycatchers. I had been told to expect them about the middle of May, their usual time of arrival, but it was not until May 29 that we found them in Miller Canyon. We heard their peculiar notes when we reached the heavy sycamore timber in the upper part of the canyon, but we followed the birds for some time before we could actually see them. They moved about very little and very deliberately, often remaining still for some time. In spite of their apparently striking color pattern, they were far from conspicuous among the foliage, as their contrasting colors blended perfectly among the lights and shadows of the leafage. The male of the pair that I shot was only wounded, and, as I held him in my hand, he pecked at my finger and spread his beautiful crest in anger; the crest when opened was nearly circular in outline and brilliant chrome-yellow in color, with a narrow black border surrounding it; it would make a fine display in courtship, as well as a striking warning in anger.

Most observers agree that this flycatcher, in Arizona at least, is partial to just such localities as that in which we found it, among the largest sycamores in the larger, wider, and deeper canyons, where it prefers to nest in the natural cavities in these big, picturesque trees. But Major Bendire (1895) states that it has been taken "among the oaks in some of the canyons near Fort Huachuca." Harry S. Swarth (1904) says of its distribution in the Huachucas:

This species does not occur in the higher parts of the range, nor is it found in the foothills. Preeminently a bird of the heavily wooded canyons, it is seen only along the streams; and all I have seen have been between 5,000 and 7,500 feet altitude. It is most abundant in Tanner Canyon, a broad, well watered canyon with a far more gradual ascent than in any of the others. It is on this account, I think, that this flycatcher occurs in it so much more abundantly than elsewhere, for besides being the longest canyon in the range, the head of it is at the lowest point along the divide; thus giving the greatest area at the altitude favored by this species of any canyon in the mountains. This canyon seems to be abundantly suited to the needs of this flycatcher for almost its entire length, and I have seen them very nearly to the head of it.

Alexander F. Skutch has contributed some elaborate notes on the typical Central American race of the sulphur-bellied flycatcher (*Myiodynastes luteiventris luteiventris*), from which I shall quote freely, as the two races evidently do not differ materially in their habits. He says that, "in Central America, as farther to the north, the sulphur-bellied flycatcher appears to be merely a summer resident, arriving from the south in early March, raising its young, and then again withdrawing southward in August or September.

"Upon its arrival from the south, this large flycatcher, easily recognized by its prominently streaked plumage and its distinctive voice, spreads over the clearings and open country from both coasts up to about 6,000 feet in the interior. It favors open groves and pastures with scattered trees, but avoids the heavy forests. Immediately it makes itself conspicuous by its shrill voice and quarrelsome habits; it is, indeed, during the mating season, one of the most quarrelsome of the Central American flycatchers. While most of the larger resident members of the family have long been mated and in many cases have occupied their breeding territory for a considerable period, the new arrivals must hurriedly pick their mates and their nest sites, which they do with much noisy bickering and many loud disputes among themselves. They must find suitable cavities for their nests, ready made and unoccupied, and there may not always be enough of these to go around."

Nesting.—The sulphur-bellied flycatcher does not reach southern Arizona until sometime during the latter half of May, or sometimes not until the first week in June. My companion, F. C. Willard, told me that we might look for it at any time after May 15, but we did not actually see it until May 29; Mr. Swarth (1904) says that his earliest date is May 19; O. W. Howard (1899), who has probably collected more eggs of this species in Arizona than anyone else, thinks that it generally does not arrive before the first of June. The birds are probably paired when they arrive, and they are apparently in no hurry to start nesting; consequently they are among the latest of the birds to breed; Mr. Howard says: "I do not know of any eggs being taken before the first of July and I found a nest with young just hatched on August 28, 1899."

In the Huachuca Mountains the nest is invariably, so far as known, built in a natural cavity in a large sycamore at heights varying from 20 to 50 feet above ground; the cavity normally selected is a knothole, where a large branch has broken off and the cavity rotted out to varying depths. This flycatcher does not like to incubate its eggs in darkness, but prefers to be able to see out while it is sitting on its eggs; so, if the cavity is a deep one, it is filled up to within an inch or two of the opening with nesting material. The opening is generally large enough to give the bird free access, and, to the advantage of the egg collector, it is usually large enough to admit a man's hand. If the cavity is a deep one, the lower part of it may be filled in to the desired height with twigs, bits of bark, or other rubbish, on top of which the peculiar and characteristic nest is built. This nest proper is almost always entirely made of the petioles and midribs from the dry leaves of the walnut tree, but sometimes there is an admixture of pine needles or fine, stiff weed stems. These petioles are rather stiff and somewhat curved; and they are so arranged that the curve conforms to the somewhat circular shape of the nest or the cavity. There is no soft lining, to make a comfortable bed for the young, but the finer leaf stems are smoothly laid in the hollow of the nest. The diameter of the inner cup of the nest is 3 or 4 inches, and the outside measurements conform to the size of the cavity. The same cavity seems to be used year after year, probably at least one of the pair returning each year to the old abode. The female gathers the material and builds the nest, without any help from the male, except that he follows her about and perches near the nest hole, giving her encouragement with the nearest approach to pleasing notes that he can produce. Mr. Willard told me that the busiest time for nest building is early in the morning, about 6 o'clock, but that the work goes on spasmodically all day; it took about eight days to build one of the nests that he watched. He said that the female collected twigs from a walnut tree, breaking them off, but the midribs of the walnut leaves were picked up from the ground.

The nesting habits of the Central American race seem to be very similar to those described above. Mr. Skutch says in his notes: "The nest is placed in a cavity in a tree, preferably in a dead trunk standing isolated in a clearing. An old woodpecker's hole may be used; but at other times the bird prefers an open niche or hollow formed by decay, and scarcely sheltered from the sky. The nests I have seen have varied in height from 11 to 45 feet. The nest is built by the female alone; but the male is most attentive to her while she works, frequently following her on her trips to fetch material, and sometimes picking up fine twiglets with which he toys, then drops." He watched a female that seemed to have difficulty in deciding which of two available cavities to occupy and was filling both with twiglets alternately. "When she flew toward one or the other of the cavities that she was filling, her mate often hurried ahead of her and came to rest on the sill of the doorway, or clung to some convenient projection close beside it, where he sang a low, hurried, twittery, somewhat harsh-voiced song, while his partner arranged the material within."

Eggs.—The three or four eggs laid by the sulphur-bellied flycatcher are very handsome and richly colored. The eggs are ovate and only slightly glossy when fresh. The ground color varies from white to rich creamy buff. They are profusely spotted and blotched, usually over the entire surface, with rich reddish browns, "chestnut," "auburn," or "liver brown," with underlying spots of different shades of "Quaker drab" or "lavender"; Bendire (1895) mentions one that is marked with dark "pansy purple." The measurements of 26 eggs average 25.9 by 19.0 millimeters; the eggs showing the four extremes measure 27.9 by 19.3, 26.0 by 19.9, and 24.0 by 18.0 millimeters.

Young.—Mr. Skutch (MS.) writes: "Once in Guatemala I stuck in the entrance of a low nest a twig that bore on its exposed end a little wad of cotton soaked with vermilion paint. The flycatchers succeeded in pulling it out and carrying it away, but while doing so both of the pair stained with vermilion the pale yellow plumage of their under parts. The female acquired the heavier marks, by which she could readily be distinguished from her mate. I watched this nest from concealment for nearly seven hours, during which the female alone incubated the eggs. Her sessions in the nest varied from 5 to 32 minutes; the longest was during a heavy shower; but during the following morning, which was fair, she sat once for 29 minutes continuously, and twice for 26 minutes. Her recesses ranged from 1 to 21 minutes. The average of 15 sessions was 17 minutes; of 15 recesses, 8.5 minutes. As others of the kind that I have watched, she could be seen from the ground while she sat warming her eggs; but probably because of the lowness of her nest, only 11 feet above the ground, she was shier than most and flew out if she saw me still a long way off. The male sometimes came to stand in the doorway of the cavity and look in at the eggs, but he never sat on them. Indeed, although I have made careful studies of the nest life of many kinds of American flycatchers, including the related noble flycatcher, I have never seen the male incubate or brood.

"The eggs hatched after 15 or 16 days of incubation. The newly hatched nestlings bear a rather copious, long, dusky down. Their bills are yellow inside and out. I have not been able to watch the care and development of the young; but undoubtedly both parents feed them."

Henshaw (1875) introduced this flycatcher to our fauna by collecting a pair of adults with their three young in the Chiricahua Mountains of southern Arizona in 1874, of which he wrote:

I obtained a pair of old birds, together with three young, August 24. These, though indistinguishable in size and perfection of plumage from the adult pair, were still the objects of their solicitous care, and were dependent on them for food. Indeed, their presence might have remained unnoticed by me, had I not been greeted, as I entered the mouth of one of the deep, narrow cañons intersecting the mountains in every direction, by the shrill notes and angry cries of the old birds, who hovered in the air at a short distance, or flew restlessly from tree to tree, endeavoring to distract my attention from the young, till taking the alarm, they flew over into an adjoining ravine, where soon after I found the whole family assembled, the old birds having immediately rejoined their charges.

A somewhat different behavior of the young was noted by Mr. Willard, who says in his notes that "on hearing certain notes from the adults, the young 'freeze' and remain motionless as long as the old birds keep up the noise."

Plumages.—The "newly-hatched nestlings," referred to above by Mr. Skutch (MS.), bore a "rather copious, long, dusky down." Mr. Ridgway (1907) says that the young, in juvenal plumage, are "similar to adults, but upper parts more strongly tinged or suffused with brownish buffy, middle and greater wing-coverts and distal secondaries edged or margined with cinnamon-buff, and yellow crown-patch more restricted."

Mr. Swarth (1904) says: "The concealed yellow crest of the old bird is lacking, the feathers of the crown merely having their bases pale saffron, not sharply defined and hardly apparent at a casual glance; and in the very young birds even this feature is almost entirely absent. Also, the dark median stripe of the rectrices is more narrow than in the adult."

The yellow crown patch is hardly noticeable in some of the young birds that I have examined, the yellow of the under parts is paler than in the adult, and in very young birds the whole plumage is softer and more blended and the bill is smaller.

I have not seen enough fall and winter specimens to come to any definite conclusion as to subsequent molts and plumages, but the following remarks by Dickey and van Rossem (1938) indicate that, even in El Salvador, these flycatchers leave for the south before the postjuvenal or the postnuptial molt of adults is complete: "A juvenile and an adult taken, respectively, July 28 and 30, have just started the body molt. In two adults taken August 15 and 16 the body molt is virtually complete, but curiously enough neither in these, nor for that matter in any of the six molting specimens, have any of the old, worn remiges or rectrices been replaced. It seems likely that *Myiodynastes*, like certain other migratory flycatchers, * * * has a midwinter wing and tail molt which occurs months after the new, fall, body plumage has been acquired."

I have seen a young male, collected in Panama on November 20, that was molting its tail feathers.

Food.—Very little has been published on the food of the sulphur-bellied flycatcher. Of two stomachs examined by Mr. van Rossem (Dickey and van Rossem, 1938), one contained insects exclusively, and one moths and small berries. He says that "many birds are stained with the purple juice of an undetermined fruit or berry." Mr. Skutch says in his notes: "The sulphur-bellied flycatcher subsists upon insects, which it snatches from the air in the spectacular fashion characteristic of it family, and small fruits and berries of various kinds."

Behavior.—Mr. Swarth (1904) writes:

This species, though a handsome, strikingly marked bird, and at times an exceedingly noisy one, is yet so shy and retiring, that, far from being conspicuous, a person unfamiliar with the habits of the species might collect for weeks in a region in which it abounded and not know that there were any around. Frequenting as they do, the tops of the tallest trees along the canyons, which are thickly covered with foliage at the time these birds arrive, a far brighter colored bird might easily escape observation; and their colors, though striking, blend exceedingly well with the surrounding vegetation, they are by no means easy to see; the more so that they frequently sit perfectly motionless for a considerable length of time. It has happened more than once, that, hearing the familiar note in some tree top, I have watched, sometimes for half an hour, endeavoring to see the bird; scanning, as I supposed, every twig on the tree, only to see it finally depart from some limb where it had been sitting, if not in plain sight, at any rate but very imperfectly concealed.

In Arizona this flycatcher does not seem to be so aggressive toward hawks and crows as are the kingbirds, perhaps because its eggs and young are better protected in the hollows in which it nests. But a pair that had been robbed previously by jays fought a red-shafted flicker for its hole for over a week and finally drove it away and took possession of the hole; the flycatchers' victory was gained more through perseverance than fighting ability, for the flicker could drive either of the flycatchers away at will; but the other one would fly down to the hole, compelling the flicker to return and drive it away. The reason that the flicker fought so long was evident when the flycatchers' nest was removed later and three fresh eggs of the flicker were found in the hollow beneath it.

Referring to the Central American race, Mr. Skutch says in his notes: "The sulphur-bellied flycatchers, especially the male, harried other birds that flew about in the vicinity of their nest. They frequently pursued the quetzals that were feeding nestlings near by, worrying them a good deal but doing them no harm. One morning the male flycatcher attacked a pigeon (*Columba subvinacea*), which was harmlessly eating berries in the top of a tall tree at some distance from the nest, and drove it from its meal. When a pair of little Dow's tanagers (*Tangara dowii*) alighted in the dead tree just in front of the nest, the flycatcher darted upon one of them. But the little bird held its ground, screaming shrilly, while the bigger assailant hovered menacingly above it. Seeing that the tanager could not be bullied into retreating, the flycatcher took no further heed of it; and for some minutes the pair of tanagers and the pair of flycatchers perched amicably close together in front of the latter's nest."

Voice.—The same careful observer says that the sulphur-bellied flycatchers are "very active and noisy, especially during the mating season. Then they pursue one another, calling *p' p' p' pe-ya*, *p' p' p' pe-ya* in shrill, petulant voices. This remains the call uttered in flight even after they are happily settled and brooding their eggs. When they call from a perch they remind me of a withered and somewhat ill-tempered old woman calling her grandson, *Weel-yum*, *Weel-gum*, in thin, high, querulous tones. With this utterance, too, they protest intrusion at their nest; but not satisfied with merely vocal protestations, they make spirited darts at the head of the intruder.

"One would hardly expect that a bird with such a thin and petulant voice could really sing; but in the gray dawn, during the breeding season in April and May, the male mounts to some lofty perch and delivers a song of appealing beauty. Frequently he chooses as his singing perch the topmost dead twig of a tall tree standing in a clearing or at its edge; and here, his dark form conspicuous against the brightening sky, a hundred feet above the dew-laden earth, he pours forth his dawn-song for many minutes together. *Tre-le-re-re*, *tre-le-re-re* he repeats tirelessly over and over, in a soft, liquid, almost warbling voice. One morning I timed a bird that sang for 17 minutes without a pause. During the hours of full daylight this appealing song is rarely heard, and then chiefly when the flycatcher is excited by a rival of his own kind, or else in brief snatches, at the beginning of the nesting time, as if he were practicing. What a contrast between this soft, cool, pellucid dawn-song and the high-pitched, strained, excited notes that the bird utters at other hours of the day."

Others have noted similar vocal utterances in Arizona, the morning and evening song and the loud, shrill, screeching call note, which sounds much like the squeaking of a wagon wheel that needs greasing. Mr. Willard thought that the song sounds much like the most musical notes of the cowbird; and Mr. Swarth (1904) says: "Though noisy their vocabulary is limited and I have never heard but the one shrill call from them, a note hard to describe but very much in the style of the familiar two-syllabled whistle of the Western Flycatcher (*Empidonax difficilis*). Of course the volume is infinitely greater than with the little *Empidonax*, but they resemble each other to this extent, that I have known a person familiar with the Sulphur-bellied Flycatcher to mistake a *difficilis* near at hand for the larger flycatcher in the distance."

Field marks.—The sulphur-bellied flycatcher is so unlike any other North American bird that it could hardly be mistaken for anything else; the white throat and pale yellow abdomen, streaked with black, and the rufous tail, with median dusky stripes, are quite distinctive. And, as it is oftener heard than seen, its characteristic voice will identify it.

Fall.—Although this is one of the latest arrivals in the spring and a late breeder, it apparently leaves for the south as soon as the young are able to follow. We have few Arizona records later than August 24. Even in Costa Rica, Mr. Skutch saw very few after the middle of July, and he has not seen it anywhere in Central America between September and February; and the latest date he gives me is August 9. Evidently the winter home is somewhere in South America.

DISTRIBUTION

Range.—Southern Arizona and Central America south to Ecuador and Peru. Resident except in extreme north.

Breeding range.—The sulphur-bellied flycatcher breeds **north to** southern Arizona (Pima Cañon, Cañon de Oro, and the Chiricahua

and Huachuca Mountains) ; Chihuahua (Cajon Bonito and Carmen) ;
Tamaulipas (Golindo and Ciudad Victoria); and Yucatan (Val-
ladolid). **East** to Yucatan (Valladolid); eastern Honduras (La
Ceiba and Segovia River) ; eastern Nicaragua (Eden and Rio Escon-
dido) ; Panama (Panama City and Obaldia) ; Venezuela (Magdalena
Valley) ; and western Brazil (Marabitanos and Borba). **South** to
Brazil (Borba) ; and southern Peru (Artilleros, Monterico, and
Perene). **West** to Peru (Perene, Nauta, and Yurimaguas) ; western
Costa Rica (Puntarenas and Santa Rosa) ; El Salvador (Miraflores
and La Libertad); western Guatemala (Escuintla, Patulu Island,
and Hacienda California) ; Oaxaca (Tehuantepec and Juquila);
Guerrero (Dos Arroyos) ; Jalisco (Mazatlan and Altemajec) ; Nay-
arit (Acaponeta) ; Sinaloa (Plomosas and Presidio) ; Sonora (San
Rafael) and southern Arizona (Santa Rita Mountains and Pima
Cañon).

Winter range.—The species appears to be resident throughout most
of its range north to southern Sinalao (Presidio and Plomosas) ; and
central Tamaulipas (Ciudad Victoria).

The range as outlined is for the species, which has been separated
into at least two subspecies. The race known as *Myiodynastes luteiventris swarthi* is the one found in southern Arizona. The southern
limits of its range are at present uncertain, and in the extreme southern part of the range there is also some confusion regarding the
species.

Migration.—Early dates of arrival in Arizona are: Paradise, May
17; Tombstone, May 18. A late date of fall departure from that
State is Tombstone, September 20, abnormally late.

Egg dates.—Arizona: 13 records, June 20 to August 15; 8 records,
July 2 to 15, indicating the height of the season.

Mexico: 1 record, June 15.

MYIARCHUS CRINITUS BOREUS Bangs

NORTHERN CRESTED FLYCATCHER

PLATES 10, 11

HABITS

"The voice of one crying in the wilderness," a disembodied sound,
a loud, striking challenge note, coming from somewhere in the woodland treetops, greets us early in May in New England, and we know
that one of our showiest and most fascinating birds has arrived with
the vanguard of the migrating hosts. He is oftener heard than seen,
for, although he is handsomely colored, his soft colors blend well into
his surroundings in the forest trees, as he sits motionless on some lofty

branch and proclaims possession of his chosen territory over which he has ruled for successive seasons. If we watch carefully he will betray his presence, and we shall see the flash of reddish brown in his wings and tail, or his yellow under parts, as he sallies forth to snap up an insect or to drive some intruder out of his domain.

Probably the crested flycatcher was originally a forest-loving bird. It still shows a fondness for what remnants of our forests are left in southern New England, though it seems to prefer the more open portions, the edges of clearings and woodland glades, and the borders of the woods. It is seldom found in the depths of extensive forest areas. Since civilized man has cleared away so much of our forested land and has made improvement thinnings in most of the remainder, this flycatcher finds fewer cavities there in which to build its nest and has learned to adapt itself to living in more open situations, in old orchards, in isolated trees in open lots and even about human habitations.

Dr. Samuel S. Dickey writes to me: "In the backwoods of West Virginia, where I have given attention to the species for a period of 30 odd years, it inhabits stands of original timber, such as white oak, tulip poplar, white linden, sugar maple, white ash, beech, and sour gum. It delights to frequent the edges of cranberry glades, at altitudes up to 3,000 feet at least. Pairs enjoy rugged cliffs of the deeply entrenched creeks and rivers. Others flit across abandoned fields, given over to wild apple bushes, saplings, and Jersey scrub pines."

Courtship.—Dr. Dickey contributes the following notes on this subject: "As early as April 25 their characteristic call notes could be heard from the open windows of our house. After quite an indication of springtime activity, lasting fully a week, males were seen to clash in more distant areas of the field, where the natural habitats of pairs overlapped. They then would draw up close to one another, over fences and bushes, expand their wings, spread their tails, and dart rapidly at each other. They then tore some feathers from breasts, held fast with their claws, and tossed and tumbled toward the ground. There were periods, too, when they would actually dally in the grass, and, after seemingly biting with beaks, would separate and flit off in the direction of their respective nesting grounds. However, such maneuvers did not last long. The remainder of their excessive ardor was directed to their females.

"Males were seen to dash from high dead limbs after females. They would glide around and around walnut trees, enter undergrowth of yellow-locust poles, and then emerge along fencerows. The female was seen to escape the encroachment of the male. She would swerve upward, dart toward an abandoned cavity of the redheaded woodpecker, and make a hurried retreat inside the orifice.

Then she would turn around and thrust her playful, comic countenance out of the hole. The male would hover close to the orifice and then return to the high perch from which he started his sally. Such maneuvers were repeated over and over again until mating was completed; whereupon the female proceeded to gather material for nest construction and take it into the woodpecker's cavity."

The impression seems to prevail that passerine birds select new mates each year, but evidently this is not always the case with the crested flycatcher, which seems to show great attachment to its home territory and, sometimes, to its former mate. Raymond J. Middleton, of Norristown, Pa., has demonstrated by banding that one pair has remained constant for three years: he says (1936): "This pair of birds was taken on each of the years 1934, 1935 and 1936 in the nesting box while feeding half-grown young on the dates above given, being mated together for three consecutive years. A147214 is now at least 8 years old."

Nesting.—The crested flycatcher is a hole-nesting species and will use almost any cavity that is large enough to hold a rather bulky nest and that has an opening large enough to admit of easy access. It seems to show a preference for natural cavities in trees but has probably always used to some extent the abandoned holes of the larger woodpeckers, such as the flicker, the pileated woodpecker, the red-headed woodpecker, and the red-bellied woodpecker. But since it has become adapted to civilization, it does not seem to fear the presence of man and has learned to nest in a variety of man-made structures, often near human dwellings. Nesting boxes erected for purple martins or other birds are most commonly used in such locations, but nests have been recorded also in hollow logs attached to buildings, hollow posts, an old wooden pump, an old lard bucket, a stove pipe or open gutter-pipe, or any old tin can or box of proper size and suitably located.

In my home territory, in southeastern Massachusetts, we have always associated crested flycatchers with old orchards, where we often saw or heard the birds as we were driving by or while exploring them as favorite nesting sites for various birds. They seemed to prefer orchards that were partially surrounded by deciduous woods or were located on the outskirts of the woods; but often they were found in orchards remote from such cover. They had no use for young or well-pruned orchards; old neglected orchards with dead trees, full of natural, rotted-out cavities were their choice. All the nests that we have found, with one exception, were in apple trees in such orchards. The cavities selected were in the trunks or main branches; sometimes these were vertical or somewhat slanting, and sometimes nearly or quite horizontal.

When open at the top or at the end of a horizontal branch, the cavities were sometimes very deep, even 2 or 3 feet, but such cavities were filled with trash and nesting material so that the nest was usually not more than a foot or so from the opening. The only nest that was not in an apple tree was in a dead stub, about 7 feet high and open at the top, in some swampy woods. The orchard nests ranged from 6 to 11 feet above the ground, as such old apple trees are not large. We found that these flycatchers formed strong attachments to their homes and returned to the same orchards year after year; one pair was found nesting in the very same cavity that it had occupied nine years earlier; I had not visited the locality in the meantime, but I like to think that it may have been occupied during at least some of the intervening years.

Dr. Dickey tells me that, in West Virginia, "a small weather-worn knot hole is a favorite breeding site, while abandoned orifices of squirrels are used too." M. G. Vaiden reports from Mississippi that he has found nests in various situations: in an old woodpeckers' hole in an immense chinaberry tree; in a yard on a farm, about 50 feet from the ground; in a natural cavity in a black locust, 28 feet up in the trunk; in a natural cavity in an ashleaf maple; and in "tall cypress trees in the brakes of the territory."

Bendire (1895) says that natural cavities are preferred, "where such are obtainable, even should these be much more extensive than are really needed, as instances are known where openings in hollow limbs fully 6 feet deep have been filled up with rubbish to within 18 inches of the top before the nest proper was begun. Both sexes assist in nest-building, and it takes sometimes fully two weeks before their task is completed. The finishing and lining of the nest is generally completed by the female."

In addition to the few trees named above, nests of the crested flycatcher have been found in natural cavities or woodpecker holes in various oaks, ashes, maples, birches, pines, and cedars, as well as in beech, chestnut, tulip, pear, tupelo, sycamore, cottonwood, and locust trees, and probably in others. Dead stubs in the woods have probably been old favorites, and even woodpecker holes in telegraph poles have been used. Nests have been reported at various heights above ground, from 3 feet in low stumps or prostrate trees up to 70 feet in large trees; but probably most of the nests are located below 15 or 20 feet.

Bendire (1895) says: "The nesting cavities selected are ordinarily from 18 to 30 inches deep and others are considerably deeper, while occasionally one is quite shallow. The inner cup of the nest varies from 2¾ to 3½ inches in diameter and from 1½ to 2 inches in depth." Dr. Dickey (MS.) writes: "Nests that I have examined consisted of loose foundations of leaves of such deciduous trees as sugar maple, white oak, scarlet oak, walnut, and apple. They were intermixed with petioles, skeleton leaves, animal hair (from dog, cat, cattle, pig, rabbit, and horse), chicken and other poultry feathers, bark fibers from trees, hemp, rootlets, pieces of cord string, and strands from ropes."

Many and varied are the materials used in the construction of the nest and the filling in of the cavity, to which the size of the nest must conform.

Prof. Maurice Brooks writes to me: "The crested flycatchers that have built in a box erected for them, near French Creek, W. Va., have given striking evidence of a sense for color. At this time we are keeping on the farm chickens of two breeds, Barred Plymouth Rocks and Rhode Island Reds. The flycatchers have utilized feathers in constructing the nest. On the outside of the nest we counted 17 Plymouth Rock feathers, but not a single Rhode Island Red feather. On the other hand, the lining of the nest was composed almost entirely of Rhode Island Red feathers, not a single gray feather from the Plymouth Rocks being used. The nest contained three pieces of snakeskin, evidently that of the pilot blacksnake (*Elaphe obsoleta*)."

In addition to the materials mentioned above, the following have been found in the nests of this flycatcher: Large quantities of grass and pine needles, a few small twigs, feathers of grouse, owls, and hawks, a rabbit's tail, woodchuck fur, seed pods, bits of bark, cloth, and paper, pieces of onion skin, Cellophane, paraffined or oiled paper, bits of eggshells, and pieces of horse manure.

I have left until the last the consideration of the use of the castoff skins of snakes about which so much has been written. Almost everyone who has written anything about the crested flycatcher has touched on this subject. There can be no doubt that such old skins are often, perhaps generally, found in the nests, though they are usually found in small pieces and are often entirely lacking or replaced with something else of similar texture. Fully 25 percent of the nests that I have personally examined have contained no pieces of snakeskin or any similar material. Mr. Vaiden tells me that "from a total of 37 nests examined in the past 30 years, snake skins have been found in only 14." On the other hand, Prof. Brooks (MS.) says of one nest: "The birds had evidently been unable to find the pieces of snakeskin, which they are accustomed to place in their nests, but in this case they had substituted three pieces of the yellowed outside skin of an onion. This is the only nest I have ever seen that did not contain at least one piece of snakeskin. I have identified the sloughed skin of the pilot blacksnake (*Elaphe obsoleta*), the black racer (*Coluber constrictor*), the common watersnake (*Natrix sipedon*), and one of the little green snakes. I saw in a nest a piece of snake-skin that bore the unmistakable checkerboard pattern of the house-snake (*Lampropeltis triangulum*)."

In my experience the snakeskin is usually found in small pieces, more or less imbedded in the body of the nest or in the lining, but in some cases it is conspicuously displayed on the rim or left hanging in a long strip outside of the cavity. This has led to the oft-repeated theory that it is used as a "scarecrow" to frighten away predatory mammals, birds, reptiles, or other enemies. Frank Bolles (1890) was evidently convinced of the truth of this theory by the following circumstantial evidence, of which he writes:

In one instance, at Tamworth, New Hampshire, I found a nest with one egg in it but with no snake skin visible. I found it about 7 A. M. one beautiful day in early July, 1888. I touched the egg and handled the nest slightly. Shortly before sunset I looked a second time into the hollow limb where the nest was placed, and was much surprised, in fact somewhat startled, by what I saw. Forming a complete circle about the egg, resting, in fact, like a wreath upon the circumference of the nest cavity, was a piece of snake skin six or seven inches long. The part which had encased the head of the snake was at the front of the nest and was slightly raised. It may not be wise to found a theory upon a single fact, but from the moment I saw that newly acquired snake skin, placed as it was, I made up my mind that the Great Crested Fly-catcher uses the skin to scare away intruders.

He had a somewhat similar experience at the same nesting site the following year, which still further convinced him. His experience was suggestive, but not convincing. Charles L. Whittle (1927) has published an interesting article on this subject, from which I quote as follows:

Since many species of birds have their young and eggs destroyed by snakes, and since old birds at nesting time are greatly concerned when a snake is seen near their nests, as I have often observed to be the case, it seems obvious that if such species, and presumably the Great-crest does not escape their depredations, in seeking nesting material recognized snakes' sloughs as suf-ficiently snake-like to act as scare-crows to other birds, or other animals, they would themselves be too much alarmed on discovering the sloughs to use them in nest building. Hence two corollaries appear to be justified: (1) that since birds gather snakes' sloughs, they do not associate the flimsy, lifeless material with their former wearers; and (2) that they themselves, not recognizing that sloughs resemble snakes, do not employ them in nest building as scare-crows, but in the same manner that birds occasionally use fragments of birch bark, leaves, strings, newspaper, etc., as nesting material.

The above argument seems convincing. Many other birds use cast-off snakeskins as nesting material, some very extensively. I have a nest of the eastern blue grosbeak in my collection that is almost completely covered with several long pieces of such skin wound about its exterior, in which case it may have been used as decoration or camouflage. But in most instances it is probably used as convenient and desirable nesting material. The fact that the crested flycatcher has been found so often using such material as onion skins, thin, greasy, or waxed paper, paraffine paper, or strips of Cellophane suggests that either these bright shiny substances attract their attention, or that they, like snakeskins, furnish a certain degree of resiliency, or perhaps ventilation, in the nest. Mr. Bolles (1890) noted that fresh pieces of skin were brought in from time to time during incubation, which may indicate that the birds appreciate the value of such light, springy, and airy material in the close confines of the nesting cavity.

Eggs.—The crested flycatcher has been known to lay anywhere from four to eight eggs; five seems to be the commonest number, six eggs are frequently found, and the larger numbers are very rare. I have heard of only one set of eight, which is in the collection of the Academy of Natural Sciences of Philadelphia. The eggs are mostly ovate or short-ovate, occasionally elliptical-ovate or elongate-ovate. They are only slightly glossy. The eggs of the genus *Myiarchus* are all handsomely and peculiarly marked, those of this species being usually more heavily marked than those of the other species. The ground color varies from creamy white or "cream color" to "cream-buff", or "pinkish buff", rarely "vinaceous-buff." The eggs are usually quite uniformly covered with the peculiar markings, but sometimes these are somewhat concentrated at one end. The markings consist of a few irregular or elongated blotches and streaks and scratches or fine hair lines, as if made with a pen, of "claret brown," "liver brown," or other browns, and various shades of drab, purple, or lavender. The measurements of 50 eggs average 22.6 by 17.2 millimeters; the eggs showing the four extremes measure 23.8 by 17.8, 23.6 by 18.0, and 20.6 by 15.2 millimeters.

Young.—The incubation period has been stated as 13, 14, or 15 days, by different observers. Bendire (1895) says: "As a rule but one brood is reared in a season, and incubation lasts about fifteen days; the female attends to these duties almost exclusively, but is not a very close setter, and it is not uncommon to find addled eggs in the nests of this species. An egg is deposited daily until the set is completed."

A. Dawes DuBois says in his notes: "The voices of the young remind me of the peepers (tree frogs) heard in early spring, though more subdued. They keep up these calls when they think they hear their parents coming. The parents usually came together with food, and one sat on a nearby branch while the other went into the nest. When the first one came out the other entered the nest and gave its supply of food to the young. The usual food for the young consisted of medium-small insects, but occasionally one of the parents would bring a large miller or a small butterfly, and give it, wings and all, to the young. The parents took excrement from the nest and flew off with it." Dr. Ira N. Gabrielson (1915) writes: "The striking thing in the feeding, at least to us, was the large percentage of larvae fed. They comprised the largest single item of food, being 21.15% of the total. Grasshoppers under two heads in the tables, were 12.50%; spiders, 6.73%; moths, 6.97%; unidentified, 26.20%; red admirals, 3.12%; flies, 3.60%; beetles, 4.08%; hymenoptera (bees and wasps), 4.32%; and the remainder, 11.30%, were miscellaneous insects." Henry Mousley (1934) spent about 30 hours watching a nest of young crested flycatchers, during the first half of July 1932. Referring to the food brought to the young by the male, he says: "On four occasions, he brought a butterfly which I easily recognized as the Silver-bordered Fritillary (*Brenthis myrina*), an insect having a spread of wings of nearly one and three-quarter inches, which will give you some idea of what the young had to put up with." He also saw the male bring "a ripe wild raspberry" and the female bring "the second large soft green caterpillar." On July 2, when he estimated that young were three or four days old, "the young were fed thirteen times in three hours, ten times by the female, and three by the male." On subsequent occasions, the young were usually fed at intervals varying from 10 to 13 minutes, though once he saw them fed 10 times in one hour. The average rate of feeding for the whole period was once in 11.25 minutes. The male fed 42 times and the female 118. He estimated that the young left the nest when they were about 18 days old, which agrees exactly with Ora W. Knight's (1908) figure.

The findings of other observers differ: Mrs. Margaret Morse Nice (1931a) gives 12 days as the altricial period, Dr. Gabrielson (1915) says 12 to 13 days, and Dr. Dickey tells me that the young remain in the nest three weeks.

Dr. Gabrielson (1915) says:

We saw no evidence of regurgitation. * * * During the study we saw the parents carry away the excreta 41 times and devour it only once. Much of it was undoubtedly removed during our absence from the blind, but there must have been much of it devoured while the birds were concealed from our view in the nest.

The nestlings were very noisy and restless. They kept up a constant peeping from the first day. On July 7 [when about six days old] one or more of them began to utter a loud clear call or whistle, "*twee-et*," which was occasionally answered by the parents from a distance. * * *

On July 8 the nestlings began to climb restlessly about in the nest. * * * They crawled part way up the sides of the cavity and fell back to the bottom again. * * * Several times on July 9 they fell out of the nest and started away through the grass. * * * Whatever the cause of this action they quieted down after July 10 and remained in the nest until July 14.

After the young had left the nest, Mr. Mousley (1934) discovered them, with their parents, "about three hundred yards to the north of the nest. They were in a tangle of small birch-trees, willows, and other shrubs, the intervening spaces being covered with brushwood and very long grass, making it impossible to get about in a hurry, but forming a wonderful get-away for young birds." Dr. Thomas S. Roberts (1932) says: "After the young leave the nest the entire family keeps together for some time and goes roving about in the upper reaches of the forest, the young, which closely resemble their parents, being fed in part by the old birds while learning to care for themselves. They are all for the most part silent at this time, only an occasional, long-drawn *wheep* from the adults and weaker calls from the young announcing their presence."

Plumages.—The young are hatched naked and blind, but they soon become scantily clothed in grayish natal down, which adheres to the tips of the juvenal feathers and does not entirely wear away until the young birds leave the nest. The pattern of the juvenal plumage is essentially like that of the adult, but the colors are somewhat duller, the upper tail coverts are "cinnamon-rufous," the outer webs of the tail feathers are edged with rusty, or "cinnamon-buff," the median and greater wing coverts are broadly tipped with "cinnamon-rufous," the buffy edgings of the primaries are broader than in the adult, and the tertials are edged with pale buff. The sexes are alike in all plumages, though the colors of the adult female may be slightly duller.

Dr. Dwight (1900) says that the first winter plumage is "acquired by a partial postjuvenal moult beginning by the middle of August, which involves the body plumage, wing-coverts and tertiaries (apparently), but not the rest of the wings nor the tail, young birds becoming practically indistinguishable from adults." This plumage is apparently fully acquired before the birds migrate, but just when the juvenal remiges and rectrices are molted does not seem to be known. Dickey and van Rossem (1938) say: "The only immature bird collected (April 18) has retained its juvenal remiges and rectrices from the previous year. It is renewing the 7th primary in each wing, in addition to an extensive renewal of the contour plumage. The retention in this single specimen of the juvenal wing and tail feathers is probably an abnormality, for the other species of *Myiarchus* which occur locally change these at the postjuvenal molt."

Dr. Dwight (1900) states that the first nuptial and the adult nuptial plumages are acquired by wear, but I suspect that we may find evidence of a partial prenuptial molt at both ages when sufficient material is available. Adults have a complete postnuptial molt beginning early in August and completed before the birds go south.

Food.—Prof. Beal (1912) examined the contents of 265 stomachs of the crested flycatcher, and found it divided into 93.70 percent animal and 6.30 percent vegetable matter. "Beetles constitute 16.78 percent of the food, and of these 0.24 percent are useful species." He identified 52 species of beetles in the food, representing 12 different families, among which such harmful species as the cotton-boll weevil, the strawberry weevil, and the plum curculio were noted. Hymenoptera (bees and wasps) amount to 13.69 percent; the destructive sawflies were found, but only one worker honeybee; a few parasitic species were noted, but the proportion was not large. Diptera (flies) amount to only 3.06 percent. Hemiptera (bugs) constitute 14.26 percent of the diet. Orthoptera (grasshoppers, crickets, and katydids) seem to be favorite food; the average for all the months is 15.62 percent, but it ran as high as 23.18 percent in September. Lepidoptera (butterflies and moths) are the largest item in the food, 21.38 percent; caterpillars were found in 73 stomachs and adult moths and butterflies in 48, an unusually large percentage of adults. The remainder of the animal food was found to consist of dragonflies, lace-winged flies, and a few other insects, 4.14 percent; spiders, 4.03 percent; and a few eggshells; and "three stomachs contained the bones of a lizard (*Anolis carolinensis*)."

The vegetable food, 6.30 percent, consists of small wild fruits, such as mulberry, pokeberry, sassafras, spicebush, blackberry, raspberry, chokecherry, wild bird cherry, Virginia creeper, wild grape, cornel, huckleberry, blueberry, and elderberry.

Since the above bulletin was published, A. L. Nelson has sent us the results of the analysis of six more stomachs and a compilation of all the material in the economic food notes files of the Biological Survey. Among these I find mentioned a number of butterflies and moths, both adults and larvae, including the destructive leopard moth; also June bugs and cicadas.

He quotes from a letter from Fred E. Brooks as follows: "While making some observations recently upon the grapevine rootborer (*Memythrus polistoformis*) in a vineyard in Upshur County, W. Va., I noticed that the insects were being preyed upon by crested flycatchers (*Myiarchus crinitus*). The adult of this insect is a dayflying moth and the birds would catch them as they flew about the vines." He also quotes the following from W. D. Doan (1888): "In eastern Pennsylvania its food consists largely of the following insects: *Anisopterix pometaria* and *A. vernata*, *Pieris oleracea* (Oleracea Butterfly), *P. rapae* (the imported cabbage butterfly), *Colias philodice* (sulphur butterfly), corn worm (*Gortyna zeae*), house fly (*Musca domestica*), white-lined house fly (*Tabanus lineola*), stable fly (*Stomoxys calcitrans*), red ant (*Formica sanguinea*), field cricket (*Gryllus abbreviatus*), mosquito (*Culex taeniorhynchus*), and red-legged locust (*Caloptenus femur-rubrum*), besides large numbers of beetles."

As may be seen from the above analyses, the food habits of the crested flycatcher are almost wholly beneficial; the harm done by the taking of a few predatory insects is far offset by the long list of injurious species that it destroys. It should be encouraged to live and increase its numbers in our orchards and gardens by leaving natural cavities in otherwise worthless trees and by putting up boxes for nesting places.

Although most of the insects named above are probably taken on the wing in true flycatcher fashion, some of its food is secured in other ways. Early in the season, before the trees are in full leafage, much of the flycatcher's food is taken on or near the ground, or from crevices in the bark of trees, or from crannies in rail fences or fence posts. Dr. Dickey (MS.) says: "When food is scarce, during a season of drought, cold, or unusual rainfall, crested flycatchers will visit haystacks, crannies of log barns, and even open doors of hay lofts, to obtain stray insect life that harbors in such nooks. Again it will deign to pass to the bases of clumps of saplings, stir a mass of leaves or a spider web, and obtain some choice tidbit."

Francis H. Allen tells me that he observed "two feeding among the foliage of trees and shrubs, flitting about much as vireos would, and not at all in characteristic flycatcher fashion." He "saw them pick insects off the leaves when hovering before them, like kinglets."

Dr. Gabrielson (1915) writes of its hunting methods:

The greatest variety of food was secured in true flycatcher fashion, i. e., by watching for passing insects and darting after them from the chosen perch. * * * Some of these if not captured in the first dash, were not pursued further, but others, notably butterflies and moths, were followed until secured. * * *

The second method was somewhat different, although the insects were still taken while the flycatchers were on the wing. This method was to hang on rapidly beating wings before a leaf or branch and pick the insects from it. * * *

The third method was a variation of the first. The Great Crests sat on a low branch until they saw an insect in the grass, when they would drop to the ground and secure it. * * * When they missed the insect, they never hopped or ran along the ground, but rose into the air and dove down into the grass again. One watched catching a grasshopper near the foot of the nest tree went through this performance several times before the prey was finally secured.

Behavior.—In a general way the behavior of the crested flycatcher is much like that of the eastern kingbird; in its erratic dashes after prey it is distinctly like other flycatchers, and its hovering flight on rapidly vibrating wings reminds us of the kingbird; its flight is swift, as it must be to secure such lively insects as dragonflies; when not hunting it often sails from one tree to another on motionless wings and spread tail, after the well-known manner of the blue jay. It is quite intolerant and aggressive toward small birds that approach its nest or even enter its territory; but it does not seem to care to attack crows and hawks, as the kingbird does, perhaps for the obvious reason that its eggs and young, often deep down in a hollow, are less accessible to these large birds.

Dr. Gabrielson (1915) writes:

A Chickadee, Downy Woodpecker, Red-headed Woodpecker, and Flicker came to the tree at various times. The woodpeckers were driven away by the Great Crests, but they paid no attention to the Chickadee. * * *

A Cowbird came into the nest tree while the female was in the nest, sneaked to the nest opening and looked in. What she saw was evidently not reassuring as she quickly backed away and flew off.

A squirrel crossing the glade was vigorously attacked and made to scamper for refuge to the nearest tree. Once safely there he turned and expressed his opinion of the Great Crest in shrill and violent language.

The most vicious performance which I witnessed was an attack on an immature Bronzed Grackle. He blundered into the nest tree while the male was sitting on one of the topmost branches, and had hardly settled himself when he was struck a violent blow from behind and sent sprawling to the ground. He lay there squawking for a few moments and then started to fly away. Hardly had he lifted himself from the ground when another blow on the back of the head caused him to turn a complete somersault into a small bush. He crawled out on the side opposite the nest and flew away without being further molested. The Great Crest used both beak and wings in the attack and the second blow took several feathers out of the grackle's head.

As we enter the nesting territory, even while we are some distance from the nest tree, the birds set up a loud outcry; and, as we draw nearer, they become more excited, flying about nervously, uttering their loud alarm or challenge notes, with crests erected and bills snapping. But when we have actually discovered the nest, they are apt to withdraw in silence, or sit and watch us quietly. They seldom offer to attack a human intruder, though Dr. Gabrielson (1915) says: "When the nestlings were taken out of the nest on July 13 they made a great fuss and the parents answered them for a few moments. This noise soon ceased and the adults, particularly the female, made a desperate attack on our party, flying about our heads and at our faces. Finally, even this stopped, and the female alighted on a branch about fifteen feet away and kept silent watch of the proceedings."

Voice.—Its voice is one of the most prominent characteristics of the crested flycatcher, as it is so much oftener heard than seen. It has been referred to as a harsh squeak, but I have never heard it make a sound that could be called harsh or squeaky; its commonest note seems more like a loud, rather musical whistle, suggesting, in quality at least, the latter part of the "bob-white" call of the quail. To me it sounds like *whoit-whoit-whoit*, or *whuit-huit*, repeated several times. It has also been written *whip-whip-whip*, or *whit-whit-whit*, or *whuir-whuree*, or *puree*. The alarm note is a loud, less musical *wheep*, or *queep*, of great carrying power, audible at a long distance. Both notes are emphatic and immediately attract attention. Forbush (1927) mentions a very different loud note, *queer-queer-queer-queer*, as rendered by J. A. Farley.

Francis H. Allen (1922) says that this flycatcher has what we must call a genuine song:

[This] is chiefly an early morning performance, but may be heard, too, at other times of the day. Like the kingbird's it is a long, indefinite song or series of songs, but it has nothing of the hurried character of the former. Indeed, it is one of the most leisurely songs I know, for there is a rest of two seconds or more after each phrase. In its simplest form the song is a repetition of the phrase *queedle* over and over again indefinitely, but each alternate *queedle* is of different character from the one that immediately precedes it. The first time I studied the song I found numbers 2, 4, 6, etc., to be about a fifth on the musical scale higher than numbers 1, 3, 5, etc.; or, rather, this was true of the first syllable of each phrase—the *quee*. The *dle* part was perhaps a third lower than the *quee* in numbers 1, 3, 5, etc., and about an octave lower in numbers 2, 4, 6, etc. To indicate the difference in pitch I am in the habit of rendering the song *coodle, queedle, coodle, queedle, coodle, queedle*, etc. The *dle* part always being on the same pitch, the inflection of the alternate *queedles* has the effect of a finality in discourse. I think the difference in pitch between the *coodles* and *queedles* is not always as great as a fifth, for, not being a musician, sometimes I have had to listen rather intently to detect it.

Mr. Allen goes on to explain certain variations in the song in the same paper, to which the reader is referred. And Mrs. Nice has referred to it in three papers (1928, 1931a, and 1931b), in which she treats mainly of the time at which she has heard the "morning twilight song" at different places; most of the songs were heard between 5:00 and 5:30 A. M., but once as early as 4:24 on June 11. The songs were given at the rate of from 28 to 30 notes a minute. "The length of these songs surprised me, one lasting 28 minutes and two others more than 35" (1931b).

Eugene P. Bicknell (1885) writes: "In July the voice of this bird begins to fail, and a silent-period is nearly approached, if, indeed, it be not actually reached, in trying summers. During this time of semi-silence the usual utterance is a single note, which is often faint, and with a mournful intonation as it sounds at slow intervals among the high trees of the woods. Towards the end of August there is noticeable on the part of the birds an attempt to regain their earlier vocal prowess, but they soon return to the low note which they learned in July. This is their farewell, and is in strange contrast to the harsh outcry with which they came upon the scene."

After reading this manuscript, Francis H. Allen writes to me: "I have been accustomed to write what is perhaps the commonest note of the crested flycatcher as *k'wheek*. Apparently others, who render it as *wheep*, have not noticed the introductory *k'*, but I feel quite sure that it is commonly used. Another very common note is the rattling cry, *creep* or *cr-r-r-reep*, as I put it. As a matter of fact, this note is really polysyllabic; that is, the rapidly repeated *ee* sound has somewhat the effect of a rolled *r*."

Field marks.—In outline the crested flycatcher is larger than the kingbird, having a more decided crest and a much longer tail; its coloring is entirely different, an olive-brown back, pale gray breast and throat, yellow abdomen, and conspicuous wing bars. The Arkansas kingbird, now often seen in the Eastern States, shows some of these colors, but its wing bars are not conspicuous, it has a black tail, with white outer webs of the lateral feathers, and the cinnamon-rufous in the wings and tail, so conspicuous in the crested flycatcher, is lacking. In flight, the cinnamon in the wings and tail of the latter species shows up very plainly as one of the best field marks. Its voice is, of course, quite distinctive.

Enemies.—Probably one of the worst enemies of this and other hole-nesting birds is the European starling, which has increased so rapidly that it is appropriating all the available nesting cavities. The crested flycatcher is possibly more than a match for it and can hold its ground, but I have no evidence on the subject one way or the other. On June 8, 1941, W. G. F. Harris and I hunted through an old orchard in Raynham, Mass., where a pair of crested flycatchers had been in the habit of nesting for a number of years. The orchard was overrun with starlings; large numbers of young starlings were flying about among the trees, in full juvenal plumage, evidently the products of first broods, and their parents were laying their second sets of eggs, one of which we found.

When we located the flycatchers' nest, we were surprised to find that it contained six eggs of the flycatcher and one egg of the starling. The flycatcher was on the nest. Incubation had started in three of the flycatcher's eggs, the other three and the starling's egg being perfectly fresh. It would seem that the starling had probably laid its egg in the flycatcher's nest before the latter's set was complete and during the interval when she was off the nest. This illustrates the keen competition for nesting sites.

Snakes and squirrels probably destroy some eggs and young, though the flycatchers could doubtless drive away the latter. Mr. DuBois tells me the following snake story: "There were two or three boys in our town who made frequent expeditions to the country. One day, while exploring a farmer's orchard, we came to an apple tree worthy of investigation because of a cavity in a main limb. It was my turn to climb up; and when I reached the level of the hole and peered into it, I could see the head and coils of a large snake. With the aid of a stick the snake was soon dislodged and was killed by the boys on the ground as soon as it had made the descent. A huge enlargement of its body invited further investigation. We opened it and found an adult great crested flycatcher that had been recently swallowed whole. The bird had been caught at her own nest. I cannot vouch for the identification of the snake. As I recall it now, we called it a bullsnake."

Winter.—Although stragglers have been observed in late fall and winter in South Carolina and even New England, most of the northern crested flycatchers leave their more northern summer homes in late September or early October and spend the winter in Mexico, Central America, and northern South America. Dickey and van Rossem (1938) say that, in El Salvador, it is "fairly common in fall, winter, and spring on the coastal plain and in the foothills and mountains up to 3,500 feet. This species is most numerous in open woods in the lower foothills and is least so along the coast. Extreme dates of arrival and departure are October 25 and April 13."

DISTRIBUTION

Range.—The United States and southern Canada west to the Great Plains. Winters south to Colombia.

Breeding range.—The crested flycatcher breeds **north** to southern Manitoba (Carberry, Portage la Prairie, and Winnipeg); southern Ontario (Gargantua, Lake Nipissing, and Ottawa); southern Quebec (Montreal and Quebec City); and central Maine (Houlton and Patten). The **eastern** boundary of the breeding range is the Atlantic seaboard south to Florida (St. Augustine, New Smyrna, and Flamingo). **South** to Florida (Flamingo, Cape Sable, Seven Oaks, and Pensacola); southern Mississippi (Bay St. Louis); southern Louisiana (New Orleans, Thibodaux, and Grand Coteau); and southeastern Texas (Houston, Victoria, and Kerrville). **West** to eastern Texas (Kerrville, San Angelo, and Commerce); Oklahoma (Wichita Mountains Refuge, Minco, and Arnett); central Kansas (Pratt, Hays, and Stockton); eastern Nebraska (Hastings and West Point); eastern South Dakota (Yankton, Sioux Falls, and Faulkton); eastern North Dakota (Wahpeton, Fargo, Grafton, and the Turtle Mountains); and southwestern Manitoba (Treesbank and Carberry).

Winter range.—The winter range extends **north** to, probably rarely, southern Texas (Brownsville); and southern Florida (Punta Rassa, Fort Myers, and Fort Lauderdale). **East** to southeastern Florida

(Fort Lauderdale, Miami, and Tavernier); probably eastern Cuba (Guantanamo); and Colombia (Bonda, Puerto Valivia, and Rio Frio). **South** to Colombia (Rio Frio and Novita); Panama (Obaldia, Quebrada, and Divala); and Costa Rica (San Jose and Bolson). **West** to Costa Rica (Bolson and Tenorio); El Salvador (Volcan de Conchogua and Rio San Miguel); Guatemala (Gualan and Quirigua); Oaxaca (Santa Efigenia and Tapanatepec); and Veracruz (Motzorongo).

The range as outlined is for the species, which is separated taxonomically into northern and southern races. The northern form (*Myiarchus crinitus boreus*) occupies the entire breeding range except the peninsula of Florida and the coastal region north to southern South Carolina, which are occupied by the southern crested flycatcher (*M. c. crinitus*).

Spring migration.—Early dates of spring arrival are: Northern Florida, March 8. Southeastern Georgia, March 21. South Carolina—Charleston, April 10. North Carolina—Raleigh, April 9. Virginia—Variety Mills, April 24. District of Columbia—Washington, April 20. New Jersey—Morristown, May 3. New York—Shelter Island, May 1. Connecticut—Hadlyme, May 4. Massachusetts—Stoneham, May 3. Vermont—St. Johnsbury, May 10. Maine—Avon, May 9. Quebec—Montreal, May 10. Louisiana—New Orleans, March 12. Tennessee—Athens, April 9. Kentucky—Eubank, April 13. Indiana—Bloomington, April 18. Michigan—Plymouth, April 25. Ontario—Ottawa, May 5. Iowa—Hillsboro, April 20. Illinois—Chicago, April 27. Minnesota—Minneapolis, May 4. Texas—Refugio County, March 13. Kansas—Manhattan, April 25. Southeastern Nebraska, May 2. Manitoba—Aweme, May 24.

Fall migration.—Late dates of fall departure are: Manitoba—Aweme, September 11. Kansas—Onaga, October 2. Minnesota—Minneapolis, October 10. Illinois—Chicago, September 18. Iowa—Keokuk, September 17. Ohio—Wauseon, October 2. Mississippi—Ariel, October 15. Quebec—Montreal, September 4. Maine—Winthrop, September 13. Massachusetts—Waverly, September 29. District of Columbia—Washington, September 29. North Carolina—Raleigh, October 16.

Casual records.—The crested flycatcher has been recorded on a few occasions outside of its normal range. Among these is a specimen taken in the Moose Mountain district, Saskatchewan, on June 20, 1924; one was reported from Eastend, Saskatchewan, on May 24, 1933; one was collected at Windsor, Colo., on August 17, 1911; and one was obtained at Douglas, Wyo., on June 14, 1896.

Egg dates.—Florida: 23 records, March 14 to June 11; 12 records, May 5 to 25, indicating the height of the season.

Georgia: 11 records, May 5 to June 27.

Illinois: 14 records, May 25 to July 10; 8 records, June 9 to 22.

Massachusetts: 25 records, May 28 to June 26; 13 records, June 7 to 15.

Ontario: 3 records, June 14 to 28.

Texas: 12 records, April 11 to July 21; 8 records, April 19 to May 26.

MYIARCHUS CRINITUS CRINITUS (Linnaeus)

SOUTHERN CRESTED FLYCATCHER

HABITS

Because the Linnaean name for this species is based on Catesby's description of a bird *supposed* to have come from South Carolina, and because South Carolina specimens are *supposed* to belong to a southern race, the above name is now restricted to the crested flycatchers of Peninsular Florida, which are supposed to range north along the Atlantic coast to southern South Carolina. This seems rather far-fetched, as we have no definite type locality given for Catesby's bird, which he said "breeds in Carolina and Virginia."

This is what Dr. E. W. Nelson (1904) had to say on the subject: "As first pointed out by Mr. Bangs, the Great Crested Flycatchers of southern Florida are readily distinguished from birds occupying other parts of its range by the much greater size of their bills. This character appears so constant and is so marked that it seems to be worthy of recognition by name, although not accompanied by any other equally well marked differences. Unfortunately the birds from the Carolinas are most like those from New England, so that Mr. Bangs in his *Myiarchus crinitus boreus* (Auk, XV, p. 179, April, 1898) renamed the type form. The name afterwards given by Mr. Howe to the bird of southern Florida must therefore be recognized."

Reginald Heber Howe, Jr. (1902), proposed to call the southern Florida bird *Myiarchus crinitus residuus*, saying: "The main and a very sufficient character of separation given by Mr. Bangs, 'the swollen bill' of the southern, as contrasted with the small and slender bill of the northern bird, is very marked even in a comparison between northern and southern Florida examples."

Mr. Ridway (1907) relegated both new names to synonymy and said: "After carefully comparing breeding specimens from Florida with those from more northern localities I am unable to find differences sufficient, in my judgment, to warrant their subspecific separation."

If the South Carolina birds are to be included in the southern race, it seems likely that the range of this subspecies *might* be extended along the Gulf coast westward to Louisiana and perhaps Texas. Dr. Oberholser (1938) says: "The Southern Crested Flycatcher is a fairly common summer resident, from March 12 to October 2, in southeastern Louisiana, wherever woodlands or sufficient trees produce a suitable habitat."

In Florida, according to Arthur H. Howell (1932), the crested flycatchers "inhabit a variety of situations—open pine forests, cypress swamps, hammocks of oak or cabbage palmetto, the custard apple forest on the shores of Lake Okeechobee, and the black mangrove swamps near Cape Sable."

In a general way the haunts and habits of the southern crested flycatcher are similar to those of its northern relative, due allowance being made for the difference in environment. It nests in similar situations but uses local material; Major Bendire (1895) mentioned a nest, taken near San Mateo, Fla., that "was placed in a hole in the side of a rotten stump in low, flat pine woods, and was composed of dry cypress leaves, pine needles, grasses, sphagnum moss, dead leaves, bunches of hair, snake exuviae, strips of cypress bark, weeds, grass roots, palmetto fiber, and feathers; it was lined with bunches of hair, feathers, strips of cypress bark, and pieces of snakeskin."

Charles J. Pennock mentions, in some notes he sent me, the attempts of a pair to build a nest in a stovepipe that served as chimney for a building occupied by a colored "Auntie." The stovepipe ran out horizontally and then turned upward. One morning the old lady lighted a fire, had trouble with the draft, and was "fairly smoked out." A peck or more of nesting material was taken out of the pipe.

Although the northern crested flcatcher does not habitually attack crows and hawks, H. H. Bailey (1925) says that birds of the Florida race of the species "are of great benefit in helping to drive off the crows and hawks bent on catching the farmers little chicks, and I have seen them pursue these intruders for some distance, in company with a kingbird."

The eggs of the southern crested flycatcher are like those of its northern relative, and do not seem to differ much from them in size. The measurements of 40 eggs average 22.4 by 17.6 millimeters; the eggs showing the four extremes measure 23.8 by 19.0, 20.6 by 17.5, and 21.4 by 15.9 millimeters.

MYIARCHUS TYRANNULUS MAGISTER Ridgway

ARIZONA CRESTED FLYCATCHER

HABITS

The Arizona crested flycatcher is the largest of the North American species of the genus *Myiarchus*. It is similar in general appearance to our familiar crested flycatcher of the Eastern States, though Ridgway (1907) says of this and the Mexican crested flycatcher, "color of upper parts much grayer (the pileum browner), gray of throat and chest and yellow of abdomen, etc., much paler, and inner webs of rectrices with a broad stripe of dusky grayish brown next to shaft." It is a bird of western Mexico, southern Arizona, and southwestern New Mexico.

It is a conspicuous and noisy member of the interesting aggregation of hole-nesting birds that have found congenial nesting sites in the deserted nest holes of the Gila woodpecker and Mearns's gilded flicker. We found this flycatcher fairly common in the region lying east of the Santa Cruz River and south of the Catalina Mountains in southern Arizona. Here the land is flat or slightly rolling, arid, sandy, and stony, even up into the foothills of the mountains. For the most part it is treeless, except for a few scattered paloverdes, low-growing mesquites, and a scanty undergrowth of creosote bushes. But scattered all through it one sees the towering straight or many-branched trunks of the saguaros, or giant cacti, standing like sentinels on the desert plains and on the lower slopes of the foothills. Here and there are scattered clumps of several species of cholla, with flowers of varied colors, an occasional barrel cactus, with its bulky interior filled with moisture for the thirsty traveler, and various low-growing cacti, with flowers of brilliant hues. All these are attractive enough when the desert bursts into bloom in the spring, but they seem to form a strange habitat for a crested flycatcher, which we in the East have learned to regard as a forest-loving bird.

Nesting.—In just such places as that described above, the Arizona crested flycatcher comes to build its soft nest in some one of the many vacant holes that the woodpeckers have made in the saguaros. Their neighbors, in addition to the two home-building woodpeckers, may be saguaro screech owls, the tiny elf owls, cactus wrens, desert sparrow hawks, purple martins, or ash-throated flycatchers. Nearly every giant cactus has one or more woodpecker holes in it, and some of the larger ones have half a dozen or more of them, at heights ranging from 15 to 30 feet above ground, though many of them are unoccupied.

The sap of the saguaro hardens around the inside of the cavity, making an ideal bird box, which holds its shape for a long time. In this cavity the flycatchers make a soft, warm nest of hair, fur, feathers, etc., usually with bits or large pieces of cast-off snakeskin. Cow's hair, often obtained from dead cattle and smelling badly, seems to be one of the principal materials; but I have seen in the nests tufts of badger fur, rabbit fur, the white tail tuft of a cottontail rabbit, a wad of feathers from a white-necked raven, various other feathers, and once a piece of skin from a Texas nighthawk, with a foot and a bunch of feathers attached to it. The center of the nest is usually lined with a felted mass of the softer fur.

All the nests I have seen, and most of those I read about, have been located in the saguaros, as described above. But W. E. D. Scott (1887) mentions a nest that he found in the Catalina Mountains at an elevation of about 4,000 feet, that was "built in a deserted Woodpecker hole in a dead sycamore stub." And A. J. van Rossem (1936) reports seeing two pairs of these birds "carrying nesting material into natural cavities in tall cottonwoods," 10 or 12 miles from the nearest giant cactus.

Eggs.—Three to five eggs are laid by the Arizona crested flycatcher. They vary from ovate to short-ovate and are only slightly glossy. They are typical crested flycatcher eggs, much like those of our familiar eastern species, but usually less heavily marked. The creamy-buff ground color is more or less completely marked with the peculiar elongated blotches, spots or scrawls, as if made by an erratic pen, of various shades of brown, purple, or lavender, "liver brown," "claret brown," and shades of "brownish drab" prevailing, with underlying markings of shades of "Quaker drab." In some eggs the smaller half is very sparingly marked; and in some others the whole surface is uniformly covered with very minute brownish dots. The measurements of 43 eggs average 24.1 by 18.2 millimeters; the eggs showing the four extremes measure 26.2 by 19.1, 26.0 by 20.0, and 21.8 by 16.3 millimeters.

Plumages.—I have not seen any very young birds, but birds in the fully developed juvenal plumage are similar to adults, though having the upper parts browner (less olivaceous), especially on the pileum, which is "sayal brown," and on the upper tail coverts, which are largely "ochraceous-tawny"; the outer webs of the rectrices are broadly margined with "ochraceous-tawny" (not narrowly margined with grayish brown, as in the adult) ; the outer webs of the primaries and secondaries are more extensively margined with "ochraceous-tawny" than in the adult; the greater and median wing coverts are broadly margined with "cinnamon-buff"; the gray areas and the yellow areas of the under parts are paler than in the old birds.

The postjuvenal molt, which apparently includes the contour plumage only, probably occurs before the birds leave for the south and produces a first winter plumage that is practically adult. What June and July birds I have seen are in decidedly worn plumage; and October and November birds in fresh plumage indicate that the annual complete molt occurs in August and September. I have seen an adult female molting both wings and tail on September 4.

Food.—We have practically no positive and accurate information on the food of this flycatcher. It is said to consist largely of beetles, but doubtless includes many other kinds of flying insects, which it captures in the air, and perhaps some wild berries and fruits. All flycatchers eat some useful insects, but a much larger number of harmful ones. They may therefore be classed as more beneficial than otherwise. There was a large apiary near the saguaro plains where we found these flycatchers breeding, but we heard no complaint from the owner that the flycatchers destroyed any numbers of bees.

Behavior.—Like its eastern relative, the Arizona crested flycatcher is a noisy, active, quarrelsome bird. Although not so bold and aggressive in its attacks on larger birds as are the kingbirds, it indulges in many squabbles with its smaller neighbors for the possession of nesting sites and is always ready to pick a fight with one, if it comes too near. However, it seems to live and rear its young, perhaps in a state of armed neutrality, among the various other species that occupy the ready-made homes in the saguaros. Against the intrusion of human beings loud vocal protests are made and considerable agitation is shown, but the flycatchers are too shy actively to attack the intruder; they generally remain at a safe distance and make up in noise for what they lack in courage.

Voice.—They have a variety of loud, harsh notes, which reminded me of similar calls of the eastern bird. One of the commonest notes sounds like "come here, come here," strongly accented on the last syllable; a loud, clear, whistling note is also often heard, which is not unmusical.

DISTRIBUTION

Range.—Southern Texas and Arizona, and Mexico south to Honduras and El Salvador. Not regularly migratory.

The range of our two races of crested flycatcher extends **north** to southeastern California (Bard); central Arizona (Big Sandy Creek, Fort Verde, Roosevelt Lake, and Paradise); and southern Texas (Corpus Christi). **East** to Texas (Corpus Christi, Mercedes, and Fort Brown); Tamaulipas (Soto la Marina and Tampico); Yucatan (Merida and Temax); Quintana Roo (Cozumel Island); and Honduras (Roatan Island and Truxillo). **South** to Honduras (Truxillo, La Ceiba, and San Pedro); El Salvador (Acajutla); Guatemala (Puebla); Chiapas (San Benito); Oaxaca (Santa Efigenia and Tehuantepec); and Guerrero (Coyuca). **West** to Guerrero (Coyuca and Acahuitzotla); Jalisco (Guadalajara); Nayarit (Tres Marias Islands and Santiago); Sinaloa (Escuinapa and Mazatlan); Sonora (Ortiz and Nogales); and southeastern California (Bard).

The range as outlined is for the two races that extend into the United States. The Arizona crested flycatcher (*Myiarchus t. mag-*

ister) is the form found in southern Arizona and western Mexico, while the Mexican crested flycatcher (*M. t. nelsoni*) occupies the range from the Rio Grande Valley in Texas south through eastern Mexico to Honduras and El Salvador.

Casual records.—Although it seems probable that this bird may occasionally nest in southwestern New Mexico, the only definite record of occurrence is of a specimen taken on June 12, 1876, on the Gila River about 20 miles south of old Fort West.

Egg dates.—Arizona: 23 records, April 4 to July 17; 13 records, May 24 to June 15, indicating the height of the season.

Texas: 67 records, March 30 to June 17; 33 records, April 28 to May 24.

MYIARCHUS TYRANNULUS NELSONI (Ridgway)

MEXICAN CRESTED FLYCATCHER

HABITS

This Mexican flycatcher extends its range from eastern Mexico into the United States only in the lower valley of the Rio Grande in southern Texas. It is somewhat smaller than the Arizona crested flycatcher but is like it in coloration. It is larger and paler in color than our eastern crested flycatcher but much like it in general appearance and behavior. We found it to be a fairly common bird in Cameron County, frequenting the open country about the ranches, but George B. Sennett (1879) says that "although found in the chaparral and low, stunted growth of mesquite, yet its home is emphatically in the heavier growths of timber, such as exist above Hidalgo."

Nesting.—The only nest that I saw near Brownsville was in a cavity in a fence post about 5 feet above ground; our guide, Capt. R. D. Camp, had taken a set of five eggs from this nest a short time previously. Dr. Herbert Friedmann (1925) had a similar experience with it; he says: "This bird is more a bird of the open than the Crested Flycatcher of the northern states, and nests commonly in fence posts bordering open fields. * * * Two nests, each with five eggs, were found. Both were in stumps used as fence posts."

Mr. Sennett (1879) says: "They nest in hollow stubs or abandoned woodpeckers' holes, at a height varying from five to twenty feet. The nests are lined with a matted felt consisting of soft strips of bark, feathers, hair, and wool, with sometimes bits of snakeskins intermingled. They begin to lay early in May, the number of eggs in a clutch being five or six. When sitting, the birds are not very timid, and, upon being flushed from their eggs, do not fly to any great distance, and soon return to the nest upon the intruder's retreat. On May 16 I flushed a male from a nest and six eggs, a circumstance making it probable that the male assists in the duties of incubation."

There are six sets of eggs of the Mexican crested flycatcher in my collection, all taken by F. B. Armstrong near Brownsville, Tex., between April 28 and May 14; one of the nests was in a hole in a fencepost, and the others were in cavities in trees at heights varying from 5 to 12 feet above ground.

A nest in the Thayer collection was taken from an old woodpecker's hole, 10 feet up in a pine tree, in British Honduras; it was made of palmetto fibers, feathers of a dove, and a piece of snakeskin. Snakeskin, apparently, does not figure so largely in the nests of this flycatcher as it does in the nests of our northern crested flycatcher.

Eggs.—The Mexican crested flycatcher lays three to six eggs, five being the commonest number. The eggs are practically indistinguishable from those of the Arizona crested flycatcher; they are much like the eggs of the northern crested flycatcher, but usually somewhat less heavily marked. The measurements of 50 eggs average 22.9 by 17.3 millimeters; the eggs showing the four extremes measure 25.2 by 20.1, 19.8 by 17.0, and 21.8 by 16.5 millimeters.

The molts and plumages, food, and general habits of this bird do not differ materially from those of other crested flycatchers. It is only a summer resident within our borders, which forms the northern limit of its range, arriving early in April and departing during the latter part of September.

MYIARCHUS CINERASCENS CINERASCENS (Lawrence)

ASH-THROATED FLYCATCHER

PLATES 12, 13

HABITS

The ash-throated flycatcher is widely distributed in western North America, breeding as far north as Washington (rarely), as far east as central Colorado, and thence southward into Mexico, from northern Lower California eastward to Tamaulipas. Over much of this area it is a common bird, and in some regions it is really abundant. Major Bendire (1895) remarked that climatic conditions do not seem to affect it to any extent, for it is as much at home in the mountain fastnesses of the southern Sierra Nevadas, at an altitude of 9,000 feet, as in Death Valley, probably the hottest place in the United States. He found these birds rather common in Arizona and said of their haunts there: "Their favorite haunts were the denser mesquite thickets in the creek bottoms, oak groves along hillsides, and the shrubbery in canyons leading down from the mountains, but I also saw them occasionally on the more open plains covered with straggling mesquite trees and patches of cholla and other species of cacti." We found it at various places in Arizona, in the canyons and foothills of the Huachuca Mountains, as well as on the washes below the canyons; it was fairly common about Tombstone, and we found it nesting in the woodpecker holes in the saguaros on the dry plains near Tucson.

In the Lassen Peak region of California, Grinnell, Dixon, and Linsdale (1930) found these flycatchers "on the low ground along the Sacramento River, * * * in sycamores, in valley oaks, in live oaks, and about dead trees or hollow snags close to the river. On the higher ground to the eastward, * * * the birds noted were on the rocky mesa clothed scantily with grass and with a few scattering blue oaks and clumps of buck-brush. At other places in the eastern part of the section this flycatcher was seen in or around junipers, willows, or sage-brush."

Nesting.—Major Bendire (1895) has a lot to say about the nesting of the ash-throated flycatcher, which is worth quoting:

The nests are usually placed in knot holes of mesquite, ash, oak, sycamore, juniper, and cottonwood trees, as well as in cavities of old stumps, in Woodpecker's holes, and occasionally behind loose pieces of bark, in the manner of the Creepers. On two occasions, near Tucson, I found the Ash-throated Flycatcher using abandoned nests of the Cactus Wren, and Mr. A. W. Anthony found them nesting in the dry blossom stalks of the yucca and *Agave americana* in southwestern New Mexico.

The Ash-throated Flycatcher nests at various heights from the ground, rarely, however, at greater distances than 20 feet. The nest varies considerably in bulk according to the size of the cavity used. Where this is large, the bottom is filled up with small weed stems, rootlets, grass, and bits of dry cow or horse manure, and on this foundation the nest proper is built. This consists principally of a felted mass of hair and fur from different animals, and occasionally of exuviae of snakes and small lizards; but these materials are not nearly as generally used as in the nests of our eastern Crested Flycatcher—in fact, it is the exception and not the rule to find such remains in their nests. Among about fifteen nests of this species examined by myself I only found it in three cases. As nearly as I have been able to observe, I think the female does most of the work on the nest, but the male follows her around while in search of material, and apparently guards and sings to her. I have known a pair of these birds to finish a nest in one day. * * *

It is surprising how little space is really required by them in which to rear a family. The inner cup of a well-preserved nest of this Flycatcher, placed behind a loose piece of bark of an old cottonwood stump, measures about 2½ inches in diameter by 2 inches in depth. The walls of this nest are composed exclusively of cattle hair, which is well quilted together and forms a fairly strong felt. The base is formed of dry grass roots, and it was placed between the soft inner and the outer bark of the tree, which kept it intact and held it firmly in position.

Of six nests that Mr. Willard and I recorded in Arizona, one was in a hole in a post 4 feet from the ground; one was in an old cactus wren's nest 8 feet up in a mesquite bush; one was in a cavity in an old mescal stub, open at the top and only 3 inches deep; another was in the dry stalk of a mescal, probably an old woodpecker's nest, 7 feet from the ground; and the other two were in the abandoned nests of woodpeckers in the saguaros. All these nests were made of hair and fur from cattle, deer, or rabbits, with a little dry grass in the foundation; they made soft, warm beds for the young; none of them contained snakeskins. Mr. Willard also recorded two very low nests in holes in mescal stalks; one was 3 feet and one only 2½ feet from the ground. There is an entirely different type of nest in the Thayer collection in Cambridge, collected by W. W. Brown, Jr., in Sonora; it was 5 feet from the ground in a hole in a tree and was made entirely of dead, gray grasses and was lined with finer grasses and hairs; it looks as if the flycatchers had appropriated an old nest of some other species.

The ash-throated flycatcher nests in a great variety of situations in addition to those named above; evidently there are not enough natural cavities in trees or old woodpecker holes available to satisfy its requirements; so it is forced to select any opening it can find that is large enough to hold its nest, often on or near human habitations or in man-made structures, such as a drain pipe from the eaves of a house, an old tin can or pot, a hole in a fencepost, a bird box, empty mail box, or any other boxlike opening. If the cavity is small, the nest is squeezed into it; if it is too large, the extra space is filled in with rubbish. Sometimes the birds show more energy than good judgment in the selection of a nest site. Fred Gallup (1917), of Escondido, Calif., hung an old pair of overalls on a line to dry; a pair of these birds began carrying in nesting material through a hole in one leg, but it fell out at the bottom of the leg as fast as they carried it in; they kept at the hopeless job for about an hour, until Mr. Gallup tied up the bottom of the leg. They finally succeeded in filling up the leg with material, lined the nest with feathers, and raised a brood of young.

Wilson C. Hanna (1931) tells a remarkable story of a pair of these flycatchers that built a nest and raised a brood of young "in the boom of a gasolene engine shovel which had been in operation almost every day in loading clay." The nest was "down three feet in a cavity on the underside of the boom and well out toward the end." The boom moved, of course, with every shovelful of clay. The site may have been selected when the shovel was not in operation, and the birds were courageous enough to stick by their eggs or young. The most remarkable part of the story is that, although the birds may have chosen what they thought was a quiet spot, they did not profit by their experience, or did not mind the disturbance, for they returned the next season and raised a brood of young in the same cavity.

Mrs. Bailey (1928) writes: "A peculiar nesting site was found by Mr. Ligon at the old Miller ranch on the Pecos—a four-inch exhaust pipe six feet long standing at an angle of about thirty degrees, coming from the cylinder of an abandoned oil engine. The pipe was smeared inside with the black fuel oil softened by the heat and the parent bird which flew from the nest, that was about twelve inches down inside the pipe, to a mesquite bush on a bank above, was so black that Mr. Ligon had difficulty in recognizing it."

Major Bendire (1895) says: "I am inclined to believe that it not infrequently dispossesses some of the smaller Woodpeckers, like *Dryobates scalaris bairdi*, of its nesting sites, as I have found its nests on two occasions in newly excavated holes, the fresh chips lying at the base of the tree, showing plainly that they had only recently been removed."

Mrs. Wheelock (1904) writes: "It has been caught nesting in newly formed cavities prepared by both the Texas and Gairdner woodpeckers, and in one case at least I know the woodpeckers were at work on the hole when driven away by usurpers. The battle raged vigorously at intervals for a whole day. No sooner had the Flycatchers settled the affair and begun to line the nest with rabbit fur, than the woodpeckers returned to the fray; during the temporary absence of the bandits they scratched out every bit of the unwelcome material, and prepared to reoccupy their home themselves. But as always, the fiercer temper of the Flycatchers prevailed over the brave resistance of the woodpeckers, and after repeated defeats they surrendered."

Eggs.—The number of eggs laid by the ash-throated flycatcher varies from three to seven, but the larger numbers are very rare; the usual set consists of four or five. The eggs are usually ovate, occasionally elliptical-ovate, and they have very little gloss. They are of the *Myiarchus* type but rather more sparingly marked than those of the Mexican crested flycatcher. The ground color varies from creamy white to pale "cream color," or rarely to a more pinkish shade. Some eggs are rather lightly streaked longitudinally with fine hair lines, some are marked with heavier lines or elongated splashes, and a few are spotted or irregularly blotched with the colors common to other species of the genus, various browns, purples, and drabs. The measurements of 50 eggs average 22.4 by 16.5 millimeters; the eggs showing the four extremes measure 24.4 by 17.3, 23.9 by 17.8 and 20.3 by 15.2 millimeters.

Young.—Major Bendire (1895) writes: "The female, I think, attends to the duties of incubation exclusively, which lasts about fifteen days. She is not a close sitter, and often leaves the nest for hours, especially during the heat of the day, but remains close by. The young are fed on the soft portions of insects, and leave the nest in about two weeks, following the parents about for some time before they are able to care for themselves."

Professor Beal (1910) reports that a nest with four young "was observed for eight and one-half hours and 119 feedings were noted, or an average of 14 feedings per hour. Both parent birds took part in the feeding until the female was unfortunately killed after the first hour of feeding on the morning of June 27." During this hour, from 5.15 to 6.15 A. M., there were 28 feedings. "At practically the same hour the next morning, June 28, the male bird alone was able to feed only 16 times. However, the young did well, and left the nest that afternoon." He estimated that "each of the young birds must have been fed about 49 times every day, or 196 insects in all."

Mrs. Wheelock (1904) tells of the activities of a male in feeding a brood of three young; at first, while the young were small and naked, he swallowed the insects, "and flew almost immediately to feed the young by regurgitation, but as they grew older he carried raw food to the nest. Often he alighted on the tree near the tiny doorway and by pulling off the wings and legs prepared the soft parts of the insect to be eaten by his nestlings. From the amount of food consumed one would imagine nothing smaller than young owls inhabitated the nursery. Twenty-two grasshoppers were taken in less than half an hour, making more than seven apiece. The nestlings being so small, this seems an appalling amount to be crammed into those tiny throats; but it evidently agreed with them, for they grew at a surprising pace, and on the sixteenth day they were well prepared for their début."

For several days after the young had left the nest, she watched the female teaching the young to catch their own food. "She brought a small butterfly and lit a little above and in front of one of the young. She fluttered out toward him holding the insect in her bill, then she released the latter so that it flew lamely down just in front of the eager baby. * * * The lesson was repeated with variations at intervals all day. Three days after this he was catching flies for himself, although still following the mother about and begging with quivering wings for the larger insects he saw her seize, and too often getting them."

Plumages.—The nestlings, which are hatched blind and naked, soon become clothed in the juvenal plumage, which is darker above and paler below than in the adults. Ridgway (1907) says that the young are "essentially like adults, but pileum cinnamon-brown or wood brown, rectrices cinnamon-rufous with a median streak of grayish brown, uper tail-coverts strongly tinged with cinnamon-rufous, outer webs of remiges mostly buffy cinnamon-rufous, other wing-markings tinged with cinnamon-buff, and yellow of under parts much paler (yellowish white)."

Apparently, young birds have a postjuvenal molt in August and September, involving the contour plumage but not the wings and tail; this produces a first winter body plumage like that of the adult; the remiges and rectrices are evidently molted the following spring and summer.

Practically all the June and July adults that I have seen are in much worn plumage; and October birds all seem to be in fresh plumage. This indicates a complete molt during August and September; I have seen one adult, taken September 21, that is in fresh body plumage but is molting both wings and tail. There is probably a very limited prenuptial molt, about the head and throat, in young and perhaps in old birds.

Food.—Professor Beal (1910), reporting on the contents of 80 stomachs of the ash-throated flycatcher, says:

Though an orchard bird, it seldom eats any cultivated fruit, but confines its diet largely to insects, most of which are either injurious or neutral. * * * Animal food amounts to 92 percent and vegetable to 8 percent for the season. * * * One stomach taken in September held 44 percent of elderberries, which is exceptional. * * * Of the animal food, beetles, almost entirely of harmful species, amount to 5 percent. * * * Bees, wasps, and a few ants (Hymenoptera) amount to 27 percent. * * * Bugs (Hemiptera) aggregate about 20 percent of the food of the ash-throat, which is in the largest showing for that order of insects yet found in the food of any flycatcher. * * * While many of these are taken upon the wing, probably some are picked from plants. One bird was seen on a mustard plant feeding upon the plant lice, which completely infested the plant. One stomach was entirely filled with tree hoppers and two with cicadas. * * * Flies (Diptera) amount to about 14 percent and were eaten in nearly every month. * * * Caterpillars were found in 20 stomachs and moths in 7. Together they amount to 19 percent of the food. This shows that caterpillars are a favorite article of food with this bird, and proves that it does not take all of its food on the wing. While no stomach was entirely filled with caterpillars, one contained nothing but moths. Grasshoppers formed about 5 percent of the food, and were mostly taken in May, June, and July. One stomach contained nothing else. As they do not often come within reach of flycatchers, these insects must be especially sought for. Various other insects and spiders amount to a little more than 3 percent. * * *

Vegetable food was found in 9 stomachs. Of these, 5 contained remains of elderberries; 2, bits of other small fruit; and 2, skins which might have been those of cultivated varieties.

Bendire (1895) adds to the vegetable food the berries of a species of mistletoe that grows abundantly in southern Arizona; and Dr. Beebe (1905) saw it devouring many varieties of small fleshy fruits, when insects were scarce. Mrs. Bailey (1928) says that "five taken near an apiary contained no honey bees, but one contained 24 robber flies, an enemy of the honey bee."

Behavior.—Grinnell and Storer (1924) write:

The Ash-throated Flycatcher resembles the Western Kingbird in general form and tone of coloration, but differs unmistakably in habits and demeanor. It has none of the aggressive, belligerent actions which characterize the kingbird, but attends to the business of catching insects in a pleasingly quiet manner. Unlike many of the Flycatcher tribe, the Ash-throat does not often return to the same location after sallying forth to capture an insect, but usually moves on to a new perch, evidently preferring to *go after* its prey rather than passively wait for the latter to chance by. Often, when taking flight for but a short distance, the bird retains the upright posture of its body, and with its tail drooped and slightly expanded flutters from one perch to the next. Nor is it so restricted in home range as the kingbird. Most flycatchers, the kingbird included, are wont to remain in a restricted area after once being established for the season, but the Ash-throat seems to be more enterprising and ranges widely over the brushlands.

That this flycatcher is sometimes more aggressive than the above remark indicates is shown by the following observations by Dr. Beebe (1905):

As the two ravens rose at our approach, one of these flycatchers appeared from a field beyond and, kingbird-like, gave a thrashing to first one and then the other, descending with his full force upon head and back and more than once sending fluffs of black to the ground.

When both ravens had disappeared, the flycatcher returned and instantly gave his attention to a Western Red-tailed Hawk. Uttering his loud *che-hoo! che-hoó!* the brave little creature dashed at the bird of prey, striking blow after blow, the hawk meanwhile never attempting to retaliate, but making every effort to escape from his small tormentor. Thus early in our trip the Ash-throated Flycatcher established a reputation for bravery which it always sustained.

Voice.—Major Bendire (1895) says that "its principal call note is a clear 'huit, huit,' a number of times repeated, which sounds very much like the ordinary call of the *Phainopepla;* it also utters some low, whistling notes which are not at all disagreeable to the ear."

Florence A. Merriam Bailey (1896) says: "Their calls closely resemble those of the eastern Great-crest, *M. crinitus.* Some are like *quir'r'r', quirp'* and *quir'r-rheá.* The bird also says *hip, hip, ha-wheer,* the *hip* emphasized with a vertical flip of the tail, the *wheer* with a sidewise dash. The Flycatcher has besides a low call of *hip* and *ha-whip.*"

Field marks.—In general appearance the ash-throated flycatcher most closely resembles our common eastern crested flycatcher, but the two are not likely to occur in the same region. A flycatcher resembling our eastern bird, but much paler in coloration, with a large, brown, bushy head, a conspicuous white throat, and a long, reddish brown tail, perching in an upright posture on some low tree or bush, is sure to be this species. It is smaller than the Arizona crested, as well as paler, and larger than the olivaceous flycatcher. The two western kingbirds have black or dark brown tails and brighter yellow under parts, as well as gray breasts.

DISTRIBUTION

Range.—Western United States and Mexico south to Guatemala.

Breeding range.—The ash-throated flycatcher breeds **north** to central Washington (North Yakima); northeastern Oregon (Weston); northern Utah (Salt Lake City); western Colorado (Grand Junction and Naturita); New Mexico (Santa Fe and Roswell); and west-central Texas (San Angelo). **East** to western Texas (San Angelo, Mason, Kerrville, San Antonio, and Losoya Crossing); and probably southwestern Tamaulipas (Miquihuana). **South** to probably southwestern Tamaulipas (Miquihuana); Durango (Rio Sestin); southern Sinaloa (Rosario); and southern Baja California (Miraflores). **West** to Baja California (Miraflores, La Paz, San Fernando, and Cocopah); California (El Cajon, Santa Barbara, Hayward, Nicasio, Ravensdale, and Edgewood); western Oregon(Klamath Falls, Rogue Valley, Prineville, and Twickenham); and Washington (probably The Dalles and North Yakima).

Winter range.—In winter the species is found **north** to rarely southwestern Arizona (Yuma); northern Sonora (Sonoyta and Pozo de Luios); southwestern Chihuahua (Duranzo); and Yucatan (Chichen-Itza). Note: This species has been detected casually in winter north to the southern point of Nevada (Fort Mohave and Searchlight). **East** to Yucatan (Chichen-Itza); eastern Guatemala (Rio Dulce and Gualan); and central Costa Rica (La Palma). **South** to central Costa Rica (La Palma and Puntarenas); El Salvador (Libertad and Barre de Santiago); western Guatemala (San Lucas, Lake Atitlan, and Sacapulas); southern Oaxaca (Tapanatepec and Chivela); southern Guerrero (Coyuca); and southern Baja California (Miraflores). **West** to Baja California (Miraflores and La Paz); western Sonora (Tesia and Guaymas); and rarely southwestern Arizona (Yuma).

The range as outlined is for the entire species, which has been separated into two subspecies. Most of the range is occupied by the typical race, *Myiarchus c. cinerascens*, the Lower California flycatcher (*M. c. pertinax*) being found only in the southern part of the peninsula of Baja California.

Spring migration.—Early dates of arrival are: Texas—San Antonio, March 10. New Mexico—Carlisle, April 16. Colorado—Pueblo, May 12. Arizona—Huachuca Mountains, April 9. Nevada—Pahrump

Valley, April 29. California—Death Valley, March 24; Berkeley, April 9; Red Bluff, April 25. Oregon—Klamath Basin, April 6. Washington—Yakima, May 13.

Fall migration.—Late dates of fall departure are: California—Berkeley, September 30; Flintridge, October 11. Arizona—Tombstone, September 26. Colorado—Mesa County, August 22; El Paso County, September 28. Texas—Bonham, October 17. New Mexico—Silver City, November 20.

Casual records.—A pair was seen on May 24, 1925, at Tacoma, Wash., which is north of the range as now known. One also was reported as seen near Libby, Mont., on September 4 and 5, 1924. A specimen was collected at Cheyenne, Wyo., on June 6, 1896, while two have been taken in Colorado east of the mountains, one at Gaume's Ranch in Baca County, on May 25, 1905, and the other in the Clear Creek Valley, near Denver, on September 17, 1911.

Egg dates.—Arizona: 22 records, May 6 to June 26; 12 records, May 21 to June 23, indicating the height of the season.

California: 79 records, April 12 to July 5; 41 records, May 25 to June 11.

Lower California: 8 records, March 13 to July 11.

Texas: 5 records, May 19 to June 4.

MYIARCHUS CINERASCENS PERTINAX Baird

LOWER CALIFORNIA FLYCATCHER

HABITS

The Lower California race of the ash-throated flycatcher is found in the Cape region of this peninsula, and thence northward to about latitude 30°, where it intergrades with the northern form.

William Brewster (1902), referring to the characters of this subspecies, states: "My specimens from the Cape Region differ rather constantly from those from western Mexico and the United States in having longer as well as usually stouter bills. They are also almost invariably grayer above, especially on the crown and nape, and less yellowish on the abdomen, crissum, under tail coverts, and flanks. The grayish on the nape is often so pronounced as to form an obscure but noticeable band or collar. In autumnal plumage the abdomen, flanks, crissum, and under tail coverts are primose yellow, the back faintly tinged with olive, the light edging of the secondaries and wing coverts slightly olivaceous; otherwise this plumage does not differ materially from that of spring."

He says further: "Its favorite haunts are arid, cactus-grown plains in the low country near the coast, but it also frequents thickets, where they are to be found."

I have seen no nests or eggs of this flycatcher and can find nothing in print to indicate that it differs materially in any of its habits from the well-known northern race of the species.

The measurements of 12 eggs average 22.99 by 17.12 millimeters; the eggs showing the four extremes measure **24.3** by 17.0, 22.8 by **17.6, 21.3** by 17.2, and 22.6 by **16.7** millimeters.

MYIARCHUS TUBERCULIFER OLIVASCENS Ridgway

OLIVACEOUS FLYCATCHER

HABITS

This is a small, pale race of Lawrence's flycatcher and is the smallest member of the genus found within our borders, not much larger than our common eastern phoebe. Its range extends from southern Arizona southward through western Mexico to Oaxaca. Several other races are found in other parts of Mexico and Central America. When first found by Frank Stephens in the Santa Rita Mountains of Arizona, it was supposed to be Lawrence's flycatcher, the type race of the species, but it proved to be a decidedly smaller form, with grayer upper parts, the rectrices only slightly, if at all, edged with pale cinnamon, and generally paler.

The olivaceous flycatcher is now known to be a fairly common summer resident of the Upper Sonoran Zone among the various mountain ranges of southern Arizona and New Mexico, especially on the oak-covered hillsides. As to its haunts in the Huachuca Mountains, Harry S. Swarth (1904) writes:

Though during the summer months the Olivaceous Flycatcher is found in considerable numbers through the lower parts of the mountains; still from its retiring habits, its mournful, long drawn, note is heard far more often than the bird itself is seen. Seldom venturing into open ground, it loves the dense, impenetrable scrub oak thickets of the hillsides better than any other place, though also found along the canyon streams wherever the trees grow thick enough to prevent the sun from penetrating. It seldom ascends the mountains to any great height, 7500 feet being about the upward limit of the species, and it is most abundant below 6000 feet. They breed down quite to the mouths of the canyons, and on one occasion during the migration I secured one in a wash over a mile from the mountains. This, however, is quite exceptional. These flycatchers begin to arrive early in April, the first noted being on April 6, but it is a week or ten days later before they are at all abundant.

Nesting.—Mr. Swarth (1904) says that "they seem to disappear during the breeding season, and though really very abundant, their plaintive note, heard occasionally from some dense thicket is almost the only evidence that the birds are still around. Consequently not a great deal is known of their breeding habits. All the nests I have seen, some six or eight, all told, were built at a considerable distance from the ground, from twenty to fifty feet. They seem to breed rather late, as Mr. Howard secured a set on June 17, 1902, and on July 25 I shot a young bird which had only just left the nest."

O. W. Howard (1899), who has probably had more experience with the nesting habits of this flycatcher than anyone else, says that although the birds are numerous, the nests are very hard to find; in four season's collecting he was unable to take an egg. One nest with young was placed in a natural cavity in an ash tree about 20 feet from the ground. Another nest was placed only 15 feet up in a dead oak stump in a deserted woodpecker's hole; this was deserted before the set was complete. "The nest was composed almost entirely of rabbit's fur with a few tail and wing feathers of jays sticking upright around the outer edge." He mentions a nest found by F. C. Willard in a natural cavity in a sycamore 40 feet from the ground.

Evidently Mr. Howard was more successful later on, for there are two of his sets, with the nests, in the Thayer collection in Cambridge; these were collected in the Huachuca Mountains on June 13, 1901, and June 8, 1902. The first nest was only 4 feet from the ground in a natural cavity in an oak stump where the tree had been broken off; it is a small nest made of fine grasses and lined with still finer grasses and hairs. The other one was in a deserted woodpecker's hole in a dead upright limb of an oak, 20 feet from the ground; it consists of a bulky mass of plant rubbish, straws, grasses, weed stems and tops, dead leaves, strips of bark, cow's hair, fur, and feathers; in the center it is lined with a felted mass of rabbit's fur.

Eggs.—Four or five eggs constitute the usual set for the olivaceous flycatcher. The eggs are usually ovate and only slightly glossy. The ground color is pale "cream color" or creamy white, and the markings, in various shades of browns, purples, and pale drabs, are much like those on the eggs of the ash-throated flycatcher but are rather finer. They are the smallest and, on the average, the most lightly marked of any of the eggs of the North American species of this group. The measurements of 43 eggs average 19.6 by 15.2 millimeters; the eggs showing the four extremes measure **21.6** by 16.2, 21.0 by **16.4, 17.6** by 14.4, 18.2 by **14.2** millimeters.

Plumages.—I have not seen any very young birds, but what immature birds I have examined are browner on the upper parts than adults and the under parts are paler; the abdominal region, which is light yellow, "primrose yellow" in the adult, is almost white or faintly tinged with yellow; the greater and median wing coverts, secondaries and tertials are broadly edged with "cinnamon"; and the rectrices are much more broadly edged on both webs with "cinnamon-rufous" than in the adult. I have not been able to trace the postjuvenal molt. June and July adults that I have seen are in much worn plumage; and October adults all seem to be in fresh plumage. I have seen adults in full molt as early as August 9, and others molting wings and tails as late as September 20; evidently the complete annual molt occurs in August and September.

Food.—Very little has been recorded on the food of this flycatcher, but Cottam and Knappen (1939) have reported on the contents of three stomachs taken in Arizona in June and July. "The evidence seems to indicate that any moving insect of small size is acceptable, one bird having ingested some twenty kinds of insects." The list includes a grasshopper, termites, an ant-lion, mayflies, treehoppers, miscellaneous bugs, including assassin bugs, leafhoppers, spittle bugs, a squash bug, beetles, including wood-borers and weevils, snipe flies and other Diptera, moths, bees and wasps, and spiders. Diptera formed the principal item in each stomach, amounting to 30.66 percent of the total food; Lepidoptera came next, 16 percent (eight individuals, seven adults, and one larva comprised 27 percent of one meal); treehoppers made up 14 percent; and bees and wasps figured 9 percent.

We found the olivaceous flycatcher fairly common in the upper parts of the canyons in the Huachuca and Chiricahua Mountains, but it is such a shy, retiring species, oftener heard than seen, that we learned practically nothing about its habits. Nothing of consequence seems to have been published about its behavior. Mr. Swarth (1904) says: "They begin to leave [the Huachuca Mountains] as soon as the young have attained their growth, being about the first of the summer residents to move south. Their numbers decrease rapidly after the end of July, and by the middle of August there were practically none left in the mountains. I saw no more, and supposed that they had all left, until September 3, when I came onto a pair of the birds feeding several young. This was right at a place where Mr. Howard had secured a set of eggs earlier in the season, and I have no doubt that, as neither of the parent birds were shot, they reared another brood and were correspondingly delayed in leaving."

DISTRIBUTION

Range.—Southern Arizona and western Mexico.

The olivaceous flycatcher is found **north** to southern Arizona (Santa Rita Mountains, Tucson, Santa Catalina Mountains, and Paradise); and northwestern Chihuahua (Cajon Bonita). From this region it ranges **south** through Sonora (Saric, Guaymas, and Agiabampo); western Durango (Rio Sestin); and Sinaloa (Escuinapa); to Guerrero (Coyuca and Acapulco). Over most of this range it is resident but it apparently withdraws from Arizona during the winter season. At this time it is found **north** to central Sonora (Oposura).

To the east and south of the range as outlined several other races have been described, and the species itself is found south at least to Ecuador and Colombia.

Migration.—In Arizona it has been recorded as arriving at Paradise on April 18, at Tombstone on April 21, and at Pinery Cañon in the Chiricahua Mountains, on April 29. A late date of fall departure from Tombstone is September 20.

Casual records.—Two specimens were taken on the west side of the San Luis Mountains, southwestern New Mexico, on July 13 and 17, 1892, and on May 11, 1883, one was collected at Fort Lyon, Colo.

Egg dates.—Arizona: 14 records, May 17 to June 23; 8 records, May 29 to June 13, indicating the height of the season.

Mexico: 8 records, April 6 to May 25.

SAYORNIS PHOEBE (Latham)

EASTERN PHOEBE

PATES 14–16

HABITS

CONTRIBUTED BY WINSOR MARRETT TYLER

Spring.—The phoebe arrives in New England from its winter quarters in the Southern States and Mexico about the time of the spring equinox, or a little later in a backward season, when most of the harsh weather is behind us. He comes in full song, flitting alone about his breeding ground. On his northward migration the bird follows so closely the awakening of insect life that we may look for him as soon as we see the little insects beginning to fly about in warm, sheltered corners or around the outbuildings of a farm, and often we find him already here.

The phoebe is a gentle little bird, dull in plumage with scarcely a field mark. He is light and easy on the wing, making swift, adroit turns and twists and sudden tumblings, something like the graceful silent dodging of a butterfly, and when he alights airily on his perch, his tail keeps swaying loosely, almost as if blown by the wind. He is tame, yet reserved in manner. Althea R. Sherman says of him in her notes, "Phoebe is exceptionally correct in his behavior—without a fault."

He will stay with us for six months or more. He and his mate may settle in a busy, noisy farmyard, or perhaps far away in some remote, rocky glen, but wherever they nest they will spend a peaceful summer, giving little heed to their neighbors, seemingly happy, contented, and self-sufficient, devoting themselves to the care of their family.

Courtship.—The courtship of the male consists merely in following the female about on the wing, singing and calling to her, but with no posturing or display, unless we consider the flight song a form of display. The singing of the male, which is incessant on his arrival, becomes less frequent as the birds pair off. I have watched a pair for an hour late in April without hearing a song. On such occasions, however, the male is attentive to the female; he makes long flights after her as she moves about and hovers near her in the air, and he sometimes utters a softly whistled note, *too-lit*, suggestive of a horned lark's flight call.

Miss Sherman (MS.) speaks of a bird "calling unavailingly from the roof of an old barn all summer" and of "a wooer, unsuccessful for a month, perhaps more, which tried to coax a female to an old nest. He went to the nest, calling *phoebe* and giving a peculiar rattling note, and when he succeeded in getting the female to come to the nest, he changed to a lower, softer sound. One day both sat together on the edge of the nest and when she slipped into it he uttered a rasping twitter."

Nesting.—The flycatchers as a group build their nests in a variety of situations—on high branches, in holes in trees, or on the ground—but each species holds fairly closely to its customary site except when circumstances force a departure from their usual habit—for example, when the kingbird breeds in treeless regions. The phoebes, however, exercise a wider range of choice in selecting their nesting site. This is notably true in the case of the eastern phoebe, doubtless because of its intimate association with man.

Originally phoebes built chiefly in rocky ravines, where their nests often rested on a firm support and were partially sheltered from above, but the birds quickly adopted man-made structures, such as barn cellars and sheds, and perhaps oftenest of all they build their nests, little mounds of green moss and mud, under bridges and trestles, which afford ideal protection. Indeed, our bird is often called the bridge pewee. Mr. Bent (MS.) cites a case of a pair of phoebes, or their successors, that built under the same bridge for 30 years. "Formerly," he says, "the trolley cars rumbled over this bridge twice an hour all day, but the birds were not disturbed, or at least not discouraged, by the noise."

The following list shows a diversity of nesting sites and indicates how great an influence man has had on the breeding habits of the phoebe: Arthur C. Bent (MS.), "around the socket of an electric lamp, partly supported by the wire"; Rev. J. J. Murray (MS.), "under overhanging roots at the top of a roadside bank"; Laidlaw Williams

(MS.), "nest built upon a hook hung within a well"; A. D. DuBois (MS.), "under a bridge, stuck to vertical face of concrete"; Albert W. Honywill, Jr. (1911), in a woodpile; George Miksch Sutton (1936) "on a strip of wallpaper that sagged from the ceiling"; E. D. Nauman (1924), 5 feet down in a well, which was planked over except for a 12-inch opening, 10 feet above the water; E. C. Hoffman (1930), in an abandoned farmhouse to which the only entrance was a 2-by-4 inch opening in a broken windowpane (one young bird left the nest, flew to the window and escaped); Emerson A. Stoner (1922), in an air shaft of a coal mine, 7 feet below the surface of the ground. Lester W. Smith writes us of a nest "on a portable cider mill."

One of the likeliest places to look for a phoebe's nest is in the sheds on a farm, the old back sheds where abandoned carriages have been backed in, and the farm machines—rakes and hay-cutters—stand waiting for the harvest. Here, on the inside beams, sometimes within reach from the ground, we often find a number of nests, all but one more or less dilapidated, and some mere relics of nests built long ago. Generally they rest on a beam, although rarely they may be stuck onto the wall. I remember one nest that was attached to a beam slanting at a 45° angle, one side built up twice as high as the other so that the top of the nest was level. Another common situation is on the porch of an abandoned farmhouse, over the door, perhaps, on the narrow wooden ledge. But even in inhabited houses phoebes sometimes build in a retired back piazza, and in a few cases the presence of people passing in and out all day has not driven them away. It is not uncommon, however, even at the present time, to find a nest placed in a niche on a rocky cliff far from any house.

A. Dawes DuBois (MS.) describes a typical nest as "composed of mud, dry grass, weed and grape-vine fibers; lined with finer fibers and hair; and covered outside with moss." Of the moss, a constant component of phoebe's nests, Dr. Samuel S. Dickey reports (MS.): "I have observed such species as *Mnium stellaria, Funaria* sp., *Polytrichum* sp., *Hypnum cristatum,* and *H. dendroides.*"

Knight (1908) states: "Nest building requires about thirteen days, though I have known exceptionally of a nest being built in seven days." Bendire (1895) says that phoebes' nests "vary considerably in shape as well as in the manner of construction. If attached to the side of an overhanging rock, it is necessarily semicircular, and mainly composed of mud pellets mixed with moss, a little grass, and occasionally a few feathers, somewhat resembling the nest of our well-known Barn Swallow. If placed on a flat beam, or rafter, or on top of a post, it is circular, and sometimes but little or no mud is used in its construction." He gives the dimensions of a nest, built "underneath the roots of a partly overturned tree," as

"4½ inches in outer diameter by 4 inches in height, the inner cup being 2½ inches across by 1¾ inches in depth."

Occasionally phoebes repair their nest for use during a second breeding period (Clinton G. Abbott, 1922), but often they build a new nest, making the change, doubtless, to avoid the parasites that usually overrun their nests. In rare cases they superimpose their second nest on the old one. Wilbur F. Smith (1905) tells of a nest five stories high, measuring 9 inches, and Richard C. Harlow (1912) reports: "Near Pine Grove [Pennsylvania] in an old ore furnace, a nest of Phoebe was found with six distinct stories." Bendire (1895) says: "Occasionally they build a new nest on the top of the old one, and this is sometimes done to get rid of Cowbirds' eggs that may have been deposited by these intruders, but ordinarily they do not appear to object much to such additions, and care for them as faithfully as if they were their own."

Frederick C. Lincoln (1926) records an instance in which man proved to be an unfavorable factor to a pair of breeding phoebes. He says:

The nest was located but five feet from an electric light that apparently was frequently burned to a late hour. * * * The electric light naturally attracted many night-flying moths, which the adult Phoebe would catch throughout the evening to feed her single offspring. This bird soon died and the second set of eggs was laid. All five of these were successfully hatched and the same procedure was again followed. The young were kept literally stuffed with moths, the parents frequently continuing feeding as late as midnight. All of these young died when they were about half fledged.

It is possible that the diet of moths alone may have been wholly or mainly responsible, but it seems more probable that the continuous feeding had the effect of upsetting the normal daily digestion, with fatal results.

Joseph Janiec (MS.) reports that in a nest he had under observation five eggs were deposited in five successive days.

Eggs.—[AUTHOR'S NOTE: The eastern phoebe may lay anywhere from three to seven or even eight eggs, but five eggs form the usual set and the extremes are rare. The eggs are usually ovate, and they have very little or no gloss. The color is pure white, and usually all the eggs in the set are immaculate, but often one or more eggs in a set are sparingly marked, chiefly about the larger end, with small spots of dark or light brown. The measurements of 50 eggs average 19.0 by 14.7 millimeters; the eggs showing the four extremes measure **20.6** by 14.7, 20.1 by **15.2**, **16.5** by 14.2, and 17.0 by **13.2** millimeters.]

Young.—Althea R. Sherman, who has given close study to the nesting of the phoebe, says in her notes: "Sixteen days is the average time of the incubation period." She found in a series of 10 nests that the period averaged slightly longer in the first brood than in the second—16⅗ and 15⅔ days, respectively. She continues: "The incubation periods of the phoebe vary more than do those of any other species that I have studied closely. Not only do the first nestings average a little longer incubation period—about one day—but the time the young stay in the nest is longer by one day than in the case of the second nesting. The female alone incubates." Frank L. Burns (1921) gives the period of nestling life as 15 to 16 days. During this time both parents bring food to the young birds.

The nestling phoebe has no difficulty in leaving the nest when it is normally placed, but, as we have seen above, the nest may be built below the ground (in a well or shaft), under a bridge across a stream, or in an inclosure such as a room in a house, from which it is difficult for a fledgling to reach the open air. In such cases the little birds find themselves in a precarious situation; they may be forced to fly upward or to some distance to avoid a stretch of running water, or to direct their flight through a tiny opening. Nevertheless, they appear to extricate themselves generally without mishap, although they spend little more time in the nest than young wood pewees do that need only to walk out on the level branch by the nest or to flutter to a branch below.

Fledgling phoebes are pretty little birds, prettier when they leave the nest than they will ever be in their lives again. They are not dingy like their parents; their backs are pale olive, and their wings are crossed by two distinct buffy bars. They are alert and active even on the day they leave the nest and flit to a branch and perch without awkwardness, twitching their short tails ably with the downward sweep characteristic of the adults. They utter frequently a little peeping note—*tereep* or *trree*—feebler, longer and less sharply pronounced than the *chip* of the old birds.

Dr. Dayton Stoner (1939a) made a detailed study of the development of young eastern phoebes and found the incubation period to be 16 days; the young, he says, reach "near-adult size within a period of 17 days."

Plumages.—[AUTHOR'S NOTE: The natal down of the young phoebe is "mouse gray" or "light drab," which only scantily covers the feather tracts. When the bird is four or five days of age the flight feathers begin to grow, followed soon by the contour plumage, with the natal down adhering to the tips of the juvenal feathers. The juvenal plumage resembles that of the adult, but the upper parts are browner, "clove brown" on the crown and nape and "olive-brown" on the back, wings, and tail; the feathers of the lower back, rump, and upper tail coverts are broadly tipped with "cinnamon-buff"; the greater and median wing coverts and the rectrices are tipped with "cinnamon-rufous," and the secondaries and tertials are edged with yellowish white; the under parts are yellowish white, with brownish "olive-gray" on the sides of the throat and breast.

A partial postjuvenal molt begins about the middle of August, involving all but the primaries, secondaries, and rectrices and producing the first winter plumage, which is much like that of the fall adult. The young bird in this plumage is greener above and yellower below, "primrose yellow," than in the juvenal plumage; the wing coverts are now narrowly edged with yellowish white. This plumage is apparently worn all through the winter and spring, with no change except by wear and fading.

The postnuptial molt of both young and old birds occurs in August and September, before the birds migrate, and is complete. Fall and winter birds are more olive above and yellower below, "primrose yellow," than spring birds, and the light wing edgings are tinged with yellow; all these bright colors fade or wear away before spring. The sexes are alike in all plumages.]

Food.—F. E. L. Beal (1912), in a comprehensive examination of 370 stomachs found the food of the phoebe "to consist of 89.23 per cent of animal matter to 10.77 of vegetable. The animal portion is composed of insects, with some spiders and myriapods, a gordius, and one bone of a tree frog. The vegetable part is made up of small fruits or berries, with a few seeds, all of them probably of wild species." The more important results of Beal's investigation are quoted below:

Useful beetles, consisting of tiger beetles (Cicindelidae), predaceous ground beetles (Carabidae), and ladybirds (Coccinellidae), amount to 2.68 per cent. Other beetles, belonging to 21 families that were identified, make up 12.65 per cent. They appear to be eaten very regularly in every month, but the most are taken in spring and early summer. May is the month of maximum consumption with 23.67 per cent. Beetles altogether amount to 15.33 per cent, which places them as second in rank of the items of animal food. The notorious cotton-boll weevil (*Anthonomus grandis*) was found in 6 stomachs taken in the cotton fields of Texas and Louisiana * * *

In the phoebe's diet Hymenoptera stand at the head, as is the case with most of the flycatchers. They are eaten with great regularity and are the largest item in nearly every month. A few of them are the useful parasitic species, which are, however, offset by quite a number of sawfly larvae, which are very harmful insects. * * *

The maximum amount of Hymenoptera was taken in August, when they aggregated 39.66 per cent. They constituted the entire contents of 7 stomachs, and were found altogether in 225, which would seem to establish these insects as the favorite food of the phoebe. In bulk they amount to 26.69 per cent of the yearly diet.

Diptera aggregate 6.89 per cent, and are a very constant, though small element of the food. * * * Hemiptera (bugs) seem to be sought for rather more than flies, as they were found in 151 stomachs, but only one was entirely filled with them. Very curiously these were leaf hoppers (Jassidae), lively little creatures that live on grass and leaves and jump like fleas. * * *

Orthoptera (grasshoppers and crickets) form 12.91 per cent of the phoebe's food. * * *

Lepidoptera (moths and caterpillars) are eaten much more regularly than grasshoppers, but not in such large quantities. They amount to 8.86 per cent of the food of the year. * * *

Spiders constitute quite a steady article of the phoebe's diet. Ticks and millepeds also are eaten. None of these creatures can be taken when they are on the wing, as they can not fly, but spiders may sometimes be picked up when they are sailing through the air upheld by their gossamer threads or they may be found on the top of a tall reed as the bird flies past. But ticks and millepeds must be taken from the ground or some other surface. The aggregate of these creatures for the year is 5.78 per cent. * * *

The vegetable food of the phoebe may be placed in two categories, fruit and seeds. Fruit amounts to 4.99 per cent. * * * Of small wild berries 17 species were identified, besides a number of seeds, but nothing of any economic value was found. * * * The great bulk of the vegetable food was taken in the fall, winter, and early spring months. * * *

Among the stomachs examined were those of four newly hatched nestlings, which merit passing notice. The stomachs contained no vegetable matter whatever, but were completely filled with insects and spiders.

Professor Beal concludes his examination of the phoebe's food with the following pleasant summary: "It seems hardly necessary to say anything in favor of a bird already firmly established in the affections of the people, but it may not be amiss to point out that this good will rests on a solid foundation of scientific truth. In the animal food of the phoebe there is such a small percentage of useful elements that they may be safely overlooked; while of the vegetable food it may be said that the products of husbandry are conspicuous by their absence. Let the phoebe remain just where it is. Let it occupy the orchard, the garden, the dooryard, and build its nest in the barn, the carriage house, or the shed. It pays ample rent for its accommodations."

B. H. Belknap (1938) reports that phoebes were very active in destroying tent caterpillars in the moth stage during an infestation by these insects at Albany, N. Y. He says: "In view of the fact that the female moth is a little larger and somewhat more showy in flight than the male, the object of the chase was more than likely to be a female. In any event, the dozens of tent caterpillar moths devoured daily represented literally thousands of tent caterpillar eggs not laid."

The phoebe captures most of its food while both bird and insect are on the wing, as we might infer from the preponderance of flying insects composing its diet. As we watch a bird fly out from its perch in pursuit of an insect—sometimes such a small one that we ourselves cannot see it—the phoebe impresses us with its lightness and agility in the air, although guiding itself with admirable precision. Its flight is a soft, butterflylike fluttering, with abrupt, short turns—quick as a flash. We hear the click of the bill as it closes on the insect, sometimes high in the air, sometimes low down. When feeding over calm water, the bird almost meets its reflection—two birds that almost touch, then fly apart again without breaking the surface, or sometimes they do touch, leaving a little ripple.

We often find phoebes near water. We may meet a pair of them quietly flitting about a pond, keeping near together as they feed. One may toss itself high in the air toward an insect, floating in an upward dive, and then drift down again, its tail high, as if blown upward by the bird's descent. All the motions are graceful and airy with little apparent force. Again, a bird may hover deftly before a branch, holding itself upright and stationary, balancing with waving wings, or one may make a wide sweep through the air and snap up a dragonfly, or stoop above a shrub and pluck off a berry in passing.

As we see the phoebe in the field, the dragonfly appears to be a more prominent article of food than Professor Beal's examinations of stomachs indicate. Phoebes eat medium-sized dragonflies frequently and feed them to their young.

Behavior.—The phoebe has lived so long and so familiarly in our farmyards that we have come to look on it, not as a wild bird, but as a member of the happy community that makes up rural life—the pigs in their sty, the hens in their coops, the horses and cows in the barn, and the phoebe in the back shed. Busy all day catching insects, unobtrusive, never noisy, it is popular with the farmers. They all know the phoebe as, over and over, it calls out its name. It is a pleasanter neighbor than the robin, for it does not burst into a distressing panic whenever you come near its nest. It has rather the nature of the chipping sparrow, another member of the family on a farm. Both birds have a quiet reserve combined with a capacity for hard work, not unlike the New England farmer himself.

Audubon (1840) held the phoebe in high esteem. He spent much time studying a pair of birds that bred in "a small cave scooped out of the solid rock by the hand of nature." He goes on: "Several days in succession I went to the spot, and saw with pleasure that as my visits increased in frequency, the birds became more familiarized to me, and, before a week had elapsed, the Pewees and myself were quite on terms of intimacy." Audubon studied this pair throughout their first breeding period, and to prove true his supposition that the young birds returned to their birthplace, he became America's first bird-bander. He continues: "When they were about to leave the nest, I fixed a light silver thread to the leg of each, loose enough not to hurt the part, but so fastened that no exertions of theirs could remove it. * * *

"At the season when the Pewee returns to Pennsylvania, I had the satisfaction to observe those of the cave in and about it. There again, in the very same nest, two broods were raised. I found several Pewees' nests at some distance up the creek, particularly under a bridge, and several others in the adjoining meadows, attached to the inner part of sheds erected for the protection of hay and grain. Having caught several of these birds on the nest, I had the pleasure of finding that two of them had the little ring on the leg."

The following two quotations illustrate the tameness of the phoebe, or at least its disregard for the presence of people. By such behavior, as well as by the choice of nesting sites, the bird shows an adaptability to man-made changes in its environment.

Clinton G. Abbott (1922) tells the story of a pair of phoebes that had built their nest on the veranda of his summer home before he and his family moved into it for the season on May 15. At this time the nest contained five eggs. The female bird was alarmed at first, but "within a week," he says, "she had succeeded in completely readjusting herself to the new conditions. From her original shy and timid self, she was metamorphosed into quite a different type of bird, stolidly remaining seated upon her nest regardless of sudden noises or the movements of people. * * * Persons—even whole tea parties—were ignored, except that once or twice we thought we detected a tone of annoyance in the Phoebe's voice upon finding a favorite chair occupied!"

The following remarkable incident is related by H. H. Brimley (1934):

On November 27 I was on a deer stand in Onslow County, N. C. The air was rather warm for the time of year and mosquitoes were quite noticeably in evidence, though not particularly aggressive. I was in a standing position with my rifle under my arm, the barrel pointing downwrd, and I had my hands clasped in front. A faint fluttering of wings caused me to look down, and I saw a Phoebe (*Sayornis phoebe*), a bird frequently known by us as Winter Pewee, trying to alight on my rifle barrel. Failing to secure a firm grip on the smooth surface of the metal, the bird slid down the barrel until the front sight was reached, where it secured the grip desired, and there it perched.

It showed no sign of fear or nervousness and in a few seconds flew up and picked a mosquito off my hands, which were not more than a foot distant from its perch. Then, it picked others off the front of my coat, off my sleeeves, and several more off my hands, meanwhile perching indiscriminately on my hands, sleeves, and gun barrel, though seeming to prefer the last.

Finally, the Phoebe discovered that my face seemed to be attracting more mosquitoes than any other part of my person so he transferred his attention to that part of my anatomy, and found a new perching place on the top of my hunting cap.

In picking mosquitoes off my face, the sharp points of the bird's bill were noticeably felt at every capture, and it was the irritation caused by a succession of these pricks that finally caused me to dispense with its attentions. * * *

When I decided to end the incident, I found a difficulty in doing so. I had presumed that any decided movement on my part would drive my little friend away, but this bird was not of the scary kind. * * * He continued to perch on my head and pick mosquitoes off my face even after I had started to move around in an effort to discourage his attentions. But my face was beginning to feel somewhat inflamed from the frequent pecking to which it had been subjected, so I called it a day and told the Phoebe to stop pestering me.

Frederic H. Kennard says in his manuscript notes under date of March 31, 1908: "A phoebe flew through our open bedroom window this morning while we were at breakfast and made himself at home catching flies and perching on the back of the rocking chair."

William Brewster (1936) describes thus the phoebe's method of bathing:

June 17, 1905. About noon today I saw a Phoebe bathing in a small pond in the Berry Pasture. It flew from a dead branch about fifteen feet from the pond and eight feet above the ground, striking the surface with its *breast* and with such force as to make a rather loud sound as well as to send heavy ripples rolling to my part of the pond. This action was repeated three times at short intervals.

There can be no question that the bird was bathing and not picking up floating insects, for each time it returned to its perch I could see that it was dripping wet. After freeing its plumage from most of the water by a vigorous shake or two, it would preen its feathers for a few moments and then take another dip.

Bendire (1895) says: "While generally of an amiable disposition toward other birds, often nesting in close proximity to the Barn Swallow, Robin, and Chimney Swift, it will not allow any of its own kind to occupy a site close to its own, fighting them persistently until driven off, and should one of the earlier arrivals presume to appropriate its old nest, war is at once declared."

The most characteristic habit of the phoebe—one by which it may be recognized at a glance—is its manner of moving its tail. The bird no sooner alights than its tail begins to sway, first a downward sweep, then a recoil, which often carries the tail above the starting point. The swings are wide and are often made toward one side or the other, giving a wagging effect. The motions, repeated several times, are rather slow, not at all like the nervous twitchings of many birds' tails.

Voice.—The voice of the phoebe is distinctive as he sings, pronouncing *phoebe*, or better perhaps *wheepy*. There is in the voice an aspirate quality roughened a little by a rolling 'r' sound, very different from the voices of the other genera of flycatchers. There is none of the pure musical tone of the wood pewee, none of the rasp of the alder flycatcher, and no hint of the arrogant shout of the greatcrest. The phoebe's song is uttered emphatically to be sure, and it is sharply accented at the start, but it is never loud: it is, in fact, only a force-

ful, whistled whisper. It may either rise or, more commonly, fall in
pitch at the end, and the second syllable is often doubled—*wheep-
pi-pi*—a little vibratory quality running through it all. The first
syllable, sweeping upward in pitch, receives the accent. The bird
sings throughout the summer and well into the autumn.

Often in spring the phoebe gives his song from the air. He
launches out from his perch repeating his chip note rapidly several
times, and flies about in a seemingly distracted manner while he utters
his song over and over, extending it sometimes with a long series of
be-be-bes. Dr. J. J. Murray (MS.) writes: "I have several times
heard what might be called the flight song of the phoebe, always late
in March, at the time when nesting is just beginning. On one such
occasion the bird flew almost straight up into the air for about 50
feet; then, with tail fully spread and wings fluttering, circled and
dived, all the while uttering a series of quick, sharp whistled notes
resembling its ordinary call; and finally returned to a perch near
where the flight began."

Early in the morning before daylight, a time that some of the flycatchers devote to a special song heard only in the twilight, the
phoebe redoubles his singing; song follows song with scarcely a pause
between them, and when two or three are within hearing, a perfect
chorus of singing ensues. Horace W. Wright (1912) says of the early
morning singing: "The song is usually continued without much pause
for an hour or more."

Albert R. Brand (1938) finds that the approximate mean average
pitch of the phoebe's song is at the rate of 4,300 vibrations a second,
about the same as the eastern meadowlark.

Of the minor notes, the commonest is a sharp, clearly cut *chip*, resembling a call note of the swamp sparrow. Francis H. Allen (MS.)
says of it: "The ordinary *chip* note of the species is sometimes
followed by a sort of echo on a lower pitch and purer in tone, as
chip-pi. The *pi* is not nearly so loud and emphatic as the *chip;* it
suggests a mechanical after-effect not deliberately sounded." The
chip note is used as a simple exclamation, or as a note of concern when
we come near the nest.

I have also heard a quarrelsome *tree-tree-tree-tree*, given in a rapid
series; a soft note, *tree-oo*, falling in pitch; and a rather musical
whistled *treet*, sometimes doubled to *sereet*.

Olive Thorne Miller (1892) says: "Whenever a phoebe alighted on
the fence he made a low but distinct remark that sounded marvelously
like '*cheese-it*.'"

Field marks.—The phoebe is a flycatcher of midsize with a dark
bill, practically no wing bars (except in juvenal plumage), and no
eye ring. Olive Thorne Miller (1892) speaks of him as "the loneliest

of its kind * * * whose big chuckle-head and high shoulders gave him the look of an old man, bent with age."

Ralph Hoffmann (1904) says: "The *sideways sweep of the tail* is a characteristic action by which the bird may always be identified; in the old birds the absence of wing bars also serves to distinguish it from the Wood Pewee. Young birds have dull wing bars, but they cannot refrain long from making a suggestive movement of the loose-hung tail."

Enemies.—Perhaps the most serious of the phoebe's enemies are the parasites that often infest the nests and debilitate or kill the young birds. Manley B. Townsend (1926) speaks of a nest "containing four newly hatched young." "A week later," he says, "on examining the nest, I found only the desiccated bodies of the young birds. The nest was swarming with parasitic insects." Lewis O. Shelley (1936a) adds his testimony on this subject: "The first nestings are invariably pretty free from parasitic pests, but second nestings may be literally overrun with mites and possible third broods will often be forced prematurely into leaving the nest. I am of the opinion that mites invariably prevent Phoebes from raising a third brood."

Harold S. Peters (1933) found the mite *Liponyssus sylviarum* in the plumage of a phoebe sent to him by P. A. Stewart from Ohio.

Frederic H. Kennard (MS.) adds the raccoon to the phoebe's enemies. He says: "May 2, 1925. A raccoon broke up our phoebe's nest on the post of our woodshed last night. Eggs and nest lay several feet from the bottom of the post this morning. I had always supposed a screech owl was guilty in past years, but on making a close examination today, I found claw marks and a hair from a coon's belly stuck to the bark of the post."

William Brewster (1936) describes thus a dramatic incident in the life of a phoebe:

A male Pigeon Hawk suddenly appeared from we hardly knew whither and with the speed of an arrow glided on set wings, on a slightly declining plane, directly at the Phoebe.

That trustful little bird, swaying at ease on his slender perch, seemed so wholly unconscious of his fearful peril that we all thought him lost, but when the Falcon was within a foot of him he did the only thing that could possibly have saved him, viz. dropped like a ripe fruit nearly to the ground and then started directly for the barn cellar. The Hawk overshot him scarce more than four feet and, stopping and turning about with truly marvelous quickness, followed and overtook him before he had gone three yards but the Phoebe doubled short and abruptly and the little Falcon, apparently disgusted at his ill success, darted off down the hill-side towards the eastward, giving us a fine view of his ashy-blue back. Only a few minutes later the Phoebe was back on the same perch again. The whole episode was most impressive—happening as it did, at what might be called the very threshold of the Phoebe's home and during a rarely beautiful and peaceful May afternoon.

George Nelson tells me that at his home in Lexington, Mass., house wrens sometimes interfere with the breeding of his phoebes by flinging the newly hatched birds from the nest. The phoebe is "one of the very commonest foster parents of the young Cowbird. In regions where both species are common, fully 75 percent of the nests contain eggs of both kinds," according to Friedmann (1929).

Fall.—In the latitude of Boston, Mass., we see the phoebe well into October, the time when the killing frosts come. It is a characteristic bird of the late fall migration, when the sparrows are passing through and the hermit thrushes are arriving from the north. Marked by its usual quiet demeanor, we meet it along the sunny southern borders of woodlands and meadows. It is nearly silent—although on warm days it may sing even now—and almost always alone.

Manly Hardy (1885) tells of a phoebe lingering late in the season at Brewer, Maine. "On Nov. 23 (1884)," he says, "when the snow here was six inches deep, and the Penobscot River frozen over above the dam, a Phoebe came into my garden and remained a long time. As it was Sunday I did not shoot him, but there is no doubt as to his identity, for my daughter and I stood within a few feet of him and watched him catch insects over a smoking manure heap."

Winter.—Bent and Copeland (1927) found the phoebe "a common winter resident about villages and in hammocks" near St. Petersburg, Fla. I have met the bird many times on the east coast of Florida during February and March. At this season it is a silent bird, rarely singing, giving only its *chip* note occasionally. Bradford Torrey (1904), speaking of the vicinity of Miami, Fla., says: "Phoebes have sung much less of late than they did in January. Then they seemed to find existence a perpetual jubilee."

Maurice Brooks writes to Mr. Bent: "Prof. E. R. Grose, State Teachers College, Glenville, W Va., reports that during the winter of 1930 a phoebe spent the winter on the college campus, feeding during the colder periods on the berries of Japanese ivy. This is, as far as I am aware, the only West Virginia winter record for the species."

There are, however, several records of phoebes wintering in more northern States, viz: F. Clement Scott (1934), in New Jersey, February 4 and 5, temperature 20° "with about 8 inches of snow on the ground"; John H. Tompkins (1928) at Babylon, N. Y., on February 6; and Miss Carol Jones (1922) at Bennington, Vt., on February 1.

Milton P. Skinner (1928), speaking of the winter birds of the North Carolina sandhills, says: "During the winter these birds [phoebes] are almost always solitary, seldom even two birds being near each other. Nor are they seen with any other species."

DISTRIBUTION

Range.—North America east of the Rocky Mountains.

Breeding range.—The eastern phoebe breeds **north** to southwestern Mackenzie (Old Wrigley, Lake St. Croix, and Hill Island Lake); northern Saskatchewan (Methye Portage and Sandfly Lake); Manitoba (probably Norway House and Gypsumville); southern Ontario (Indian Bay, Kenora, North Bay, Algonquin Park, and Sulphide); southern Quebec (Montreal, Hatley, and Mont Louis); Prince Edward Island (Charlottetown); and Nova Scotia (Sydney). The **eastern** boundary of the breeding range extends southward from this point along the coast to southeastern Virginia (Cobbs Island). **South** to southeastern Virginia (Cobbs Island and Richmond); North Carolina (Raleigh and Charlotte); northwestern South Carolina (Spartanburg and Seneca); northern Georgia (Atlanta); northern Alabama (Long Island and Florence); northern Mississippi (Iuka); Arkansas (Bertig, Magazine Mountain, and Rich Mountain); northeastern Oklahoma (Tulsa); and east-central New Mexico (Santa Rosa). **West** to east-central New Mexico (Santa Rosa); western Oklahoma (Kenton); Kansas (Stockton); eastern Nebraska (Red Cloud and Lincoln); eastern South Dakota (Yankton, Armour, and Dell Rapids); North Dakota (Bowman); Alberta (Morrin, Edmonton, and Battle River); and western Mackenzie (Fort Simpson, Two Island Indian Village, and Old Wrigley).

Winter range.—The winter range extends **north** to central Chihuahua (Chihuahua City); northeastern Texas (Houston and Gainesville); probably rarely southeastern Oklahoma (Caddo); northern Alabama (Leighton); occasionally eastern Tennessee (Knoxville); western North Carolina (Asheville); and casually the District of Columbia (Washington). Note: In mild winters the phoebe has been detected **north** to northwestern Arkansas (Rogers); southern Illinois (Olney); southern Ohio (Columbus); southern Vermont (Bennington); and southern New Hampshire (Milford). The **eastern** boundary of the normal winter range extends southward along the Atlantic coast from the District of Columbia (Washington) to Florida (Daytona, Miami, and Royal Palm Hammock) and rarely Cuba (Habana and Guantanamo Bay). **South** to rarely Cuba (Guantanamo Bay) and Oaxaca (Tapanatepec). **West** to Oaxaca (Tapanatepec and Cuicatlan); Morelos (Morelos); Mexico (San Antonio Coapa); Nuevo Leon (Monterrey); and Chihuahua (Chihuahua City).

Spring migration.—Early dates of spring arrival are: New Jersey—New Providence, March 8. New York—Ballston Spa, March 20. Connecticut—Hartford, March 13. Massachusetts—Beverly, March 12. Vermont—St. Johnsbury, March 22. Quebec—Montreal, April

10. New Brunswick—Scotch Lake, April 18. Missouri—St. Louis, March 3. Illinois—Chicago, March 12. Ohio—Oberlin, March 14. Michigan—Petersburg, March 10. Ontario—Strathroy, March 19. Kansas—Onaga, March 13. Iowa—Keokuk, March 12. Minnesota—Lanesboro, March 22. Alberta—Edmonton, April 19. Mackenzie—Fort Simpson, May 14.

Fall migration.—Late dates of fall departure are: Alberta—Belvedere, September 7. Manitoba—Oak Lake, October 13. Minnesota—Minneapolis, November 3. Illinois—Chicago, November 10. Iowa—central part, October 28. Ontario—Point Pelee, October 18. Ohio—Hillsboro, October 30; Toledo, November 15. Missouri—Concordia, October 30. Quebec—Montreal, October 8. New Brunswick—Scotch Lake, October 8. Massachusetts—Danvers, October 26. Rhode Island—Providence, October 27. Connecticut—Hartford, October 30. New Jersey—New Providence, November 5.

Casual records.—There are four for the phoebe in Colorado as follows: A specimen taken at Fort Lyon on April 20, 1884; one collected in Pueblo County on April 5, 1896; another in the Clear Creek Valley, between Denver and Golden, on September 17, 1911; and the last near Denver, in Adams County, on April 26, 1933. Two specimens were collected at Paradise, Ariz., on October 8, 1918, and on August 16, 1919, while a third was taken at Blue Point on October 8, 1933. One was obtained a San Fernando, Calif., on February 14, 1901, and another at Moss Beach, near Pacific Grove, on March 7, 1913. Sight records have been reported from Pullman, Camas, and Yakima, Wash., but there appears to be no specimen collected in that State.

Egg dates.—Alberta: 5 records, May 15 to June 3.

Illinois: 28 records, April 1 to June 16; 14 records, April 30 to May 26, indicating the height of the season.

Massachusetts: 34 records, April 24 to June 20; 18 records, May 8 to June 13.

Texas: 5 records, April 11 to May 12.

Virginia: 6 records, April 27 to June 10.

SAYORNIS NIGRICANS NIGRICANS (Swainson)

BLACK PHOEBE

PLATES 17, 18

HABITS

CONTRIBUTED BY ROBERT S. WOODS

In the more verdant portions of the valleys and coastal plains of the Southwestern States the attractive black-and-white figure of the black phoebe is a familiar sight. No other western flycatcher, except possibly *Empidonax trailli*, is so partial to the vicinity of water; it may confidently be looked for by the side of almost any marshy lowland pond or under the sycamores bordering the lower reaches of the mountain streams. The black phoebe is one of the most domestic of all western birds and is frequently seen about barnyards, sweeping over city lawns, or even hunting in the artificial canyons of downtown Los Angeles. Undoubtedly this is one of the species that have thrived and increased with the settlement of the Southwest. Peaceable and unobtrusive, free from annoying habits, and eminently beneficial in its diet, the black phoebe is one of our most valuable birds.

This species and the vermilion flycatcher are the only members of their family that may be considered substantially nonmigratory throughout most of their ranges within the United States. Their seasonal movements appear to be more in response to local conditions than to any general migratory urge. In many parts of southwestern California the black phoebe is the one resident flycatcher, Say's phoebe and Cassin's kingbird occurring mainly as winter visitants, and the remaining species as summer visitants or migrants. The black phoebe is only sparingly distributed over the interior or more arid portion of its territory, because of the scarcity of its preferred types of habitat.

Although the black phoebe is for the most part a bird of the lower altitudes in California, it is reported by various observers to nest occasionally at elevations of 4,000 to 6,000 feet, and Major Bendire (1895) writes: "Dr. Edgar A. Mearns, United States Army, informs me that he found a pair breeding at the reservoir from which the town of Tombstone derives its water supply, in Millers Canyon, Huachuca Mountains, southern Arizona, on July 31, 1894. This is located in the Douglas spruce zone (*Pseudotsuya* [sic] *taxifolia*), at an altitude of about 8,000 feet." The general withdrawal of the majority of the birds from the valleys or plains into the foothill canyons in spring, as noted by Bendire in southern Arizona, is undoubtedly represented to a certain extent throughout the entire range, but only in a limited degree on the Pacific slope.

Courtship.—In keeping with its unostentatious demeanor, the courtship practices of the black phoebe are not such as to readily attract attention, but Ralph Hoffmann (1927) states that "in the mating season the male often makes a song flight, fluttering about in the air, repeating *ti-ti-ti* for a few seconds and then slowly descending."

Nesting.—With respect to the localities frequented by nesting black phoebes, the words of Grinnell and Storer (1924) may well be quoted:

Black Phoebes are not distributed locally with the regularity observed in shrubbery-inhabiting birds such as Wren-tits or Brown Towhees. The peculiar nesting requirements of the phoebes probably account for this lack of uniformity in their distribution. They must have sheltered faces of rocks or wooden

walls against which to place their nests, and these sites must be within carrying distance of some source of the mud used in nest construction. Such sites
are widely and irregularly scattered. The building of bridges over creeks and
the maintenance of stock barns with watering troughs near by have probably
increased the population of these birds in the country as a whole.

The following account of nesting sites chosen by black phoebes in
the vicinity of Fresno, Calif., together with dates of nesting, by
John G. Tyler (1913), is typical of descriptions by various observers
in both California and Arizona:

Nests of these birds are sometimes fastened to the walls of deserted cabins,
and occasionally a pair will build in an old well if they can gain entrance,
such nests being from six to fifteen feet below the surface of the ground. The
most common nesting sites, however, are the large stringers of bridges, where
the nest is securely fastened above the water. I have never known this species
to choose a place where there would be support for the bottom of the nest, as
the Eastern Phoebe is said to do. Our bird attaches its wall pocket to the vertical surface of a plank, and so securely is it fastened that it will often break
apart rather than give way. This species often nests on the faces of rocks in
the hills, but such sites are almost entirely wanting in the Fresno district. I
have found one or two nests fastened to the partly dead trunk of some large
tree, but it is safe to say that nine out of ten birds choose the protection
afforded by bridges, where mud is easily secured, and horsehairs as well, for
these two ingredients enter largely into the construction of the nest. * * *
I have found eggs nearly ready to hatch on April 5 and fresh ones June 15,
so the nesting period may be said to extend from March 1 to July 1, with
probably two broods reared in a season, in some cases at least.

Several other writers have mentioned wells as nesting sites, 4 or 5
feet below the surface. As an instance of the birds' fearlessness in
attending their nests despite disturbances, Florence A. Merriam
Bailey (1896) wrote: "April 30, 1898, I found three eggs in the nest
of a Black Phoebe five feet down in a deserted well. Before the eggs
hatched, a pump was put down the well and water pumped up every
day, but the birds did not desert the nest." Illustrating the same
characteristic, Milton S. Ray (1906) says, "About Visalia I noticed
about half a dozen nests with eggs placed in sluice boxes through
which the water coursed uncomfortably close to the mud-made domi-
cile."

John McB. Robertson (1933) describes the unusual occurrence of a
black phoebe's nest in a willow tree leaning over a watercourse near
Artesia, Calif.: "One tree, about eight inches in diameter, had a dead
limb on its lower side extending downward at a sharp angle, and on
the end of this was a typical mud nest of the Black Phoebe, containing two young birds about a week old. The parent birds were nearby. Shreds of willow bark had been used with the mud and fringed
the outside of the nest; the lining was of bark and hair. The nearly
horizontal trunk of the tree formed a shelter about six inches above
the nest which was about three feet from the water." The black

phoebe's habit of building its nest beneath some sheltering projection has been noted by ornithologists as far back as Dr. J. G. Cooper (1870), who stated that the nest is "stuck against a wall or sometimes on a shelf, beam, or ledge of rock, but always under some protecting roof, often under a bridge." Concerning the nest itself, Major Bendire (1895) says:

Mud seems to enter largely into the construction of its nests, and I believe is invariably used. These are located in similar situations to those of the two preceding species [*Sayornis phoebe* and *S. saya*]. It is equally attached to a locality once chosen for a nesting site; and instances are recorded where four clutches of eggs have been laid in one season the three previously laid having been taken. Two broods are generally reared in a year and perhaps three. The exterior of the nest consists of small pellets of mud mixed with bits of dry grass, weed fibers, or hair, and somewhat resembles that of a Barn Swallow; the outer mud wall is carried up to the rim. Inside it is lined with weed fibers, fine roots, strips of bark, grass tops, hair, wool, and occasionally feathers. If their eggs are taken, they generally lay another set within two weeks. A nest now before me, taken by Mr. H. W. Henshaw, at Santa Ysabel, California, on April 28, 1893, measures 5 inches in outer diameter and 3½ inches in height; the inner cup is 2¾ inches in diameter and 1¼ inches in depth, and is lined with plant fibers and fine grass tops.

Concerning the nesting of this species at Escondido, Calif., James B. Dixon (MS.) confirms the observations of others as to the nature of its nesting sites and also remarks that "they establish themselves and return for years to the same place. There has been a pair nesting on our old home place to my knowledge since 1900, and they are still there. Probably a good many generations have lived there during the 40 years, but there has never been a season that a pair did not show up and raise one and quite often two broods of young. Nests are of mud and lined with fine dry grass tissue or animal hair. Horsehair is often used and sometimes results tragically, as they hang themselves in the loops of horse hair woven into the mud and lining." Similarly stressing the attachment of individuals to their own particular nesting sites, F. B. Jewett (1899) writes:

My observations have been confined to one pair of birds which have nested on my barn for some eight years past. While I cannot state positively that it has been the same pair during the entire term I am led to believe that such is the case. During the first two or three years the birds changed the site of their nest frequently, probably owing to some disturbance, for afterwards when I guarded them against interference they chose a site which they have occupied ever since. * * *

Both birds assisted in the construction of the nest, one working while the other kept watch. Both also incubated, dividing the work equally, as nearly as I could judge. In most cases the eggs were laid on consecutive days, incubation commencing immediately after the laying of the last egg. * * *

The birds have used the same nest for four years, tearing out the old lining and replacing it with new at the beginning of each season and mending places that had been broken.

Contrary to Mr. Jewett's experience, George Oberlander (1939) mentions the building of two nests entirely by the females, successive mates of the same male.

Several instances have been reported in which newly built black phoebes' nests have been used by other birds. Wilson C. Hanna (1933) found two nests that had apparently been seized by house finches (*Carpodacus mexicanus frontalis*) in spite of the presence of the rightful owners, while Harold M. Holland (1923) describes the joint use and alternate occupancy of a nest by phoebes and house finches. This latter nest was later found to have been deserted by both, though containing six eggs of the phoebe and five of the finch. Emerson A. Stoner (1938) reports finding a nest containing three phoebe's eggs below, then a scanty lining of fine hairs, and above that an egg of the dwarf cowbird (*Molothrus ater obscurus*) and three of the western flycatcher (*Empidonax difficilis*).

Eggs.—Mr. Tyler (1913) remarks that "it is interesting to note that when four eggs constitute a set there are generally three that are unmarked and one that is quite heavily spotted with red dots on the larger end but when there are five in the set the additional egg nearly always has just a few very fine spots like dust. My observations show that nearly always the spotted egg is the last one to be deposited. If that is the rule, then should a set of seven or eight eggs happen to be laid we might expect one or two specimens as heavily spotted as a kingbird's egg."

The measurements of 50 eggs average 18.7 by 14.4 millimeters; the eggs showing the four extremes measure **20.3** by **15.2, 17.3** by **14.2,** and 17.8 by **13.2** millimeters.

Young.—The incubation period for two sets of five eggs each has been recorded by Mr. Oberlander (1939), counting in each case from the laying of the last egg. In one nest three eggs were hatched, the first after 17 days, the others on the following day; in the second case, three eggs were hatched on the fifteenth day and another the next day. In this second nest, the four nestlings remained about 21 days, which in the light of the following account indicates considerable variability in the length of this period.

Mr. Jewett (1899) states that the young "remained in the nest on an average about two weeks, or until it was too small for them," and that "three broods were generally reared in each year, the first and second usually consisting of five, and the last of four birds. The youngsters never remained long after they had been turned adrift, usually disappearing on the third day " Regarding the condition and care of the young, Mrs. Irene Grosvenor Wheelock (1904) says:

For some unexplained reason the nest of this species, like that of Say phoebe and the Eastern phoebe, is infested with innumerable insects, which frequently cause the death of the young. This seems strange in the case of birds that splash in the water so much as do these. One of the first lessons taught the young is the delight of a bath in an irrigation ditch; to this wholesome recreation they are initiated when about five weeks old.

* * * At first the feeding is done by regurgitation, but when five days old the nestlings are fed on fresh insects.

As soon as they are ready to fly the male takes entire care of them, leaving the patient mother to repair the old nest and undertake the bringing up of a second family. He teaches the young to catch food on the wing, just as the Arkansas and Cassin kingbirds teach theirs, and as I believe all flycatchers do,—by releasing a maimed insect in the air just in front of the hungry little one, who, forgetting fear, instinctively darts out to catch it.

The black phoebe chanced to be the object of a series of experiments by Mr. and Mrs. Eric C. Kinsey (1935), the results of which, while not necessarily applicable to this species alone, are of considerable interest to students of behavior. A brood of fully feathered young was substituted for a newly hatched brood, and the latter for a set of fresh eggs. After a few minutes of uncertainty, one of the birds, believed to be the male in each case, began feeding the young, which soon appeared to have been accepted unreservedly. Again, according to Mr. Kinsey—

We shifted young several days old into a nest containing young on the point of leaving. The latter brood was placed in the nest containing the eggs, the eggs were placed in the nest which formerly contained the young several days old. A watch was maintained, in turn, upon each nest with the following results. The partly grown young were accepted in lieu of the older group, the older group was accepted, after slight hesitation, by the adults with the eggs; however, the eggs were, so far as we could tell, never brooded by the pair from which the half grown young were taken. All of these nests were subsequently visited some three or four days later and all of the respective broods of young were being cared for by their foster parents just as though they were of their own hatching.

For the safety of its nest and young the black phoebe evidently relies solely upon the inaccessibility of its nesting site. Accordingly it does not, like birds which practice concealment, endeavor to avoid being seen near the nest, but on the contrary, may use the nearest convenient perch as a base for its hunting activities.

Plumages.—[AUTHOR'S NOTE: The sexes are alike in all plumages. Young birds in juvenal plumage, in June and July, are much like the adults in color pattern, but the darker parts are sootier; the feathers of the lower back, hinder scapulars, rump, and upper tail coverts are indistinctly tipped with pale brown; the wing coverts are tipped with cinnamon or light rusty; and the white of the abdomen is suffused with brownish along the border. This plumage is worn but a short time; I have seen young birds molting the contour plumage, but not the wings and tail, as early as July 25. In first winter plumage, young birds can be distinguished from adults by the juvenal wings, which are apparently retained during the first winter. Adults and one-year-old young birds have a complete molt in July and August.]

Food.—The most comprehensive accounts of the food habits of the black phoebe are those of Prof. F. E. L. Beal (1910 and 1912), who asserts that "this bird eats a higher percentage of insects than any flycatcher yet studied except the western wood pewee." For the earlier report there were available 333 stomachs, which showed 99.39 percent of animal matter to 0.61 percent of vegetable. Ground beetles (Carabidae), ladybirds (Coccinellidae), and tiger beetles (Cicindelidae), all presumed to be useful, made up 2.82 percent. Other beetles, all more or less harmful, amounted to 10 percent, consumed throughout the year. Hymenoptera, principally wild bees and wasps, contributed 35 percent, rising to 60 percent in the month of August. No trace of a honeybee was found.

Hemiptera, including plant lice and a number of aquatic species, formed 7 percent of the total. Flies (Diptera) ranged from 3 percent in August to 64 percent in April, averaging 28 percent. Grasshoppers and crickets supplied only about 2.5 percent, moths and caterpillars 8 percent, miscellaneous, principally dragonflies, with some spiders, 6 percent. In his second report, which differs little from the first in percentages, Professor Beal adds that ants "for a short time in midsummer constitute quite a notable part of the food."

In the stomachs of 24 nestlings, tabulated separately from the adults, "no great difference was apparent in the kind of food eaten nor in the relative proportions," except that the percentage of vegetable matter was a trifle higher. Regarding the consumption of vegetable matter, Professor Beal (1912) says: "It is not at all improbable that this species and many others seldom or never take vegetable food intentionally. In many cases the vegetable substance found in the stomachs is mere rubbish accidentally picked up with insects. Bees and wasps often light on berries to suck the juice, and a bird making a quick snap at such an insect might take berry and all." However, Prof. A. J. Cook (1896) found pepperberries in the stomachs of nearly all black phoebes killed in winter (presumably in southern California).

In towns and cities much of the bird's hunting is done by skimming low over lawns, where it seems to capture a goodly number of fair-sized moths, probably adults of the highly injurious cutworm.

Mr. Oberlander (1939) found that indigestible portions of large insects were regurgitated in the form of spherical or conical pellets usually 7 or 8 millimeters in diameter. A pellet was ejected nearly every night except in rainy weather, and with undetermined frequency during the day. He also mentions an item of diet which is unusual in birds of this family: "The birds sought to catch the inch-long minnows which dimpled the water surface in the evening. They were not often observed to be successful, but daring attempts to snatch these choice morsels were made. Once, when a phoebe sighted a minnow, such a desperate attempt was made to get it that the bird dipped its head in the water nearly to the eyes. When a wiggling fish was caught, it was held in the bill, and tapped against the perch in the same manner described for large insects until finally, with gulping, the fish was swallowed."

Behavior.—Cooper (1870) speaks of the eastern phoebe as the "exact analogue" of this species in habits and of the similarity in their cries, resemblances that were noted also by Bendire (1895). The black phoebe is eminently solitary in its disposition, and aside from mated pairs in the breeding season, or the resulting family parties, it is always seen alone, a condition that seems to prevail by mutual consent and without much bickering. Averse though the phoebe is to the society of its own kind, in its contacts with other birds it shows no trace of the aggressiveness that has earned for its family the name Tyrannidae and that is at times demonstrated even by its near relative, Say's phoebe. To what extremes it has carried its pacifism is indicated by the previously cited instances of the usurpation of its nests by house finches. The occasional sight in fall or winter of a black phoebe hunting in close proximity to a Say's phoebe suggests that its aversion to companionship must be directed solely toward those of its own particular species.

The two western species of *Sayornis*, though occupying large portions of California and Arizona in common, ordinarily choose strikingly different types of habitat. While *saya* frequents open, often more or less barren, country, *nigricans* prefers the vicinity of streams or ponds, irrigated fields, well-watered lawns and gardens, or the neighborhood of buildings and barnyards. Rarely seeking the tree-tops, the black phoebe usually perches instead on the shaded lower branches, on fences, stones, or other low objects, but seldom on the ground itself. It is fond of taking up its station at the edge of a pool and darting out over the water, occasionally bathing by dipping its lower parts beneath the surface in passing. Its flight is rather soft and mothlike, punctuated by the sharp snap of the bill as it captures—or misses—its insect prey. Mr. Oberlander (1939) has called attention the fact that—

Pursuit flights are usually downward from the perch level so that most insects are taken from or within a few inches of the lawn, water, or weed patch below the perch. * * * This capture of insects below the perch level seems to be correlated with their greater abundance there. Observed insects slowed down in flight when they hovered about patches of weeds and lawn.

This apparently makes for ease of catching. On bright days I found it diffi-cult to see insects when they were flying high in the glaring light, yet when they chanced to fly between me and a dark background, sighting became easier. Hence, it seems probable that the increased visibility of insects sighted below the phoebe's perch might be another factor influencing the number of flights directed downward.

A complete reversal of this flight direction came with the approach of dark-ness. * * * In the dim light the insects could not be sighted near the ground among the long shadows, yet hundreds were visible in the air above. Of 117 flights noted during the approach of darkness, 79 per cent were directed upward. The number of upward flights increased with increasing darkness as shown in a recorded sample of flights.

Usually no hesitation is apparent before an insect is pursued. Yet the directness of flight as well as the usual capture of but a single insect on each trip indicates that the insect is sighted from the perch rather than while flying. Where the bird feeds over a lawn, insects are sighted again and again from ten to twenty feet from the perch; the bird flies directly to the exact position on the lawn where the insect is located, plucks it up without the slightest indication of uncertainty and returns to the perch.

Where slow flying insects are numerous in the air, two insects are occa-sionally taken in one flight, the second presumably sighted in the air but probably after the first, since a series of insects, regardless of their abundance, never was taken without the bird returning to its perch.

When insects were scarce, on rainy days and early on cold mornings, hover-ing was noted over grassy areas or alongside walls. Evidently these hoverings were a means of sighting insects hidden in the tall grass. These hovering flights are not the regular method of feeding, nor does the bird regularly range away from the perch in this manner; instead, they seem to be an adapta-tion to unfavorable conditions where it is necessary to search out insects that can not be sighted from the perch, and less frequently for detecting in-sects already sighted from the perch but lost from the bird's view after flight began.

The method of searching from a perch not only enables the phoebe to locate insects with a minimum of energy, but the rejection of unsuitable species apparently occurs before leaving the perch. Common yellow cabbage butterflies (*Eurymus*) frequently were observed to pass slowly in jerky flight, unmolested, within a few feet of a phoebe. * * *

In late spring and early summer the adult phoebes were observed to pass up dozens of checkered fritillary butterflies (*Argynnis*). * * * The young phoebes fresh from the nest were not so selective as their parents in foraging. On three occasions the young were observed to snatch a fritillary and swallow it, gulping down the large wings as well as the body. These incidents in-dicate that the older birds become conditioned against this insect. * * *

The majority of insects caught were small enough to be swallowed in flight. But when forms larger than the house fly were taken, the catch was carried back and tapped against the perch several times as if to smash and kill it before swallowing. * * *

After catching large insects the birds exhibit noticeable preference for substantial flat perches against which they can break up their prey.

The two western species of *Sayornis* are sufficiently distinct in their manner of flight to enable them to be distinguished when seen only in outline, as in looking toward the setting sun. The flight of the present species, when not engaged in pursuits, is comparatively direct and businesslike, lacking the airy, butterflylike indecision that often marks that of its relative.

Voice.—Almost constantly the black phoebe announces its presence by a crisp *tsip*, or, interchangeably, a more prolonged *chee*, accompanied by a jerk of the tail when perching or repeated with more vivacity while on the wing. Though rather plaintive in tone, these notes are by no means so mournful as those of the less voluble Say's phoebe or the pewee. The only other common utterance of this species is its song, if such it may be called. This consists of an indefinite repetition of two pairs of notes, rendered by Mr. Hoffmann (1927) as *ti wee, ti wee*, the one pair usually having an upward inflection and the alternate pair a downward inflection. This song is sometimes heard in fall as well as in spring.

Field marks.—In strong contrast to many of the Tyrannidae, the black phoebe is one of the most easily recognized of all our birds. The blackish throat and breast at once distinguish it from any other North American flycatcher, while its mannerisms and flight, together with its color pattern, will prevent confusion with birds of any other family.

Enemies.—Little information is available regarding any possible enemies of this species. Its nonterrestrial habits and the nature of its typical nesting sites remove it from the sphere of most walking or crawling predators, and its flight, though not swift, is so well controlled that it would seem a difficult victim for the sharp-shinned hawk. Any destruction by man would be purely wanton and without a shadow of excuse.

Roland Case Ross (1933) tells of a black phoebe killed by a honeybee's sting in the roof of its mouth, and he speculates on the possibility that the scarcity of the species in certain localities might be due to disastrous experiences with this unaccustomed form of prey.

Winter.—An excellent description of the winter habits of the black phoebe, as observed in Monterey County, Calif., is furnished by Grinnell and Linsdale (1936):

> The black phoebe is one kind of resident bird which was conspicuously more numerous in winter than in summer at Point Lobos. A count made in early January placed the number of individuals stationed on the area at eight. In the early nesting season only one pair was found.
>
> Each bird possessed a certain series of perches which marked its location for a large part of the winter. It remained close within this circuit and thus avoided close contact and conflict with other birds of the same or closely related species. These home ranges with which we became well acquainted through frequent observation were mainly in two types of situation—along the rocky shore, and within but close to the margins of pine woods or chaparral areas.

In the pines the birds usually perched within a few feet of the ground, on the dead stubs or limbs where these were shaded by the main crown of the tree. At certain times a bird would perch on or close to the top of a low tree; and some individuals showed a distinct preference for fence posts and the top wires of fences for perches. Most of these places were alike in providing situations where the birds could be within shadows most of the time. The more exposed (to light) perches were most likely to be used on cloudy days. Possibly this mannerism was more in the nature of a coloration-concealment device than a direct response to the warmth of the sun. The perches chosen also were alike in providing protection from strong, or cold, wind.

DISTRIBUTION

Range.—Western United States, Central and South America; nonmigratory.

The range of the black phoebe extends **north** to northern California (Benbow and Baird); southern Nevada (Alamo); Arizona (Grand Canyon and Salt River Bird Refuge); southern New Mexico (Cooney, Chloride, and Carlsbad Bird Refuge); and southwestern Texas (Spring Creek and San Angelo). **East** to western Texas (San Angelo and Pecos River High Bridge); Honduras (Ceiba); Nicaragua (Banbana); Venezuela (Colon); Bolivia (Concepcion); and northwestern Argentina (Orillas del Rio Lavallen and Anfama). **South** to northern Argentina (Anfama) and Peru (Huanuco). **West** to Peru (Huanuco); Ecuador (Zamora, Bucay, and Esmeraldas); western Colombia (San Jose and Medellin); Costa Rica (San Jose and Carrillo); El Salvador (Libertad and San Salvador); western Guatemala (San Geronimo and Duenas); Baja California (San Lucas, Triunfo, San Fernando, and Todos Santos Island); and California (San Clemente Island, Santa Barbara, Berkeley, Point Reyes, and Benbow).

The range as outlined is for the entire species, which has been separated into several subspecies or geographic races. Only one of these, the typical race (*Sayornis n. nigricans*), is found commonly in the United States. The San Quintin phoebe (*S. n. salictaria*), of northern Baja California, also comes into southern Arizona. The southern part of Baja California is occupied by a third race, the San Lucas phoebe (*S. n. brunnescens*).

Casual records.—One of these birds was recorded from the Umpqua Valley, Oreg., in 1857, and another from Salem, Oreg., in July 1879. Neither of these records has been satisfactorily traced.

Egg dates.—Arizona: 8 records, April 16 to June 26.

California: 126 records, March 17 to August 15; 64 records, April 15 to May 16, indicating the height of the season.

Lower California: 25 records, April 6 to May 27; 13 records, April 22 to May 9.

SAYORNIS NIGRICANS SALICTARIA Grinnell

SAN QUINTIN PHOEBE

HABITS

This race of the black phoebe is found in the Upper Austral Zone of northern Lower California, from about latitude 30° northward to the United States boundary; also in southern Arizona in the valley of the Colorado River.

In describing and naming it, Dr. Grinnell (1927) said that it is only slightly smaller than the well-known black phoebe of California, but the "general tone of color of dark parts blacker, more slaty, less brown. Color of dorsum close to 'dusky neutral gray' of Ridgway. * * * This marked slatiness of color tone is apparent not only on the forward lower surface, head and back, but also pertains to the remiges and rectrices, especially their concealed portions in closed wing and tail. This the *blackest* of the races of the Black Phoebe."

He said that it is largely "restricted associationally to willows, which plants, of course, as a rule in an arid country mark the near vicinity of water or at least the presence of damp ground-surface."

It apparently does not differ in its habits from the northern race, as A. W. Anthony wrote to Major Bendire (1895) that it breeds "wherever water is found, building under the eaves of adobe houses when near human habitations, and on the sides of ledges along streams in the unsettled parts."

The eggs are apparently indistinguishable from those of the species elsewhere. The measurements of 30 eggs average 18.91 by 14.56 millimeters; the eggs showing the four extremes measure **21.0** by 15.1, 19.5 by **15.3**, and **18.0** by **13.9** millimeters.

SAYORNIS NIGRICANS BRUNNESCENS Grinnell

SAN LUCAS PHOEBE

HABITS

The black phoebes of the Cape region of Lower California have been given the above name by Dr. Joseph Grinnell (1927). As compared with the upper California race, he calls this subspecies "similar but browner in general tone of all dark areas, bill broader, and wing slightly and tail decidedly shorter. Color of dorsum close to 'fuscous' of Ridgway"; whereas, in the California bird, the dorsum is "close to 'chaetura drab' of same plate." William Brewster (1902) called attention to the "comparatively faded, brownish coloring" of the Cape bird but thought it might be due to adventitious bleaching.

But Dr. Grinnell (1927) said that "this is certainly not the case, for a considerable number of October and November examples, thus in new autumn plumage, show the character to be an innate one. In this connection it should go without saying that in one-molt birds like *Sayornis* only early fall plumages should be used for the determination of finely manifested color characters."

The San Lucas phoebe ranges north in Lower California to about latitude 29°.

J. S. Rowley writes to me: "The habits of this phoebe are identical with those found to the northward. From Miraflores and Todos Santos, I took several sets, one containing four eggs, three containing three eggs, and one with two eggs, all full complements. All these nests were typically made of mud and were saddled to the side of a boulder near water."

There are two sets of eggs, with the nests, of this phoebe in the Thayer collection in Cambridge, taken by W. W. Brown, Jr., at Comondu on May 5 and 7, 1909; each nest contained three eggs. One nest was on the side of a shack near a brook, and the other was on a crossbeam against the wall in a shack, also close to a brook. The nests are typical of the species, made externally of small pellets of mud, mixed with dried grass stems and fine shreds of inner bark, and lined with finer pieces of similar materials.

The eggs are indistinguishable from those of the species elsewhere. The measurements of 37 eggs average 19.24 by 14.69 millimeters; the eggs showing the four extremes measure **20.6** by 14.8, 18.9 by **15.4,** and **17.8** by **13.6** millimeters.

SAYORNIS SAYA SAYA (Bonaparte)

SAY'S PHOEBE

PLATES 19, 20

HABITS

This large phoebe, clad in pleasing shades of gray and brown, sharply contrasted with its black tail, replaces throughout a large portion of western North America our familiar eastern phoebe, which it resembles in many of its haunts and habits. It is as much at home among the western ranches as our eastern bird is about our New England barnyards, equally fond of human company, and often building its nest on or about, or even in, the rancher's buildings. It is a wide-ranging species, breeding as far north as central Alaska and as far south as northern Mexico. It is a summer resident only in the northern portion of its range, where it is one of the earliest arrivals in the spring, but it is found all winter in southern California, Arizona, and New Mexico.

Say's phoebe is a bird of the open country, the prairie ranches, the sagebrush plains, the badlands, the dry, barren foothills, and the borders of the deserts, where it can forage widely over the stunted vegetation, or perch on some low bush or tall weed stalk to watch for its insect prey. But it is also found in the mouths of canyons or rocky ravines, perched on some commanding boulder as a watchtower. It has no special fondness for watercourses, or for rich agricultural lands, and is seldom seen in heavily timbered regions. As the deeply shaded retreats are more favored by the somber-hued black phoebe, so are the open, sunny places more suited to this sandy-colored species; perhaps each is less conspicuous in its normal habitat.

Nesting.—I made my acquaintance with Say's phoebe in southwestern Saskatchewan, where we located five pairs of these birds about the ranches in 1905 and 1906. A nest was found, on May 30, 1905, on a rafter under a bridge, but we could not reach it. Another nest was discovered on a shelf under the eaves of a station house; it held two fresh eggs on June 5. A nest examined on June 10, 1905, was built on a shelf under the roof of a cattle shed; there was a foundation of mud, on which was constructed a pretty nest of soft, fine grasses, lined with cow's hair and woolly substances. Other nests were observed in Saskatchewan, and in 1922 in Arizona, all of which were placed on or in deserted ranch buildings. The above locations seem to furnish the favorite nesting sites for this phoebe, but it also nests in a variety of other situations.

Major Bendire (1895) says that at Fort Lapwai, Idaho, "they generally arrived during the third week in March, the males preceding the females about a week, and nest repairing or building commenced about the latter part of this month. I have taken a full set of eggs, containing small embryos, on April 17, 1871. Here they nested mostly under the eaves of outhouses and stables; but one pair selected the plate or rail over the main door of my quarters, and another a corner on the hospital porch. In this vicinity I also found a pair occupying an old Cliff Swallow's nest attached to an overhanging ledge of rock in Soldiers' Canyon, on the road to Lewiston, Idaho, and another in a very unusual position in the same canyon, in an old Robin's nest, placed in a syringa bush, about 4 feet from the ground." He says further:

Besides the various localities already mentioned in which Say's Phoebe has been found nesting, burrows of Bank Swallows are also occasionally occupied. Ordinarily mud is not used in the construction of their nests; which are rather flat structures; the base usually consists of weed stems, dry grasses, moss, plant fibers of different kinds, wool, empty cocoons, spider webs and hair, the inner lining being generally composed of wool and hair alone. A well-preserved nest, now before me, from the Crooked Falls of the Missouri, Montana, taken by Mr. R. S. Williams, June 3, 1889, measures 5½ inches in outer diameter by 2¼ inches in height, the inner cup being 2½ inches by 1¼ inches in depth. This is a compactly built structure, the materials composing it being well worked together, and it is warmly lined with cattle hair.

Before the advent of man-made structures, the primitive nesting sites of Say's phoebe were evidently on shelves or in crevices of rocky cliffs, protected from the weather by overhanging rock, on ledges in caves, in natural cavities in trees, or in holes in vertical or overhanging banks. Many pairs still nest in such situations, especially in uninhabited regions and in the far north. There is a set of eggs in my collection that was taken from a hole in a bank, well sheltered from rain, as the rim of the nest was flush with the face of the bank. But after the coming of man, it did not take the birds long to learn to take advantage of the new, and often more secure, nesting sites offered. Abandoned mine shafts and old wells took the place of caves. Harry S. Swarth (1929) says: "At our camp on the Ashburn Ranch [southern Arizona] a pair of Say Phoebes had a nest in a well, built in a crevice in the dirt wall about 15 feet down. This is a favorite nesting site with the species in this region and I have seen a number of nests similarly placed, in wells or in mine shafts." R. T. Congdon has sent me a photograph of a nest located 15 feet under ground, at the bottom of the well-like shaft of an old irrigation flume, where an inverted siphon formerly carried water under a railroad track. Another of his photographs shows a nest in a lard pail inserted in the stovepipe hole of a chimney in a deserted forest cabin. While in Arizona, in 1922, Frank Willard photographed a nest in an old mail box on a post by a roadside.

As indicated above Say's phoebe occasionally appropriates the nest of some other species, sometimes driving away the rightful owners. E. S. Cameron (1907) writes from Montana:

In May, 1895, a pair took possession of a Barn Swallow's nest in the stable and forced the rightful owners, which were renovating it, to build an entirely new one affixed to a beam. In 1904, a pair of Say's Phoebes nested below the eyrie of the Golden Eagles and were unmolested. Another pair which, in 1906, built in a hole near the Prairie Falcon's eyrie (on one of the highest buttes along the Yellowstone) were killed by the latter for their young. In May, 1907, a still more remarkable site chosen by these flycatchers was the old abode of a Cliff Swallow; one of several nests situated above a wolf-den in a huge sand rock. The den was inhabited by a she-wolf with her six pups, and the birds were exposed to constant disturbance, both from these animals and from men who suffocated the young wolves with a pitch pine fire. The she-wolf escaped with one ten-weeks-old pup and intermittent efforts were made to trap her at the den. Nevertheless the flycatchers did not desert their nest."

Eggs.—Say's phoebe lays ordinarily four or five eggs, sometimes as few as three, and very rarely as many as six or seven. The eggs vary from ovate to short-ovate and have little or no gloss. They are usually pure white, but occasionally one or more eggs in a set may have a few small spots of reddish or dark brown. The measurements of 50 eggs average 19.5 by 15.1 millimeters; the eggs showing the four extremes measure **21.6** by 15.5, 19.1 by **15.6** and **18.0** by **14.0** millimeters.

Young.—Major Bendire (1895) says: "Incubation lasts about twelve days; the young are fed entirely on insects, mainly on small butterflies, which are abundant about that time, and they are ready to leave the nest in about two weeks, when the male takes charge of them, the female in the meantime getting ready for a second brood." Apparently two broods are generally raised in a season throughout most of the range of Say's phoebe, and in the southern portions often three. Mrs. Wheelock (1904) writes:

Incubation lasts two weeks, and although the male does not brood he sits all day long on a lookout near by. The newly hatched young are naked except for a slight gray fuzz on their saffron skin. Until six days old their eyes are closed by a skinny membrane, and during this time they are fed by regurgitation. They mature very rapidly, and in two weeks have their feathers well in order for their first attempts to fly. Up to this time the father bird has diligently fed and guarded both them and the mother, coming to the nest every two or three minutes with butterflies in his bill. But as soon as they are ready to try their wings, he assumes full charge, teaching them to fly and to catch insects on the wing in true flycatcher fashion.

She says that, while incubating on the second set of eggs, the female is seldom fed by her mate, "but, since the days grow warmer, leaving oftener and for longer intervals to forage for herself. When the second family is ready to fly, she takes charge of it unless the necessity of rearing a third brood should compel her to desert them; and then, from somewhere, the hitherto unnoticed male appears, to assume care of them."

Plumages.—I have seen no very young birds of this species, but the young bird in full juvenal plumage is essentially like the adult, except that the upper parts are browner, and the greater and median wing coverts are broadly tipped with "cinnamon" or "cinnamon-buff," forming two distinct bands. The sexes are alike in all plumages.

Adults have a complete postnuptial molt during the latter part of July and in August; I have seen adults in full molt as early as July 27 and as late as September 2; and I have seen an adult male in full fresh plumage on August 31. I have seen a number of birds, taken in January, February, and March, in which the contour plumage about the head and neck is much worn; and others, taken in May, in wholly fresh plumage; this would seem to indicate a partial prenuptial molt; two specimens, taken on March 30 and May 17, show signs of body molt in progress. As adults and young are almost indistinguishable after the postjuvenal molt in fall, these may be young birds.

Food.—Professor Beal (1912) reports on the contents of 11 stomachs of Say's phoebe, taken during every month in the year. Animal food made up 99.78 percent and vegetable food 0.22 percent of the whole. Beetles of the three most useful families, Cicindelidae (tiger beetles), Carabidae (predaceous ground beetles), and Coccinellidae (ladybirds), amount to 5.95 percent of the food. "This," he says, "is a surprisingly large percentage to be eaten by a flycatcher." Other beetles, either harmful or neutral species, amount to 9.72 percent. Hymenoptera seem to be the largest item of food, 30.72 percent. "They are mostly bees and wasps, with a few ants. No honeybees were found." Hemiptera (bugs) amount to only 4.45 percent, but Diptera (flies), "mostly of the families of the house fly, the crane fly and the robber fly," are more popular, amounting to 16.67 percent. Caterpillars were found in 17 stomachs and moths in 19. "Here for the first time is found a flycatcher that eats more of adults (moths) than it does of the larvae (caterpillars)." Grasshoppers and crickets occurred in 48 stomachs and amount to 15.36 percent of the food. "Dragon flies, spiders, millepeds, and a few sowbugs, together amount to 4.79 percent of the food, and make up the remainder of the animal quota. * * * The vegetable food of Say's phoebe can be dismissed with a few words. It consists of seeds of elder (*Sambucus*) contained in 3 stomachs, nightshade (*Solanum*) in 2, a single seed of a fig in 1, seeds of tarweed (*Madia*) in 1, and rubbish in 4. Thus it has no economic importance."

Bendire (1895) says: "I have repeatedly seen it catching good-sized grasshoppers on the wing, as well as different species of beetles, flies, moths, and butterflies. It has a habit similar to the Owls of ejecting the indigestible portions of its food in the shape of pellets. My attention was drawn to this fact by observing several such lying on the porch of my quarters at Fort Lapwai, Idaho, where a pair of these birds nested over the door."

Claude T. Barnes writes to me about a Say's phoebe that arrived in the foothills above Salt Lake City, Utah, on February 21, 1939, in the midst of a blizzard of great intensity: "For a week I observed it daily, pitying its fluffed loneliness, as it sat on limbs heavily laden with snow. At last in its hunger it came to a Boston ivy (*Parthenocissus tricuspidata*), which a western robin had appropriated; and, thereafter, it would flutter, as occasion permitted, to the seeded berries of the ivy, only to be driven away forthwith by the equally doleful robin. During the time that I observed it, the phoebe never made any sound; and, on account of the almost constant snowfall, the ivy was apparently its sole subsistence."

Behavior.—Major Bendire (1895) writes:

Its general habits and actions resemble those of the eastern Phoebe; like it, it is one of the earliest spring migrants to return from its winter haunts, and it is equally attached to its old home, to which it regularly returns from year to year. It appears to be much more tolerant in its disposition toward other members of its kind than the Phoebe, as I have found several pairs breeding within 100 yards of each other apparently in perfect harmony. Its manner of flight is also similar. * * * I consider it a more restless bird than the Phoebe, if that is possible; for it is never idle, but constantly darting back and forth from its perch after passing insects, which form the bulk of its food and of which it never seems to get enough.

In its favorite haunts in the flat, open spaces, Say's phoebe flits about over the stunted vegetation with rather powerful wing strokes in its somewhat zigzag flight. It does not ordinarily fly high and favors rather low perches, on some small bush, tall weed stalk, or low rock, seldom higher than a fencepost. Thence it sallies forth to seize its flying prey with a loud click of its bill, or to pick up some lowly insect from the ground, and returns to its perch with a flick of its black tail. It is not at all shy and shows its confidence in the human race by living about the ranches and placing its nest on occupied dwellings.

Laurence M. Huey (1927) discovered an interesting night-roosting habit of this phoebe:

Camp was established near an old adobe ruin, where, after a few days, a Say's Phoebe was noticed early in the morning and again at sunset, with precise regularity. This occasioned some speculation, and not until it was discovered that the bird had a chosen roosting site nearby, were its actions explained. * * * On the evening when the phoebe's secret was discovered, an inspection tour was made about 10:30. * * * Before entering the doorless doorway of the old building, a glance upward revealed, almost directly over the doorway, an old nest of a Black Phoebe and protruding from the edge of the nest was a bird's tail. A closer look showed the preemptor to be none other than the regular-twice-daily-occurring Say's Phoebe. Blinded by the light, the bird did not flush, but tried to crouch more closely into the cup of the nest.

Every night thereafter the phoebe was looked for and always found. On two occasions, when the weather had turned decidedly cooler, the bird resorted to a niche in the wall a few inches above the nest; otherwise it was always in its favorite spot.

Voice.—The ordinary call note of Say's phoebe is quite unlike the note of the eastern phoebe; it is a soft, plaintive *phee-eur*, rather sweet but somewhat melancholy in tone; it is often accompanied by a twitching of the tail and a raising of the crest. Ralph Hoffmann (1927) says: "The Say Phoebe in the mating season utters repeatedly a swift *pit-tsee-ar*, finally fluttering about in the air repeating a rough trilling note. Even in the winter this mating song is occasionally heard."

Field marks.—Its flycatcher behavior, its grayish-brown back, and its reddish-brown belly, together with its conspicuous black tail, make this phoebe easily recognizable. Its flight, as referred to above, is characteristic.

Range.—Western North America, including northwestern Mexico; accidental east of the Great Plains.

Breeding range.—Say's phoebe breeds **north** to central Alaska (White Mountains and Circle); western Mackenzie (Fort McPherson, Fort Simpson, and Hay River); southern Saskatchewan (Johnstone Lake, Regina, and Indian Head); and southern Manitoba (probably Oak Lake, and Aweme). **East** to southwestern Manitoba (Aweme); North Dakota (Charlson and Stutsman County); South Dakota (Tuttle and possibly Vermillion); Nebraska (Valentine, Greeley, and Lincoln); western Kansas (Gove and Coolidge); western Oklahoma (Kenton); southeastern New Mexico (Carlsbad); and Chihuahua (Rio Sestin). **South** to southern Chihuahua (Rio Sestin) and central Baja California (San Bartolome). **West** to Baja California (San Bartolome, Valladareo, and Sierra San Pedro Martir); California (Escondido, Ventura, Pleasant Valley, and Red Bluff); eastern Oregon (Warner Valley, Malheur Lake, and Brogan); eastern Washington (Grand Dalles, Yakima, and Cheney); British Columbia (Chilliwack, Clinton, Glenora, and Atlin); southwestern Yukon (Fort Selkirk); and central Alaska (White Mountains).

Winter range.—The winter range extends **north** to central California (Berkeley, Fresno, and Death Valley); central Arizona (Salt River Reservation); and central New Mexico (Albuquerque). **East** to New Mexico (Albuquerque, San Acacia, and Carlsbad); central Texas (San Angelo, Laredo, and Brownsville); Tamaulipas (Matamoros); and Veracruz (Jalapa and Orizaba). **South** to southern Veracruz (Orizaba); Puebla (Puebla and Chapulco); Durango (Lerdo); and southern Baja California (San Jose del Cabo). **West** to Baja California (San Jose del Cabo, La Paz, Cerros Island, and Todos Santos Island) and the coastal region of southern California (San Diego, Santa Barbara, and Berkeley).

The range as outlined is for the entire species, of which two subspecies are currently recognized. The typical race, *Sayornis saya saya*, occupies the entire range except for Baja California, which is the general range of the San Jose phoebe, *S. s. quiescens*.

Spring migration.—Early dates of spring arrival are: Oklahoma—Kenton, March 20. Kansas—Ellis, March 18. Nebraska—Antioch, March 29. South Dakota—Lacreek, April 4. North Dakota—Bismarck, April 10. Manitoba—Treesbank, April 6. Saskatchewan—Eastend, April 22. Colorado—Boulder, March 27. Wyoming—Cheyenne, April 7. Montana—Terry, April 5. Utah—Linwood, April 15. Alberta—Edmonton, April 22. Mackenzie—Fort Simpson, May 4. Eastern Oregon—Enterprise, February 28. Eastern Washington—Grand Dalles, February 12. British Columbia—West Summerland,

February 25. Yukon—Fortymile, May 5. Alaska—Mount McKinley, June 5.

Fall migration.—Late dates of fall departure are: Alaska—McCarthy, August 21. Yukon—Fort Selkirk, August 13. British Columbia—Okanagan Landing, November 7. Washington—College Place, September 30. Oregon—Fort Klamath, September 20. Alberta—Glenevis, September 19. Montana—Fortine, September 8. Wyoming—Laramie, October 3. Utah—St. George, October 26. Colorado—Grand Junction, October 4. Saskatchewan—Eastend, September 19. North Dakota—Charlson, September 24. South Dakota—Great Bend, September 18. Nebraska—Ashby, October 3. Oklahoma—Black Mesa country, Cimarron County, October 1.

Casual records.—On several occasions Say's phoebe has been recorded east of its normal range, although several of the alleged occurrences are without complete data. One was collected at Stotesbury, Mo., sometime previous to 1907; two specimens are said to have been collected by Robert Kennicott, at West Northfield, Ill., previous to 1876; according to Kumlein and Hollister one was taken at Racine, Wis.; a specimen was taken at or near Godbout, Quebec, on October 19, 1895; and one was collected at North Truro, Mass., on September 30, 1889. In the north, a specimen was obtained at Point Barrow, Alaska, on May 27, 1932.

Egg dates.—California: 44 records, March 7 to June 16; 22 records, April 1 to May 6, indicating the height of the season.

Colorado: 16 records, May 7 to June 26; 8 records, May 20 to June 16.

Idaho: 6 records, April 14 to June 30.

New Mexico: 12 records, April 4 to July 14; 6 records, April 27 to June 1.

North Dakota: 8 records, May 31 to July 9.

SAYORNIS SAYA QUIESCENS Grinnell

SAN JOSE PHOEBE

HABITS

Dr. Joseph Grinnell (1926) described this race of Say's phoebe as similar to the well-known northern form, "but tone of coloration paler, this paleness being in the direction of ashy gray rather than light brown." The seven specimens on which this name is based all came from a very limited "area in northwestern Lower California on the Pacific drainage from the Sierra Pedro Martir west to the seacoast. Life-zone chiefly Upper Sonoran." All the localities, where the specimens were collected, "lie between latitudes 30°30' and 31°30'." As all the specimens are in full fresh annual plumage," the paleness is not due to wear or fading.

I can find no evidence that this phoebe differs in any respect in its habits from its more northern relative. It evidently breeds in the area indicated above, for A. W. Anthony told Major Bendire (1895) that he found nests of this phoebe in that region in abandoned mining shafts and prospect holes, as much as 25 feet below the surface of the ground. How much farther south, east, or north it breeds does not seem to be known. No form of Say's phoebe has been found breeding in the Cape region of Lower California, where the species seems to be only a winter visitor and rather rare at that.

The subspecific status of the birds of this species that have been taken in winter in the Cape San Lucas region evidently has not been determined.

Dr. Louis B. Bishop (1900a) described a northern race of this species, which he named *Sayornis saya yukonensis*, based on the study of 15 specimens collected in the Yukon Valley, Alaska. He characterized it as "similar to *Sayornis saya* but darker, the gray of the upper parts clearer—less scorched, with the pale edgings of the wing-coverts and secondaries narrower; the tail longer; the bill shorter and relatively broader." Mr. Ridgway (1907) relegated it to synonymy; and the A. O. U. Check-list has not yet admitted it.

The measurements of eight eggs average 20.4 by 15.5 millimeters; the eggs showing the four extremes measure 22.1 by 16.3 and 19.0 by 15.3 millimeters.

EMPIDONAX FLAVIVENTRIS (Baird and Baird)

YELLOW-BELLIED FLYCATCHER

PLATES 21, 22

HABITS

To most of us this pretty little flycatcher is known only as a spring and fall migrant. It is not often seen then, as it is a shy, retiring bird, frequenting the low, wet, swampy thickets along streams or the borders of swamps or ponds. Since it is mostly silent on migrations, its characteristic notes do not tell us of its presence, and as it generally succeeds in keeping out of sight it must be sought for diligently. In its summer home its voice betrays it, but there, also, the searcher must invade the moist, gloomy morass of some northern forest bog, beneath the shade of spruces and firs, and endure the attacks of hoards of black flies and mosquitoes, to get even a glimpse of this woodland waif.

Dr. Samuel S. Dickey, who has had considerable experience with this flycatcher, says in his notes: "In the Northern States and in Canada they are met with in the shadowy underwoods of evergreens, paper birches, and mountain ashes, where cranberries, trailing white snowberry, rare orchids, and an array of slightly emerald mosses carpet the forest floor and cover the crumbling logs." In Northumberland County, New Brunswick, he and R. C. Harlow found it in the evergreen forest, where they explored "many a little glade, beautified with an array of botanical treasures, such as the twayblade (*Listera cordata*), small green wood orchis (*Habenaria clavellata*), and green coralroot (*Choralhoriza trifida*). In the Adirondack Mountains of New York yellow-bellied flycatchers were present in mossy glades under the towering firs. Now and then a bird was routed from colonies of tripwood (*Viburnum alnifolia*), over pretty beds of the twinflower (*Linnaea borealis*), and the red-berried *Cornus canadensis*."

In the southern portions of its breeding range, which extends at least as far south as the Pocono Mountains in Pennsylvania, the yellow-bellied flycatcher finds congenial summer homes only on the mountains, at elevations of 2,500 feet or more, among the forests of spruces, firs, hemlocks, and tamaracks, and where the damp ground is carpeted with sphagnum moss.

Nesting.—The earliest accounts of the nesting habits and the eggs of the yellow-bellied flycatcher were based on wrong identification and are known to be quite at variance with present-day knowledge; the nests were said to be placed in the forks of bushes and the eggs to be white and unspotted, whereas we now know that the nests are always placed on or near the ground, or in cavities in the upturned roots of fallen trees, and the eggs are spotted.

We are indebted to H. A. Purdie (1878) for the first authentic account of the nesting habits of this species. He and Ruthven Deane were shown this nest in Houlton, Aroostook County, Maine, in 1878, by a collector named James Bradbury. He relates the incident as follows:

Mr. Bradbury informed us that he found, on June 15, a nest unknown to him with one egg. On the 18th he conducted us to the edge of a wooded swamp, and, pointing to the roots of an upturned tree, said the nest was there. We approached cautiously, and soon saw the structure and then the sitting bird, which appeared to be sunken in a ball of green moss. Our eager eyes were within two feet of her, thus easily identifying the species, when she darted off; but, to make doubly sure, Mr. Deane shot her. There was no mistake; we at last had a genuine nest and eggs of the Yellow-bellied Flycatcher. A large dwelling it was for so small and trim a bird. Built in and on to the black mud clinging to the roots, but two feet from the ground, the bulk of the nest was composed of dry moss, while the outside was faced with beautiful fresh green mosses, thickest around the rim or parapet. The home of the Bridge Pewee (*Sayornis fuscus*) was at once suggested. But no mud entered into the actual composition of the nest, though at first we thought so, so much was clinging to it when removed. The lining was mainly of fine black rootlets, with a few pine-needles and grass-stems. The nest gives the following measurements: depth inside, one and one half inches; depth outside, four and a quarter inches; circumference inside, seven and a quarter inches.

Other nests have been found in upturned roots, notably one found by Major Bendire (1895) in Herkimer County, N. Y., which he described as follows:

> The nest was placed among the upturned roots of a medium-sized spruce tree, to which considerable soil, which was entirely covered with a luxuriant growth of spagnum moss, was still attached. This perpendicular moss and fern covered surface measured about 6 by 8 feet. The nest was sunk into the moss and soil behind, about 14 inches above the ground; the entrance was partly hidden by some ferns and the growing moss around it, and, taken all in all, it was one of the neatest and most cunningly hidden pieces of bird architecture I have ever seen. I might have walked past a dozen times without noticing it. It contained four eggs, in which incubation was about one-third advanced. The entrance was nearly circular, and measured about 1¼ inches in diameter. The inner cup of the nest itself measured about 2 inches in diameter and 1¼ inches in depth. It was composed of fine grasses and a few black, hair-like rootlets and flower stems of mosses.

Nests of the yellow-bellied flycatcher in upturned roots are the exception rather than the rule; most of the nests reported have been on or near the ground, on the sides of hummocks or mounds, and well hidden in sphagnum moss or other low vegetation; some have been found under the roots of standing trees or stumps; and W. J. Brown, of West Mount, Quebec, tells me that he once found a nest "on a cliff near a trout stream." He has probably found more nests of this flycatcher than any other man, He says that it is "very abundant" in the County of Matane, Quebec, on the lower St. Lawrence, where it nests in the "thick evergreen woods on the borders of peat and blueberry bogs." He says, in his letter, that he has the records of over 200 nests found in that county, and that he locates 10 to 15 nests every year. "The nests, of course, always have moss and are lined with black plant fibres and fine rootlets. A few nests have been lined with fine, bleached grasses only."

The finest description of a nest of this species that I can find is that of a nest found by Dr. A. K. Fisher near the summit of Slide Mountain in the Catskill Mountains, N. Y., at an altitude of over 3,500 feet, quoted by Major Bendire (1895) as follows:

> The nest was built in a cavity scooped in a bed of moss facing the side of a low rock. The cavity had been excavated to a depth of 2½ inches and was 2 inches across. The opening, but little less than the width of the nest, was 9 inches from the ground, and, partially hidden by overhanging roots, revealed the eggs within only to close inspection.
> The primary foundation of the nest was a layer of brown rootlets; upon this rested the bulk of the structure, consisting of moss matted together with fine-broken weed stalks and other fragmentary material. The inner nest could be removed entire from the outer wall, and was composed of a loosely

woven but, from its thickness, somewhat dense fabric of fine materials, consisting mainly of the bleached stems of some slender sedge and the black and shining rootlets of, apparently, ferns, closely resembling horsehair. Between the two sections of the structure, and appearing only when they were separated, was a scant layer of the glossy orange pedicels of a moss (*Polytrichum*) not a fragment of which was elsewhere visible. The walls of the internal nest were about one-half an inch in thickness, and had doubtless been accomplished with a view of protection from dampness.

Prof. Daniel C. Eaton, of New Haven, very kindly assumed the task of determining the different species of moss which entered into the composition of the nest and of the moss bed in which it rested, and his investigation disclosed the fact that the mosses which abounded immediately about the nest had not been utilized as building material. * * * In addition there were found among the materials of construction catkin scales of the birch, leaves of the balsam, and fragments of the dried pinnae of ferns; but, as suggested by Professor Eaton, the presence of some of these was probably accidental. Springing from the verdant moss beds immediately about the nest were scattered plants of *Oxalis acetosella, Trientalis americana, Solidago thyrsoidea,* and *Clintonia borealis.*

William L. Bailey (1916) reports the finding of three nests of the yellow-bellied flycatcher on Pocono Mountain, Pa., by himself and by some of his friends. He says:

The nesting sites were all in little open sunny spots of wet sphagnum in the dense secluded forest of spruce, hemlock, balsam and tamarack; and all through the moss grew the wintergreen, bunch berry and occasionally the fragrant white swamp azalea. The nests were hidden in the sides of little mounds of sphagnum; only a little black flat hole was visible, which did not even look suspicious. The nest which had young was composed first of small spruce twigs, and then lined thickly with pine needles only, and set right in the sphagnum deeply cupped. As I had not flushed the bird, I poked my finger into it for investigation before I knew it to be a nest. Mr. Stuart's nest, which contained eggs, was simply lined with pine needles.

A nest found by Dr. Dickey in New Brunswick was a rather loose affair, "a weave of pretty sprays of sphagnum, the Knight's crest moss (*Hypnum dendroides*), some dark rootlets, a few culms of *Juncus*, and sedge (*Carex disperma*), and was nicely lined with the brown needles of the red pine (*Pinus resinosa*)."

Eggs.—W. J. Brown tells me that the usual number of eggs is three or four; in over 200 nests he has found only 25 sets of five. The eggs vary from ovate to short-ovate and are practically lusterless. The ground color is pure, dull white. There are sometimes a few small blotches, but more often fine dots sparingly scattered over the egg, or more or less grouped about the larger end, rarely concentrated into a wreath. These markings are in various shades of brown, "cinnamon-rufous," "walnut brown," "pinkish cinnamon," or "cinnamon-buff"; Bendire (1895) says: "Occasionally a specimen shows a speck or two of heliotrope purple." The measurements of 50 eggs average 17.4 by 13.4 millimeters; the eggs showing the four extremes measure **18.6** by 14.0, 18.1 by **14.7**, **16.0** by 13.0, and 17.0 by **12.3** millimeters.

Plumages.—Dr. Dwight (1900) says that the natal down is "brownish olive-green." He says of the juvenal plumage: "Upper parts, sides of head and throat, an obscure pectoral band, and lesser wing coverts olive-green, the crown feathers centrally darker. Wings and tail deep olive-brown; median and greater wing coverts edged with rich buff yellow forming two distinct wing bands, secondaries narrowly and tertiaries broadly edged with yellowish white. Below sulphur-yellow, including the orbital ring."

C. J. Maynard (1896) says that the nestlings are "quite slaty above, and much lighter below [than fall juvenals], being nearly white, and the darker areas are slaty." Dickey and van Rossem (1938) write:

The postjuvenal plumage (the "first winter plumage") is, as supposed by Dwight, not fully acquired until very late in the fall. A specimen taken October 8 is still largely in juvenal feather ventrally, while one taken November 30 still shows many juvenal feathers on the lower throat. This is the last part of the body plumage to be replaced. The juvenal remiges and rectrices are retained until April. In early April there commences a complete spring molt (the "first prenuptial") which involves the entire body and the replacement of the old, worn juvenal wing and tail feathers. A specimen collected April 6 has just commenced this molt and six others, taken between April 22 and April 30, represent every stage to its completion.

The adults vary a good deal in the time of completion of the fall molt. * * * In the present adult series, two specimens taken December 1 and 7, respectively, have nearly finished the body molt. The wing molt is extremely slow and, starting as it does about the time the last of the new body plumage has been acquired, takes most of the winter and early spring. * * * Thus the wing replacement fills in, roughly, the time interval between the winter (postnuptial) and spring (prenuptial) molts. The complete, spring, body molt of the adults (and juveniles) begins in late March and is finished by the end of April."

Food.—Professor Beal (1912) examined the stomachs of 103 yellow-bellied flycatchers, in which the contents were practically 97 percent animal matter and 3 percent vegetable matter. He says that "its bill of fare includes insects of a number of species which are injurious to garden, orchard, or forest, as the striped squash beetle, several species of weevils, tent caterpillars, and leaf rollers." Beetles amount to 16.53 percent, with less than 2 percent of useful species. He continues:

Hymenoptera amount to 46.25 percent of the food and were found in 81 stomachs. Of these, 48 contained ants, which amounted to 13.42 per cent of the whole. * * * This bird is probably the greatest eater of ants of any of the flycatchers and stands near the head in the eating of Hymenoptera in general. * * * Hemiptera were found in 33 stomachs only, and amount to 4.16 per cent of the food. * * * Diptera were contained in only 29 stomachs, but amounted to 14.89 per cent. * * * They belonged to several families, including the house fly, horsefly, and the long-legged crane fly. * * * Lepidoptera were found in 28 stomachs, of which 4 contained the adult moths and 24 their larvae or caterpillars. * * * Spiders are eaten by this bird to a greater extent than by any of the other flycatchers. They amount to 8.52 per cent of the food and are taken quite regularly through the season. Beginning with 2.21 per cent in May they gradually increase to 14.28 per cent in September. Hymenoptera alone stand higher in the food of that month. With the exception of certain nestlings no other species of bird yet studied shows so high a percentage of spiders in its food, though wrens and titmice and some warblers approach it.

He reports that "the vegetable food consists of a few small fruits, none of which are of domestic varieties, a few seeds of poison ivy, some cedar foliage, some scales from a bud, and rubbish. The poison ivy is the only thing of any interest and that was found in only one stomach."

Dr. J. M. Wheaton (1882) noticed a pair of birds, which he afterward identified as yellow-bellied flycatchers, "feeding on some elm saplings. Alighting near the bottom of the trunk they hopped from one to another of the alternate twigs, ascending spirally. Meantime they gathered their food, which I soon discovered to be small black ants."

Gardner P. Stickney wrote an interesting letter to Professor Beal about some unusual feeding habits of this flycatcher, which was afterward published (Stickney, 1923). During the latter part of September there was a heavy, wet snowstorm at his camp in northern Wisconsin, and after the storm many birds came to feed on the berries of the mountain-ash trees:

In the afternoon of the day following the storm, two yellow-bellied flycatchers appeared among the other birds in these small trees and seemed to be very fond of the mountain-ash berries. Instead of handling the berries as the other birds did, the flycatcher would pick a berry and crush it between its mandibles, getting out the pulp and dropping the skin, rather perfectly clean, to the ground. There were so many of the birds of various sorts and the berries were going so rapidly, that I detached four or five bunches of the berries and took them into camp.

It took only two or three days for the birds to otherwise entirely denude the trees of the berries and after the last berry had been picked from the trees, I noted two flycatchers hopping around on the ground and picking up the berries which had been dropped as the various species were feeding in the trees. The flycatchers were very tame and it occurred to me that I might feed them with some of the berries which I had previously picked and had in camp. Working very carefully in an hour or two I had these two yellow-bellied flycatchers on my knee, picking the mountain-ash berries from between my thumb and fore-finger. It was a very delightful experience and it was interesting to see how thoroughly the flycatchers would clean the berries, eating everything but the skins, which they invariably dropped to the ground. The birds stayed around for two or three days, in fact as long as I had any berries to feed them, and then disappeared. The last day that they were with us was bright and sunny and they spent most of their time fly-catching, but would occasionally come back to me and take a berry.

Behavior.—Mr. Maynard (1896) gives a very good account of the normal behavior of this flycatcher on its migrations, as most of us are likely to see it, in the following words:

Yellow-bellied Flycatchers are most decidedly, of all the genus, the true children of the shade, for they are seldom found elsewhere than in the thickest swamps. Even in these secluded retreats, they avoid the tops of the bushes, keeping well down in the dense foliage, often perching within a foot of the ground. Alder swamps which are so filled with undergrowth that it is difficult to force one's way through them, are the favorite resorts of these Flycatchers. It is extremely difficult to detect the presence of these little birds in such places, not only on account of the luxurious vegetation, but principally because they are extremely quiet, the only note which they utter during the migrations being a plaintive *pea* given only at intervals and, so low as to be inaudible a few yards distant. I have frequently entered a swamp in which I was certin some of these Flycatchers had taken refuge and have, at first, been unable to find a single specimen, but upon remaining quiet for a moment, I would hear the low *peas* in all directions. Guided by the sound of the nearest, I would proceed cautiously in its direction and after a moment's search, would see the bird as he sat on some low twig, occasionally launching outward for a short distance to catch a passing insect which his keen eye had informed him was especially palatable. As long as I remained perfectly still, the Flycatcher would pursue his vocations but upon my making the slightest movement, he would observe me and, giving a quick, upward flirt of his tail, would flit silently but with marvelous celerity among the brown stems of the alders, and skillfully wending his way through the labyrinth of twigs, vines, and leaves, he would almost instantly disappear.

In its summer home this flycatcher is equally shy and retiring, not easily approached and oftener heard than seen. It can be lured from its shady retreat by the well-known imitation of a squeaking mouse or of the cries of a bird in distress, a trick so often used by ornithologists to call birds into the open. The flycatcher comes eagerly enough, together with all the other small birds within hearing, but he eyes the intruder only momentarily before he discovers the deception and dashes back into cover; he cannot be so easily fooled again.

Voice.—The vocabulary of the yellow-bellied flycatcher is not elaborate or particularly musical, but it is quite distinctive and its few notes are easily recognized. Bendire (1895) writes: "Its call note is a low, plaintive 'peeh peeh,' the last part more emphasized; another, an alarm note, sounds like 'turri turri'; the same note I put down the previous season as 'trehe-eh, trehe-eh,' with the remark that it reminded me somewhat of the sound produced by sliding a finger over a violin string."

Dr. Dwight (Chapman, 1912) says: "The song is more suggestive of a sneeze on the bird's part than of any other sound with which it may be compared. It is an abrupt *psĕ-ĕk'*, almost in one explosive syllable, harsh like the deeper tones of a House Wren, and less musical than the similar but longer songs of the Alder or the Acadian Flycatcher. It is hardly surprising that the birds sing very little when we see with what a convulsive jerk of the head the notes are produced. Its plaintive call is far more melodious—a soft, mournful whistle consisting of two notes, the second higher pitched and prolonged, with rising inflection, resembling in a measure *chŭ-ē-ē'-p.*"

There are almost as many interpretations and renderings of this bird's notes as there are descriptions of them. L. M. Terrill (1915) writes:

Owing to the ventriloquial quality in the voice of this Flycatcher it was some time before I discovered the bird, probably the female, perched on a branch two feet from the ground. Its alarm was sharper and more abruptly ended than the call notes, sounding somewhat like the syllables "pee-wheep"; the first note suggesting the Wood Pewee and the latter the Alder Flycatcher. The last note commenced with a rolling and ended with a grating sound, as if the bird had snapped off the sound by suddenly closing its mandibles, accompanied by a tail and bodily twitching that indicated considerable effort.

Hearing the usual call notes one would not suspect much effort, in fact, their ordinary notes, "peeh-peeh," the latter slightly accented and prolonged, have none of the explosiveness of "alnorum," but are peculiarly soft, drowsy, and effortless.

Various other renderings of the call note are: *Chee-weep, chee-weep; phee-i; puh-eé; peá;* or *too-weé.* There seems to be some confusion, or difference of opinion, among observers as to which note is the alarm note, which the song, and which the call note; and I shall not attempt to differentiate between them. Francis H. Allen (MS.) thinks that "Dwight's rendering of the song, in Chapman's Handbook, is very good when it is heard near at hand; heard farther off, the song sounds more like *killink.*" Forbush (1927) says that "on its breeding grounds a song is attempted with very indifferent success, given by Dr. Hoy as *pea-wāyk-pea-wāyk* several times repeated; it is soft and 'not unpleasant.'" Dr. Glover M. Allen (1903), on June 19, in New Hampshire, "heard one of these birds give a peculiar flight song, just after sunset. It flew slantingly upward for some twenty feet and repeated a number of times alternately its ordinary '*pu-ee*' and '*killick.*'"

Field marks.—This is a small flycatcher, olive-green above and more decidedly yellow below, including the throat, than any other small flycatcher in eastern North America; in fall, when some other flycatchers are somewhat yellowish below, this character is not so prominent, though it is yellower then than the others; and it has a conspicuous yellow eye ring.

Fall.—The yellow-bellied flycatchers begin to leave their northern breeding grounds during the last half of August. They usually pass through Massachusetts between August 25 and September 25, with a few scattering late dates up to December 6. They frequent the same kind of dense, swampy thickets as in spring and can be seen only by the most ardent observers who are willing to look for them in their still, shady retreats.

Taverner and Swales (1907) say that, in Ontario, "this species seems to start on its southward migration about the middle of August," but, at Point Pelee, "others come in before the earlier arrivals leave, and many linger until well into September."

Winter.—In El Salvador, Dickey and van Rossem (1938) report this flycatcher as "fairly common in fall, winter, and spring throughout the Lower Arid Tropical Zone and, locally, in the lower edge of the oak-pine association in the Arid Upper Tropical. Extreme elevations are sea level and 3,600 feet. Dates of arrival and departure are October 8 and April 30.

"The yellow-bellied flycatcher, while confined to levels below 3,600 feet, was, during the proper seasons, fairly numerous and evenly distributed. Although found in all sorts of woodland it shows preference for thin, open undergrowth beneath heavy forest. * * *

"It was noticeable that the winter population remained fixed, and there was little or no local shifting once the winter quarters were selected. Each individual had its own particular patch of shrubbery where it could be seen or heard at all times of the day."

DISTRIBUTION

Range.—North and Central America east of the Rocky Mountains; accidental in British Columbia and Greenland.

Breeding range.—The yellow-bellied flycatcher breeds **north** to southern Mackenzie (probably Taltson River and Soulier Lake); Ontario (north shore of Lake Superior); Quebec (Lake Timiskaming, Lake Mistassini, Godbout, and Anticosti Island); and Newfoundland (Port Saunders). **East** to Newfoundland (Port Saunders, Nicholsville, and 3-pond Barrens); Nova Scotia (Dartmouth); New Brunswick (Grand Manan); Maine (Ellsworth and Auburn); southern New Hampshire (Mount Monadnock); southeastern New York (Slide Mountain); and northeastern Pennsylvania (Mount Pocono). **South** to northern Pennsylvania (Mount Pocono); southern Ontario (probably London and Listowel); southern Wisconsin (Bark River and Albion); northern Minnesota (Lake Itasca and Moose River); southern Manitoba (probably Winnipeg, probably Portage la Prairie, and Oak Lake); probably southern Saskatchewan (Johnston Lake); and central Alberta (Edmonton and Glenevis). **West** to Alberta (Glenevis, Belvedere, probably Lake Athabaska, and probably Smith's Landing) and Mackenzie (probably Taltson River).

Winter range.—During the winter season these birds appear to be concentrated in eastern Mexico **north** to central Tamaulipas (Rio

Martinez); and **south** through other Central American countries to Panama (Divala, Cava, and Veragua).

Spring migration.—Early dates of spring arrival are: Florida—St. Marks, May 2. Georgia—Atlanta, May 6. Virginia—Lynchburg, May 9. District of Columbia—Washington, May 9. New Jersey—Milltown, May 10. Massachusetts—North Amherst, May 8. New Hampshire—Monadnock, May 18. New Brunswick—St. John, May 19. Quebec—Lake Mistassini, June 2. Louisiana—Bayou Sara, April 26. Mississippi—Biloxi, April 30. Tennessee—Athens, April 25. Kentucky—Lexington, May 1. Missouri—St. Louis, May 8. Illinois—Chicago, May 11. Ontario—Ottawa, May 19. Iowa—Grinnell, May 14. Minnesota—Lanesboro, May 19. Alberta—Lake Athabaska, June 3.

Fall migration.—Late dates of fall departure are: Alberta—Camrose, August 26. Manitoba—Shoal Lake, September 7. Minnesota—Minneapolis, September 28. Wisconsin—Milwaukee, September 25. Iowa—Sigourney, September 25. Ontario—Toronto, October 12. Michigan—Detroit, September 27. Ohio—Hillsboro, October 6. Illinois—Rantoul, October 1. Missouri—St. Louis, October 5. Mississippi—Biloxi, October 16. New Brunswick—St. John, September 4. Quebec—Montreal, Sepember 24. Maine—Portland, September 21. Massachusetts—Harvard, September 23. New Jersey—Morristown, October 9. Virginia—Lynchburg, October 9. North Carolina—Piney Creek, October 2. South Carolina—Porchers Bluff, October 8. Florida—Leon County, October 11.

Casual records.—An adult male was collected at Pike River, in the Atlin District of British Columbia, on August 3, 1914, and another was taken at Hazelton in this Province on July 24, 1913. Two specimens were taken in 1853 at Godthaab, Greenland, and another was obtained at sea off Cape Farewell, in September 1878.

Egg dates.—Alberta: 3 records, June 8 to 20.

Maine: 11 records, June 1 to July 12.

New York: 10 records, June 10 to July 4.

EMPIDONAX VIRESCENS (Vieillot)

ACADIAN FLYCATCHER

PLATES 23, 24

HABITS

CONTRIBUTED BY BAYARD HENDERSON CHRISTY

This small flycatcher is hardly to be identified by sight alone. So closely indeed does it resemble the related species *trailli* and *minimus* that even the most skillful field observers cannot certainly distinguish between them.   The greater then is the need for dwelling upon characteristics of range, habitat, mode of nest building, and notes, for in each of these *virescens* is peculiar, and by one or another of them, in the summer season at least, identification may be made sure.

This is a bird of the Austral Zones; *trailli* and *minimus* are of more northerly range; and it is only northward of Mason and Dixon's line that there is overlapping.   Southward of that line *virescens* is of the three the only breeding species.

E. trailli is a bird of alder thickets, of willow-grown stream margins; *minimus* is found in orchards and in open, bush-grown places, cultivated and waste; but *virescens* is found in the forest: in cypress swamps, in heavily wooded bottomlands, in the depths of wooded ravines.

Of its habitat in Florida, Williams (1928) quotes from Herbert L. Stoddard this—"About four other pairs inhabit a half mile of this strip of swamp and I find the bird a fairly common summer resident in similar situations along small water courses in northern Leon County."   Kopman (1915), writing of the bird in Louisiana, says it is found "in swampy woods of every character.   It is evenly distributed throughout the wet wooded lands of the fertile alluvial region, and occurs wherever there are river swamps and creek bottoms in other sections."   In the Okefenokee Swamp, of Georgia, say Wright and Harper (1913), the bird "finds a congenial haunt in the gloom of the cypress 'bays,' where one often hears its note as he paddles along the narrow runs.   It also frequents the hammocks and the cypress ponds."   In the Great Dismal Swamp, of Virginia, according to Daniel (1902), it is "not uncommon along the margins of the inlets, notably where the foliage forms a canopy over the water."   In western North Carolina, wrote Brewster (1886), "everywhere below 3000 feet this Flycatcher is a very common species, inhabiting all kinds of cover, but occurring most numerously in rhododendron thickets bordering streams, where its abrupt, explosive note of *wicky-up* could be heard at all times of the day."   Schorger (1927), writing of conditions in Wisconsin, says, "The essential requirement of the Acadian Flycatcher appears to be a large tract of undisturbed timber.   The typical habitat is a deep, well-wooded ravine having a rocky stream bed, which is usually dry.   It may also be looked for in the heavy timber of the river bottoms and in tamarack swamps in the southern portion of the state."   W. E. Saunders (1909), who discovered the bird in Ontario, wrote, "About fifty miles southeast of Detroit and only a few miles from Lake Erie there was formerly an immense black ash swamp, portions of which are still in existence, and it was in these, where the mosquitoes were of sufficient quantity to feed a large number of Flycatchers, that I found the Acadians." Simmons (1925) characterized its habitat in central Texas as "open glades in timber near running water, and along creeks; wooded ravines; rarely in heavy woodlands, usually in dry, deciduous second-growths along water courses; Spanish oaks overhanging creek valleys in hills; wooded roadsides; deep, shady woodlands or second-growth forests, watered by small streams, and with either little or much undergrowth and vine-tangle."

In my own territory, of the eroded ravines of southwestern Pennsylvania, the bird is abundant in its peculiar habitat. It is found along streams in wooded ravines. The timber is almost wholly of hardwoods; and it is where the pendent lower branches of great beeches overhang the small streams that these birds are most likely to be found.

It is because of the remarkable likenesses between the small flycatchers of the genus *Empidonax* that the early ornithologists, having nothing to go by and necessarily feeling their way, gave confused and misleading accounts. Wilson (1810), for instance, characterized the Acadian flycatcher excellently well. (He called it the small green-crested flycatcher.) He manifestly knew the bird in the field; but he added to his account a description of nest and eggs that certainly belongs, not to *virescens*, but to *minimus*. Audubon (1831) first described *trailli*, and the brothers Baird (1843), *minimus*. It remained for Henshaw (1876), Wheaton (1882), Brewster (1895), and Oberholser (1918) to complete accurate analyses of the three species.

It is, indeed, in consequence of confusion that *virescens* carries today the inappropriate vernacular name, *Acadian* flycatcher. The name was applied initially to a bird taken in Acadia, that is to say in Nova Scotia. The bird, so taken, was held to be the type specimen, until, in the light of fuller knowledge, the truth appeared that this species never reaches Nova Scotia, and that the bird first called *acadicus* must have been of one of the other species. The technical name *acadicus* was thereupon changed to *virescens;* but, in English, Acadian flycatcher had become too well established and has not been supplanted.

Spring.—Returning from their wintering grounds in northwestern South America, the birds begin to enter their breeding range early in April and by the middle of May have completed their migration. S. S. Dickey, of Waynesburg, Greene County, Pa., writes (MS.): "After their return from winter quarters the males are quite combative. They may then be seen chasing one another up and down the courses of the ravines. So intent are they that, heedless of a man's presence, they will dash up almost into his face. While they primarily fight off other males of their own species, I have noticed that they attack and drive vireos, tanagers, and warblers from their nesting territories."

Nesting.—E. trailli and *minimus* both build cup-shaped nests, resting in crotches between upstanding shoots or twigs of such trees of low growth as alder and sumac; *virescens* swings her nest hammock-wise between horizontally spreading twigs. It is a frail, shallow basket of fine, dry plant stems or other fibrous strands, hung by its rim between slender forked twigs. In the region of my own observation (southwestern Pennsylvania), great beech trees stand along the narrow bottomlands of the wooded ravines; the lower branches droop and spread out horizontally toward their leaf-bearing tips; and it is in these places that the flycatchers commonly place their nests. They may hang them at a height of 8 or 10 feet, and often directly over some pool in the course of the stream.

At the time of nest building (the last week in May) the canker-worms (*Paleacrita vernata*) are in their heyday. Their threads hang everywhere through the woods, and as one presses through the undergrowth they cling to one's perspiring forehead; the caterpillars themselves, "measuring worms," caught upon the clothing, climb upward to one's neck. In the hanging threads of silk the falling withered staminate flowers of forest trees and the scales of opening buds are caught. These airy festoons the flycatchers catch up and carry to their nests. In my region, at least, the wild silk so gathered forms an ever-present, and, as I judge, an essential nest-building material. The strands are collected in such quantity as to form a web, and this web commonly swathes the supporting crotch. It spreads upon the nest rim and in it the ends of the frail and loose vegetable fibres are enmeshed. The effect is that the structure, of flimsiest appearance, is in fact adequate to outlast its usefulness. I quote from my own notebook of observations made in Allegheny County, Pa.: "May 30. Acadian flycatchers nest-building at the tip of a pendent lowest branch of a large beech, about 15 feet up, and immediately above the stream. Nest seemingly all but finished. The birds paid little attention to my presence. One, the female presumably, seemed to do all the building. The male called at intervals, and once or twice it flew up as the female came in to the nest. The female kept calling too, more frequently than the male, and continued even when she was in her nest. Her call note is softer than the male's and less emphatic. Once when she was in the nest I heard a reiterated chattering note, and occasionally (also from the nest) a softer, whistled note with falling inflection. When the male approached, the notes of one or perhaps of both birds were softer and more musical.

"As the female repeatedly flew in I thought once or twice that I detected material in her bill, but, standing in shadow and looking upward toward a bright sky, I could not be sure. Generally I could see nothing. Seated within her nest, the bird kept reaching over the rim and seemed to be engaged in drawing strands of material (silk?) over the rim of the nest inward. This was the oft-repeated and, indeed, the chief action of the bird at the nest. Once or twice she remained for a minute or more, but ordinarily her visits were of shorter duration."

When the building is done, threads of silk with the entangled blossoms are commonly left, streaming from the rim of the nest. The appearance from without is lacking in symmetry; it is altogether casual; and the nest on that account must commonly be overlooked by marauders. It would be an interesting matter of inquiry what the value of wild silk as nest-building material may be to such birds as the wood pewee, the hummingbirds, the gnatcatcher, and the vireos.

I collected the nest about which I have spoken, after the brood had flown from it, and took it to Dr. O. E. Jennings, botanist of the Carnegie Museum, asking him to be good enough to analyse the materials. This he has generously done and has given me the following report:

"The main body of the nest is composed of a tangled mass of very slender herbaceous stems and branches mostly averaging between $\frac{1}{4}$ and 1 mm. in diameter. Many of these are specked with decay fungi and were probably dead when gathered. It is noteworthy that no grass leaves were used in the nest.

"There are a few strips of grapevine bark, but these were apparently of not much importance and are not much interwoven. There are, however, a few coiled grapevine tendrils (small ones) which by their coiled and twisted character help to hold the loose nest together. The main binding material in the nest consists of cobwebs to which the various small twigs and also bits of dead leaves of various kinds are sticking. The small slender twigs making up most of the framework evidently come from woodland herbs rather than those growing out in the open, the woodland habitat also being indicated by the thin nature of the bits of leaves. It seems rather likely that such woodland plants as sweet cicely and honewort are the source of this material, and it is also noteworthy that no grass leaves are involved, grasses being scarce in shaded woodlands."

Here is evidence that the bird during nesting season confines itself to a narrow territory. And yet, somewhat at variance with this, is a report from Prof. Maurice Brooks (MS.), of Morgantown, W. Va., who writes of finding nests woven of poverty grass (*Danthonia spicata*), a material that must have been brought more than a quarter of a mile, from the open fields where it could be found. Additional descriptions of nests *in situ* follow:

S. F. Rathbun writes (MS.) of a nest in New York, on the shore of Lake Ontario, saying, "The other day I found a green crested fly-catcher's nest [=*virescens*] when at the lake. It was built on the horizontal fork of a maple sapling at a height of 12 feet. It was made of very small hemlock twigs, interwoven with the maple blossoms one sees so plentifully now. That was the body of the nest, and to its under part the birds had attached, by using spiders' web or fila-ments, a loose string of the maple blossoms, until it formed a pen-dulous mass, 8 inches or more long, that came conically to a point; and this would sway in every breath of wind. The nest was a very beautiful affair."

W. E. Saunders (1910), describing a nest taken in southwestern Ontario, says, "The nest is composed of fine grasses and rootlets bound together on the outside by what appears to be caterpillar web. The well known habit of this species of making the nest appear like an accidental bunch of drift, by the addition of loose flowers of alder, walnut or oak, is varied in this instance by the substitution of a large number of bud scales, apparently of beech. The nest is, as usual, shallow, the cavity measuring ⅞ of an inch deep, by 1½ inches wide, while the external measurements are 5 x 2."

Elliott Coues (1880) was one of the first to make clear the distinc-tion between this species and *E. trailli*. He quotes from a corre-spondent, Otto Widmann, who had sent him specimen nests of both, and had written, saying, "I have seen many nests of *acadicus* [= *virescens*] in the woods, as they are easily found, hanging in conspicuous places between 12 and 25 feet above ground. They were all made alike, the only difference being that some were more difficult to collect than others, hanging on slender limbs far from the trunk of the tree." Dr. Coues then describes the Acadian nests. These Mr. Widmann had taken in St. Louis County, Mo. He continues:

The two nests of this species [*acadicus*] are *strikingly* different from the three of *trailli* in structure, in material, and in position. They appear to have been taken from long, slender, horizontal branchlets, in the *horizontal* forks of which they rest. They are *shallow* nests—in fact, rather saucer-like than cup-shaped, some 3½ inches across outside, by less than 2 inches in depth; the cavity over 2 inches across the brim, by scarcely 1 inch in depth. They are very light, "open-work" structures, so thinly floored that the eggs may have been visible to one looking up from below; and the walls, though more com-pact, still let daylight through on all sides. These nests, in short, may be compared to light hammocks swung between forks. Each is composed almost entirely of long walnut (*Carya*) aments, which, drooping in slender sprays from all sides, give a tasteful, airy effect to these pretty structures. There is a slight lining in each case of slender grass-stems and still finer rootlets, loosely interlaid in every direction on the bottom, rather circularly disposed around the brim. These specimens were taken June 13 and 18, 1879, in hickory woods, at altitudes of 10 and 15 feet.

Natural historians, in their thoughtlessness, are apt to speak slightingly of the Acadian's nest-building ability; but they produce no evidence of inadequacy; and, when attention is given to the matter, her nest will be seen to be, as such products commonly are, a perfect piece of artistry, utterly sufficient to her need. As for durability, Prof. Brooks writes (MS.), "The nests are highly deceptive in appearance. They are so loosely woven as frequently to make possible the counting of the eggs from below. Despite this apparent looseness, the nests are resistant to winter storms, and it is not unusual to find solidly suspended nests in early spring, before the leaves have appeared. I saw one nest which lasted through two winters, being still fairly firm at the beginning of the third season."

The material that, caught in caterpillar thread, hangs from the nests is commonly, as has been said, small faded flowers or bud scales; it may be bits of bark, the dust of wood-borers, or whatever litter sifts through the woods.

In the South, Spanish moss is a frequently used nesting material. Stockard (1905), writing of conditions in Mississippi, says, "Two nests taken in Adams County were very interestingly constructed, being composed entirely of Spanish moss woven between the prongs of small elm forks. A surplus of moss was used so that long beards or streamers of it hung down for a length of eighteen inches below the actual nest. This arrangement gave the exact appearance of ordinary bunches of this gray moss hanging from the branches. Both nests would have been passed unnoticed but for the fact that the birds flew off as I passed under the limbs."

Williams (1928) says of a nest collected in Leon County, Fla., that it was composed "entirely of Spanish moss and imbedded in a cluster of that epiphyte."

I have before me the records of 55 nests, 49 collected in the North, 6 in the South. Nearly all were placed in the lower branches of large trees; a few in saplings, shrubs, and bushes. Of 44 northern nests, 20 were in beech trees, 8 in maples, 3 in hemlocks, 2 in hickories, 2 in white oaks; and 1 each in elm, locust, linden, walnut, elder, wild apple, red haw, hazel, and witch-hazel. Of the 6 southern nests, 2 were hung in cypresses, 2 in water oaks, and 1 each in elm and sweet gum. In southwestern Pennsylvania the nest is commonly placed in a beech; but in South Carolina Wayne (1910) has found it to be "invariably built in the forks of a dogwood tree (*Cornus florida*)"; and Brimley (1889), writing of nests found in the vicinity of Raleigh, N. C., says that "about half of them are placed in small dogwoods, the balance being in post oak, water oak, sweet gum, birch and tulip poplar trees, sweet gum being second favorite to dogwood."

In height the average of 44 nests was 10½ feet. Thirty-three were between 8 and 20 feet; one was as low at 3½ feet; two were as high as 25. Twelve overhung water—a flowing stream or a cypress swamp.

Eggs.—[AUTHOR'S NOTE: The Acadian flycatcher lays two to four eggs to a set, usually three. They vary from ovate, the commonest shape, to elliptical-ovate and have very little or no gloss. The ground color varies from creamy white to buffy white. They are sparingly marked with small spots or minute dots, generally mainly near the larger end of the egg, of different shades of brown, such as "liver brown" or "ferruginous," the darker colors being commoner. The measurements of 50 eggs average 18.4 by 13.8 millimeters; the eggs showing the four extremes measure **19.8** by 14.5, 20.1 by **14.7, 16.5** by 13.7, and 16.8 by **12.7** millimeters.]

Young.—The period of incubation, as observed by Harold M. Holland (MS.), of Galesburg, Ill., is 13 days.

Butler (1897) makes record, *fide* V. H. Barnett, of a young bird that could not fly well, shot in Warren County, Ind., on September 25. This, of course, is abnormally late.

Plumages.—[AUTHOR'S NOTE: The small, nearly naked nestling is sparingly clothed in whitish down, which adheres to the tips of the juvenal plumage. A half-grown nestling before me has a "light brownish olive" crown, with pale buff edgings; the back and rump are similar, with the widest edgings on the rump; the median and greater wing coverts on the half-grown wings are broadly tipped with "light ochraceous-buff"; the secondaries are narrowly and the tertials more broadly edged with the same color; the chin is white, the chest tinged with olive-gray, and the abdomen is white to yellowish white. In an older bird, taken in August and fully grown, the edgings on the upper parts have worn away and the plumage has a greener cast; the under parts are all white, except for an obscure olive-gray pectoral band.

Apparently a partial postjuvenal molt of the body plumage, mainly after the birds have left for the south, produces a first winter plumage which is not very different from the above. Adults probably have a complete postnuptial molt after migrating, but, for lack of winter specimens, we do not know what molts take place during the winter and early spring.]

Food.—Prof. Beal (1912), of the Biological Survey, examined the contents of the stomachs of 100 Acadian flycatchers "collected in 14 States, the District of Columbia, and Canada, and from April to October," and found "97.05 per cent of animal matter and 2.95 per cent of vegetable." His more detailed report is as follows:

Animal food.—Beetles are eaten to the extent of 13.76 per cent of the whole food. Of these 1.66 per cent are of the three prominently useful families (Carabidæ, Cicindelidæ, and Coccinellidæ). The others were of more or less harmful families and include such well known pests as spotted cucumber beetle (*Diabrotica 12-punctata*), rose beetle (*Macrodactylus subspinosa*), rice weevil (*Calandra oryzæ*), and a scolytid. Beetles were found in 76 stomachs and were eaten quite regularly till October, when none were taken.

Wasps, bees, and ants amounted to 39.93, or practically 40, per cent of the bird's food, and are eaten so regularly that no month's consumption falls much below the average. They were found in 76 stomachs, or 84 percent of all, and four were entirely filled with them. Ants were contained in 29 stomachs and parasitic species in 13, but some of the latter may have been overlooked, owing to their broken condition. Hymenoptera as a whole are the largest item of animal food with this as well as most other flycatchers. Flies (Diptera) amount to 8.15 percent of the food, and are not taken as regularly as Hymenoptera and in October are not eaten at all. They were noted in 39 stomachs and were the sole contents of 1. Most of them are of the housefly family, but a few long-legged crane flies were found in 5 stomachs. Bugs are eaten still less than flies. They amount to 6.03 percent, but are not taken very regularly and not at all in October. They were contained in 29 stomachs and consisted of such families as the leaf hoppers, tree hoppers, stinkbugs, and assassin bugs.

Orthoptera were found in 1 stomach taken in Florida in April and 2 collected in Pennsylvania in September, but the percentage in each of these 3 stomachs was so great that the amount for the whole season is 6.38 per cent of the food, or more than the last item. The contents of the Florida stomach could not be determined further than that they were orthopterous, but the contents of the other 2 were identified as *Œcanthus niveus*, the snowy tree cricket, known in some places as the August bird. As these creatures are rather nocturnal in their habits and not much given to flying at any time, it is rather surprising to find that a flycatcher had nearly filled its stomach with them.

Moths, in both the adult and larval form (caterpillars), are second in importance in the animal food. They are taken pretty regularly in every month, but with some falling off in July. The amount for the whole season is 18.87 per cent. They were found in 38 stomachs, of which 31 contained caterpillars and 8 held moths; 3 contained no other food. No special pest was observed among them. A few miscellaneous insects, such as dragon flies, scorpion flies, and a few insects not identified, amount to 0.99 per cent, and have no special interest. Spiders and millepeds were eaten in moderate quantities from April to August. They amount to 2.94 per cent and complete the quota of animal food. As usual, many of them were the long-legged harvestmen or daddy longlegs. . . .

Vegetable food.—Fruit was found in 5 stomachs and vegetable refuse in 1. There were a few seeds of blackberries or raspberries, and these were the only things that could have been the product of cultivation. The rest was wild fruit of no economic value.

Summary.—The habits of the Acadian flycatcher do not lead it to the garden or orchard, and its food has little direct economic interest. It does not catch many useful insects and, as it does not prey upon any product of cultivation, it may well be considered as one of those species whose function is to help keep the great flood of insect life down to a level compatible with the best interests of other forms of life.

It is a well-established result of observation that an animal and its environment are accommodated to each other as hand and glove;

and accordingly it is a fair inference that if only we had eyes to see we should discover some peculiarity—in food supply, in all likelihood—that would afford us the reason why this little bird is not widely and generally distributed but found only in restricted areas of specialized sort.

Prof. Beal's report, informing though it is, does not afford certain guidance in following that inquiry. The food found in a bird's stomach is broken and macerated. Here and there a particular insect may be recognized, but for the greater part it is hardly possible to go beyond family or genus; and, since the genera of insects are widely distributed, one could hardly hope to discover a predominance of forms characteristic of forested ravines. One clue, however, lying in Prof. Beal's data, is this: The score that *virescens* makes in the consumption of Lepidoptera is exceptionally high. With *difficilis*, Lepidoptera make up 5.68 percent of the total consumption; with *trailli*, 7.73 percent; with *minimus*, 7.27 percent; but with *virescens* Lepidoptera make up 18.87 percent of the whole, and the creatures are largely consumed when in larval state. Among the Lepidoptera, *Paleacrita vernata*, the common canker moth, is worthy of note, because it abounds in the haunts of *virescens* and because, in the northern latitudes at least, the silk that its caterpillar produces is, as has been noted, an invariable and seemingly an indispensable item of this bird's nest-building material. These caterpillars are abundant late in May and early in June; in July they are gone; but after a month or more those of another species succeed them. The indications are quite definite that, in the North at least, the cankerworm fills an important place in the economy of the Acadian flycatcher.

Behavior.—Generally speaking, *virescens* is a very inconspicuous bird. It must be sought for to be seen. Throughout the summer it does not leave its woodland habitat, and while nesting its territory may be no more than one or two hundred yards across. It may ordinarily be found perched in deep shade, less than 20 feet from the ground, and well beneath the canopy of foliage. It flits its tail as it calls, and it hawks its prey in the manner common to all flycatchers. Todd (1940) says that it is shy and suspicious and contrives to keep well out of sight; and adds that "like other flycatchers, it is a solitary bird; each pair, after having settled for the season, has its own definite territorial limits beyond which it does not pass unchallenged." Langille (1884), writing from western New York, says:

In a shadowy part of the woods, where young hemlocks are thickly interspersed, I hear sharp, quick notes, *pee-whee, quee-ree-ee*, which I at once recognize as those of the Small Green-crested Flycatcher (*Empidonax acadicus*), a very common summer resident of our upland woods. I look sharply into the shadows for some time before I get sight of it. It is perched on a dead limb, near the base of a small hemlock; and always accompanies its note with a quick jerk of the tail. Like the rest of the Flycatchers, it sits still on its perch and waits for its prey; and when that prey appears, be it beetle, fly, or moth, it darts quickly after it, cutting a smooth curve, which is sure to intercept it, and seizing it with a sharp click of the mandibles. With its quick, well-directed movement, the broad gape of its deeply cleft mouth and tangle of bristles on either side of it, there is but a slim chance of escape for its victim.

In the matter of confidence as against timidity, testimonies are various and inconsistent. Brownell (1887), writing from Plymouth, Mich., remarks that "the sharp chirp of the female often repeated was an infallible guide to its nest." Brimley (1889), from Raleigh, N. C., says that "they leave [the nest] so quietly and unobtrusively on the approach of man as to make it next to impossible to find the nest by flushing the bird." Similarly, Wayne (1910), writing from Charleston, S. C., says, "When the birds are building nests or in cubating their eggs they are always extremely shy, and leave the nest long before a person has approached within twenty-five yards of its location." Stockard (1905) writing from Adams County, Miss., calls this "a species with most retiring habits. The nest may be found and removed without the birds having made their appearance or the slightest sound." He describes two nests particularly and adds that he would have failed to find them, "but for the fact that the birds flew off as I passed under the limbs." Brewster (1886), writing from western North Carolina, says, "It is one of the tamest and least suspicious of the small Flycatchers, but owing to its retiring disposition, and habit of sitting perfectly motionless among the foliage, it is much oftener heard than seen." Sutton (1928) tells of an experience in Crawford County, Pa.: "By closely watching a female on May 26 I found a nest just ready for eggs partially suspended from a long, swaying beech limb. On May 30 the nest was complete. On June 3 it held three eggs. This nest, which was over twenty feet from ground, was secured by making a huge tripod from three saplings bound together at one end. Though the branch was considerably shaken and swayed the female would not leave until I touched her." I have myself stood within 10 feet of a bird upon her nest in a low sapling. And no doubt it is true that the stage to which incubation has advanced is a factor in the equipoise between behaviors that we call timid and courageous.

It has sometimes been intimated that the bird breeds in colonies (see, for instance, Porter, 1907), but this, no doubt, is a hasty conclusion, drawn from the fact that within the now greatly restricted areas of suitable habitat it ordinarily is common. In the course of half a mile along a woodland stream I have found as many as six pairs resident, but have discovered no evidence of interdependence between them. In the same territory pairs of ovenbirds were as closely spaced, and were far more abundant, since they were settled over the slopes of the ravine, while the flycatchers were confined to the margin of the stream.

Coues (1880) quotes Widmann, who, speaking of conditions in St. Louis County, Mo., says of *acadicus* (=*virescens*) that, while found in the forests only, it there is "very abundant—that means, one pair to every few acres."

Dr. Sutton writes (MS.) from Bethany, W. Va.: "The Acadian Flycatcher may be more active in the late summer than at any other season. Watchful as it is for passing insects, it must now devote part of its time to the keeping of its plumage free of the webs encountered in capturing its prey. I have at this season watched the bird on its perch, fluttering wings and tail, running its bill rapidly along its feathers, scratching its face with its foot, and so casting off these shreds of silk."

If it were possible to rename the bird, it should be called *Empidonax sericiferens*, the silk-bearing *Empidonax*.

Voice.—The oft-repeated cry when on the nesting ground is of two syllables run into one: abrupt, startling, a hiccup of a song. *Ka-zeep*, one writer vocalizes it; *Wicky-up*, another. And Warren (1890) gives, as a vernacular name, *Hick-up*. Wilson (1810) wrote: "This bird is but little known. It inhabits the deepest, thick shaded, solitary parts of the woods, sits generally on the lower branches, utters, every half minute or so, a sudden sharp squeak, which is heard a considerable way thro the woods; and as it flies from one tree to another has a low querulous note, something like the twitterings of chickens nestling under the wings of the hen. On alighting this sound ceases; and it utters its note as before." (A good instance this of the Father of American Ornithology at his best in characterization.)

Bradford Torrey (1896) speaks of "the petulant, snappish cry of an Acadian flycatcher." Aretas A. Saunders (1935) writes:

The song is two-syllabled, the second note higher than the first, with a slight burr in it, and a little longer and strongly accented, like "*ka-zeep.*" This is repeated at short intervals through the breeding season.

The call note is a simple "*peet,*" and the bird also sometimes produces a series of short musical notes all on one pitch, "*we-we-we-we-we,*" which resemble strongly the sound produced by the Mourning Dove in flight, and are supposed by many to be made by the wings. The bird sometimes produces this sound when in flight, as the Dove most frequently does.

Of this sound, supposed by many to be made by the wings, Brewster (1882a) wrote, "They [the Acadian Flycatchers] had another note also which was much like the whistling of wings. I afterwards satisfied myself that this sound was a vocal one."

An excellent record of the calls of this bird is that of Simmons (1925) : "A single *peet* or *spee*, with a twitch of the tail; or a louder *pee-e-yuk*, uttered with wings trembling and bill pointed up. A soft, murmuring, whistling call, rarely heard, uttered during a short, fluttering flight. A loud, quick, and emphatic, or again, faint and fretful, *tshee-kee*, *tshee-kee*, or "what-d'-see, what-d'-ye-see." A short *wick-up* or *hick-up*, followed by a harsh, abrupt *queep-queep*. A short *whoty-whoty*. Calls frequently uttered." S. S. Dickey wries (MS.) : "They are markedly noisy just before twilight, when the woods are still."

The song period, as given by Baerg (1930) for Arkansas, is from May 1–June 4 to July 26–August 28.

Field marks.—As has been said, the bird must be identified in the field by habitat and note (or by nest structure, if a nest be found). As to appearance, here are the baffling findings of those best qualified to speak. Griscom (1923), for example, says:

The identifiability in life of the species of *Empidonax* is a matter to which Messrs. W. DeW. Miller, J. T. Nichols, C. H. Rogers and the writer have given special attention. Collecting has proved that in spite of the greatest care, it is *impossible* to be absolutely certain in separating the Acadian, Alder and Least Flycatchers by color characters even in the spring. In the fall plumage it is out of the question, the determination of museum skins often being very critical. It is quite true that extremes in size or highly plumaged individuals can often be named with approximate certainty, but even here collecting has proved a low percentage of error. * * * They are exasperating birds. Every spring and fall I see individuals which I am convinced are one or the other, but all too rarely will they open their mouths and sing their names.

Sutton (1928) writes: "In the fall this species [*virescens*] is almost impossible to identify, save by collecting specimens. However we are positive of three records * * * (identification through call-note.) * * * This call-note might be written 'weece', energetically uttered."

Todd (1940) writes: "It is necessary to discard or at least to question the accounts of the earlier authors, who more or less confused this with other species * * *. As a matter of fact, once one has become familiar with their haunts and habits and in particular with their call notes, it is probably easier to distinguish them in life than in preserved specimens."

Enemies.—Occasionally, though not often, this bird becomes a host for the cowbird (*Molothrus ater*). Evermann (1889), in writing of the birds of Carroll County, Ind., said of the Acadian flycatcher that it is "one of the most frequent victims of the Cowbird"; but Friedmann (1929), after a careful review, writes: "I know of but few definite records, although many writers claim to have found parasitized nests of the Acadian Flycatcher. Twenty-two records, ranging from New England, and Pennsylvania, west to Ohio, Indiana, Illinois, and Michigan have been found. * * * Of these twenty-two no fewer than twelve come from southern Pennsylvania (Jacobs, Oologist, May, 1924, pp. 52–54)."

In qualification of Mr. Jacobs's finding, it should be observed that the nests of *virescens* were but 12 in number, in a total of 234 nests of various species that held cowbird eggs.

Bendire (1895) says: "Occasionally the Acadian Flycatcher builds a double nest—for instance, when a Cowbird has deposited an egg in one just completed, before the owner has laid in it. Mr. W. E. Loucks, of Peoria, Illinois, sends me such a record. The nest found by him contained a Cowbird's egg in the lower story and three fresh eggs in the upper one."

But for this, it does not appear that the Acadian flycatcher is a species heavily preyed upon. The number of eggs that it lays and the fact that (so far as is known) no more than one brood is normally raised in a season indicate that it is a vigorous species, with few enemies.

Miller (1915) has remarked upon the decrease in the numbers of Acadian flycatchers in the vicinity of Philadelphia; and Griscom (1923) has made similar comment upon conditions in the vicinity of New York City. The latter is at a loss to find explanation; but the former undoubtedly recognizes the true reason, in observing that in Fairmount Park, Philadelphia, where the extensive forests cover the banks of the stream with both conifers and deciduous trees, the species is apparently increasing. When the forests are cut away the bird disappears. Certainly in its proper environment it continues to be abundant.

DISTRIBUTION

Range.—Eastern United States and southern Ontario; winters in northwestern South America.

Breeding range.—The breeding range of the Acadian flycatcher extends **north** to northeastern Nebraska (West Point); Iowa (Sioux City, Woodward, and McGregor); southern Wisconsin (Prairie du Sac and Calhoun); northern Michigan (Blaney and probably Sault Ste. Marie); southern Ontario (probably Coldstream and Dunnville); northern New York (Lockport, Watertown, and Albany); and southern Vermont (Bennington). **East** to Vermont (Bennington); Massachusetts (Hyde Park); Connecticut (Danbury and Stamford); and south along the coast to southeastern Georgia (St. Marys); and Florida (Gainesville). **South** to northern Florida (Gainesville, Oldtown, Tallahassee, and Pensacola); southern Louisiana (New Orleans); and Texas (Houston and San Antonio); **West** to central Texas (San Antonio and Kerrville); Oklahoma

(Minco, Tulsa, and Copan); eastern Kansas (Wichita, Emporia, and Geary); and eastern Nebraska (Omaha and West Point).

Winter range.—During the winter season this species is concentrated in eastern Colombia (Bonda, Puerto Valdivia, Puerto Rico, and Las Lomitas) and eastern Ecuador (Cachair, Rio de Oro, and Chimbo). One was collected at Palmul, Quintana Roo, on February 9, 1910, but this may have been merely a case of exceptionally early migration.

Spring migration.—Early dates of spring arrival are: Florida—Whitfield, April 6. Georgia—Savannah, April 4. South Carolina—Mount Pleasant, April 7. North Carolina—Raleigh, April 20. District of Columbia—Washington, April 25. Pennsylvania—Waynesboro, April 25. New Jersey—Englewood, May 5. New York—Lockport, May 13. Louisiana—New Orleans, March 30. Arkansas—Helena, April 20. Kentucky—Eubank, April 18. Missouri—St. Louis, April 27. Indiana—Waterloo, May 1. Ohio—Oberlin, May 4. Michigan—Petersburg, May 8. Iowa—Hillsboro, May 5. Illinois—Chicago, May 6. Wisconsin—Milwaukee, May 11. Texas—San Antonio, April 14. Oklahoma—Norman, April 30. Kansas—Lawrence, May 2. Nebraska—Childs Point, May 6.

Fall migration.—Late dates of fall departure are: Kansas—Lawrence, September 30. Oklahoma—Norman, September 11. Wisconsin—New London, August 24. Illinois—Rantoul, October 12. Iowa—Sigourney, October 14. Ontario—Harrow, September 5. Michigan—Detroit, September 28. Ohio—Ashtabula County, October 3. Kentucky—Danville, October 26. Louisiana—Covington, October 27. New York.—Rochester, September 26. New Jersey—Elizabeth, September 29. District of Columbia—Washington, September 15. North Carolina—Chapel Hill, September 23. South Carolina—Mount Pleasant, September 22. Georgia—Athens, September 28. Florida—Tallahassee, October 9.

Casual records.—One was collected at Pine Ridge, in northwestern Nebraska, on May 26, 1900.

Egg dates.—Michigan: 7 records, May 30 to June 25.

New York: 10 records, June 5 to July 4; 6 records, June 6 to 17, indicating the height of the season.

Pennsylvania: 44 records, May 21 to June 23; 22 records, June 3 to 10.

South Carolina: 15 records, April 28 to July 8; 9 records, May 27 to June 27.

EMPIDONAX TRAILLI BREWSTERI Oberholser

LITTLE FLYCATCHER

HABITS

The western race of Traill's flycatcher, for which the above name is now accepted, is widely distributed in western North America.

The 1931 Check-list states that it "breeds from extreme southwestern British Columbia, northern Washington, central Idaho, and central Wyoming south to northern Lower California, southern New Mexico, central Texas, and Durango." It is an abundant summer resident in all suitable localities, "its favorite haunts being the willow-covered islands and the shubbery along water courses, beaver meadows, and the borders of the more open mountain parks; in such places it sometimes reaches an altitude of 8,000 feet in summer, especially in California, Colorado, and Utah," according to Major Bendire (1895).

S. F. Rathbun writes to me that the little flycatcher is a common summer resident in western Washington and has a very uniform distribution. "It is a bird of the lowlands, seldom met with at any considerable elevation; and it appears to prefer quite open places more or less overgrown with shrubs or bracken or both, the location of which is along the margin of a mixed growth of trees, mostly deciduous, with water or low ground not far away. And should any locality of this kind remain more or less unchanged, it is very apt to have a pair of these flycatchers resort to it year after year."

W. L. Dawson (1923) says that, in California, it "is a lover of the half-open situations, bushy rather than timbered, of clearings, low thickets, and river-banks. Above all, it is wedded to the lesser willows, *Salix flavescens*, *S. lasiolepis*, *S. sessilifolia*, and the rest. Unlike its congeners, it will follow a stream out into the desert, if only a few willows or cottonwoods will keep it company." He also found it in the heart of the Sierras, "though not often to altitudes above 6,000 or 7,000 feet. * * * The highest elevation at which I have ever found this species breeding is at Mammoth Camp, in southern Mono County, at an elevation of 8,000 feet."

Grinnell and Storer (1924) say: "The Traill adheres closely to the cover of thickets; it must be looked for beneath the level of the willow tops. It is thus very different in perch predilection from most of the other Empidonaces."

Nesting.—Mr. Rathbun says (MS.) that in the vicinity of Seattle, Wash., this flycatcher does not appear to arrive until about the middle of May or later. "The many records we have indicate its nesting period to be from about the middle of June until well into July. Ususally the nest is placed at a height of only a few feet in some small bush, and often in a large bracken, if any such are in the bird's territory, for this growth seems to be favored by this flycatcher. Once I found a nest in a small clump of willows standing in quite a depth of water, my attention being attracted to the nest by its size and the many dry, long grasses dangling from it. This flycatcher's nest is a neat affair, somewhat loosely made of various kinds of dried plant material, such as bracken, fibers of weed stems, coarse grasses, and bits of moss; sometimes vertically woven into the structure are a few stiff straws or weed stalks, as if for strength; next is a thin layer of soft material, which is lined very often with the reddish flower stalks from ground mosses."

L. B. Howsley, of Seattle, Wash., has sent me some elaborate notes on the nesting of the little flycatcher, from which I quote as follows: "The nesting site selected is usually the forks of the large field fern [presumably the western bracken, *Pteridium aquilinum* var. *pubescens*], so common in this section of the country. In some portions of the western part of the State, this fern grows over 6 and 7 feet high, and the plants are so thick that they constitute miniature forests. Wherever they are available, usually in an open parklike site, the little flycatcher seems to prefer it, possibly for concealment or protection, as the fronds afford both. This flycatcher is one of the commonest birds in this area, and scarcely a fern patch is without a summer resident. The height of the nest averages 3 feet, my records showing the lowest as 30 inches and the highest as 5½ feet up. The fork of the fern is always used, preference being given to a 3- or 4-way fork—this, undoubtedly, for the extra support. To the surrounding stems, the nest is securely tied with shreds of weed bark."

The building program, based on four observations, but perhaps not the invariable rule, he divides into the following operations:

"(1) The placing of a bunch of semidecayed weed bark, lint, and bleached, dead grass. This platform is used by the bird in the future construction work. (2) The bird then ties the framework of the nest to each support, starting at the bottom until halfway up. Then the top support, which will eventually be the upper edge of the outside dimension, is next tied before the balance is tied in. (3) After the skeleton has been completed, more miscellaneous foundation material is piled in the bottom, rather loosely, no special attention being paid to the ragged ends and somewhat unshapely contour. (4) When the construction has reached the proper height, the filling in of the nesting cup begins, weaving and tying in vegetable fibers until the inner cup frame has been reached. (5) At this point the rather haphazard actions of the builder lose all carelessness, and the final touches are affixed with a great deal of care and much attention to detail. The completion of the cup consumes about one-third of the total time necessary to finish a nest.

"The total construction time was 5 to 7 days; and in one instance laying started before the inside cup finish was completed. A day or so is usually spent in tucking in the stray ends and binding down the outside with what appeared to be a very fine vegetable fiber or spider web. I have a nest neatly bound with spider web.

"The materials comprising this flycatcher's nest can virtually be called standard. Almost invariably the outside is of weed bark and fiber, and soft, bleached grasses, well bound together. The cup consists of a thin layer of very fine grasses, horse hairs (where available), and the inevitable small bits of vegetable down. Sometimes the rim is decorated with a few small feathers, often lightly and neatly bound.

"One nest was placed 4 feet up in a dense tangle of thimbleberries at the bottom of a deep draw, one placed 3 feet up in a shittim bush, one in a small swamp maple 3 feet up on an open hillside, and one in a wild rose bush 3 feet up in dense brush, near a trail. Aside from these, all others observed were in ferns, in cleared brushlands or overgrown fields."

Three California nests and one Arizona nest that I have seen were all placed in upright or slanting forks of willows, securely fastened between the forks or supporting twigs, at heights varying from 4 to 6 feet above ground. The materials used were essentially the same as those mentioned above, but considerable willow cotton was mixed with the other material. Dr. R. T. Congdon has sent me some photographs of a nest in a raspberry bush in an orchard, near Wenatchee, Wash. Bendire (1895) mentions a nest, found by Dr. Clinton T. Cooke, near Salem, Oreg., that was 18 feet from the ground in the upright crotch of a slender willow. Nests have also been found in alders and blackberry bushes, sometimes as low as 1 foot from the ground. Nests placed in upright crotches are usually in the shape of inverted cones, and sometimes measure as much as 5 inches in height, though usually much less; two that I measured were only about half that height and about 3 inches in outer diameter, and the inner cavity was about 2 inches wide and 1½ inches deep. The nests are generally well and compactly built, but some are rather flimsy.

Eggs.—The number of eggs to a set varies from two to four and is usually three or four. They are ovate to short-ovate, generally the former, and they have practically no gloss. The ground color is pure dead white, creamy white, pale buffy white, or rarely has a slight pinkish tinge. Some few eggs are immaculate, but almost always they are more or less marked with fine dots, spots, or small blotches, mainly about the larger end. The markings are in various shades of light, reddish brown, such as "vinaceous-rufous" and "ferruginous." The measurements of 50 eggs average 17.8 by 13.3 millimeters; the eggs showing the four extremes measure **19.3** by 13.7 18.8 by **14.7, 15.5** by 12.7, and 16.3 by **12.4** millimeters.

Young.—Bendire (1895) says that "only one brood is raised in a season, and incubation lasts about 12 days; the young are fed on insects of various kinds, and remain in the nest about two weeks."

Mrs. Irene G. Wheelock (1904) writes: "Only the mother bird broods in the beautiful nest; the male simply straddling the edge in masculine helplessness when left in charge, looking very wise but really quite useless so far as keeping the eggs warm is concerned. In twelve days queer naked bits of bird life fill the cradle, and now the small brown master is full of importance. They are hungry; away he darts for food, but the demand is ever greater than the supply. To satisfy these four open mouths means a trip every two minutes or oftener. No time has he now for scrapping or bullying his little wife. From early morn he must hustle, snatching time for a hastily swallowed bug *en route* if he can, going hungry if he must."

Plumages.—The sexes are alike in all plumages, and the young in juvenal plumage is essentially like the adult, except that the upper parts are of a browner olive and the wing bands are buff or "cinnamon-buff." Both young and adult birds migrate southward before molting, and the young birds retain the juvenal remiges and rectrices through the winter and have a complete prenuptial molt in April, according to Dickey and van Rossem (1938). They say of the adults: "A specimen taken September 3 has just commenced the molt, while one taken on the 29th has nearly completed the body molt and is halfway through the primary molt. One of those taken February 3 is in very fresh plumage, and it is not unlikely that *brewsteri*, as in the case of the allied form, sometimes drags along with the wing molt until late in the winter. In the spring there is a complete body molt, which is finished just before the northward migration in April."

Food.—The food of the little flycatcher is essentially the same as that of the alder flycatcher, due allowance being made for the difference in the ranges of the two forms. This subject is treated more fully under the eastern subspecies, to which the reader is referred. Professor Beal (1912) says: "No special differences in the food habits have been noticed, and as many of the stomachs used in this investigation were collected before the two forms had been clearly distinguished, it is not practicable to separate them now. * * * It is evident from the nesting habits of this species that it is not likely to injure any product of industry, and the contents of the stomachs examined corroborate this observation."

Behavior.—Mrs. Wheelock (1904) has this to say about the behavior of the little flycatcher:

It is restless and energetic, flitting about among the bushes but keeping out of sight except when a too enthusiastic sally after a passing insect betrays its whereabouts. But for this and a habit it has of calling out in a fretful tone at the approach of any person, it would never be noticed, so small is it and so well concealed by the waving leaves. * * *

Although so busy, this Flycatcher is never so occupied as to miss a chance of driving another bird, great or small, away from the special clump of alders which the pugnacious mite has preëmpted for his own. When there is no one else within scrapping distance, he contents himself with scolding his mate on the nest. Apparently nothing suits him from the time the nest site is chosen until the brood is reared.

Major Bendire (1895) says: "They never remain long in one place, but move from perch to perch, snapping up insects as they fly; they are pugnacious, quarrelsome little creatures, making up in courage and determination what they lack in size. I have seen one drive a Red-shouldered Blackbird away from the vicinity of its nest, pitching down on it angrily and pecking at its head and neck in the manner of its larger relatives when chasing Crows or Hawks."

Voice.—Ralph Hoffmann (1927) gives the song as "an explosive *weeps-a-pidéea*" and says: "The vigorous four or five syllabled song, given in one utterance, with the characteristic emphasis at its close, is quite unlike that of any of the other small flycatchers. The song is occasionally shortened to the last two syllables, *pi-déea* and is often preceded by an explosive *prrit*. Besides the song the Traill Flycatcher utters constantly a sharp *whit* and, when two birds quarrel, a grating twitter."

Dawson (Dawson and Bowles, 1909) evidently thought that the notes are "not always distinctive. Particularly, there is one style which cannot be distinguished from the commonest note of the Hammond Flycatcher, *switchoo*, *sweéchew*, or unblushingly, *zweébew*, *zweébew*, *zzweet*. Other notes, delivered sometimes singly and sometimes in groups, are *pisoó; swit'oo*, *sweet*, *swit'oo; Swee*, *kutip*, *kutip; Hwit* or *hooit*, softly."

Bendire (1895) gives a slightly different version: "One of their common call notes sounds like 'queet-queet,' and the alarm note uttered when the nest is approached is something like that of Traill's Flycatcher, 'whuish-whuish.' When pursuing each other during the mating season, they sometimes give vent to a twittering note, not unlike that of the Arkansas Flycatcher, and a sharp 'quéét-quéét' is often heard while these restless little beings flit about in the low willows, or when perched on some tall weed or coarse marsh-grass stalk."

Mr. Rathbun says in his notes: "It is one of the earliest birds to begin to give its calls in the morning, and often these continue in a rapid way for two or three hours, then gradually grow less; and no matter whether the sun shines or it is raining, neither condition seems to have any effect on the frequency or the strength of the flycatcher's notes.

"In midsummer, after the little flycatcher's nesting time is over, and when many of the other birds are silent, its notes will still be heard almost each day, though more so near the close, also occasionally after sunset; and this is the case until well into August, after which they wholly cease."

Mr. Howsley contributes the following notes: "During the earlier mating period, it is common for this flycatcher to give the *pre-pe-deé* call, not *pre-pe-deér* like the western flycatcher, but a short, snappy, dead tonal sound. Later, this note changes to *zweé-beck*, just as short and snappy. This call is given altogether after mating has been completed and nesting begun, the *pre-pe-deé* not being heard afterward. In the earlier stages of courtship, several birds were heard to mix the two calls, sometimes using one, sometimes the other."

Field marks.—As stated in this paragraph under the other species, the four common, western Empidonaces are difficult to recognize in life by color or markings; they have no prominent field marks, though, as Dawson (Dawson and Bowles, 1909) says, "comparing colors, Traill's gives an impression of brownness, where the Western is yellowish green, Hammond's blackish, and Wright's grayish dusky. These distinctions are not glaring, but they obtain roughly afield, in a group where every floating mote of difference is gladly welcomed." The notes of the four species are quite distinctive, and they are commonly found in quite different types of habitat. None of the other three is likely to be seen in the low, dense, moist thickets frequented by the little flycatcher.

Enemies.—Both subspecies of *Empidonax trailli* are rather uncommon victims of the cowbirds of their respective regions, except in southern California and Colorado. Dr. Friedmann (1929) says: "Forty records have come to my attention, two from Colorado, and the rest from California."

Winter.—The fall migration must start very early in August, for Dickey and van Rossem (1938) say:

The first fall arrivals of this species appeared August 14, 1925, in the flooded forest at Lake Olomega [in El Salvador], and within a few days it became extremely common all through the undergrowth. By far the greater part were of the western subspecies, *brewsteri*. During the winter Traill's flycatchers were fully as abundant as during the fall, and as many as fifty were seen in one day in the willows and shrubbery along the San Miguel River. Out of four specimens taken at random in that locality on February 3, 1927, three were *brewsteri* and only one was *traillii*. Between April 1 and 12, 1927, during the spring migration, these flycatchers were literally swarming in the underbrush of the sandy peninsula at Barra de Santiago. Three specimens taken were all *brewsteri*.

ALDER FLYCATCHER

PLATES 25, 26

HABITS

This is the Traill's flycatcher of the older Check-lists and the bird that Audubon named for his friend Dr. Thomas S. Traill, of Edinburgh, and which he supposed represented a single species with a distribution extending from the Atlantic to the Pacific. Since the species has been subdivided, so much confusion has existed in the use of both scientific and common names for the two forms, as well as differences of opinion as to their distribution, that it seems best to abandon the common name Traill's flycatcher, revert to the old name little flycatcher for the western form, and use the above name for the eastern race; so, now we have a combination of Audubon's scientific name for the species and Brewster's (1895) proposed common name for the eastern subspecies. Some idea of the confusion that existed at the time that Brewster proposed the name alder flycatcher for the eastern race can be gained by reading his paper (1895). Both he and Ridgway (1907) regarded the birds of the Mississippi Valley region, or at least the southern part of it, as referable to the western race; and the 1910 A. O. U. Check-list concurred in this view; but the 1931 Check-list refers these birds to the eastern race. I do not feel competent to argue the case, but I have noticed that the nests and nesting sites of the birds breeding in the Mississippi Valley and adjacent States are quite different from those found east of the Allegheny Mountains and farther north and are much like those found west of the Rocky Mountains. Dr. Oberholser tells me that before he applied the name *brewsteri* to the western race he had studied a large series of these flycatchers from all over their range in North America and that there cannot be the slightest doubt that the Mississippi Valley birds belong to the eastern race.

The haunts of the alder flycatcher are not very different from those of its western relative. In the eastern and northern portions of its range, it makes its summer home in dense, low, and usually damp thickets of alders, willows, elderberries, sumacs, red osier dogwood, and viburnums, along the banks of some small stream, around the shores of a pond, or on the borders of a marsh or bog; but sometimes in the Middle West it is found breeding under suitable conditions at some distance from any water. John A. Farley (1901a), who has made a special study of this flycatcher, writes:

The Alder Flycatcher arrives in eastern Massachusetts about May 20. By the thirtieth of the month it has always reappeared on its breeding grounds.

These are bushy meadows grown (or growing) up more or less thickly with alders. The lower growth in some places consists of wild roses (*Rosa*), sweet gale (*Myrica gale* L.), and other swamp shrubbery, together with the usual mixed meadow herbage. Mingled with the alders will be young swamp maples and birches and oftentimes scattering white cedars. The whole forms a thick, at times almost choked, expanse of meadow growth. The wild roses in which the Flycatcher is so fond of nesting seem to be almost as much an essential in its summer home as the alders themselves.

Maurice Brooks writes to me: "As a breeding species in West Virginia, this bird has a rather peculiar distribution. For many years we believed it to be restricted to a few high mountain swamps and accepted this as normal behavior for a northern species. Recently Dr. George Miksch Sutton has found the birds breeding in a swamp at low elevation along the Ohio River in Brooke County. From this it would appear that lack of suitable swamps, rather than elevation, may be the restricting factor."

In Lucas County, Ohio, Louis W. Campbell (1936) found that a group of six or eight pairs had "selected a dry pasture, thickly overgrown with shrubs and small trees, as a nesting ground." The nearest water was a winding creek about half a mile to the southeast. "What grass there is is kept very low by grazing cattle but much of the ground is covered by shrubs and small trees; cock's spur hawthorne (*Crataegus Crus-Galli* L.), wild crabapple (*Malus coronaria* (L.) Mill.), black haw (*Viburnum prunifolium* L.), hazel nut (*Corylus americana* Walt.), and prickly ash (*Zanthoxylum americanum* Mill.)." A few taller trees were growing in the vicinity, and the herbaceous plants in the area were all indicative of dry soil.

Furthermore, Charles J. Spiker (1937) says: "All observations I have made on this species in the State of Iowa have taken place in dry, upland pastures, especially where there were rank growths of hazel bushes, wild crab, and hawthorn". And he suggests that such surroundings "may be fairly typical of its haunts farther west."

Spring.—On its spring migration the alder flycatcher may be seen almost anywhere, in open country, in deciduous woods, or even pine woods, as well as in the swampy thickets. It is one of the later migrants, coming when summer is near at hand, when most of the other birds have come, and when the trees and shrubs are in fresh green leafage. It passes through Massachusetts between the middle of May and the first two weeks of June, depending on the weather. Edward H. Forbush (1927) writes: "On some warm still morning in the waning of the Maytime the bird watcher notes here and there in the edge of the woods, on a pasture fence, in a small tree by the bog or even in the orchard, a small flycatcher usually on a rather low perch, sitting quite erect, silent and watchful, occasionally dashing out in pursuit of a flying insect or flitting from one point of vantage to another. This is the Alder Flycatcher in migration—quiet, watchful and discreet."

Nesting.—Mr. Farley (1901a) gives a very good account of the nesting habits of the alder flycatcher in eastern Massachusetts, as follows:

So far as I have observed, it nests invariably in a bush, selecting most often a wild rose, or clump of rose shoots or sprays—usually *Rosa carolina* L. [Footnote: I recall finding a nest once in a small shrub of meadow sweet (*Spiraea salicifolia* L.)]. The nest is often overshadowed by the alders which are scattered here and there in clumps in the bushy meadow. But it is as likely to be placed in unshaded shrubbery in the full glare of the sun. When in the open, it is more or less hid, however, by the mingled mass of wild roses, sweet gale, and other bushes rising breast-high all about it. It is often in the thickest jungle of such growth where tall, waving ferns vie in height with the predominating tangle of rose bushes that the Alder Flycatcher hides away its nest.

The height of the nest from the ground is from two to four feet. It is placed rather loosely, at times even flimsily, in an upright crotch or rather fork, or else between independent twigs that furnish a similar support. In either case the nest is suspended in a characteristic and peculiar way. I have never seen it set snugly down into a crotch after the manner of the Least Flycatcher. It is, instead, supported between twigs or prongs. It gets its chief support, as a rule, from two main shoots which often grow from the ground independently of each other, but which will be sometimes members of one bush, forming in this case a long crotch or fork. * * *

A beautiful nest that I found in 1895 in Essex County merits description because, in addition to being the handsomest structure of the Alder Flycatcher that I have seen, it is typical (although in a somewhat exaggerated way) of the general architecture of the species. The nest was three and one-half feet from the ground in a clump of the swamp rose (*Rosa carolina* L.), being one foot below the top of the bush. The nest is large, representing the extreme in size. Its inside depth is two and one-eighth inches; outside depth, three inches; outside diameter, three and three-eighths inches; inside diameter, one-half inch less. It is composed of fine grasses and strips of *Asclepias*, the latter woven into the body of the structure as well as wound about the outside and over the rim. It is deeply-cupped and thickly-walled, with rim slightly curving over and in on one side. The lining is composed of the finest of hair-like, dried, yellow grasses. A pretty effect is obtained by the use of a very delicate grass which, projecting above the rim, shows the finest of tassels. * * *

This nest has in common with all others that I have seen the usual, characteristic, loose, unfinished, even ragged, appearance outside and below. But the long grasses and especially the fibrous strips of *Asclepias* hang or string down in the present case in unusual quantity and length. Much of this reaches down six inches below the nest. Some of it extends down for one foot. A studied air of disarrangement, of negligence, of elegant confusion, is thus secured. The decorative effect is heightened by the silvery *Asclepias*, which, in addition to entering so largely into the body of the nest, causing it to shine flax-like, streams down and out therefrom in what might be termed a fibrous cascade. In greatest possible contrast to the disarranged, silvery-gray exterior is the round, deeply-hollowed interior with its exquisite yellow lining of finest grass. The excessive use of *Asclepias* in this nest is exceptional.

The above description is fairly typical, although somewhat extreme as to the amount of loose material in the lower exterior, of all the nests that I have seen or read about in northern New England and eastern Canada. In addition to the shrubs mentioned above, nests have been found in alders, willows, swamp azaleas, elders, dogwoods, hazels, a birch sapling, spicewood, wild raspberry, gooseberry, and wild currant bushes. Probably any small tree, bush, or bit of rank herbage in a suitable locality might be used. F. H. Kennard mentions in his notes a nest that he found in northern New Hampshire that was only one foot from the ground in a clump of royal fern (*Osmunda regalis*). The heights from the ground range from this extremely low level up to six feet, but three or four feet are the commonest heights. F. A. E. Starr tells me of a nest he found near Toronto, Ontario, that was in a clump of wild raspberry bushes by a roadside.

Nests found in the mountain regions of Pennsylvania and West Virginia seem to be similarly located and made of similar materials. An extra large nest, found by Dr. Samuel S. Dickey in Warren County, Pa., measured 4½ inches in outside diameter and 5 inches in height, the inner cavity being 2 inches wide and 1½ inches deep. Those that I have measured have varied from 3½ to 4 inches in outside diameter, and the body of the nest, exclusive of the loose ends, was not much more than 2 inches in height. Dr. Dickey (MS.) says of his nest: "The foundation consisted of a loose weave of stems of yellow marsh grass (*Calamagrostis canadensis*), fine panicles of grass, fine weed stems, bark strips of weeds, and gray mats of spider cocoons. It was lined with fine, dusky-colored weed stems, fine yellow grass panicles and several brown needles of the white pine (*Pinus strobus*)."

P. M. Silloway (1923) says that in the western Adirondack forest the alder flycatcher "commonly finds a site in an upright crotch of a bush or sapling, but in some instances it saddles its nest on a horizontal branch from twenty to thirty feet from the ground." Harold M. Holland writes to me from Illinois that "whereas willow and like growths may offer suitable nesting places along watercourses, on the 'prairie' it almost invariably nests in the osage-orange hedgerows. Years ago these hedges lined country roads and formed farm boundaries very extensively, and flycatchers were correspondingly plentiful." But the hedges are gradually disappearing. A. Dawes DuBois tells me that he once found a nest, in Jersey County, Ill., that was 8 feet from the ground "in a small ash tree in an orchard." He sends me the description of another nest, found on waste land north of Springfield, Ill., that was "composed chiefly of soft plant fibers and thin shreds from the outside of weed stalks, with several soft feathers matted into the walls, lined with very fine stiff plant stems and one large, soft feather, and slightly contracted at the rim. In exterior appearance, it resembles a yellow warbler's nest."

Several other observers have called attention to the resemblance of western nests of the alder flycatcher to the nests of this warbler or those of the goldfinch, whereas eastern nests have often been referred to as resembling nests of the indigo bunting or bush nests of the song sparrow; thus two very different types of nests are indicated. Three nests, found by Mr. Campbell (1936) in the upland colony in Lucas County, Ohio, were all placed on nearly horizontal branches of cockspur hawthorns, 3½ to 4½ feet from the ground; another nest was on a slanting branch of a small elm tree, 7 feet from the ground. The photographs shown in his paper indicate that these nests are quite unlike eastern nests. Many years ago Otto Widmann sent to Dr. Elliott Coues (1880) three Missouri nests of this flycatcher which he considered "identical with those of *E. minimus.*" One was taken "from an oak-tree, at an elevation of 10 feet; another, with three eggs, June 21, from an elm, at a height of 18 feet; the third, with a single egg, June 17, from an ailanthus, only 6 feet from the ground." The above data seem to indicate that the nests and the nesting sites of the alder flycatchers (or perhaps more properly Traill's flycatchers) breeding west of the Alleghenies are both quite different from those of the species breeding east of that range and in Canada.

Eggs.—The three or four eggs laid by the alder flycatcher are practically indistinguishable from those of its western relative. Mr. Farley (1901a) gives some detailed descriptions of individual eggs, one of which, he says, is "of a creamy ground color and is beautifully marked after the typical style with a fairly complete ring of pale brown blotches having darker centres, and with dark brown (almost black) round dots interspersed among the blotches, a rich effect being thus secured." The measurements of 50 eggs average 18.5 by 13.5 millimeters; the eggs, showing the four extremes measure **19.8** by 14.0, 19.3 by **14.5, 17.0** by 12.7, and 19.8 by **12.6** millimeters.

Plumages.—Dr. Dwight (1900) says that the natal down is "pale olive-brown." Subsequent plumages and molts follow the same sequence as in the western race, the changes taking place while the birds are in the their winter home, though some young birds begin the postjuvenal molt before they leave in September. Young birds in juvenal plumage are somewhat yellower on the under parts than are adults.

Food.—Professor Beal (1912) examined the stomach contents of 135 specimens of the two races of *Empidonax trailli,* taken in various parts of the country. Animal food made up 96.05 percent and vegetable food 3.95 percent.

Beetles of 65 species, all harmful species, except for a few ladybird beetles that eat plant lice and scale, amount to 17.89 percent. "Hymenoptera are the largest item of animal food, not only in the aggregate but in every month." They are mostly in the form of wasps and bees, but there are a few of the parasitic species and some ants. They amount to 41.37 percent of the food, a record exceeded by but two other flycatchers. Hymenoptera of all kinds were found in 93 stomachs and were the sole contents of one.

"Diptera, such as crane flies, robber flies, house flies, and dung flies, were found in 47 stomachs and were the entire contents of 4. They amount to 14.20 per cent of the food." Hemiptera were found in 44 stomachs and amount to 7.24 per cent. In one individual 12 chinch bugs were identified and the fragments of many more were found in the same stomach. "Lepidoptera, that is moths and caterpillars, were found in 41 stomachs, of which 18 contained moths and 25 held caterpillars, 2 containing both. The aggregate of both is 7.73 per cent." Orthoptera, made up mostly of small grasshoppers, amount to 3.91 percent and were contained in 16 stomachs. "A few odd insects, such as dragon flies and some ephemerids, were occasionally taken, and altogether amount to 2.77 per cent of the diet. A cattle tick was found in one stomach and a snail in another, both identifiable. Spiders and millepeds were eaten to the extent of 0.94 per cent and complete the animal food."

Of the vegetable food, he says: "Elderberries were found in 6 stomachs, blackberries or raspberries in 2, dogwood berries in 1, juniper berries in 1, fruit not further identified in 3, seeds unknown in 2, and rubbish in 1."

Behavior.—The alder flycatcher does not differ materially in its behavior from the western subspecies. It has been said to be a very shy bird, but it is really no more so than many of our small birds; it is more retiring than shy. During the nesting season, it spends most of its time in the dense thickets, where it nests and where it secures most of its food; in such places it keeps out of sight and is not easy to approach, as it hears the necessarily noisy movements of the observer and retires ahead of him by short flights from one low perch to another, hidden among the leafage. It comes out occasionally in pursuit of some passing insect or perches for a moment on some topmost twig to give its emphatic little song. On its arrival in spring it is much more in evidence, flying about from treetop to treetop in the open, prior to the selection of a nesting site and a mate. About its nest it is shier, or more nervous, than at other times; it is not a close sitter and can seldom be surprised on its nest, for it slips quietly away when it hears the observer coming; and it will not readily return to its nest, even to feed its young, if it knows that anyone is watching; but it will flit about in the bushes, just beyond vision, uttering its exasperating *pip-pip* scolding notes.

Audubon (1840) says: "When leaving the top branches of a low tree, this bird takes long flights, skimming in zigzag lines, passing close over the tops of the tall grasses, snapping at and seizing different species of winged insects, and returning to the same tree to alight. Its notes, I observed, were uttered when on the point of leaving the branch."

Voice.—When I first heard the distinctive and striking note of the alder flycatcher, many years ago, I recorded it at *raiz-wée* or *raiz-whéo*, the first part harsh and rasping, and the last syllable a clear, loud whistle that was rather musical. Francis H. Allen has sent me the following very good description of the notes: "Heard near at hand the song sounds to me like *wee-zeé-up*, with the *up* very faint. This is a slight modification of Dwight's rendering in Chapman's Handbook. A good rendering of the song is *vee-feél* (the German *wie viel*), as somebody else has suggested. One of the notes is an explosive *queeoo*, the *oo* at the end very short, the whole having a rasping quality. The head is thrown back and the bill pointed up for this note, as well as for the song. The ordinary call note is a liquid *pip*, which is sometimes, or perhaps often, heard at dusk."

Mr. Farley (1901a) says: "The minor notes of the Alder Flycatcher, like its harsh cry, are perfectly characteristic and unlike the notes of any other bird. They are of two sorts, the common low *pip* or *pep*, which to some ears may resemble *peep*, and the softly whistled whisper (or whispered whistle), *pip-whee* or *pip-whing*. There is an interval between the two syllables of this soft song, and the last is accented. * * * It is a faint little cry that rarely rises above the gentle rustle of the alder and maple leaves as they are stirred by the June zephyrs."

Many other renderings of the various notes have been published, but they all seem to be different interpretations of the above notes. And, of course, some of these are similar to those of the western subspecies.

Fall.—When the young are strong on the wing, during the latter part of August, the families start on their southward migration; and before the middle of September they have all departed from their summer homes. They seem to follow a southwestward course, west of the Alleghenies and through Texas into Mexico and Central America, where they join the western race in its winter home.

Dickey and van Rossem (1938) say that the alder flycatcher is a fairly common fall migrant and less common midwinter visitant to the lowlands of El Salvador. "Extremes of altitude are 200 and 1,000 feet. Dates of arrival and departure are August 25 and February 10."

DISTRIBUTION

Range.—North America, south in winter to northwestern South America.

Breeding range.—The species breeds **north** to Alaska (Nulato, Fort Yukon, and Circle); Mackenzie (Fort McPherson, Fort Goodhope, Fort Providence, and probably Fort Resolution); northern Manitoba (Brochet Lake, Cochrane River, and Norway House); northern Ontario (Moose Factory); northern Quebec (Richmond Gulf); and east-central Labrador (Esquimaux Island). The **eastern** boundary of the range extends southward along the coast from this point to northern New Jersey (Plainfield). **South** to northern New Jersey (Plainfield); probably western Maryland (Thayerville); central Ohio (Columbus and Lewistown Reservoir); central Indiana (Indianapolis); southern Illinois (Mount Carmel and Olney); northeastern Oklahoma (Tulsa); central Texas (probably Cameron and San Angelo); central Veracruz (Orizaba); southwestern Tamaulipas (Jaumave); Durango (Rio Nasas); probably northern Sonora (Nogales); and northern Baja California (Cerro Prieto, Las Cabras, and Los Coronados Islands). The **western** limits of the breeding range extend northward from northern Baja California (Los Coronados Islands) along the Pacific coast, chiefly in the mountainous regions, to Alaska (Chickamin River, probably Cordova Bay, and Nulato).

Winter range.—In winter the species is found **north** to southern Guerrero (Coyuca); Guatemala (Los Amates); Honduras (Lancetilla and Ceiba); Nicaragua (San Carlos); Panama (Gatun); northern Colombia (Mamatoca, Bonda, and Buritaca); and northwestern Venezuela (Encontrados). **East** to western Venezuela (Encontrados); east-central Colombia (Puerto Berrio, Honda, and Rio Frio); and northeastern Ecuador (Rio Suno). **South** to Ecuador (Rio Suno, Gualaquiza, Zamora, and Zaruma). **West** to southwestern Ecuador (Zaruma); Costa Rica (Bolson); El Salvador (Colima); southwestern Guatemala (San Jose and Mazatenango); and Guerrero (Coyuca).

The range as outlined is for the entire species, which has been separated into two subspecies or geographic races. The typical race, known as the alder flycatcher (*Empidonax trailli trailli*), occupies the northern part of the breeding range south to southern British Columbia, Colorado, Arkansas, Kentucky, Maryland, and New Jersey. The little flycatcher (*E. t. brewsteri*), occupies the balance of the range but in the northern portions of its range it apparently overlaps that of *E. t. trailli*. The two races appear to occupy a common winter range, except that *brewsteri* extends somewhat farther south.

Spring migration.—Early dates of arrival are: North Carolina—Raleigh, May 14. District of Columbia—Washington, May 8. New Jersey—Milltown, May 12. Massachusetts—Stockbridge, May 10. Vermont—Randolph, May 17. New Hampshire—Monadnock, May 20. New Brunswick—Scotch Lake, May 23. Quebec—Quebec City, May 25. Missouri—St. Louis, April 29. Illinois—Odin, May 3. Ohio—Oberlin, May 7. Michigan, Plymouth, May 11. Ontario—Ottawa, May 14. Minnesota—Minneapolis, May 1. Manitoba—Shoal Lake, May 9; Winnipeg, May 11. New Mexico—Carlisle, April 16. Colorado—Fort Lyon, May 9. Montana—Fortine, May 18. Alberta—Glenevis, May 22. California—Los Angeles, May 4. Oregon—Mercer, May 10. Washington—Spokane, May 15. British Columbia—Okanagan Landing, May 16. Alaska—near Lynn Canal, May 24.

Fall migration.—Late dates of fall departure are: Alaska—Fort Yukon, August 31; Sergief Island, September 3. British Columbia—Atlin, August 31; Vancouver, September 14. Washington—Spokane, September 30. California—Pasadena, September 26; Yosemite Valley, October 1. Alberta—Glenevis, August 31. Montana—Fortine, September 8. Wyoming—Laramie, September 7. Colorado—Yuma, September 10. Manitoba—Winnipeg, August 31. Minnesota—Minneapolis, September 24. Iowa—Elkader, September 20. Ontario—Ottawa, September 4. Michigan—Ann Arbor, September 13. Ohio—Hillsboro, September 23. Illinois—Chicago, September 20. Missouri—St. Louis, October 4. Mississippi—Bay St. Louis, October 18. New Brunswick—Scotch Lake, September 20. Maine—Dover-Foxcroft, September 3. Massachusetts—Dennis, September 5. Pennsylvania—Beaver, September 7. District of Columbia—Washington, September 17. North Carolina—Raleigh, September 21. South Carolina—Mount Pleasant, September 30.

Casual records.—There are numerous records of the collection of one race or the other outside the normal range but space is not available for their citation. Chapman refers without date to specimens from Embarcacion, Argentina, and a specimen was taken at Vista Alegre, Peru, on October 11, 1922. According to Reid a specimen was taken at Stocks Point, Bermuda, and preserved in the Bartram collection.

Egg dates.—California: 94 records, April 19 to July 21; 48 records, June 6 to 28, indicating the height of the season.

Colorado: 9 records, June 25 to July 30.

Illinois: 32 records, June 8 to July 9; 16 records, June 20 to 28.

Maine: 20 records, May 15 to July 1; 10 records, June 16 to 22.

New York: 19 records, June 9 to 28; 10 records, June 15 to 21.

Oregon: June 8 to July 18; 15 records, June 25 to July 2.

EMPIDONAX MINIMUS (Baird and Baird)

LEAST FLYCATCHER

PLATES 27, 28

HABITS

The familiar little "chebec," as we used to call it, is widely distributed in the Canadian and Transition Zones of eastern North America, where it is a common and well-known summer resident. In the more thickly settled regions it has become adapted to the environments of human habitations and makes itself at home in our orchards and gardens, in the shade trees about our houses, along the streets of towns and villages, and even in our city parks. It is equally at home along country roads, the borders of streams and ponds, in partially overgrown pastures, and on the edges of the woods. It frequents the more open woodlands, rather than the denser forests, but shows a decided preference for the rural countryside. On an early morning walk, in spring, along a New England country roadside, one is almost sure to see this sprightly little bird, or at least hear its familiar "chebec."

The least flycatcher extends its summer range northward in Canada as far as it can find deciduous tree growth suited to its needs, where it seems partial to thickets of balsam poplar and quaking aspen; here it also frequents groves of willows along the shores of lakes, and thickets of birches and maples along the streams.

Prof. Maurice Brooks writes to me: "Least flycatchers are summer residents of the Alleghenian division of the Transition Life Zone. In places they are abundant, their nests placed in either deciduous or coniferous trees at 10 to 20 feet from the ground. They occur in West Virginia at elevations ranging from 1,500 feet to 3,300 feet. I have never found them in the red spruce belt, although they closely approach it."

Referring to the same general region, Dr. Samuel S. Dickey (MS.) writes: "While most pairs of these birds seemingly prefer to inhabit apple orchards, some of them frequent mixed tracts of pines, spruces, hemlocks, oaks, and birches. They are heard venting their calls from thickets of alders, silky dogwood, winterberry, and mountain holly."

Spring.—Writing of the least flycatcher in Massachusetts, Mr. Forbush (1927) says:

The bird arrives in spring before the other small flycatchers. A few individuals are here ere the end of April, and when a thousand orchards burgeon with the bloom of spring, when the first misty green begins to screen the woodlands, a host of these little feathered warriors spreads over New England. At first, in migration, they are rather silent and appear wherever open spaces among the trees or along the edges of thickets gives them fly-room. At this time they may be mistaken for the Alder Flycatcher, as they may frequent alders along a brook and may even appear among the tall bushes at the edges of the meadows. Later, when the females come, the males are the most vociferous and pugnacious of their kind, and nearly every orchard resounds with their cries."

Courtship.—As soon as the females arrive courtship begins in earnest. The males are then the most active, noisiest, and most pugnacious of any of our small birds. Rival males indulge in frequent combats, fighting furiously until the vanquished is driven away among the foliage in search of hidden females, and the latter are chased about in pursuit flights through the leafage, across clearings and over the open spaces, until the successful suitor proclaims his victory with many vigorous *chebecs* and much flirting of wings and tail. The pair is then ready to select its nesting site.

Nesting.—In southern New England the least flycatcher shows a decided preference for apple trees as nesting sites; at least two-thirds of the nests that I have recorded in Massachusetts have been in apple trees in old orchards near houses; in such situations the nests have been placed on horizontal branches, usually partially supported by upright twigs, or in an upright fork of some small branch. In my egg-collecting days, old and partially neglected orchards were always considered favorable places to look for the nests of this and several other species of small birds. We have also found the least flycatcher nesting rather commonly on the pine barrens of Plymouth County, Mass., where there was a scattering growth of small or medium-sized pitch pines, with an undergrowth of scrub oaks and other underbrush; here I have found as many as three nests in a short walk. Some of the nests were placed near the ends of horizontal branches and others against the trunks of the smallest pines at heights ranging from 7 to 15 feet above ground. But other nests in southern New England and New York have been found in pear trees, maples, willows, oaks, alders, sycamores, locusts, beeches, elms, birches, sumacs, wild cherries, and others. In northern New England, nests have often been found in spruces, tamaracks, and other conifers. William Brewster (1937) says that, in the neighborhood of Umbagog Lake, Maine, "their nests are usually built in balsams, hemlocks, or other evergreen trees and commonly lined much more profusely with feathers than are those found in eastern Massachusetts."

In Canada, favorite trees are yellow and paper birches, quaking aspens, balsam poplars, and mountain-ashes, as well as the conifers mentioned above. Near James Bay, Canada, Dr. Dickey tells me that he found many nests of this flycatcher in dense growths of speckled alder, red osier dogwood, quaking aspen, balsam poplar, and various willow shrubs; the nests were 2 to 6 feet above ground, and "were adorned with mats of white hair from the husky dogs and feathers plucked last autumn from 'wavey geese' and other wild fowl."

The nests of the least flycatcher are seldom more than 25 feet from the ground, usually much less in eastern and northern localities, but A. Dawes DuBois has sent me some notes on Wisconsin and Minnesota nests that were much higher; one was about 50 feet up in an elm, one 40 or 50, and another about 60 feet from the ground. He watched the bird building the latter nest, and says: "One of these little flycatchers, after gathering some material on the ground, flew to a fallen birch, took from it some fine, thin shreds of birch bark, and mounted into the tree above me. It worked at the nest about 10 seconds, flew away and was gone two minutes, then worked at the nest 25 seconds, and was gone again for about two minutes on another material-gathering errand."

One that I saw building worked equally fast, returning to the nest with material every three or four minutes. Mr. Forbush (1927) says that the building of the nest usually occupies from six to eight days. A. A. Cross wrote to him about watching the building of a nest that was nearly finished:

About one half of the upper edge seemed literally torn to pieces, the frayed fragments projecting in all directions. The work of some robber, I thought. Such was not the case, for presently the owner appeared with her beak full of building material which, a piece at a time, she thrust into the edge of the nest, leaving the loose ends free. Watching her, I noted that she was gathering the inner bark from the dead and broken stems of last year's goldenrod. She made many trips, working rapidly, and disposing of the material as in the first case. In about 20 minutes she had finished, causing the edge of the nest to look like a miniature hedge. She then settled herself solidly in the nest, hooked her head over the edge and pivoting on her legs ironed out the rough brim with her throat, putting considerable energy into the work and working first one way and then the other. In this manner she was able to take in about one-third of the circumference of the nest before changing her position. Then readjusting herself, she continued the process until the nest was finished. This was the last step in the building of the nest.

The nest is compactly and well made, firmly settled down into a crotch or firmly attached to twigs arising from a horizontal branch, resembling in these respects certain nests of the yellow warbler or the redstart. It is deeply cupped, the walls are rather thin, hardly any material intervening between them and the supporting forks, and the upper rim is often somewhat incurved. The body of the nest is made of shreds of the inner bark of trees, shredded bark of coarse weeds, bits of string and paper, fine weed stems and grasses, the dried blossoms of weeds and grasses, thistledown and the down from cottonwoods and ferns, cotton, shreds of rope, spider webs and cocoons, and various other vegetable fibers and rubbish. The rim is neatly finished off with the finest of the fibers and grasses, firmly interwoven or pressed down into place. The cup is smoothly lined with the finest of grasses, cow's hair or horsehair, thistle, milkweed, dandelion, willow, or cottonwood down, and a few feathers. Nests that I have measured vary from 3 to 2½ inches in outer diameter; from 2 to 1¾ inches in inner diameter; in outside height from 2½ to 1¾ inches; and the inner cavity varied from 1½ to 1¼ inches in depth.

Eggs.—Bendire (1895) says that the number of eggs laid by the least flycatcher varies from three to six, usually four, and that one is deposited daily. I have never seen more than four eggs in a nest and think that any larger numbers must be rare. A set of seven, reported by Dr. Dickey (MS.), was apparently the product of two females. The eggs are ovate, short-ovate, or rounded ovate and are not glossy. They are creamy white and unmarked. The measurements of 50 eggs average 16.1 by 12.9 millimeters; the eggs showing the four extremes measure 17.8 by 12.2, 16.8 by 15.0, 15.0 by 12.7, and 15.2 by 11.4 millimeters.

Young.—The period of incubation is said to be 12 days. Probably both sexes incubate; some observers state so positively, and others are more or less in doubt about it. But certainly both parents assist in the feeding and care of the young, for both have been seen at the nest together. Often, if not regularly in the southern portion of its range, two broods are raised in a season; and sometimes the second brood is raised in the same crotch.

Ralph Hoffmann (1901) removed an empty nest after the young had left it, and says: "When the young had been out two days, and were being fed constantly by the male, I saw the female fly to the empty crotch, where the old nest had been. In a moment she repeated her visit, and when I walked to the tree, I saw the skeleton of a new nest already completed. Two days later the nest was finished. It was interesting to note that the beginning of the new series of instinctive acts involved in raising a second brood did not destroy the force of the last series, for when the nest was finished the female returned to help the male feed the first brood."

Miss Mildred Campbell has sent me some elaborate notes on her observations made at a nest of young least flycatchers in Michigan, from which I quote as follows: "The adults covered the eggs day and night. The incubating bird's body movements indicated that the eggs were being turned at intervals. Nestlings were hatched on June 29, June 30, and July 1, respectively. The pieces of shell were carried from the nest by the adults. After 24 hours the large-bellied, naked nestling began to stretch for food."

During the 11 days in the nest a young bird increased in weight from 1.42 to 8.21 grams. Down appeared on the third day, pin-feathers on the fourth and sixth days, eyes opened on the eighth day, and on the eleventh day the young bird was completely feathered.

"The nestlings were fed by regurgitation of both male and female at first. Later a diet of captured craneflies, Mayflies, grasshoppers (mainly nymphs), beetles, spiders, and harvestmen, with some light green food (unidentified) was brought to the young. The nestlings were fed in rapid succession in a more or less definite order. When no attempt was made to swallow food pressed into the young throat, the morsel was removed and placed in another gaping mouth. During my observation, the greatest interval between feedings was 54 minutes. The birds were fed heavily from 4:40 until 10:40 A. M. and from 2:00 until 6:00 P. M. A sudden increased demand for food came just before the young prepared to fly from the nest. Usually one bird was fed per visit, but several times two were fed on the same occasion.

"The adults fed the nestlings and waited for the voided 'excreta sac', which was picked up and removed from the nest. The rim of the nest became badly worn by the continuous activity, so that the young soon were able to void the 'excreta sac' over the side of the nest.

"The diurnal brooding period decreased rapidly from day to day. On very warm days the adults relieved one another during the heat of the day. One and then the other remained astride the nest, with outstretched wings, keeping the young from suffocating and from the torrid rays of the sun.

"The nestlings pushed and turned, stretched and preened their wings, and spread their tails. Much of this was done from the rim of the nest, and then the young dropped down in the nest in order to keep from going overboard during this activity. The first one left the nest during the evening of the eleventh day, the second on the morning of the twelfth day, and the third one preened and stretched on the rim and in the bottom of the nest all day. Food was caught as the parents flew by this fledgling. Finally, in the evening, this one dropped from the nest and balanced itself on a bouncing twig. Here it was fed and then led away by the parents."

Observations by Aretas A. Saunders (1938) and his students on a nest in Allegany State Park, N. Y., "show that the parent fed the young 24 times an hour, as an average, in 13¾ hours of observation. The times varied, however, with the age of the bird and the time of day, the least number being seven feedings an hour and the greatest 37. Apparently the greatest amount of feeding in an hour was in the morning between 9 and 11 o'clock."

Plumages.—The natal down is said by Dr. Dickey to be light gray (MS.). Dr. Dwight (1900) describes the juvenal plumage of the least flycatcher as follows: "Above, including sides of head, olive-brown, greener on the back, a faint ashy gray collar. Wings and tail deep olive-brown, median and greater coverts edged with pale buff forming two wing bands, secondaries and tertiaries with dull white. Below, grayish white, a smoky gray pectoral band; pale primrose-yellow on abdomen and crissum. Orbital ring dull white. * * * The species in this plumage is not so green above as *E. virescens*, but browner and very like *E. t. alnorum* from which it may be differentiated by its grayer lower parts, somewhat paler wing bands and smaller bill."

The postjuvenal molt is mainly accomplished after the young birds migrate southward, but some birds "become greener above and yellower below before they leave for the south late in August." What first winter birds I have seen are much greener above and yellower below than the adults. There is a first prenuptial molt during the winter, or early spring, which is apparently complete and produces a plumage like that of the adult.

Adults evidently follow the same sequence of molts and plumages as in *flaviventris* and *trailli*, having a postnuptial body molt in fall, after arriving in the south, a wing molt during winter, and a prenuptial body molt during the early spring, before coming north.

Food.—Professor Beal (1912) says of the least flycatcher: "It is a typical flycatcher in food habits, but like most others of the family it does not take all of its food upon the wing. The writer has seen one scrambling about on the trunk of a tree and catching insects from the bark like a creeper." In his study of the food, 177 stomachs were examined taken within the months from April to September. "The food consisted of 97.83 per cent of animal matter and 2.17 of vegetable. * * * Hymenoptera are the largest item, * * * 41.10 per cent. * * * Three stomachs were entirely filled with ants and four with other Hymenoptera. Parasitic species were eaten to the average extent of 11.66 per month. * * * This percentage is higher than is desirable."

He lists 67 species of beetles as identified in the food, but useful beetles amount to only 1.41 percent, and harmful beetles total 19.94 percent. The average for Hemiptera is 11.12 percent, for Diptera 11.34 percent, for Orthoptera 2.59 percent, and for Lepidoptera, both moths and caterpillars, 7.27 percent. "Ephemerids found in one stomach, dragon flies found in 3, and an unidentified insect in 1, make up 0.95 per cent. One stomach was entirely filled with a large dragon fly. Flycatchers are among the comparatively small number of birds expert enough to catch dragon flies on the wing, and these insects are too wary to be taken sitting. Spiders are eaten to a small extent in every month in the season * * * 2.11 per cent."

Of the vegetable food he says: "Fruit amounts to 1.83 per cent, and consists of Rubus seeds found in 2 stomachs, elderberry seeds in 2, pokeberry seeds in 1, rough-leaved cornel in 1, and fruit skins not further identified in 4. Various seeds were contained in 6 stomachs, and rubbish in 3; altogether they amount to 0.34 per cent."

W. L. McAtee (1926) says that "insects injurious to woodlands which are eaten by this flycatcher include carpenter ants, gipsy moths, click beetles, leaf beetles, nut weevils, tree hoppers, leaf hoppers, and leaf bugs." To this list might be added cankerworms, or inchworms, which it catches in the air as the worms spin down to the ground on their webs. The bird also picks off many of these and other caterpillars from the leaves while hovering in the air.

Dr. Dickey writes (MS.): "Once a flycatcher performed a singular, spiral flight, a distance of 8 yards, to pursue over a little glade a speeding beefly (*Bombycillus*). They are prone to approach spider webs and small caterpillars that dangle from silken cords. They lean out from twigs and cleanse the webs of these desiderata. They even mount high in dead branches, scan the nooks and corners, and show by their every movement that they are finding the nourishment conducive to their sprightliness. Rarely I observed a flycatcher pass close to the ground, brush almost the tops of sickle sedges, and snap some stray bug, then return to an alder branch to devour it."

Behavior.—Mr. Forbush (1927) says that the least flycatcher "is the smallest, earliest, tamest, smartest, bravest, noisiest, and most prominent member of its genus in New England," and it is certainly entitled to all these superlatives. The earliest arrivals are much in evidence, as their emphatic calls advertise their presence in our orchards and gardens, as well as everywhere in the open countryside, along roads and on the borders of the woods. One does not have to look far before he sees the trim little warrior perched upright on some bare twig, the top of a post or on a convenient wire, or even on the top of some low bush or tall dead weed stalk. He shows his tameness, or his indifference to our presence by darting out to snatch some passing insect so close to our head that we can plainly hear the snapping of his bill. He is a restless, active little fellow, far less sedate than our wood pewee, preferring to dart about among the foliage, or from one perch to another, in quest of his prey, rather than sit on one favorite perch and watch for it.

He is a sociable and friendly little fellow toward human beings. Manly Hardy wrote to Major Bendire (1895): "A pair of these birds or their descendants have nested regularly in or near my garden, usually building in a maple. These birds know me, and, what is more, I believe remember me from one year to another. They often sat on a dry twig, or on a bean pole near by, and watched me hoe, and suddenly one would dart down and catch a moth or other insect which I had disturbed, flying so close to me that I could distinctly hear the sharp snap of its bill. Then it almost invariably returned to the place it darted from to eat its prey. Both birds often came close to the window and watched my family inside."

But during the nesting season it is not so friendly toward other birds; it then becomes pugnacious and drives away from the vicinity of its nest, with vicious attacks, almost any small bird that ventures too near. Even the gentle migrating warblers that are peacefully hunting for food among the foliage are quickly put to rout. Mr. DuBois tells me that "a least flycatcher drove an English sparrow out of a tree in the orchard by flying at the sparrow while fiercely snapping its bill." But he also mentions a nest of this flycatcher that was only about 5 feet from a wood pewee's nest in the same tree. "The pewee was sitting in her nest while the least flycatcher was building."

Mr. Forbush (1927) quotes the following notes from F. H. Mosher:

A pair of Least Flycatchers had just begun their nest in an apple tree by placing some bunches of cottony material and a few strings and straws. A female Oriole, happening along, appropriated the string for her own use, and carried it away. The Flycatchers came soon after, and were very much disturbed on finding the nest materials scattered, and had quite a talk over it. In a few moments the Oriole came back for more string, when both Flycatchers flew at her and snapped their bills savagely in her face. The Oriole did not seem to mind them much, and kept on going toward the nest. When the Flycatchers found they could not scare her in this way, they both attacked her fiercely, and pulled out quite a number of feathers, keeping up quite a steady scold. The Oriole attempted to retaliate, but when she attacked one of the Flycatchers the other struck her from the other side, and several times she was knocked completely off the branch. Finally she beat a precipitate retreat, one of the Flycatchers chasing her out of sight.

After the nesting season is over it becomes much less quarrelsome and less noisy; then, according to Dr. Thomas S. Roberts (1932), "it wanders away from the woodlands and may then be found with the Sparrows and fall Warblers in dense growths of tall weeds and composites, where, with the Warblers, it finds abundant insect food while the Sparrows feast on the seeds. The top of a rigid mullein stalk or wild sunflower furnishes a suitable vantage-point from which to sally out after the flying insects that constitute the bulk of its food."

About its nest the least flycatcher is very tame and brave, often allowing a close approach and making a brave attempt at defense. Professor Brooks tells me of a remarkable case of close sitting; he and his companions sawed down a small hemlock in which one of these flycatchers was incubating its eggs; "the parent bird remained on the nest until the tree had fallen, only leaving it when" they "approached the nest." Dr. Roberts (1932) tells of one that he stroked and then lifted her off her set of heavily incubated eggs.

Voice.—During spring and early summer, or until its nesting activities absorb too much of its time and attention, the least flycatcher is a noisy bird, and its voice is heard almost constantly, especially during the early morning, when it is one of the first birds to be heard, and toward evening. But after the end of June its vocal efforts slow down, and after the middle of July it is seldom heard. By far the commonest and most characteristic note, from which one of its popular names is derived, is the emphatic 2-syllable *chebeć*, strongly accented on the last syllable, given with much vehemence, and accompanied with an upward jerk of the head and a flirt of the tail, as if asserting his independence and authority over his domain. This note is so much a part of the bird that it is often combined with some of its other notes, which are not numerous and not so often heard. Mr. DuBois writes to me that this note sounds to him more like *te-bić*, which holds true for all these flycatchers that he has heard anywhere. Others have spelled the syllables slightly differently.

It also has a short call note, or alarm note, that sounds like *whit*. Dr. Dickey writes to me that "at mating time and around disturbed nests they cause the underwoods to resound with noises that sound like *speetz* and *sperk*." Miss Campbell tells me that, at the nest she was watching, "twice, once at noon and another time in the evening, the male sang an unusually melodious warbling song and the female on the nest responded by soft murmurings and rhythmic movements." Forbush (1927) refers to "a flight-song (?) a 'twittering warble'". E. A. Samuels (1883) says that the bird sometimes changes his *chebec* note into "*chebec-trree-treo, chebec-treee-cheu.*" Dr. Chapman (1912) writes: "In crescendo passages he literally rises to the occasion, and on trembling wings sings an absurd *chebeć tooral-ooral, chebeć, tooral-ooral*, with an earnestness deserving better results." H. D. Minot

(1877) mentions "querulous exclamations (*wheu, wheu, wheu*) which are more or less guttural and subdued."

Dr. Winsor M. Tyler sends me the following notes on the voice of the least flycatcher: "After listening to the chebec's song thousands of times with the question of stress in mind, I can detect little difference in the accent of the two syllables. The song suggests to me the words *a ship*, snapped out emphatically, enthusiastically, like a mariner sighting a sail at sea. The bird sometimes varies its singing by running several songs into a quick series. At a distance this form resembles the house sparrow's reiterated *chillip*. On June 13, 1915, I saw a least flycatcher fly across a field—a distance of 75 yards—singing a jumble of notes in which his regular *chebec* occurred frequently. He alighted high in an apple tree, presumably for the night, for it was almost dark.

"The bird is an incessant singer. I have heard it repeat its song about once a second, with occasional pauses, for two hours or more. My notes state: 'On June 18, 1912, the chebec woke at 3:18 A. M. Once in the course of a minute he sang 60 times; during another 60 seconds, he sang 75 songs.' When out of doors in the night, long before day, I have heard several times a sharp *chebec* suddenly crack out of the darkness above my head."

Field marks.—The least flycatcher looks much like a small, chunky wood pewee, with a more prominent eye ring and with more extensively white under parts. Compared with the other small flycatchers of the *Empidonax* group, it is much whiter below, with hardly any tinge of yellow, and the wing bars are much whiter; furthermore, it is the smallest of the genus within its summer range. Its common note is the most distinctive of all, and its haunts and habits are quite different from those of the others, always active, brave, and conspicuous. It could hardly be overlooked or mistaken.

Enemies.—Like all other small birds this flycatcher is preyed upon by many predatory birds and mammals. G. Bartlett Hendricks (1933) tells of an attack on a least flycatcher and its young by two gray squirrels. A gray squirrel was seen to run up a tree with the adult in its mouth. It "halted half way up the tree and proceeded to eat the bird. * * * In the meantime, another gray squirrel was stalking the first bird," which was rescued.

William Brewster (1906) writes: "The last nest that was built in our garden (in 1895) was attacked by a large troop of English Sparrows when it contained young about half grown. Although both parents defended it with utmost spirit, the Sparrows succeeded in tearing away part of the outer walls of the nest, and one of them, standing on its rim, bent down and delivered several murderous but fortunately ineffective pecks at the heads of the young. In the end the Flycatchers triumphed and put the cowardly horde to ignominious flight."

Dr. Friedmann (1929) says that the least flycatcher is "an uncommon victim" of the cowbird; he found only 10 records.

Fall.—After the breeding season is over and the young are strong on the wing, old and young scatter about and mingle with the early migrating warblers and other small birds. During the last week in August the southward migration from Canada and the Northern States begins, numbers decrease during the first half of September, and by the end of that month most of these flycatchers have left their northern homes. The migration is mainly southwestward through Texas and into Mexico and Central America, where they make their winter home.

Winter.—Dickey and van Rossem (1938) say that the least flycatcher is "common in fall, winter, and spring throughout the Arid Lower Tropical Zone" in El Salvador, between September 3 and April 22. "The species is most numerous below 2,500 feet, and rare and local as high as 3,500 feet. * * * The least flycatcher occurs over the same country occupied by *Empidonax flaviventris*. The two are present in about the same numbers, relatively, and both are found in similar situations, that is to say, undergrowth in the woods, mimosa thickets, shrubbery along watercourses, or in the top foliage of low, open woods."

William Beebe (1905) reports, in Mexico, "a small, loose flock observed several times near camp in a lower barranca; the only flycatchers which seemed to remain together in any association which could be called a flock."

DISTRIBUTION

Range.—North and Central America, chiefly east of the Rocky Mountains.

Breeding range.—The least flycatcher nests **north** to southwestern Mackenzie (Willow Lake River and Fort Rae); northern Alberta (Smith Landing and the Athabaska Delta); Saskatchewan (Lac Île à la Crosse and Reindeer River); Manitoba (Oxford Lake); Ontario (Lac Seul and Moose Factory); southern Quebec (St. Joachim and Restigouche Valley); Nova Scotia (Baddeck); and Prince Edward Island (Malpeaque. **East** to Nova Scotia (Baddeck and Halifax); Maine (Machias and Portland); Massachusetts (Cape Cod); Connecticut (New Haven); New York (New York City); southeastern Pennsylvania (Hamburg); western Maryland (Accident); West Virginia (Sago and Cranberry Glades); western Virginia (Blacksburg); and western North Carolina (Piney Creek, Rock Mountain, and Waynesville). **South** to southwestern North Carolina (Waynesville); Indiana (Worthington); southwestern Missouri (Marionville); southeastern Nebraska (Falls City); and southeastern Wyoming (Cheyenne). **West** to Wyoming (Cheyenne and Careyhurst); Montana (Livingston, Bozeman, and Great Falls); Alberta (Midnapore and Jasper Park); eastern British Columbia (Peace River Block); and southwestern Mackenzie (Willow Lake River).

Winter range.—In winter this species is found **north** to rarely Durango (Tamazula); northern Tamaulipas (Matamoros); Yucatan (Merida and Chichen Itza); and Quintana Roo (Cozumel Island). **East** to Quintana Roo (Cozumel Island and Xcopen); British Honduras (Orange Walk and Belize); Honduras (Tela, Lancetilla, and **Progreso**); and Panama (David). **South** to Panama (David); Costa Rica (Nicoya); El Salvador (Puerto del Triunfo and Colima); southern Guatemala (San Jose and Patulul); Oaxaca (Santa Efigenia and Tapana); and Guerrero (Acapulco). **West** to Guerrero (Acapulco, Coyuca, and Chilpancingo); Michoacan (Apatzingan); and rarely Durango (Tamazula).

Spring migration.—Early dates of spring arrival are: Georgia—near Roswell, April 11. North Carolina—Waynesville, April 20. Virginia—Blacksburg, March 19 (exceptionally early). West Virginia—White Sulphur Springs, April 19. District of Columbia—Washington, April 20. Pennsylvania—Beaver, April 22. New York—Alfred, April 29. Connecticut—Hadlyme, April 26. Rhode Island—Providence, April 27. Massachusetts—Melrose, April 25. Vermont—St. Johnsbury, April 29. Maine—South Harpswell, April 30. Quebec—East Sherbrooke, May 6. New Brunswick—Scotch Lake, May 8. Nova Scotia—Wolfville, May 4. Louisiana—New Orleans, March 30. Mississippi—Biloxi, April 9. Arkansas—Monticello, April 24. Missouri—St. Louis, April 26. Indiana—Waterloo, April 22. Ohio—Oberlin, April 27. Michigan—Petersburg, April 29. Ontario—Ottawa, May 5. Iowa—Hillsboro, April 24. Minnesota—Lanesboro, April 30. Manitoba—Aweme, May 11. Texas—San Antonio, April 14. Oklahoma—Norman, April 30. Kansas—Lawrence, May 1. Colorado—Yuma, May 13. Nebraska—Falls City, May 2. South Dakota—Sioux Falls, April 30. Wyoming—Torrington, May 5. North Dakota—Fargo, April 21. Montana—Huntley, May 13. Saskatchewan—Indian Head, May 12. Alberta—Red Deer, May 16. Mackenzie—Fort Simpson, May 24.

Fall migration.—Late dates of fall departure are: Alberta—Glenevis, September 14. Saskatchewan—Last Mountain Lake, September 3. Montana—Custer County, August 24. North Dakota—Argusville, September 26. Wyoming—Laramie, September 19. South Dakota—Aberdeen, September 22. Nebraska—Omaha, September 24.

Colorado—Holly, September 19. Kansas—Lawrence, September 21. Oklahoma—Kenton, September 26. Texas—Somerset, October 2. Manitoba—Shoal Lake, September 25. Minnesota—Hutchinson, October 8. Iowa—Emmetsburg, October 28. Ontario—Toronto, October 4. Michigan—Detroit, October 14. Ohio—Lakeside, October 13. Indiana—Carlisle, October 12. Missouri—Jasper County, October 20. Kentucky—Bowling Green, October 21. Louisiana—Southwest Reef Light, October 23. New Brunswick—Scotch Lake, September 24. New Hampshire—Jefferson, October 3. Massachusetts—Dennis, October 16. Connecticut—Hartford, October 7. New York—Orient, October 15. New Jersey—Milltown, October 1. District of Columbia—Washington, October 1. Virginia—Lexington, September 20. North Carolina—Piney Creek, September 17. Georgia—Athens, October 12.

Casual records.—There is an old record (previous to 1884) of the occurrence of the least flycatcher at an elevation of 1,500 feet near Guajango, Peru; one was taken in March 1904 on Grand Cayman Island, between Jamaica and western Cuba; and one was recorded on June 15, 1922, at Ellis Bay, Anticosti Island, Quebec.

Egg dates.—Illinois: 12 records, June 4 to 20.

Massachusetts: 35 records, May 26 to June 17; 17 records, June 1 to 9, indicating the height of the season.

New York: 30 records, May 24 to June 27; 16 records, May 24 to June 27.

EMPIDONAX HAMMONDI (Xantus)

HAMMOND'S FLYCATCHER

PLATE 29

HABITS

Three western flycatchers of the genus *Empidonax* form a very puzzling group. The difficulty of recognizing the three species, even in hand, has doubtless led to errors in identification, which must be taken into account in considering the value of field observations. Fortunately, habitats, nesting habits, and voices often help in field identification, as will be referred to later.

As to the characters of specimens in hand, Dr. Joseph Grinnell (1914b) writes: "In spite of the largely increased extent of material illustrating this genus, the relative characters of *Empidonax griseus*, *Empidonax wrighti*, and *Empidonax hammondi* remain somewhat subtle. * * * The color differences are minute: *hammondi* is slatiest, *griseus* ashiest, *wrighti* intermediate; *wrighti* is greenest dorsally and pectorally; the outer web of outer tail-feather is distinctly white nearly to its tip in *griseus*, grayish white in *wrighti*, and but slightly paler than rest of feather in *hammondi*. The lower mandible is entirely blackish brown externally, in *hammondi*, dull or lighter brownish in *wrighti*, while in *griseus* it is blackish brown at tip and abruptly straw yellow for its basal two-thirds, brightest along the rami."

In his table of measurements, it appears that, in general dimensions, *hammondi* is the smallest, *wrighti* intermediate, and *griseus* the largest, except that *wrighti* has the longest tail.

The breeding ranges of these three species are none too well defined, and many published records may prove to be subject to correction. According to the 1931 Check-list, Hammond's flycatcher is said to breed "in Transition and Canadian zones from southeastern Alaska, southern Yukon, and southern Alberta to the Sierra Nevada, central California and Colorado."

The breeding haunts of Hammond's flycatcher are mainly at higher elevations than those of the other small flycatchers in the open forests of firs, spruces and pines. Harry S. Swarth (1922) found it, in northern British Columbia, "abundant on the upper Stikine, where it is largely a bird of the poplar woods." He does not mention finding a nest there, so his birds may not have been on their breeding grounds at the time, "the last week in May."

Samuel F. Rathbun writes to me: "Hammond's flycatcher is a more or less common summer resident of western Washington, but it is in the section known as the Olympic Peninsula that this species reaches the height of its abundance, possibly because of the forest conditions which there prevail. It seems to be partial to the somewhat open coniferous forest, though, even here, more often near their borders than in their depths; and occasionally its note is heard in the fringes of deciduous trees which sometimes grow along the edges of the conifers."

W. L. Dawson (1923) writes: "In its summer home, in Oregon and Washington, Hammond Flycatchers may be locally very common. I have seen twenty in the course of a morning's walk in early June. Fir groves, the edges of clearings, bush-clad hillsides with fallen trees scattered about, the timbered banks of streams, these are favorite places of residence."

Farther south this flycatcher breeds at much higher elevations in the mountains of California and Colorado. At one time both Dr. Grinnell and Mr. Dawson expressed some doubt as to the breeding of this species in California, but it is now a well-established fact that it breeds in the high coniferous forests of the Sierra Nevada. James B. Dixon tells me that in Mono County it seems to breed almost exclusively above 9,000 feet and that some nests were found at the topmost tree limit, very close to 11,000 feet above sea level.

In Colorado it has been recorded at elevations varying from 7,500 to 10,000 feet.

Spring.—Hammond's flycatcher is evidently one of the earlier migrants through Arizona and California, where it seems to be quite generally distributed and, at times, fairly abundant. Referring to Arizona, Mr. Swarth (1904) says: "Of the migrating birds passing through this region in the spring the Hammond Flycatcher is one of the first to put in an appearance, and about the last to leave. The earliest noted, a male, was taken on March 30; the bulk of them arrive early in April, and they remain in the greatest abundance until the middle of May, when they begin to rapidly diminish in numbers, the last being seen May 22. In the spring I found them in all parts of the mountains, but most abundantly below 6000 feet, and usually along the canyons, not far from water."

Mr. Rathbun, in his notes from Seattle, says: "This flycatcher arrives in the latter part of April. It is one of the species found in the 'bird waves' that one sees at times. On these occasions, it is in the company of warblers, kinglets, chestnut-backed chickadees, and often red-breasted nuthatches, with now and then a few others; and I have never seen a wave of our migrating woodland birds that did not have at least a few Hammond's flycatchers."

Nesting.—Mr. Rathbun says further (MS.): "On two occasions I have found its nest. In both instances it was placed in a small fork of a limb of a fir tree of medium size, and the nests were at heights of between 50 and 60 feet. The material used in each was thin strips of fibrous bark and plant fibers; the lining was of fine dry grasses and bits of dry mosses, neatly woven together." Mr. Dawson (1923) writes:

In the summer of 1906 Mr. Bowles and I found these flycatchers nesting on a fashionable hillside section of Spokane. In two instances the birds were building out in the open, after the fashion of the Western Wood Pewee (*Myiochanes richardsoni*): one on the bare limb of a horse-chestnut tree some ten feet from the ground; the other upon an exposed elbow of a picturesque horizontal limb of a pine tree at a height of some sixty feet. A few miles farther north we located a nearly completed nest of this species on the 20th of May, and returned on the 1st of June to complete accounts. The nest was placed seven feet from the trunk of a tall fir tree, and at a height of forty feet. The bird was sitting, and when frightened dived headlong into the nearest thicket, where she skulked silently during our entire stay. The nest proved to be a delicate creation of the finest vegetable materials, weathered leaves, fibers, grasses, etc., carefully inwrought, and a considerable quantity of the orange-colored bracts of young fir trees. The lining was of hair, fine grass, bracts, and a single feather. In position the nest might well be that of a Wood Pewee; but, although it was deeply cupped, it was much broader, and so, relatively flatter.

Rose Carolyn Ray (1932) published the first authentic account of the nesting of Hammond's flycatcher in the high Sierras of California.

She found the nest on June 21, 1929, "in a forest of pines and firs on the Sierran summit at an altitude of 7600 feet," in Eldorado County. It was 6 feet from the ground in a tamarack tree and "placed where several small trees and a slanting sapling came together and thereby offered concealment, but it was not woven to the limb as the nests of the Wright Flycatcher are." A few days later, her husband, Milton S. Ray, was able to identify the nest by shooting the parent bird. "The nest was rather loosely made for a flycatcher and is basically composed of dark red bark strips, together with light gray bark strips, rootlets, grasses, stems, feathers, string, cocoons and woolly substances, and thickly lined with feathers. The outside measurements are, top, 3¼ by 4½ inches; depth, 2¼ inches. The nest cavity is 2 inches in diameter by 1 inch in depth."

Grinnell, Dixon, and Linsdale (1930) report an interesting nest found by them in a lodgepole pine in the Lassen Peak region of California:

The nest is of rather unexpected construction, in that three successive years' occupation of the site is in evidence. The lowermost of the three layers that are distinguishable is hard-packed, resistant to the touch, and fitted closely into the supporting crotch and against the accessory side branch; upon this is a less compact but also weathered layer about 20 mm. thick as viewed on the exposed surface; above this is the new, current nest proper with free rim showing about 20 mm. still higher. The color of this new and superimposed cup is, in sharp contrast to that of the other two layers, light brown because of its constituent material being of unweathered but dried coarse light-brown vegetable fibers and grass stems, mixed with blackish filaments of a lichen. This latest increment is well rounded and of close texture.

Major Bendire (1895) describes several nests from different localities that were located in conifers and made of similar materials. He suggests, as several others have done, that the nests of Hammond's flycatcher more nearly resemble in shape and position the nests of the wood pewee than those of the other Empidonaces. He also mentions some nests in the National Museum, collected by Roderick MacFarlane, in British Columbia, some of which "were apparently placed in upright crotches of willows, and others on horizontal limbs close to the trunks of small conifers, at no great distance from the ground."

Although the parent birds came with these nests, there is the possibility that some of his collectors may have made mistakes, or the specimens become mixed. There are numerous other published accounts of supposed nests of this species, in upright crotches and at low elevations in small trees and bushes, that apparently resemble the nests of Wright's flycatcher. They look suspicious, in view of all that has been said above, and may be referable to the latter species.

Eggs.—Three or four eggs make up the usual set for Hammond's flycatcher. They are mostly ovate and have little or no gloss when fresh; some eggs are more elongated or shorter ovate in shape, and some have been reported as somewhat glossy. The ground color seems to vary from dull white to pale creamy white, or a deeper cream color; Mr. Ray describes the ground color of his set as a "peculiar, clear, rich, yellowish tint." A majority of the eggs are unmarked, but many are marked with minute dots, or small spots, of dark liver-brown, or lighter shades of brown. A set taken by Denis Gale in Colorado is described by Bendire (1895) in Mr. Gale's own words, as having "a decided light-yellow ground, with a slight powdering of dark specks, with larger shell markings of lavender tints." The measurements of 50 eggs average 16.8 by 12.9 millimeters; the eggs showing the four extremes measure **17.8** by 13.7, 17.6 by **14.0**, **15.2** by 13.2, and 16.6 by **12.4** millimeters.

Young.—The period of incubation is probably the same as for the other small flycatchers, about 12 days. Probably only the female incubates the eggs, but both parents feed and care for the young. Dr. Russell T. Congdon has sent me some photographs of the feeding process (pl. 29), and he says in his notes: "The taking of food by the young bird stimulates active peristalsis, and the parent is so alert to grasp her opportunity that not once was the nest soiled. After feeding the insect to one of the young the parent flycatcher would remain on the edge of the nest in a watching attitude, and after a brief period, the young bird just fed would elevate its rump, extend the tail feathers way back, and extrude the pellet (or bolus) of waste matter. This the parent immediately seized in her bill, before it could drop into the nest, and made off with it."

A family of young, "obviously just out of a nest," was watched by Grinnell, Dixon, and Linsdale (1930): "They kept at a height of about ten meters above the ground, among the branches toward the center of a lodgepole pine.

"The young birds were well feathered except that their tails were not quite of full length. Both parents were in attendance. After the old female had been collected the male alone fed the family. The young ones would, in turn, fly out after the parent and there would be a melee during which feeding apparently took place. The young birds gave a faint food call, *chlip.*"

Referring to the Sierra Nevada, W. W. Price (Barlow, 1901) says: "Late in the summer the young of this flycatcher are common in the tamarack thickets along Silver Creek and on the slopes of Pyramid Peak. They are usually associated with the young of two or three warblers and Cassin vireo. I have noted this congregation each season previous to the migration. The large scattering flocks are often miles in extent, and probably contain thousands of birds."

Plumages.—There are two quite distinct color phases, said to be independent of age or season, in the adult Hammond's flycatcher. As described by Ridgway (1907), in the white-bellied phase the pileum and hind neck are "dull deep brownish gray (nearly mouse gray), the back, scapulars, lesser wing-coverts, rump, and upper tail-coverts similar, but decidedly more olive; * * * chest and sides of breast pale gray, gradually fading on sides and flanks; rest of under parts dull white, yellowish white, or very pale primrose yellow." The yellow-bellied phase he describes as similar, "but more olivaceous (sometimes brownish olive) above, chest and sides of breast olive or buffy olive, and abdomen, etc., primrose or sulphur yellow." The two extremes are rather uncommon, most of the specimens are more or less intermediate, and the color differences may be seasonal.

Mr. Ridgway (1907) says that the young are "essentially like whitish-bellied adults, but color of upper parts grayish brown, rather than olive, wing-bands light buffy, and marginal under wing-coverts buffy."

Dickey and van Rossem (1938) write: "Hammond's flycatcher differs materially from the other visiting species of *Empidonax* in that it molts before leaving the north. We have many specimens from the United States showing all stages of the postjuvenal and adult fall molt which begins in August and is ordinarily complete by the latter part of September. In this species the juvenal rectrices (but not the remiges) are replaced with the body plumage at the postjuvenal molt. Another point of difference is that the spring molt ordinarily is not extensive. Good series of migrating spring specimens taken in various western states in April and May show varying amounts of new body feathers, particularly on the foreparts and back, but most of the plumage is that acquired at the molt of the previous fall."

The postnuptial molt evidently begins very early, for an adult male, taken on June 30, by Mr. Swarth (1922) "shows the beginning of the molt"; and an adult female, taken July 27, "had renewed a large part of its plumage." A young bird taken "August 10, shows the beginning of the molt into the first winter plumage." On the other hand, the adult molt sometimes is much delayed, for Mr. Swarth (1929) collected an adult female in Arizona, on October 3, that was "apparently just beginning the annual molt."

Food.—Practically nothing seems to have been published on the food of this flycatcher, which probably does not differ materially from that of the other members of the genus.

Behavior.—Hammond's flycatcher generally spends most of its time among the higher branches of the trees in which it lives, at greater heights than most other members of the genus. Mr. Rathbun says in his notes: "A favorite perch is among the upper branches or on the extreme top of some tall, slender, dead tree at or near the edge of an open space in the forest. From this it will make short flights after winged insects, and invariably, after returning to its perch from such excursions, it will flirt its tail once or twice before lapsing into a quiet attitude." He says that on days of heavy rain it "will be found low among the trees, at times only a few feet above the ground." Bowles and Decker (1927) say of its behavior about the nest:

The female sits very closely after incubation has commenced, so that it is sometimes necessary to lift her off the nest in order to ascertain the contents, but she is seldom or never found on the nest until the set is complete. After being flushed she is the tamest of the small Flycatchers, usually returning to the tree very soon and otherwise displaying her anxiety. The male is very watchful around the nest and will promptly drive off any other bird that comes in its vicinity, in this way sometimes showing the oologist that a nest is near at hand. We once saw a beautiful male Townsend's Warbler attacked and driven off after quite a battle, in which the dusty colored little Hammond's looked like a tramp. These birds are never at all shy in the vicinity of the nest and are usually easily approached at any time elsewhere.

Voice.—Hammond's flycatcher is said to be a rather silent species. Mr. Rathbun says of it (MS.): "The note of this flycatcher, when heard at a distance, seems to be a single one, but, when a person is close, it appears to be broken, or there seems to be a slight hesitancy after the first part. To us, the note sounds like *pee-eet*, or even *pee-zeet*, given somewhat deliberately; the *zeet* with a rising inflection, lightly accented and slightly prolonged; it then ends abruptly."

The bird also has a faint, soft call note that sounds like *pit*, or *quip*, often uttered continuously. And it has a rather distinctive song, by which it can be recognized. Grinnell, Dixon, and Linsdale (1930) describe this as follows: "One singing bird that was watched June 24, 1925, had as its singing perch the very topmost snag of a dead-topped white fir. The height of the perch was estimated at forty meters from the ground. The bird continually shifted its body, most frequently the head, from side to side. The head was thrown back simultaneously with the utterance of the notes. The complete 'song,' given over and over again with monotonous regularity, sounded to the observer as follows: *sé-put* (uttered rapidly), *tsur-r-r-p* (roughly burred), *tséép* (rising inflection)."

Field marks.—Hammond's, Wright's, and the gray flycatchers can hardly be recognized by the characters that separate the species (see the first page of this chapter), except under the most favorable circumstances and at close range. The songs, which are quite different, are the best field marks. Habitats are helpful. A small flycatcher at a considerable height in a coniferous tree, especially at the higher altitudes in the mountain ranges, is quite likely to be Hammond's. At low elevations, while migrating and in the lowlands, when the birds are mostly silent, recognition is almost hopeless.

Bowles and Decker (1927), however, observe that "in life Hammond's shows himself a dusky backed, sooty breasted, short tailed little chap, while Wright's is a gray backed, light breasted, long tailed bird, appearing decidedly the larger of the two. These characteristics may seem a trifle exaggerated here, but as seen in life they are recognizable at once. In fact, Hammond's suggests more than anything else an undersized Western Wood Pewee."

Fall.—Mr. Rathbun tells me that this species departs from Washington during September. W. E. D. Scott (1887) took specimens in Arizona from early in October until the 25th of that month. But Mr. Swarth (1904) states that in the Huachuca Mountains, Ariz., they reappeared in August, "not in the foothills and along the canyons," as in the spring, "but up in the pines, none being seen below 9000 feet. The first was seen on August 26, and from that time on, though not at all abundant, I found them in small numbers scattered through the pines along the divide." It would appear from this that the fall migration is leisurely and quite prolonged; this may be due to the fact that the fall molt is accomplished before the birds reach their winter home.

Winter.—Dickey and van Rossem (1938) found this flycatcher to be a "common winter visitant to the oak-pine association of the Arid Upper Tropical Zone of the interior mountains" of El Salvador, "rarely straggling as high as 8,700 feet in the cloud forest of the Humid Upper Tropical. Extremes of elevation are 3,500 and 8,700 feet. Dates of arrival and departure are November 21 and March 12. * * * It is probable that the date of arrival is somewhat in advance of that given above, for the birds were present in numbers on November 21, the initial day of collecting on Mt. Cacaguatique."

DISTRIBUTION

Range.—Western North America; south in winter to El Salvador.

Breeding range.—Hammond's flycatcher breeds **north** to Alaska (Charlie Creek); and central Alberta (Lesser Slave Lake). **East** to Alberta (Lesser Slave Lake and Jasper Park); western Montana (Fortine, Flathead Lake, and Sourdough Canyon); Wyoming (Yellowstone Park and Laramie); and Colorado (Northgate, Gold Hill, and Salida). **South** to southern Colorado (Salida and Fort Lewis); southern Idaho (Emigration Canyon); and central California (Yosemite Valley and Grizzly Creek). **West** to California (Grizzly Creek, Mineral, and Mount Shasta); Oregon (Little Butte Creek and Powder River Mountains); Washington (Swamp Creek, Tacoma, and Seattle); British Columbia (Nootka Sound, Hazelton, and

Atlin); southeastern Alaska (Glacier); western Yukon (Carcross and Selwyn River); and eastern Alaska (Charlie Creek).

Winter range.—In winter the species is concentrated in Central America, **north** to the southern Mexican States, as Jalisco (Barranca and Ibarra); Guanajuato (Rancho Enmedia); and Veracruz (Orizaba) and **south** to southern Guatemala (Tecpam and Volcan de Fuego) and El Salvador (Los Esesmiles and Mount Cacaguatique).

Unseasonable records are of specimens taken near Livingston, Calif., on December 20, 1918, and in Sabino Canyon, Tucson region, Ariz., on February 24, 1934. The latter specimen, with one other, was first seen on January 18 and regularly thereafter to the date of collection.

Migration.—Migration data for this and some other members of the genus *Empidonax* are frequently unsatisfactory for the reason that in most cases visual field identifications are practically worthless.

Spring.—The following appear to be early dates of spring arrival: Colorado—Fort Lyon, May 15. Wyoming—Laramie, May 13. Montana—Hargan, May 10. Alberta—Banff, May 12. Arizona—Santa Catalina Mountains, March 31. Idaho—Priest River, May 19. California—Los Angeles, April 9. Oregon—Lake Malheur, April 17. Washington—Kiona, April 21. British Columbia—Victoria, April 20.

Fall.—Late dates of fall departure are: Alaska—McCarthy, August 23. British Columbia—Huntington, October 5 (unusually late). Washington—Pullman, September 18. Oregon—Wallowa County, September 23. California—Los Angeles, October 30. Idaho—Coeur d'Alene, September 1. Utah—Beaver, September 22. Arizona—Santa Catalina Mountains, October 25. Alberta—Jasper Park, August 7. Montana—Fortine, August 25. Wyoming—Pacific Springs, September 4. New Mexico—Gallup, September 30.

Casual records.—A specimen was taken at Ciudad Victoria, Tamaulipas, on March 2, 1908, and one was collected at Crawford, Nebr., on September 17, 1911.

Egg dates.—British Columbia: 9 records, June 4 to July 7.

California: 17 records, June 12 to July 10; 9 records, June 22 to July 2, indicating the height of the season.

Colorado: 4 records, June 24 to 30.

Washington: 8 records, June 5 to July 14.

EMPIDONAX WRIGHTI Baird

WRIGHT'S FLYCATCHER

PLATES 30–32

HABITS

According to the 1931 Check-list, Wright's flycatcher "breeds in the Transition and Canadian zones from central British Columbia,

Yukon (casually), and west-central Saskatchewan south to southern California, Arizona, New Mexico, and western Texas, and east to the eastern base of the Rocky Mountains." Throughout this range it is confined mainly to the foothills and slopes of the various mountain ranges, at elevations varying with latitude.

L. B. Howsley has sent me some extensive notes on the habits of this flycatcher in Stevens County, eastern Washington, in which he says of its haunts: "The altitude ranged from 900 feet to 2,800 feet above sea level. However, this flycatcher was rarely observed above the 1,500 foot level, and most nests were found at much lower altitudes. The favorite nesting habitat seemed to be the lower, more open, and rolling slopes and the benches scattered here and there, especially those portions covered with a species of willow which grew in isolated clumps throughout the area. Logged-off lands were evidently preferred, probably on account of the open situations available."

Henry J. Rust writes to me that "this species arrives in northern Idaho about the middle of May and is found sparingly distributed over the low, brushy hillsides and partly wooded flats." Mrs. Florence M. Bailey (1928) writes: "In Colorado in summer, Mr. Henshaw found the Wright Flycatcher a bird of the mountains, frequenting deciduous trees and bushes along streams; and in Arizona he found it among the oak openings; but in the vicinity of Santa Fe he saw it on pinyon-clad hills, and at Lake Burford Doctor Wetmore found it common among junipers and pines on the dry hillsides above the gulches."

Referring to the Great Basin region, Robert Ridgway (1877) says that "it inhabits both the aspen groves and copses of the higher cañons and the mahogany woods of the middle slopes, in which places it is sometimes one of the most numerous of the smaller birds." In California, Grinnell, Dixon, and Linsdale (1930) say that in the Lassen Peak region, "we gain a picture of chaparral *with trees scattered through it*, as characterizing the average habitat." And, in the Yosemite region, Grinnell and Storer (1924) found a pair in "about an acre of dense chaparral on a flat near the stage barns. The thicket was about four feet high and comprised a dense growth of snowbush (*Ceanothus cordulatus*), green manzanita, and chinquapin. The male had a number of forage posts at the tops of some dwarfed black oaks which struggled up slightly above the general level of the chaparral; he would progress from one to another of these in rather regular succession, catching flies en route. Occasionally he would go up higher, 30 feet or so, to one of the outstanding limbs of a neighboring sugar pine or red fir, and from there he would sing."

Nesting.—Mr. Howsley says in his notes from Stevens County, Wash.: "Of some 30 nests studied, 70 percent of them were placed in willows. Alders, 'snowbush,' and 'meadowsweet' comprised the others. No nests were found in conifers, or in the numerous clumps of chokecherry. My notes show the nesting heights ranging from 4 to 7 feet. None was higher. In all cases the nest was tightly secured to the surrounding uprights. Evidence of many old nests that had survived the severe storms and snows of that region testified to the care with which they were anchored.

"The nesting material is usually composed of fine, bleached weed fibers and fine grasses, woven compactly into a neat, well-rounded cup, which is lined with very fine weed bark, grasses, often bits of vegetable down and feathers, although feathers are a rather uncommon occurrence. I have never observed horsehair or other animal hair in the structure. Sometimes, however, there seems to be considerable spider webbing used in binding the outside of the nest; and I have rarely found a nest whose outer structure showed many 'loose-ends,' except the lowest part next the crotch, which seems to be quite carelessly and slovenly laid. The nest as a whole, however, compares very favorably with other flycatchers' both in neatness and compactness."

Many fine photographs of Idaho nests, sent to me by Mr. Rust, illustrate nests in different shrubs and one in a young *ponderosa* pine, the latter resting on two small branches and against the trunk (pl. 31); all seem rather loosely constructed at the base, and one has many loose ends hanging down. He tells me that nesting begins by the middle of June and that the young are on the wing early in July. "On brushy hillsides nests are usually located in low shrubs, and on wooded flats in small pines from 7 feet up."

James B. Dixon writes to me that he has found this flycatcher nesting in the Mono Basin area of the Sierra Nevada and in the San Bernardino Mountains. "Its breeding range seems to lie between 6,000 and 9,000 feet. It is commonly found in the aspen groves of the stream beds, where it nests at varying heights from the ground. One of its favorite locations is on the so-called knees of the aspens. These knees are apparently caused by the snow bending over a young sapling, and then, the next year, the sapling grows upright again, leaving a step or knee."

There is a nest in the Thayer collection taken at an elevation of 8,000 feet in the San Bernardinos; it was placed 2½ feet up in a buckthorn bush, and the collector, H. A. Edwards, comments that this is almost the only situation in which this species nests in that vicinity; he also says that the nests are almost impossible to see on account of their buffy color. This nest is firmly and well made of shredded weed stems and grasses, light buff in color, mixed with fine vegetable fibers and plant down; the rim is plastered with spider webs, and the deep cup is lined with fine grass, cattle hair, and feathers; it measures 2¾ inches in outside and 1¾ inches in inside diameter; the inner cup is 2 inches deep, and the entire height of the nest is 3½ inches. A nest in the same locality, also in a buckthorn bush, at 7,000 feet altitude, is shown in a photograph sent to me by Wright M. Pierce (pl. 30).

Bowles and Decker (1927) say: "Any kind of deciduous bush or small tree seems about equally desirable as a nesting site, for we have found them in the following locations: red willow, birch, wild rose, alder, cottonwood, maple, and one each in fir and western yellow pine, the last two being the only instances we have seen of their using conifers. The nest is almost invariably built in an upright crotch, the only exception being the one in the fir, which was saddled on a forked limb close to the ground. About six feet from the ground is the average height, the extremes being fifteen and two feet."

In addition to the trees and shrubs mentioned above, nests of the Wright's flycatcher have been found in other species of small pines and spruces, wild plum, manzanita, hazel, dogwood, mountain mahogany, laurel, and serviceberry, and probably others.

Mrs. Wheelock (1904) tells of the part that the male plays in the construction of the nest: "One sunny day about the tenth of June, you will see him bring a bunch of plant fibre and, placing it in the chosen crotch, jump on it and pack it into place with feet and bill. He has worked hard to get it, tugging with all his little strength to loosen some of it, which is the inner bark of the willows, and chewing it back and forth in his beak to render it fine and pliable. After the first bit has been put in place the female does the shaping and weaving, while the male brings the material."

Eggs.—Wright's flycatcher lays either three or four eggs, apparently never more, and perhaps sometimes only two. They are rather short or rounded-ovate, or normal ovate, and they have no gloss. Mr. Howsley says they are much like bobwhites' eggs in shape, "being quite blunt at the larger end receding suddenly to the point." The color is dull white or pale creamy white, and, so far as I can learn, they are never, or very seldom, spotted. The measurements of 50 eggs average 17.3 by 13.2 millimeters; the eggs showing the four extremes measure 18.5 by 14.2, 17.0 by 14.5 and 15.2 by 12.2 millimeters.

Young.—Mr. Howsley says in his notes that "the period of incubation lasts 12 to 15 days, one set showing a period of 17 days from the last egg laid." In this latter case it is probable that incubation did not begin until some time after laying had ceased. Mrs. Wheelock (1904) writes:

During incubation, which lasted thirteen or fourteen days in two cases, the male was frequently found on the nest, not merely guarding but brooding. When not thus occupied, he flitted restlessly through the bushes, bringing insects to his mate, not spending one moment in idleness except to take a sunbath, and his cheery twitter could be heard all day above the music of his more ambitious neighbors. As soon as the young Flycatchers were out of the shell, he redoubled his efforts and seemed to do much more than half the feeding. For the first few days this was by regurgitation, but later fresh food was given to them. Small wonder that with four such voracious appetites to satisfy he came and went in preoccupied silence. In two weeks the babies had filled the nest to overflowing and were fairly crowded out of it. Then the trials of the father bird really began, for they tagged him from twig to twig with open mouths and quivering wings. * * * In every instance the mother helped faithfully, and in one case she alone fed a nestling almost as large as herself, at the rate of six bugs in three minutes. Sometimes she liberated one in front of him, in an effort to teach him to hunt for himself, but he was the only young Flycatcher I have ever seen refuse to try to catch an insect; he would not budge. This little comedy was played all one day, and early the next morning the worn and weary little mother was seen alone, no trace of the youngster could be found, nor did she seem to care.

In a nest watched by Grinnell, Dixon, and Linsdale (1930) the young remained approximately 18 days. Ten days later "an observer was surprised to see a bird sitting on this same nest. When disturbed the brooding bird flew to a near-by twig, and the nest was found to contain three eggs."

Plumages.—Young Wright's flycatchers, in fresh juvenal plumage, are much like the adults, except that the top and sides of the head are decidedly grayer, the wing bands are buffier, and the breast and flanks are strongly buffy, the latter becoming whiter later on. Apparently the postjuvenal molt is not accomplished until the young birds reach their winter home.

Some adults may begin the fall molt while migrating, but, as I have seen a number, taken in Mexico in September, that were still in worn summer plumage or molting the body plumage, it seems as if the molt must be mainly accomplished after migrating. Birds collected in Central America in October and November are mostly in fresh winter plumage. I have seen no evidence of a spring molt in the few specimens available, but probably there is at least a partial prenuptial molt before the birds return north, as is true of several other species of the genus.

Food.—Referring to the feeding habits of this flycatcher, Mr. Howsley says: "The parent never brought a worm to the nestlings. The food was always a very small moth. These moths were taken on the wing, after which the parent would sit for several moments on a neighboring twig and utter a few weak *cheeps* intermittently before feeding the young. However, I have often noticed the parent fluttering beneath a leaf from which she seemed to obtain insects, in the manner of a warbler or vireo, which she devoured herself. Once in a while she would also swallow the moth she had evidently gathered for the offspring, but decided, either through hunger or my too close proximity, to take no chances."

I can find no published account of the food of Wright's flycatcher, which probably does not differ materially from that of the other western Empidonaces.

Behavior.—Mr. Howsley says (MS.): "Wright's flycatcher is a friendly little soul, not altogether shy, although not allowing too much familiarity. Each pair occupies a separate territory and seems to respect the other's rights by not trespassing unless inadvertently. The only instance I noted of one territorial inhabitant resenting intrusion from his neighbor was during the period following incubation, especially the early stages after hatching. Any bird, related or not, that comes in close proximity of the nest tree, is immediately put to flight by a vigorous attack, the parent uttering a rapid scolding note during pursuit." Bowles and Decker (1927) write of the behavior about the nest:

The bird can only be closely approached when the female is on the nest, and here her actions are exceedingly unusual. She will always sit so closely that one has to lift her off the nest, but it is impossible to judge the contents by this as she is just as likely to be building as to have a complete set of eggs. A striking example of this was one nest that we found built about ten feet up in a slender alder, which the sitting bird positively refused to leave. The tree could not be climbed with safety, so we whittled it off about three feet from the ground and lowered it down. The bird "sat tight" all the time and had to be pried off the nest which, much to our amazement, was little more than half completed and absolutely empty. We then tied the two sections of the tree together and later found that the change and disturbance had not seemed to bother the bird at all, as she finished her nest and laid a set of four eggs. * * * Tame as the female is when on the nest, her actions are completely changed the instant that she leaves it, for then she is the shyest of the shy and it requires a long and cautious hiding in order to get even a sight of her afterwards. We have never seen the male indicate the presence of the nest in any way or come near it at any time.

Mr. Dawson (1923) tells an interesting story of his experience with an unusually tame and confiding Wright's flycatcher that allowed herself to be handled quite freely, and even fed from his hand.

Voice.—Mr. Dawson (1923) gives a very good account of the notes of the male, as follows:

He has an extensive repertory of notes, entirely unlike any uttered by the female, and of so varied a character as to have given rise to great confusion. The two syllabled *pewick'* or *pusek'* note, especially, is very like that of *hammondi*, although it is undoubtedly milder and less sharply accented. This note is susceptible of great variation, especially when uttered in groups of three: *Pusek'*— *pitic'—squiz' ik* ; *sit' ick—chit' ick—sue whit'* ; *pssit pewick pussee'*. It is, however, the high-pitched and resonant *whêw hit'* call which startles the woods and marks the movements of the male at the mating season. This note is essentially a mating, or seeking, call; and it is uttered successively from prominent tree-tops over a wide range of territory.

Dr. Alexander Wetmore (1920) gives a somewhat different rendering of what are apparently the same notes: "The ordinary call note was a loud *tsee-wick*, given almost as one syllable, that when heard near at hand was startlingly like the *chebec* of the Least flycatcher. At a distance however this resemblance was lost. The males had a peculiar jerky song divided into couplets with slight pauses between that may be represented by the syllables *see-wick, tsee-ee, se-wick, tsil-ly tsee-ee*."

Ralph Hoffmann (1927) writes: "The song of the Wright Flycatcher is a little more vigorous and much more varied than that of the Hammond. It is commonly built up of three notes, *psit hreek pseet*, the last note the highest, but these are often grouped in series of fours, *psit hreek psit pseet*, or otherwise varied. Even when the song is made up of a series of three, it may be distinguished from that of the Hammond by the absence of the low *tsurp* characteristic of the latter. The Wright Flycatcher also utters, particularly toward dusk, a quite different series of notes, *tee, tee, tee-hick*. The common call note, used by both sexes, is a soft *pit*, heard constantly from the bushes in which the bird nests."

Field marks.—As stated under other species of *Empidonax*, the puzzling western species of this group can hardly be recognized in the field by the characters that separate the species; it is difficult enough to recognize some of them, even in the hand. Habitats and voices are the best guides to identification, but some of the habitats nearly or quite overlap, and the various renderings of the songs and calls, even as given by some of the keenest observers, are, to say the least, a bit confusing and none too distinctive. Keen and discriminating ears and considerable field experience are evidently necessary in order to learn to recognize them. Wright's flycatcher can easily be distinguished from the western, as the latter is much yellower. The little flycatcher (*trailli*) has a very distinctive voice and lives mainly in willow and other damp thickets. Hammond's lives mainly at higher elevations and prefers coniferous forests. And the gray flycatcher lives mainly in the sage-covered plains and has a more restricted distribution.

Enemies.—Mr. Howsley found many sets of fresh eggs in July. "Some of these," he says, "may possibly have been second layings necessitated by numerous destructive forces active throughout the section. I have personally seen chipmunks eating the eggs, and snakes, weasels, and other predators, which overrun the district, no doubt account for many destroyed nests. * * * There is a high mortality among the young in this particular section due to sudden and severe wind and rain storms during the nesting season. I have found whole broods wiped out, either through being *blown* out of the nests, or drowned *in* the nests. However, I think the loss is principally caused by chilling after getting wet."

DISTRIBUTION

Range.—Western United States and Canada, wintering in Central America.

Breeding range.—Wright's flycatcher nests **north** to northern British Columbia (Atlin); Alberta (Jasper Park, Banff, and Canmore); and southern Saskatchewan (Cypress Hills). **East** to southwestern Saskatchewan (Cypress Hills); western South Dakota (Elk Mountains); eastern Wyoming (Guernsey and Laramie); central Colorado (Estes Park, Idaho Springs, Crested Butte, and Silverton); and northern New Mexico (Taos Mountains, Sangre de Cristo Mountains, and Willis). **South** to northwestern New Mexico (Willis, Santa Fe, and Fort Wingate); southern Arizona (White Mountains and Huachuca Mountains); and southern California (Big Bear Lake and Barley Flats). **West** to California (Barley Flats, Mount Pinos, Mineral, and Salmon Mountains); Oregon (Pinehurst, Fort Klamath, and Powder River Mountains); Washington (Yakima and Chelan); and British Columbia (Okanagan Landing, 168-mile House, Hazelton, Telegraph Creek, and Atlin).

Winter range.—In winter this species is found in western Central America **north** to Sonora (Tiburon Island and "15 miles southwest of Nogales, Arizona") and **south** to Guatemala (Momostenango, Finca El Soche, and Finca Sepacuite).

Spring migration.—Early dates of spring arrival are: New Mexico—Silver City, April 15. Colorado—Colorado Springs, April 25. Wyoming—Guernsey, May 9. Montana—Columbia Falls, April 24. Saskatchewan—Cypress Hills, May 31. Arizona—Fort Whipple, April 11. Utah—Gooshoot Pass, May 9. Idaho—Rupert, May 5. Alberta—Banff, May 20. California—Agua Caliente, April 8; Redlands, April 20. Nevada—Carson City, April 21. Oregon—Mulino, May 5. Washington—Kiona, April 29. British Columbia—Okanagan Landing, April 24; Atlin, May 22.

Fall migration.—Late dates of fall departure are: British Columbia—Atlin, August 17; Okanagan Landing, September 17. Oregon—Wallowa County, September 21. Nevada—Hidden Forest, September 19. California—Los Angeles, November 5; El Monte, November 7. Alberta—Jasper Park, September 12. Arizona—White Mountains, September 27. Montana—Fortine, September 8. Wyoming—Yellowstone Park, September 23. Colorado—Escalante Hills, September 5. New Mexico—Glen, September 22.

Casual records.—Wright's flycatcher is accidental west of the Cascade Range, but one was collected at Tillamook, Oreg., on May 24, 1913, and in western British Columbia it was reported from Chilliwack in April 1888 and from Hastings in April 1889. Four specimens were taken at Whitewater Lake, Manitoba, from May 15 to June 5, 1925. Several specimens have been taken in Brewster County, Tex., from April 13 to June 6, chiefly in the Chisos Mountains.

Egg dates.—California: 65 records, May 27 to July 18; 33 records, June 14 to 24, indicating the height of the season.

Colorado: 15 records, June 15 to July 27; 9 records, June 22 to July 2.

Oregon: 26 records, June 12 to July 25; 14 records, June 17 to July 4.

Washington: 28 records, May 27 to July 14; 14 records, June 15 to July 8.

EMPIDONAX GRISEUS Brewster

GRAY FLYCATCHER

PLATES 33, 34

HABITS

Although the gray flycatcher was described and named over 50 years ago, it was many years before its breeding range was discovered, and even its characters as a species were none too clearly recognized. For a full discussion of the errors and misunderstandings that occurred during the first 25 years, the reader is referred to what W. L. Dawson (1923) has to say about it in his "Birds of California." Even such an eminent authority as Dr. Joseph Grinnell (1908) wrote a long account of it as a breeder in the San Bernardino Mountains, which he afterward discovered to be an error. As late as 1915, in his "Distributional List of the Birds of California," he made the statement that "typical *griseus* does not seem to have been authentically reported in summer north of the Mexican boundary." Although much has been learned about it in more recent years, it is still one of the least known of the California Empidonaces and has the most restricted distribution. It is now known to be restricted in the breeding season to the Great Basin region of northeastern Colorado, Nevada, northern Arizona, extreme eastern California, and eastern Oregon.

William Brewster (1889) based his name for the gray flycatcher on a study of 65 specimens from Lower California and 13 from western Mexico. He described it as "nearest *E. obscurus* [= *wrighti*], but larger and much grayer, the bill longer, the basal half of the lower mandible flesh colored in strong contrast to the blackish terminal half." And in his account he elaborates more fully on its resemblance to *wrighti* and *hammondi*, which has puzzled ornithologists ever since.

The normal breeding habitat of the gray flycatcher is now known to be on the sagebrush plains, or semiarid flats overgrown with desert underbrush or junipers. James B. Dixon tells me that he found it breeding in Mono and Modoc Counties, Calif.; in both instances the type of habitat was distinctive; the birds were breeding on vast semi-desert plains, where a heavy growth of winter-killed thornbrush was the principal cover, with some sagebrush. "Out on these vast areas of brush they seem to colonize, as we located over 16 nests in one area of two square miles."

Grinnell and Storer (1924) write: "The Gray Flycatcher, when settled for the summer, is a bird of the arid Great Basin fauna. It enters the Yosemite region in the environs of Mono Lake, where our limited information suggests its restriction to the tracts of sagebrush and Kunzia where these bushes reach largest size. In this sort of 'chaparral,' the Gray Flycatcher doubtless nests, as does its near relative, the Wright, in the darker-hued, more typical chaparral of the Sierras. It is interesting to note that the Wright Flycatcher, as a breeding bird, was found to extend eastward down the slopes of Leevining Peak nearly or quite to the edge of Mono Lake; it there becomes a close neighbor of its very near relative, the Gray Flycatcher."

Spring.—Harry S. Swarth (1904) writes: "I found this species to be a common migrant in the Huachucas, more abundant than its near relative *wrighti*, and generally frequenting ground of a different character. Some specimens were taken along the various washes, but the region where they were most abundant was in the most barren of the foothill country; rough boulder strewn hills with but a scattering growth of scrubby live oaks. In such places I found them fairly abundant, that is I have seen as many as twelve or fifteen in the course of a morning's collecting; but they never ventured above the very entrance of the canyons, nor ascended the mountains at all."

Nesting.—Reliable information on the nesting of the gray flycatcher is rather scanty, owing to a number of errors in identification. James

B. Dixon, however, has sent me data on 17 nests that he has found on the Mono Flats in Mono County, Calif. He says: "The nests of these birds are the best concealed of any of the flycatchers that I have ever contacted. They usually build their nests in the bases of the dead, winterkilled thornbushes, where the bark has sloughed off and lies piled up in the crotches; and the nest is so carefully embedded in this debris as to be almost invisible. I have often seen the bird sitting on the nest and flushed it, and could then hardly believe that it was a nest. They sometimes nest in the live thornbushes and sage bushes."

Russell W. Hendee (1929) records the finding of three nests in Moffat County, in northwestern Colorado. In describing the habitat in which the nests were found, he says that away from the Snake River "the flats are covered with sage, greasewood, and rabbit brush, while the ridges are densely covered with pinyon and juniper." It was in the latter section that the nests were found, of which he says: "The nests were all built from juniper bark and lined with feathers. All of the bark used was carefully gathered from the gray and weathered outside strands, and, with the irregular outlines of the nest, served to make them surprisingly difficult to see. All of the nests were built in forks in juniper trees. The birds were rather noisy and not at all difficult to observe. They were sometimes seen among the sage bushes at some distance from the junipers."

Alexander Walker (1914) took a nest and three eggs of the gray flycatcher on June 7, 1913, in Crook County, Oreg. The nest was about 2 feet above the ground, in "the crotch of a sage-bush, on a sage and juniper flat"; the nest was "composed of small dead weed stems, plant down, hair, shreds of sage-brush bark and some grasses, quilted together and lined with wool and fine feathers." The parent bird was taken and identified by two of the best ornithologists in the country.

I have never seen an authentic nest of the gray flycatcher, even in a museum or private collection, but in what published photographs of them I have seen the nests seem to be very bulky and rather loosely constructed externally, with many loose ends projecting and giving them a ragged appearance. If this is characteristic of the species, it might help in distinguishing the nests of the gray flycatcher from those of its near relative, and, in some places, its near neighbor, Wright's flycatcher.

Eggs.—I have never seen any eggs that I felt sure were laid by the gray flycatcher. The nests found by Mr. Dixon contained either three or four eggs, and he tells me that he has never seen any markings on any of the eggs. Mr. Dawson (1923) gives the number of eggs in a set as three or four and describes them as ovate or short-ovate and pale creamy white. Mr. Walker's set consisted of three eggs, and he describes the color as creamy white. The measurements of 50 eggs average 17.7 by 13.4 millimeters; the eggs showing the four extremes measure **19.3** by 13.7, 18.3 by **14.3, 16.2** by 13.5, and 17.3 by **12.4** millimeters.

Young.—Russell and Woodbury (1941) have made a careful study of the nest life of the gray flycatcher at a nest in a juniper in Navajo County, Ariz., where a pair of these birds were raising their second brood for the season in July. According to their observations, "two broods are raised, leaving the nest in June and early August, respectively." They found that incubation lasted 14 days and that the young remained in the nest 16 days. The female alone incubated the eggs, but both parents assisted in feeding the young. During "four hours of observation, the young were fed 30 times, an average of eight minutes between food-bringing visits to the nest and an average of ten feedings per nestling. This means that each young bird was fed on an average once every twenty-four minutes. The periods between feedings were by no means regular. They varied in length from one to twenty-eight minutes and were shorter and much more regular in the early morning than toward noon. * * * So far as we could tell, the food consisted entirely of insects. It varied in size from tiny beetles to a butterfly so large that the young could scarcely take it, and included such recognizable forms as grasshopper, yellow wasp, moth, and ant-lion."

They noted that the parents shaded the young from the hot sun while they were in the nest, and fed them for an estimated period of two weeks after they left the nest.

Plumages.—Ridgway (1907) says that the young in juvenal plumage are similar to spring and summer adults, referred to below, "but wing-bands pale buff instead of gray or grayish white; upper parts brownish gray or grayish brown rather than olive; gray of chest more brownish, and white of under parts tinged with pale brownish buff." He refers to two types of coloration in adults, one with the lower parts white and one with these parts primrose yellow, but says that "they seem to be mainly seasonal, a large majority of those which are white, or very faintly tinged with yellow beneath, being spring and summer birds while those decidedly yellowish beneath were nearly all obtained in autumn or winter." Also, he says that the upper parts of adults are "more decidedly olive" in autumn and winter specimens than in spring birds.

We do not seem to know much about the postjuvenal molt of young birds, but adults, apparently, at least begin to molt, if they do not wholly accomplish it, before they go south. Mr. Swarth (1904) says that all the adults he collected in August were "in worn, abraded plumage, many of them in the midst of the autumnal moult with hardly enough feathers to cover them."

He also suggests a "slight prenuptial moult," saying: "Specimens taken during February and the early part of March are in fresh, unworn plumage, soft and blended in appearance and with a considerable olivaceous on the dorsum. Those secured at the end of March and throughout April have the feathers rather worn and abraded, the upper parts dull grayish with a few new olivaceous feathers showing on the back. Specimens taken in May present a bright, fresh appearance, with the upper parts olivaceous with but a few of the old gray feathers remaining, and with considerable yellow on the abdomen."

These remarks would seem to indicate a partial prenuptial molt at variable times late in winter or in spring. In the series that I have studied March and April adults are in worn plumage, or undergoing body molt, while the late April and May birds are in fresh plumage.

Food and behavior.—Except for the food of the young described by Russell and Woodbury, nothing seems to have been published on the food of this flycatcher, which probably does not differ materially from that of other small flycatchers. Nor is its behavior essentially different from that of its near relatives. Ralph Hoffmann (1927) writes: "It shares this domain [sage-bush plains] with the Sage Thrasher and the Brewer Sparrow; the latter glean their food from the ground or the bushes themselves, but the Gray Flycatcher, perched on the top of a tall sage-bush, watches the air for its prey. When a Gray Flycatcher is started, it dives from its perch and in its flight keeps well down among the brush. Its song is more emphatic and less varied than either the Hammond's or the Wright's. It has only two elements, a vigorous *chí-wip* and a fainter *cheep* in a higher pitch. These two notes are used in a variety of combinations, but when once they are heard, the Gray Flycatcher can be instantly recognized. The call notes are a sharp *whit*, like a Traill's, and a liquid *whilp* which passes into a gurgling note, similar to that of several of the other small Flycatchers."

Only through long-continued field experience, practice, and close observation can one expect to learn to recognize in life these four small western flycatchers. The gray flycatcher is the largest and grayest, but only at close range and under most favorable circumstances can the characters mentioned on the first page of our account of *hammondi* be recognized.

DISTRIBUTION

Range.—Western United States and Mexico.

Breeding range.—The gray flycatcher has been so hopelessly confused with the closely related Wright's flycatcher, which it greatly resembles, that it is almost impossible to separate the breeding ranges of the two species, except where breeding birds have been collected.

For the present, therefore, the breeding range of the present species may be provisionally outlined as follows: **North** to Oregon (Paulina Mountains and Malheur Lake); and northern Colorado (Majors Sandwash and Mount Zirkel). **East** to Colorado (Mount Zirkel, Hot Sulphur Spring, probably Fountain, and Bondad); and northeastern Arizona (Segi Canyon). **South** to northern Arizona (Segi Canyon, Navajo County, and Grand Canyon); central Nevada (Arc Dome and Cloverdale); and east-central California (Inyo County). **West** to northeastern California (Inyo County, Mono County, Redrock, and Alturas) and central Oregon (Paulina Mountains).

Winter range.—During winter the gray flycatcher is found **north** to southern California (Pasadena and Furnace Creek Ranch); Arizona (Rillito Creek); and Chihuahua (Chihuahua City). **East** to Chihuahua (Chihuahua City); western Durango (Villa Ocampo and Tamazula); western San Luis Potosi (San Luis Potosi); Guanajuato (Irapuato); Hidalgo (Tulancingo and Irolo); and Puebla (Chalchicomula). **South** to Puebla (Chalchicomula); Mexico (Mexicalcingo, Coapa, and Chimpalpa); and Jalisco (Bolanos and La Barca). **West** to Jalisco (La Barca); Baja California (San Jose del Cabo, Triunfo, and La Paz); and California (Potholes, Salton Sea, El Monte, and Pasadena).

Spring migration.—Early dates of spring arrival are: Colorado—Mack, April 24; Two Bar Ranch, April 27; Fountain, May 3. Nevada—Smoky Valley, April 26. Oregon—Harney County, April 28.

Fall migration.—Late dates of fall departure are: Oregon—Harney County, August 29. Nevada—Toyabe Mountains, September 4; Santa Rosa Mountains, September 9; Smoky Valley, September 11. Colorado—Escalente Hills, September 5.

Egg dates.—Arizona: 2 records, May 29 and July 4.

California: 46 records, June 4 to July 10; 23 records, June 7 to 18, indicating the height of the season.

Colorado: 3 records, May 31 to June 4.

Oregon: 2 records, June 7 and 15.

EMPIDONAX DIFFICILIS DIFFICILIS Baird

WESTERN FLYCATCHER

PLATE 35

HABITS

This flycatcher was formerly called the western yellow-bellied flycatcher and was at one time treated as a subspecies of our eastern *flaviventris*; but it is now recognized as a distinct species, as it belongs to a group of Empidonaces having a different wing formula; it also has a longer tail than *flaviventris* and differs from it in some of its habits.

The western flycatcher is widely distributed in western North America, chiefly from the Rocky Mountains westward, from Alaska to southern California and Texas, with other races in Lower California and Mexico.

Unlike the yellow-bellied flycatcher, which it superficially resembles, it is not especially partial to the coniferous forests in the breeding season, but is much more generally distributed and nests in a much greater variety of situations. S. F. Rathbun says in his notes on this species: "This small flycatcher is a common summer resident of western Washington. I have found it from the Cascade Mountains to the Pacific Ocean, and its distribution appears to be general, although it seems to occur much oftener in the lower country. It is a bird to be looked for in woods mostly deciduous, where maples, alders, and dogwoods grow, for it shows little fancy for the evergreen growth; and in the former it is usually found in the vicinity of low ground. A good place to find this flycatcher is in some quiet glen, especially if such has a trickling little stream, and here its note will be frequently heard. But wherever it occurs, the place is apt to be a quiet one, for it is a bird that seems fond of the stillness of the woods.

"Usually it arrives in this section near the close of April. It begins to nest by the first of June and continues to do so throughout the month, but on occasions its nest will be found in July."

Nesting.—Mr. Rathbun writes (MS.) on this subject: "This flycatcher does not seem at all particular in its choice of a nesting place, but from our experience the locality selected is invariably within the woods. I have found its nest at various heights, from 3 to 30 feet, although the latter was exceptional, and in all kinds of spots. At times the nest will be placed among the roots of an upturned tree, again on the top of a low stump, or in the crotch of a very small tree; and once I found its nest in the crown of a devil's-club (*Echinopanax horridum*), a showy shrub bearing countless spines and little prickles irritating to the skin, a most unusual place for a bird's nest. And the localities favored are usually in a somewhat retired part of the forest, near low ground. Oftener than not green and dry mosses represent most of the nest, and invariably it is lined with fine strips of shredded bark. On one occasion I found a nest of this bird that was so beautiful I will describe it. It was placed at a height of 15 feet in the main crotch of a little alder tree growing near the edge of a bit of swampy ground not far within the border of the woods. All the material used in this nest, except its lining, was ground moss of a rich dark-green color, enough being used to fill the small crotch to a height of about 5 inches. The inside of the nest was a round depression in the quite level top surface of the moss, it having a diameter of 1¾ inches with a depth of 1¼. This depression was lined entirely with fine shreds of inner bark of the cedar, the texture of which was very soft. It was neatly, very smoothly woven, showing a high degree of skill; and, because of its reddish color, it made a beautiful contrast to the dark-green moss. The inside rim of this nest was a perfect circle; never in any bird's nest have I seen one so flawless; and it seemed to me that the maker of this particular nest had at least some sense of artistry."

Major Bendire (1895) mentions a number of quite different nesting sites. C. A. Allen, of Nicasio, Calif., wrote to him:

I have found its nests in all sorts of situations; sometimes in a small tree, placed in the upright forks of the main stem; again on the side of the stem, where a small stub of a limb or some sprouts grew out; or in a slight 'cavity in a tree trunk; against an old stump or root which had been washed down during a flood in the middle of a stream; among curled-up roots near the water, etc. I have found a number of nests, when fishing for trout, by flushing the bird from under a bank; and on stooping down and looking I found the nest nicely concealed by the deep-green moss, such as covered the surrounding stones. They always use this particular kind of moss, no matter where the nest is placed. Occasionally they nest in deserted woodcutters' huts, in outbuildings near cover, and a friend of mine has some large water tanks in the woods back of his house, where for nineteen consecutive years these birds have built under the covered roofs of these tanks. I know of no place in this locality where they do not breed, excepting in very open country. * * *

In Belt Canyon, Montana, on July 6, 1889, Mr. R. S. Williams found the Western Flycatcher nesting in a narrow fissure of limestone, about 7 feet above the base of the wall. A nest observed by Mr. A. W. Anthony, near Howardsville, San Juan County, Colorado, on June 25, was placed on a ledge of rock, about 10 feet above a wagon road, and looked like a large ball of green moss, with a neat little cup in the center, lined with cow and horse hair. * * *

Mr. A. M. Ingersoll reports finding a nest of this species at the bottom of a hole 5 inches deep, made by a Red-shafted Flicker in a live oak; nests have also been taken in piles of driftwood, on beams under bridges, etc. * * *

The nest is composed of weed stems, dry grasses, plant fibers and down, strips of the inner bark of the redwood, fine rootlets, dead leaves, and bits of moss. It is usually lined with finer materials of the same kind, and occasionally with horse and cattle hair or a few feathers. The outside of the nest is usually coated with green moss when obtainable, but some nests before me show no trace of this in their composition. They are generally placed not far from water, but there are exceptions to this. A well-preserved nest now before me, * * * measures 4 inches in outer diameter by 2 inches in height. The inner cup measures 2¼ inches by 1¾ inches deep.

That this flycatcher does not always nest in the solitude of the forests is shown by the nests reported by Dr. Grinnell (1914a) on the Berkeley Campus: "In one case a brood was reared in a nest ensconced in a niche 18 feet above the ground in the side of an oak trunk near the Faculty Club. In another instance, the nest was built in a fern basket on a porch at 2243 College Avenue. This site was but five feet from a frequently used door, and it was only two feet from the porch-light which shone into the nest on frequent occasions in the evening without appearing to disturb the birds. On May 17 (1908) this nest held four eggs, and two young were successfully reared. In 1909, the same site was chosen, but the nest was subsequently deserted."

Nesting sites on beams in unoccupied buildings are often used year after year, until a whole row of old nests may sometimes be seen. Nests built in trees may be placed on any suitable support, in a natural crevice, or behind loose slabs of bark, such as occur on redwoods, alders, or eucalyptus trees. These birds are not only very much attached to favorite nesting sites, especially in buildings, but are sometimes very persistent in their attempts to raise a brood. Joseph Mailliard (1881) tells of a pair that nested every year in the shed covering his tanks. One season he took five nests with eggs from this same pair of birds in the same shed. The first nest was taken on May 15 and the last on July 6. Thus five nests were built and 21 eggs were laid by this pair of birds in a little over two months.

Denis Gale says in his notes that he has known one pair of these birds to use the same nest for three years in succession and another pair for four, repairing the old nest or partially rebuilding it.

Eggs.—Three or four eggs are ordinarily laid by the western fly-catcher, usually four and very rarely five. The eggs vary from ovate to short-ovate or even rounded-ovate. They are practically lusterless, with a dull white or creamy white ground color. The markings consist of spots or small blotches, usually concentrated about the larger end, but on some eggs there are minute dots or small spots scattered more or less evenly over the whole surface. These markings are in shades of bright reddish brown, "cinnamon-rufous," or the lighter shades of "buff-pink"; occasionally there are a few faint spots of lavender. Bendire (1895) says that "the spots are, as a rule, coarser and heavier" than on the eggs of the yellow-bellied flycatcher. The measurements of 50 eggs average 16.8 by 13.1 millimeters; the eggs showing the four extremes measure 18.8 by 13.7, 18.0 by 14.2, and 15.2 by 12.5 millimeters.

Young.—The period of incubation is said to be 12 days. Whether both sexes incubate does not seem to be known, but both parents assist in the feeding of the young and probably in brooding them also. Professor Beal (1910) says that the food of the young shows no marked difference from that of the adults. "The young in one nest were fed 24 times in an hour. Owing to the nest's location the number of nestlings was not ascertained. If there were four, as is probable, and the feeding was continued fourteen hours, each was fed 84 times during the day."

Plumages.—I have seen no very young specimens of the western flycatcher. The sexes are alike in all plumages, and in juvenal plumage the young birds are much like the adults, but much browner above, and paler yellow or buffy below; the wing bands are "cinnamon-buff" or "ochraceous."

The molts of this species apparently correspond to those of the yellow-bellied flycatcher, to which the reader is referred. The post-juvenal molt occurs long after the young birds have left for the south and the prenuptial molt is accomplished before the birds return in the spring, consequently molting specimens are scarce in collections. What few specimens are available seem to indicate that young birds molt the body plumage late in the fall and have a complete prenuptial molt in late winter or spring. Adults seem to follow a similar sequence, with perhaps a renewal of the flight feathers during the winter. More winter specimens are needed to trace these molts.

Food.—Professor Beal (1910) examined 141 stomachs of the western flycatcher, and his "analysis gives 99.28 percent of animal food to 0.72 percent of vegetable." Of the animal food this bird appears to eat more ladybird beetles than does any other flycatcher, more than 7 percent of the food in August, but an average of only $2\frac{1}{3}$ percent for the season. He says:

Other beetles amount to nearly 6 percent, nearly all harmful, the exception being a few ground beetles (Carabidae).

Hymenoptera form the largest constituent of the food of this as of most other flycatchers. They amount to over 38 percent. * * * No honeybees were identified. * * * Hemiptera (bugs) amount to nearly 9 percent of the food. * * *

Diptera amount to a little more than 31 percent of the whole food. * * *

Lepidoptera, in the shape of moths and caterpillars, amount to about 7 percent for the year, and were found in every month except March. They appeared in 36 stomachs, of which only 7 contained the adult insects—moths—and 29 the larvae or caterpillars. * * * A few unidentified insects and some spiders make up the remainder of the animal food—about 6 percent.

Vegetable matter was found in 16 stomachs, though some of it could not properly be called food. One stomach contained seeds of Rubus fruit (blackberries or raspberries) ; 7, seeds of elderberries; 1, the skin of an unidentified fruit and a seed of tarweed (*Madia*) ; while 6 held rubbish. The Rubus fruit might have been cultivated, but probably was not.

Theed Pearse writes to me that he saw one of these flycatchers capture a good-sized fly on a branch, hold it on the branch with its feet, and tear it with its beak. Most of its food is probably captured on the wing.

Behavior.—There seems to be nothing peculiar in the behavior of the western flycatcher, as compared with the other small flycatchers of the *Empidonax* group. Both parents are devoted to the defense of their home and family, and the male stands guard nearby while the female is incubating and drives away any other birds that venture too near the nest. Mr. Pearse tells me that the flight is hesitating, like that of the kingbird. I am not acquainted with the bird in life.

Voice.—Mr. Pearse (MS.) mentions an alarm note similar to the *tsip* alarm note of warblers, and another alarm note that resembles the call of Harris's woodpecker, but he writes the usual note as *pisint*.

Ralph Hoffmann (1927) writes: "From under live oaks in a canyon, from deciduous trees near a stream or even from shady plantations about dwellings from April to July a single sharp note, *pee-ist*, like the expiration of wheezy breath, catches the ear of an attentive listener. * * * Besides the *pee-ist* note, almost but not quite two syllables, the Western Flycatcher utters a low *whit*. In the breeding season the male repeats, often for long periods from the same perch, three syllables which constitute his attempt at song, *ps-séet ptsick*, and after a slight pause *sst*."

Grinnell and Storer (1924) write:

On the morning of June 3, 1915, a Western Flycatcher was watched as it sang and foraged among the big-trunked incense cedars and huge mossy boulders on the north side of the Yosemite Valley, at the foot of Rocky Point. The greenish yellow of the bird's upper plumage and its yellowish under surface were the only sight characters available, but the call note and song were both distinctive. The former was a single high-pitched, even piercing, *sweé ip* or *tweé it*; less often a fainter *peet* was uttered. The song goes *seé rip*, *sip*, *seé rip*, or sometimes *seé rip*, *sert*, *sip*, *seé rip*, and is repeated over and over again, often so continuously that the pauses between songs seem no greater than the intervals between the constituent notes. The syllables were given in varying order, and often the single combination, *seé rip*, was uttered over and over again. While singing, this bird was perched on various twigs and branches 10 to 20 feet above the ground. The song is to be heard most often in May and early June, but as late as July 30 a bird was heard in full summer song.

Bendire (1895) quotes C. A. Allen as saying: "Its song consists of a soft, low note. It shows much distress when its nest is taken, uttering then a low, wailing note, like 'pee-eu, pee-eu' and frequently flutters about the person taking it and snapping its mandibles together."

Field marks.—The small flycatchers of the *Empidonax* group are very difficult to distinguish in the field by color characters alone. The western flycatcher is more olivaceous above and more extensively yellowish below, with a much brighter shade of yellow, especially on the belly, than any of the small western species; but there are three other western species that are more or less yellowish on the under parts, *trailli*, *hammondi*, and *wrighti*, all of which closely resemble *difficilis* in other respects. Fortunately the habitats of the four species are somewhat different. Ralph Hoffmann (1927) says on this point: "The Traill Flycatcher is very similar in appearance to the Western, but though often found in the same general region, is restricted to willow thickets and to bushy places in wet mountain meadows. * * * The Western Flycatcher, though it also affects the neighborhood of streams, demands for its hunting a certain amount of open space in the shade of tall trees of mixed growth; neither the Western nor the Wright would be found regularly in the dense willow thickets which the Traill prefers." Wright's flycatcher is more likely to be found at higher elevations on the mountain slopes, and Hammond's frequents the tall coniferous forests. But the best character by which these species may be distinguished is the call note or song, which the keen ear of a good observer can learn to recognize, as the notes of the four are quite different; when the birds are not singing, identification is often almost hopeless.

Fall.—Mr. Rathbun says in his notes from Seattle: "After the end of the breeding season, when its young are on the wing, this flycatcher appears to move about the country, for it is apt to be heard or seen almost anywhere, even at times in the cities and towns. This movement is but preliminary to its fall migration, which takes place in September, and after late in this month the species is no longer seen."

DISTRIBUTION

Range.—Southeastern Alaska, British Columbia, and Western United States and Mexico.

Breeding range.—The western flycatcher breeds **north** to southeastern Alaska (Sitka and Juneau); east-central British Columbia (Salmon River Forks); Montana (Belt River Canyon); northern Wyoming (Yellowstone Park and Sheridan); and probably western South Dakota (Box Elder). **East** to probably western South Dakota (Box Elder); southeastern Wyoming (Wheatland); Colorado (Estes Park, Goldhill, and Hancock); New Mexico (Twining, Sacramento Mountains, and Guadalupe Mountains); and western Texas (Chisos Mountains). **South** to southwestern Texas (Chisos Mountains); Chihuahua (Pinos Altos); Nayarit (Isabela Island); and northern Baja California (San Fernando). The **western** limit of the breeding range extends northward from northern Baja California (San Fernando and Vallecitos), along the coasts and islands of California, Oregon, Washington, and British Columbia, to southeastern Alaska (Forrester Island, Ketchikan, and Sitka).

Winter range.—The winter range of the western flycatcher is in western Mexico from southern Baja California (La Paz and San Jose del Rancho) and southern Sonora (Chinobampo, Tesia, and Alamos) **south** to southern Guerrero (Coyuca and Chilpancingo) and Oaxaca (Pluma).

The range as outlined includes the two North American races of this species. The typical western flycatcher (*Empidonax difficilis difficilis*) is the form found in the United States, Canada, and Alaska, while the San Lucas flycatcher (*E. d. cineritius*) is confined to Baja California. An additional nonmigratory race is found in central and southern Mexico.

Spring migration.—Early dates of spring arrival are: New Mexico—Apache, May 12. Colorado—Littleton, May 11. Wyoming—Wheatland, May 15. Montana—Fortine, May 14. Arizona—Tucson, March 24. California—Berkeley, March 12. Oregon, Weston, March 20. Washington—Tacoma, March 24. British Columbia—Courtenay, March 31. Alaska—Ketchikan, May 6.

Fall migration.—Late dates of fall departure are: Alaska—St. Lazaria Bird Reservation, September 30 (unusually late). British Columbia—Courtenay, September 9. Washington—Seattle, September 17. Oregon—Coos Bay, September 20. California—Pasadena, October 10. Wyoming—Yellowstone Park, September 16. Colorado—Colorado Springs, September 19. Arizona—Huachuca Mountains, October 1.

Egg dates.—British Columbia: 4 records, June 16 to July 3.

California: 113 records, April 10 to July 27; 57 records, May 6 to June 17, indicating the height of the season.

Colorado: 16 records, June 4 to July 23; 8 records, June 19 to 27.

Baja California: 3 records, June 21 to July 27.

Washington: 13 records, May 20 to July 25; 7 records, June 6 to 22.

EMPIDONAX DIFFICILIS CINERITIUS Brewster

SAN LUCAS FLYCATCHER

HABITS

The Lower California race of *Empidonax difficilis* breeds in the mountains of the Cape region of that peninsula and from there northward to the Sierra San Pedro Martir. It was described and named by William Brewster (1888) as a new species, based on a series of some 25 specimens collected by M. A. Frazar at La Laguna. He describes it as "most nearly like *E. difficilis* but with the general coloring much duller, the upperparts with scarcely a tinge of greenish, no decided yellow beneath, excepting on jugulum and abdomen; wing-bands brownish white." He says of it elsewhere (1902):

The St. Lucas Flycatcher is resident in the Cape Region, where it is not uncommon. Mr. Frazar found it in the greatest numbers in the Sierras de la Laguna in May and early June. He also obtained specimens at San José del Rancho in July and at La Paz in February and March. Mr. Bryant has taken it at Comondu, and San Benito and Santa Margarita Islands, while on San

Pedro Martir Mr. Anthony found it "very common all over the mountain, especially along the streams and in the willows. It was evidently nesting" at the time of his "visit in May, but no eggs were taken." He also states that it occurs sparingly near the mine and about the mission at San Fernando, where he thinks it nests "in the thick mesquite growth." It probably replaces *E. difficilis* in the breeding season throughout the greater part of Lower California.

J. S. Rowley has sent me the following notes on his experience with the San Lucas flycatcher: "While camped at La Laguna, atop the Sierra de la Laguna, from May 23 to 28, 1933, I took three sets of four eggs and one set of three eggs, all sets being practically fresh. At this camp a small creek had running water in it at this date, and several pairs of these flycatchers were nesting here. So far as I could see, the nesting habits are the same as the northern race; these nests were all placed behind climbing vines on rocks or in rotted parts of trees, and not more than a few feet from the ground. These little flycatchers were the first to sing in the early dawn and the last to sing at night, seeming to never tire of their liquid-sounding song."

The eggs are apparently indistinguishable from eggs of the species found in California. The measurements of 12 eggs average 17.2 by 13.3 millimeters; the eggs showing the four extremes measure 18.0 by 13.3, 17.3 by 14.3, and 16.3 by 12.3 millimeters.

EMPIDONAX FULVIFRONS PYGMAEUS Coues

BUFF-BREASTED FLYCATCHER

PLATE 36

HABITS

This pretty little flycatcher is the tiniest of the Empidonaces that occur in North America. It reaches the northern limit of its distribution across our southwestern border in the mountains of southern Arizona and New Mexico. It is a northern race of a Mexican species and is described by Ridgway (1907) as "similar to *E. f. fulvifrons*, but decidedly smaller and color of upper parts either darker or grayer."

We found this flycatcher to be rather rare in the upper parts of the canyons in the Huachuca Mountains, Ariz., at elevations between 6,500 and 8,500 feet. Its favorite haunts seemed to be the steeply sloping hills that rose gradually from the bed of the canyon and were covered with a scattering growth of tall pines and small oaks, and with an undergrowth of shrubs. It was not seen in the lower parts of the canyons. Harry S. Swarth (1904), referring to the same region, says:

The Buff-breasted flycatcher is one of the rarest of the regular summer visitants to these mountains, and as it is a small, inconspicuously colored bird, and in my experience rather shy and difficult to approach as well, it is a species that is most easily overlooked. It arrives in the Huachucas about the middle of April, and all the migrating birds I have taken have been along the base of the mountains, where they were usually sitting in low bushes or weeds. * * *

On May 26, 1903, I found these flycatchers breeding near the head of Tanner Canyon in such a way as to almost indicate a "colonizing" tendency, for I found seven or eight pair breeding within a radius of about a quarter of a mile, and three or four of these were within a hundred yards of each other. This may have been due, however, to the exceptionally favorable nature of the ground; for it was different from most of the region thereabout in that the canyon opened out into a considerable area of low, rolling hills, covered with a scattering growth of large pines.

Nesting.—We did not succeed in finding any nests of the buff-breasted flycatcher while I was in the Huachuca Mountains, but my companion, Frank C. Willard, found two nests after I left; one nest was destroyed by jays or squirrels, but he collected a set of three eggs from the other, a typical nest in a pine tree. He had had considerable experience with this bird in past years and published an interesting article (1923b) on it, from which I quote as follows:

Early in June, 1897, I was climbing the last slope up to the main ridge of the Huachuca Mountains in Cochise County, Arizona. During a pause for breath, a small bird flitting about among some young pine trees five or six feet tall caught my eye, and a few moments of observation convinced me that it was another of the numerous strangers to me. This was my first year in the West, and nearly every day was bringing new acquaintances. While I was debating the probable identity of this flycatcher, as its actions and appearance betokened it, the bird dropped to the ground, picked up a fine rootlet, and flew up into one of a group of tall pines about seventy-five yards away. * * * A week later I was standing near the pine tree where the bird had been last seen. A club thrown among the branches flushed her, as the alarm note she uttered announced; but I could not see her, nor could any sight of the nest be obtained. Strapping on my climbers, I was soon astride the first branch forty feet up. A careful scanning of all the nearby branches failed to reveal the nest, and I stood up and clasped my arms around the trunk of the tree preparatory to climbing higher. Something soft gave under my hand and I knew without looking that it was the nest. Hastily climbing to the branch above, I looked down into a small, deeply hollowed cup, snug against the trunk and saddled on a short stub about three inches in diameter. Another longer stub was a few inches almost directly above it. Three cream colored eggs were the contents. While I was busy packing them, the female came close, scolding vigorously.

On May 23, 1907, he found a nest in the same region in a small white oak on a steep mountain side. "The nest was twelve feet up from the ground, saddled on the lower prong of a fork, the upper prong forming a protective overhang." He found several other nests that year, all of which were similarly located, with a protective prong, stub or branch above the nest. He says of the composition of the nest: "The nest is composed largely of lichen-like leaves, dark gray in color, of a small low growing weed. These are held together with cobwebs. The lining is of fine grass, rootlets, and hair, with a few feathers near the rim, which is slightly incurved like a humming-bird's nest. It is rather insecurely fastened with cobwebs to the branch on which it rests. In appearance, it is much like a Western Gnatcatcher's (*Polioptila caerulea obscura*) nest."

R. D. Lusk (1901) writes of a nest that he found in the Chiricahua Mountains, Arizona:

Very early one morning, June 16 I saw the female fly repeatedly from the ground on the hill-side to the same limb of a large sycamore about which they had spent much time. * * * The female did all the work. The nest was placed in an inclined fork among the thick branches, pretty well up, about 35 feet. It was well-constructed, compact, deep, of dried grasses, a few vegetable fibers, plenty of spider's silk and into the lining were woven a few bright feathers. Two nests found this last season also contained several bright feathers, one of them, bright yellow ones of the Audubon's Warbler (*D. auduboni*), a blue one of the Chestnut-backed Bluebird (*Sialia m. bairdi*) and a barred feather of the Whip-poor-will (*Antrostomus v. macromystax*) fluttering on the edge of the nest.

He also found this flycatcher "breeding in a virgin forest of pines and firs, among the trees surrounding a little 'park', or treeless, open space, of which there are many in these mountains." It was near the summit of the range. "But not one of the nests found in this upper location was built upon a limb, but all against the trunks of the trees, 20 to 35 feet from the ground, in two cases in the angle of a short dead stub and in two cases with only a tiny jutting piece of bark for support or a slight depression caused by a wound in the tree. To this there was one exception * * * where it was located far out on the limb of a large fir."

There are five nests of the buff-breasted flycatcher in the Thayer collection in Cambridge, all collected by Virgil W. Owen in the Chiricahua Mountains. One was in an oak tree on the edge of a bank above a stream; it was in a slanting crotch and was partly supported by another branch. Two were in sycamores: one only 9 feet from the ground, saddled on a small, dead branch close to where it joined a live limb, and partly supported by dead twigs; the other well out on a limb and well concealed in a bunch of leaves, 25 feet from the ground. The other two were in pines: one 15 feet up and 12 feet out on a limb, in a fork and against a bunch of needles; the other 45 feet from the ground and 10 feet out from the trunk near the end of a small limb.

These nests are all similar to those described above, neatly and compactly made and deeply cupped; the material consists of various plant fibers, cottony substances, fine grasses, fine rootlets, weed blossoms, seed heads, and bits of dried leaves, all bound together and to the branch with spider webs; they are smoothly lined with the finest grasses, plant down, horsehair, and more or fewer feathers. All five of the nests are more or less profusely decorated with what look like pale gray lichens or spider cocoons, but these are probably broken bits of the small leaves of a low-growing weed referred to by Mr. Willard. They are all about the same size, 2½ to 3 inches in outside diameter, 1½ to 1¾ in inside diameter, 1¼ to 1½ deep inside, and about 2 inches in outside height.

Eggs.—The buff-breasted flycatcher lays ordinarily either three or four eggs. Mr. Lusk (1901) says that, of the sets of which he has records, 50 percent were sets of three, 30 percent sets of four, and 20 percent sets of two eggs each. All five sets in the Thayer collection contain four eggs each; and Mr. Willard found one set of five eggs. What few eggs I have seen are ovate, without gloss, and plain creamy white and unmarked. The measurements of 30 eggs average 15.5 by 11.9 millimeters; the eggs showing the four extremes measure **16.0** by 12.4, 15.6 by **12.7, 14.7** by 11.5, and 15.6 by **11.3** millimeters.

Plumages.—Young birds in juvenal plumage are much like the spring and summer adults, but they are browner above, and the median and greater wing coverts are broadly tipped with dull buff or "cinnamon-buff," instead of grayish or buffy white.

Fall and winter adults are more richly colored than spring and summer birds, the upper parts more buffy, instead of "hair brown," the wing bands more suffused with light buff, the chest often tawny-buff, and the throat and belly pale yellow and buff, but there is much individual variation. Mr. Swarth (1904) says: "The darkest colored one I have, a female, has the breast deep ochraceous buff, with the throat and abdomen but little paler; while a rather large sized male in fresh unworn plumage, has the upper breast yellowish buff, fading to pale yellowish on the throat and abdomen, almost white along the median line."

I have seen no specimens in actual molt, but adults in June, July, and August are in more or less badly worn plumage. Probably, as with certain other Empidonaces, the postjuvenal and postnuptial molts take place after the birds have retired to their winter homes. There is probably a prenuptial molt of at least some of the body plumage in the spring, for Mr. Swarth (1904) states that "specimens taken in April frequently have a few new feathers scattered over the back."

Food.—The only published account of the food of the buff-breasted flycatcher that I can find is the following recent report by Cottam and Knappen (1939), which I quote in full:

From the limited data available, it apparently like others of its kin feeds on those insects that are most available. The contents of the single stomach (from Huachuca Mountains, Arizona, June 17, 1922) available for laboratory analysis, suggests that the bird is a beneficial species. Fragments of the following insects were noted: five beetles (including one *Bembidion* sp., one *Agrilus* sp., one *Pachybrachys* sp. and one *Rhynchophora*), 6%; one short-horned grasshopper (*Acrididae*), 1%; true bugs, including leafhoppers (*Cicadellidae*), chinch-bugs (*Lygaeidae*) and big-eyed bugs (*Geocoris* sp.), 13%; more than forty ants of three genera (*Formica* sp., *Myrmica* sp., and *Solenopsis* sp.), 65%; and undetermined hymenopterous fragments, 15%.

Behavior.—The general behavior of the buff-breasted flycatcher is evidently not very different from that of other small flycatchers, but practically all we know about it comes from egg collectors and relates to behavior about the nest. Mr. Willard (1923b) says:

In leaving a nest built high up from the ground, the female drops almost straight down to the brush below and does not make a sound for some time. The male seems to give the signal to her and he does most of the scolding, flying from tree to tree and endeavoring to lead the intruder away. When the nest is discovered, and the birds realize it, the female becomes very bold and alights on the nest where she remains, frequently until nearly touched. The male takes his departure about this time.

During the nest building the male stays pretty close to the nesting tree, but offers no assistance except to drive off intruding birds. He is good at that and can also put big squirrels to flight. The female works persistently and rapidly, but the nest requires a lot of material and she often takes ten days to build it. She sits on the nest for short periods before the eggs are laid, and also as the eggs are laid, but does not seem to make a real business of it until the set is complete.

Mr. Lusk (1901) tells of a male that seemed to show intelligence in leading him away from the nest: "The male had a plan for frustrating the hunter which he worked diligently and as I have noticed it in several, I take it to be characteristic. Each time as I approached the location of the nest, he came out some distance to meet me and began calling and occasionally scolding in a certain locality, thus leading me to believe the nest was somewhere in that vicinity. Once, however, I waited until long after sunset in the vicinity of two large pines, near which the soft *pit*, *pit* of the female, as I felt sure it was, answering the male had suddenly ceased the day before. Meanwhile the male was persistently, for two long hours, insisting that all his interests were in the vicinity of a tall, leaning pine a hundred yards distant, to which point he had come to meet me day after day." He found the nest the next day in one of the pines in which he had heard the female, and from which the male had tried to draw him away.

Voice.—He describes the note of the female on the nest above, and says: "Every now and then [evidently at other times] the soft *pit*, *pit* of the two, as they kept good account of each other's whereabouts, was varied by the *Chicky-whew* of the male." Mr. Willard (1923b) describes the alarm note as *quit-quit*, or again as *quit-quit-quir-r-r*.

Enemies.—Squirrels probably destroy some eggs and young of this flycatcher, as the birds have been seen to drive away these animals. Long-crested jays are common in the same region with the flycatchers, and Mr. Willard thinks that they undoubtedly rob some nests. Mr. Lusk blames the jays for some damage and mentions lizards as possible enemies. We found sharp-shinned hawks nesting in the same vicinity, which are well-known enemies of all small birds. These little flycatchers are adept at concealing their nests and very courageous in their defense.

Field marks.—Its posture and actions mark it as a flycatcher. It is the smallest of its tribe in the region where it lives, except for the tiny beardless flycatcher, which is said to appear more like a verdin than a flycatcher. Its general color is buff, quite unlike any other North American flycatcher, and the warm buff of its breast is particularly noticeable when facing the light; it seems to glow with warmth in the sunlight.

DISTRIBUTION

Range.—Arizona, New Mexico, and Central America south to Honduras.

Breeding range.—The breeding range of the buff-breasted flycatcher extends **north** to central Arizona (Fort Whipple) and New Mexico (Inscription Rock). **East** to western New Mexico (Inscription Rock, Apache Canyon, and Fort Bayard); Veracruz (Las Vigas); and Honduras (Cerro Cantoral). **South** to Honduras (Cerro Cantoral); El Salvador (San Jose del Sacare); and southern Guatemala (Duenas, Lake Atitlan, and Quezaltenango). **West** to western Guatemala (Quezaltenango); Morelos (Cuernavaca); Nayarit (Tepic); western Durango (Cienega de las Vacas); western Chihuahua (Pinos Altos and Pacheco); and Arizona (Santa Rita Mountains, Santa Catalina Mountains, and Fort Whipple).

Winter range.—The species is resident in the southern part of the range but migratory in the United States. During the winter season it is found **north** to southern Sonora (Tesia and Guiricoba); southern Chihuahua (Durazno); and northeastern Puebla (Huehuetlan).

The range as outlined is for the entire species, which has been separated into several geographic races. Only one of these (*Empidonax fulvifrons pygmaeus*) is found in the United States, ranging south in summer from Arizona and New Mexico to Durango, and in winter to Michoacan and Morelos.

Spring migration.—Early dates of spring arrival are: Arizona— Santa Rita Mountains, April 5; Chiricahua Mountains, April 12. New Mexico—Fort Bayard, April 16; Silver City, April 26.

Fall migration.—Data indicative of the fall movement are not satisfactory, but late dates of departure appear to be: New Mexico—

Apache, August 18.   Arizona—Seven-Mile-Hill, September 12; Huachuca Mountains, October 18.

Egg dates.—Arizona: 17 records, May 23 to July 16; 9 records, May 26 to June 17, indicating the height of the season.

MYIOCHANES PERTINAX PALLIDIVENTRIS (Chapman)

COUES'S FLYCATCHER

PLATE 37

HABITS

This, the largest species of the genus, is another one of those Mexican species that find the northern limit of their summer range just across our southwestern border in Arizona and New Mexico.   The breeding range of the whole species covers most of Mexico, but that of our race covers only the northwestern part of that country, Sonora, Chihuahua, and Durango to Nayarit.   Our race is decidedly grayer above and paler beneath than the type race, the abdomen being dull white or yellowish white; hence the name *pallidiventris*.

We found Coues's flycatcher, which reminded us in some respects of its smaller relative the wood pewee, fairly common in the Huachuca Mountains, Ariz., at elevations between 7,000 and 9,000 feet.   Harry S. Swarth (1904) says that "during the breeding season these birds are to a great extent restricted to the higher parts of the mountains, being most abundant from 8,000 to 10,000 feet; though I have seen one or two nests as low as 7,000 feet."

One of their favorite haunts was in Stoddard Canyon, a branch of Ramsay Canyon, where the land rises in steep, rough slopes toward the summit, about 9,000 feet.   Scattered over the slopes are tall bull pines and various medium-sized oaks, with an undergrowth of oak scrub and various bushes.   Here long-crested jays were far too common for the welfare of small birds' nests.   Mearns's woodpeckers carried on their showy courtships in the tops of the tallest pines, hepatic tanagers nested in the pines with the buff-breasted and Coues's flycatchers, and the rare Grace's warblers concealed their nests too well for us to find them.   Here, too, we found the nests of the Rocky Mountain nuthatch, the lead-colored bushtit, the black-throated gray warbler, and the spurred towhee.   Bird life was plentiful on these rather open, sparsely wooded slopes.

Spring.—Mr. Swarth (1904) writes: "During the summer months this flycatcher is one of the characteristic birds of the pine regions of the Huachucas, where if not seen, it can at least be heard almost everywhere.   It is one of the first of the summer residents to arrive, and one was heard calling as early as March 29th.   The usual time of arrival is the first week in April, and during this month they can be found generally distributed over all parts of the mountains; while I have taken specimens, evidently migrating birds, quite at the base of the range, as late as May 25th, though others were found breeding at an earlier date."

Nesting.—Mr. Swarth (1904) says: "In the choice of a nesting place they show a marked preference for the conifers, the nest being usually built at a considerable distance from the ground, on some limb affording a wide, uninterrupted outlook, but there again no hard and fast rule can be laid down, as I have seen nests built in maples in the bottom of a canyon, not twenty-five feet above the ground, and nearly hidden by the luxuriant foliage. I have seen birds beginning to build in the middle of May, and eggs can occasionally be found until at least the middle of July."

We found a pair of Coues's flycatchers building a nest 40 feet from the ground on a horizontal fork of a high branch of a rather small bull pine, on May 12, 1922, in Stoddard Canyon. The birds were still working on the nest on the 26th; my companion Frank Willard collected this nest, with the four eggs that it contained, on June 4. He says that the male tried for some time to drive the female back to the nest after he had collected it. Mr. Willard's notes contain the data for three other nests found in the same locality; two of these were saddled on horizontal forks of oaks, one 25 and one 15 feet from the ground; the other was 40 feet up and 25 feet out from the trunk near the tip of a long branch of a large spruce, at an altitude of 8,000 feet. I have a set in my collection, taken by R. D. Lusk, at an altitude of 7,000 feet in the Chiricahua Mountains, Ariz., on July 12, 1899; the nest was placed 15 feet above ground in a horizontal fork of a sycamore beside a stream.

There are six nests of this flycatcher in the Thayer collection in Cambridge; four of these had been placed in horizontal forks of pines, at heights ranging from 10 to 38 feet; one was 30 feet from the ground in a maple; and the other must have been beautifully concealed and camouflaged, as it was built 6 feet out from the trunk and 35 feet from the ground on a horizontal limb of a red spruce that was heavily covered with large loose lichens and usnea; these lichens had been so thoroughly worked into the exterior of the nest that it must have been almost invisible except from above.

Coues's flycatcher builds a beautiful nest, very uniform in pattern and in materials used, strongly suggesting a glorified wood pewee's nest. Practically all the nests are built in the horizontal forks of branches, though occasionally one is placed on a horizontal branch where it is partially supported by upright twigs. It is very compactly made and firmly plastered to the branch and the prongs of the fork with plenty of cobwebs, so that the center of the nest is often unsupported in the space between the prongs. The upper rim is very firmly and smoothly finished, usually forming a nearly or quite perfect circle. In the nests that I have seen the body of the nest is firmly made of fine grasses, fine weed stems, shreds of weed stalks, a few bits of dry leaves, and flowering grass and weed tops. The exterior is profusely decorated, or camouflaged, with large and small pieces of lichens, often selected to match those naturally growing on the branch, in black, gray, or pale green colors, all securely bound on with cobwebs. The deeply hollowed interior is smoothly lined with the finest of bright yellow grasses and the slenderest grass tops. The yellowish-buff interior is sharply defined against the lichen-covered exterior, making a pleasing contrast and a pretty picture.

The nests that I have measured vary from 4 to 5 inches in outside diameter; the inside diameter seems to be quite uniformly 2¾ inches, and the depth of the cavity about an inch and a half; the external height varies from 2 to 3 inches. All the nests that I have seen conform very closely to the same pattern. Samuel B. Ladd (1891) gives the following good description of the composition of a nest: "The body of the nest seems to consist of the web of some spider intermingled with the exuviae of some insect, fragments of insects, and vegetable matter, such as staminate catkins of *Quercus emoryi* and a pod of *Hosackia*, some leaves of *Quercus emoryi* and *Q. undulata*. The interior of the nest is made up of grasses, principally of two species of *Poa*, also some fragments of a *Bontelona* and a *Stipa*."

Eggs.—Three or four eggs generally constitute the set for Coues's flycatcher. These are ovate and practically lusterless. The eggs that I have seen have a dull white or creamy white ground color, which is sparingly marked, mostly near the larger end, with small spots or dots of different shades of brown, sometimes very dark, sometimes paler and rarely reddish brown, with a few scattering small spots of shades of Quaker drab. Some writers have compared them to eggs of the olive-sided flycatcher or the wood pewee, but I have never seen any that bore the slightest resemblance to either of these. Bendire (1895) says that "the shell is frail and without luster, of a rich cream tint, and is sparingly spotted, principally about the larger end of the egg, with different shades of chestnut, ferruginous, and lavender."

The measurements of 50 eggs average 21.1 by 15.8 millimeters; the eggs showing the four extremes measure **23.5** by 15.9, 2.17 by **16.9**, and **18.8** by **15.0** millimeters.

Young.—Mr. Swarth (1904) says: "On July 23, 1902, I secured a young bird which had just left the nest but was as yet hardly able to fly, and two weeks later broods of young, attended by the parents could be seen everywhere. After the young had left the nest, a general movement toward a lower altitude began, and by the middle of August young and old could be found quite commonly along the canyons, and in the groves of live oaks at the mouths of the same."

Plumages.—Young birds in fresh juvenal plumage differ from adults in being darker and more olivaceous above; the pileum is much darker than the back, instead of nearly uniform with it, as in the adult, varying from "sepia" to "mummy brown", in rather sharp contrast to the color of the back; the upper tail coverts are more or less broadly tipped with "cinnamon" or buffy; the median and greater wing coverts are broadly tipped with "cinnamon"; the light edgings of the secondaries and tertials are tinged with pale buff; the abdomen, under tail coverts, and sometimes the throat are "light ochraceous-buff" or pale buffy yellow in sharp contrast against the dark gray of the sides and chest.

I have not seen any molting specimens and can only guess that the molts are accomplished mainly after the birds migrate southward. Mr. Swarth (1904) says: "An adult female taken August 24, 1902, just commencing the autumnal moult, has most of the plumage so worn and faded as to have lost all distinctive coloring, but on the upper breast and on the dorsum the new feathers are just beginning to appear."

Food.—I can find no definite information on the food of Coues's flycatcher, which probably does not differ materially from that of the other flycatchers of the region where it lives; it apparently lives on any kind of flying insects that it can find, as it can repeatedly be seen darting out into the air in pursuit of them from its perches in the trees. Living as it does, so far away from human habitations, its food habits cannot be of much economic importance.

Behavior.—Mr. Henshaw (1875) writes: "Each pair apparently takes possession of a large area, and allows no intrusion of their kind within the limits. Having spent a few moments in one spot, the bird makes a hurried dash, and in a few moments its voice can be just distinguished, as it is sent back from afar in answer to the mate near by. A short interval elapsing, it will suddenly re-appear from among the trees, and, with an exultant whistle, settle firmly down on some perching place, giving short, nervous jerks of its long tail, and turning its head quickly here and there, every motion betraying the nervous activity of its nature. These sudden erratic flights from point to point are quite characteristic of the bird."

O. W. Howard (1904) gives the following interesting account of this flycatcher in the role of a protector of its own nest and those of other small and less aggressive birds:

The Coues flycatcher is a lively, wide-awake fellow, and while sitting on his lofty perch he keeps a sharp lookout for any of his numerous enemies who may venture too near his dwelling place. The moment a jay, hawk, squirrel or snake makes its appearance, the flycatcher leaves his perch and pounces upon the intruder, at the same time giving the note of alarm which never fails to bring the female to the scene. Then there is a snapping of beaks, and a regular whirl of wings and tails about the unwelcome visitor, who is forced to leave the locality faster than he came.

With all his warlike proclivities, the Coues flycatcher has another quality—that of attracting friends—which is equally strong. Among the more timid birds he numbers a host of friends who seem to be conscious of the existing bond, and very readily take advantage of it.

He refers to a habit, noticed on several occasions, that various small birds have formed of building their nests in close proximity to the nest of the flycatcher, for the protection afforded by this aggressive fighter. On one occasion, he states—

on the same limb, not more than four feet apart, was a nest of the Coues flycatcher and one of the hepatic tanager, with a nest of a plumbeous vireo not more than twenty feet from the others. All these nests contained full sets of eggs, showing that nest building had been carried on at the same time in all three cases. Naturally we wondered how these three pairs of birds, including the belligerent flycatcher, could get along in perfect harmony, building their nests and sitting on their eggs side by side. * * *

On many occasions, in seasons following, I found nests of various warblers, vireos, tanagers, and other birds in close proximity to nests of the Coues flycatcher. Once by using a small cloth scoop on the end of a pole I took a set each of Coues flycatcher and a black-fronted warbler, without changing my position in the tree. Another time I took a set of olive warbler and a set of black-fronted warbler from the same tree, and a set of Coues flycatcher from a tree not more than fifteen feet distant. In these, as well as in many other instances, I had the opportunity to learn the reason for these family gatherings. In the locality where my observations have been made, the smaller and more peaceable birds suffer great loss from snakes, squirrels, and jays. Probably the most bitter enemy of the smaller birds is the long-crested jay, who is continually in search of their nests. When the jay locates a nest, his call-note brings as many as half a dozen of his hungry comrades to the scene, and under a feeble attack from the parent birds, the eggs or young, as the case may be, are carried off or devoured on the spot. Many times, even, the nest is torn into shreds. All this, however, does not occur when there is a nest of the Coues flycatcher in the vicinity, for upon the first alarm, the flycatcher comes to the rescue, and the would-be assailant is forced to leave. This wholesale slaughter seems to teach these much imposed upon species to seek the protection of the more independent flycatcher.

Voice.—The only note that we heard from this flycatcher sounded somewhat like the note of the wood pewee, but louder, more forceful, and perhaps more musical—*pe-wée-ee*, the middle syllable strongly accented and the last prolonged on a slightly higher key. The notes are said to be like those of the olive-sided flycatcher but are readily distinguishable. Mr. Swarth (1904) says: "As with *borealis* the male bird is fond of getting in some elevated position, usually the extremity of a dead limb at the top of some tall pine or fir, and remaining there for hours, uttering at frequent intervals its loud, characteristic call. In character and tone this call is quite similar to that of *borealis*, but the notes differ. The local name for the species, derived from its cry, is Jose Maria (pronounced Ho-say Maria, with the second syllable of the last word drawn out and emphasized), a far better translation of the sounds than is the case in many similar instances."

Alexander F. Skutch has sent me some notes on the Central American race of this species, in which he says that "during the breeding season, Coues's flycatcher mounts in the gray dawn to some lofty perch, and for many minutes repeats incessantly, in a rather dry voice, a simple little phrase which sounds like *Fred'rick fear, Fred'rick fear.*"

Field marks.—This is a rather large flycatcher that looks like an overgrown wood pewee and acts like an olive-sided flycatcher. Its notes are characteristic. The wide and light-colored mandible, brownish upper parts, the whitish chin, and the extensively gray breast and sides should serve to distinguish it.

It gives the impression of a decidedly and uniformly gray bird, with no very marked contrasts; the pale yellowish white on the belly is not very conspicuous. The olive-sided flycatcher is much darker, with a sharply contrasted white abdomen and conspicuous white tufts on the rump. The wood pewee is much smaller and more olivaceous, less grayish.

Fall.—Mr. Henshaw (1875) says: "By the latter part of September, many individuals had passed to the southward; but, at Mount Graham, at this time the species was still present. I noticed them on several occasions on the outskirts of the flocks of Warblers and Nuthatches, which were moving slowly onward. They appeared to be migrating in their company, forming as it seemed to me, a very incongruous element in these sociable gatherings. Their call notes at this time were given almost as incessantly as during the summer."

Winter.—Mr. Skutch's notes refer to the type race of the species, but its habits are doubtless similar to those of our more northern race. He says of its winter habits: "At the close of the breeding season, the families of Coues's flycatcher break up; and the birds live singly through much of the year. On the Sierra de Tecpán I usually found a single Coues's flycatcher—never more than one—in each of the motley flocks, composed of a great variety of resident and migratory warblers and other small birds, which roamed through the oak woods during the months when few birds nested. The oft-repeated *wic wic wic* of the crested, gray flycatcher, voiced briskly as it returned from the aerial sallies that it undertook from one of the more exposed branches, added variety to the chorus of mingled notes raised by the flock."

DISTRIBUTION

Range.—Arizona and Mexico south to Honduras; casual in western New Mexico.

The range of this species extends **north** to central Arizona (Fort Whipple, Mogollon Mountains, and White Mountains). **East** to eastern Arizona (White Mountains, Fort Apache, and Chiricahua Mountains); Chihuahua (Pinos Altos and Jesus Maria); Tamaulipas (Realito and Montelungo); Veracruz (Jalapa); British Honduras (Augustine); and Honduras (Mount Cacaguatique). **South** to southwestern Honduras (Mount Cacaguatique and Los Esesmileo); southern Guatemala (Tecpam and Volcan de Fuego); Guerrero (Chilpancingo); and Nayarit (San Blas). **West** to Nayarit (San Blas); western Durango (Arroyo del Buey); Sinaloa (Chinobampo and Alamos); Sonora (Oposura and Saric); and Arizona (Huachuca Mountains, Santa Rita Mountains, Santa Catalina Mountains, and Fort Whipple).

During the winter season the species apparently withdraws from the United States and is then found **north** to northern Sinaloa (Alamos) and central Tamaulipas (Montelungo).

The typical subspecies, known as Swainson's flycatcher (*Myiochanes pertinax pertinax*), occupies the southern part of the range north to central Mexico, while Coues's flycatcher (*M. p. pallidiventris*) is found in northwestern Mexico and in Arizona.

Migration.—Early dates of arrival in Arizona are: Sabino Canyon, March 23; Santa Rita Mountains, April 5. Late dates of departure are: Mount Graham, September 24; Tombstone, September 26.

Casual records.—In 1876 the species was recorded twice in New Mexico. It was reported as seen on June 21 in the Zuni Mountains east of Agua Frio, and on July 16 a specimen was taken at Fort Bayard. The Colorado record for this bird (Auk, vol. 4, p. 264, 1887) was probably erroneous and is rejected.

Egg dates.—Arizona: 33 records, May 12 to July 13; 17 records, June 10 to 25, indicating the height of the season.

Mexico: 9 records, May 27 to June 27.

MYIOCHANES VIRENS (Linnaeus)

EASTERN WOOD PEWEE

PLATES 38–40

HABITS

CONTRIBUTED BY WINSOR MARRETT TYLER

Spring.—The long spring migration is drawing to a close. The hardy adventurers of March have settled here in New England for the summer or have passed farther northward. In April the hordes of sparrows swept through the country, and early in May the orioles came back to us from South America. The rush of warblers has mainly passed now, but the last of the blackpolls are marching through, and the northern thrushes, the oliveback and the graycheek, the rear guard of the migration, are hiding in the shadows.

It is at this time of the year, when spring is in full bloom, when the countryside is brilliant green and the forest leaves are almost summer size, that the wood pewee calmly takes his place among the big trees of our woodlands, the shade trees of our streets, and, if the trees be tall, even in our gardens. His slow, sweet, quiet, three-note song tells us that he is here, hidden among the leaves, although the bird remains for the most part so high up in the thick foliage that we may not catch a glimpse of him for weeks unless we look sharply —not perhaps until the young are fledged and descend from their lofty nest and begin to wander about with their parents.

All the way on its journey from the south, the wood pewee has loitered behind the hurrying migrants, leisurely delaying its home-coming, and now, at last on its breeding ground, it finds many of its neighbors with broods already hatched, engaged with the turmoils of parenthood.

Courtship.—The wood pewee seems to have no well-marked ritual in its courtship behavior. He does indeed break away from his characteristic calm and becomes more animated during the short nuptial season, flying about more rapidly than usual and engaging in lively, weaving chases among the branches. Such pursuits, how-ever, apparently constitute, as is the case with many of the smaller birds, the only courtship display. Audubon (1840) says: "During the love season, it often flies, with a vibratory motion of the wings, so very slowly that one might suppose it about to poise itself in the air. On such occasions its notes are guttural, and are continued for several seconds as a low twitter."

Dr. Samuel S. Dickey has contributed to Mr. Bent, in careful, extensive notes, the result of his long study of the wood pewee. These notes will be quoted repeatedly hereinafter. Of courtship he writes: "During the mating period they are unusually vivacious. They tweek their wings and agitate their tails and spring prettily forward. It is no uncommon sight to see two males in combat. They draw up to each other, hover an instant in a clearing, and then in close contact they fall downward together, but before they reach the ground they usually swerve to one side. With squeaking out-cries they continue the chase until one bird, tiring of the contest, takes shelter in some distant tree. When a male has found a female to his liking, he pursues her in and out of the avenues between the trees. She will then sometimes disappear into the midst of the body of a tree and leave him hovering in bewilderment close by."

Speaking of the period of courtship, Dr. Thomas S. Roberts (1932) says: "The male Wood Pewee has, besides the usual *pee-a-wee*, a rapid chattering utterance, delivered as he pursues the female among and over the tree-tops; also, at such times, a few full, sweet notes, almost as though he were about to warble a song and suggesting a phrase from that of the Ruby-crowned Kinglet. This was heard on one occasion (June 20) just at sundown as a pair of Pewees that had a nest near by were indulging in most ardent expressions of devotion, accompanied by aerial evolutions so rapid as to make it difficult to follow them with the eye."

Nesting.—The nest of the wood pewee is a dainty little structure, harmonizing so closely with the surroundings that our eye may easily pass along the limb to which the nest is bound without detecting it. The nest seems tiny for the size of the bird, sits close to the branch—the bottom thin, the walls low and thick—and the outside is sheathed with bits of lichen.

The site of the nest is generally on a small limb, often dead and patched with lichens, commonly at a height of about 20 feet, in or near a level fork well out from the trunk of the tree.

Bendire (1895) states that the bird "shows a decided preference for open, mixed woods, free from underbrush, and frequents the edge of such as border on fields, clearings, etc., either in dry or moist situations," and that "an average and typical nest of the Wood Pewee measures 2¾ inches in outer diameter by 1¾ inches in depth; the inner cup is about 1¾ inches wide by 1¼ inches deep."

Arthur C. Bent writes in his notes: "Most of the nests that I have seen have been on horizontal, lichen-covered limbs of old apple trees in orchards, or on dead limbs of pitch pines in the Plymouth woods." The Plymouth woods is a dry, tangled wilderness, extending over many square miles in southeastern Massachusetts, overgrown with pitch pines and scrub oak and interspersed with small ponds.

Dickey (MS.), whose investigations were largely conducted in Pennsylvania, gives a long list of trees in which he has found wood pewees' nests. It includes oaks (white, red, and black), sugar maple, black walnut, yellow locust, elm, apple, and pear, generally in specimens of large growth. He has found a nest in a flowering dogwood tree only 8 feet above ground. He says that willows are used rarely, but he speaks of one nest in a partly dead willow tree five feet out from the main stem. Another nest was "in a stalwart sycamore, six feet through at the butt, in a horizontal fork 45 feet aloft and 18 feet out from the main bole."

Ira N. Gabrielson (1922) describes a nest "saddled on a long straight limb of an elm perhaps fifteen feet from the ground and about the same distance from the trunk of the tree. The only foliage on this branch was a spreading spray of leaves several feet beyond the nest. One would think that a nest so located would be easily discovered but such was not the case. While conspicuously located it was cunningly woven onto the branch and so thoroughly covered with lichens that I could scarcely believe it was a nest even after seeing the bird alight upon it. From below it looked to be simply a lichen-covered knot or a small fungus growth upon the limb and only after we were on a level with it did it seem at all conspicuous."

A. Dawes DuBois, describing in his notes a deserted nest, says: "Its inner lining consisted chiefly of stiff, curved, two-branched, wirelike stems resembling the fruit stems of the basswood tree—some of them 2 inches long. There were about 70 of these. There were also long, hairlike stems of plant fibers, other coarser stems, shreds of weed bark, some 9 inches long, a piece of spider cocoon, and a 3-inch piece of string. At one spot, near the center, the branch itself served as the bottom of the nest. The body of the structure was built of similar but coarser materials. No hair was used in this nest. The outside was well covered with lichens, firmly held in place by cocoon silk."

DuBois also stresses the point that, owing to the situation of the wood pewee's nest—i. e., directly on the bark of a horizontal limb and often not supported in a crotch—the nest must be fastened to the bark. This necessary anchorage is secured by the bird while building who "repeatedly wipes her bill from side to side along the limb, making the materials adhere to the bark."

Bendire (1895) says: "The inner cup of the nest is usually lined with finer materials of the same kind, and occasionally with a little wool, down of plants, a few horsehairs, and bits of thread," and he examined "a unique nest of this species, taken * * * from a horizontal limb of an apple tree, about 8 feet from the ground. * * * This nest, which is well preserved, is exteriorly composed entirely of wool. * * * It is very sparingly lined with fine grass tops and a few horsehairs, while a single well-preserved apple leaf lies perfectly flat and exactly in the center and bottom of the nest."

Ora W. Knight (1908) reports that the male "does not seem to do any active work, either at nest building or assisting in incubation, but I have however seen him feed the female more or less frequently while she was sitting."

The wood pewee appears to become attached to a group of trees and returns sometimes year after year to build its nest on the same branch. Katie Myra Roads (1931) gives an instance of this habit when she reports: "For thirty-five years a Wood Pewee's * * * nest has been placed in the same fork of an elm tree about forty feet from the ground."

Eggs.—Major Bendire (1895) says: "From two to four eggs are laid to a set, generally three, and sets of four I consider rare." He describes them as follows:

The eggs of the Wood Pewee vary in shape from ovate to short or rounded ovate; the shell is close-grained and without gloss. The ground color varies from a pale milky white to a rich cream color, and the markings, which vary considerably in size and number in different sets, are usually disposed in the shape of an irregular wreath around the larger end of the egg, and consist of blotches and minute specks of claret brown, chestnut, vinaceous rufous, heliotrope, purple, and lavender. In some specimens the darker, in others the lighter shades predominate. In very rare instances only are the markings found on the smaller end of the egg.

The average measurements of seventy-two eggs in the United States National Museum collection is 18.24 by 13.65 millimetres, or about 0.72 by 0.54 inch. The largest egg of the series measures 20.07 by 13.97 millimetres, or 0.79 by 0.55 inch; the smallest, 16.51 by 12.95 millimetres, or 0.65 by 0.51 inch.

Young.—The young pewees, generally three in a brood, grow rapidly and soon overfill their little nest. However, in color they match the surrounding bark and lichens so closely that they remain inconspicuous even when, almost ready to fly, the three of them are in plain sight from below, crowded together on a nest that was none too big to accommodate their parent.

Dickey (MS.) indicates how the young birds prevent themselves from falling out of the nest. "When I attempted to take them from the nest," he says, "they resisted with more strength than one would have supposed they possessed. They grasped the lining of the nest with their claws and pulled it out as I lifted them up."

Burns (1915) gives the incubation period as 12 to 13 days, Bendire (1895) as "about twelve days," and Dickey (MS.) says: "The eggs were incubated for a period of exactly thirteen days in six nests I had under observation."

A. Dawes DuBois gives in his notes an account of the nest life in a family he watched closely. He says: "On the day of hatching, the single nestling was only a bit of animated fuzz, but by the evening of the next day it had apparently grown to twice its original size—an odd little creature with tufts of whitish gray down on its back and head. When the nestling was four or five days old it was brooded only part of the time. The feeding was done very quickly. The parent brought what appeared to be a small moth; the nestling's head went up, instantly the food went in, the head dropped back, and the parent brooded, all in a second or two without any ceremony. Two days later the nestling was well feathered. Occasionally it stretched and flapped its wings. While being fed it never uttered any sound that was audible to me. The feeding continued to be a very matter-of-fact, well-regulated business; the young one opened its mouth only at the auspicious moment, and the food was quickly gulped down. Excreta were swallowed until the nestling was four or five days old; later they were carried away and discarded.

"During its thirteenth and fourteenth day the nestling was occupied chiefly in stretching up on the edge of the nest, flapping its wings, looking down at the ground or out through the trees, or watching a butterfly if one came near. It fluttered, stretched, dozed, and took nourishment by turns. Occasionally it almost toppled from the edge of the nest but seemed to have no thought of taking a walk on the branch. But the next morning the youngster ventured out to a distance of about 2 feet, and later, purposely dislodged by a parent, I thought, it fluttered to the ground. From here it struck out on its own account, almost reaching the eaves of a low building 30 yards down the slope before again fluttering to the ground."

Bendire (1895) says: "The young leave the nest in about sixteen days, and are cared for by both parents." Knight (1908) gives the period of nest life as "about eighteen days after hatching." Mr. DuBois's bird left on its fifteenth day.

Dr. Thomas S. Roberts (1932) states: "The young, when first out of the nest, sit huddled together in a row, waiting to be fed and voicing their impatience in a plaintive *squeak*, like a mouse in distress."

Plumages.—[AUTHOR'S NOTE: In the early stages of the juvenal plumage the feathers are soft, fluffy, and blended, but they appear firmer in September with the beginning of the postjuvenal molt. In the juvenal plumage the upper parts are "olive-brown" but much darker on the pileum, the feathers of the crown and rump being narrowly edged with buffy brown; sometimes the entire upper parts have these faint edgings, and sometimes the feathers of the nape are faintly edged with ashy gray; the median and greater wing coverts are tipped with "light ochraceous-buff," forming two distinct wing bands; the central and posterior under parts are "pale primrose yellow," abruptly defined against the "olive-gray" sides of the throat and flanks, with an indistinct pectoral band of olive-gray.

A postjuvenal molt, probably incomplete, begins early in September and evidently is not wholly finished before the birds go south. Whether the wings and tail are molted at this time or later in fall or winter does not seem to be known. Dr. Dwight (1900) says that the first winter plumage "resembles closely the previous dress, but grayish instead of brownish tinged above, the edgings and collar lost and the new wing-bands grayish."

Apparently young birds become practically adult during their first winter or the following spring, perhaps by a complete or partial prenuptial molt.

Adults have a complete postnuptial molt, beginning late in August or in September but chiefly accomplished after the birds have migrated. They may have a partial prenuptial molt before they come north, but we have no specimens to show it.]

Food.—Waldo L. McAtee (1926) states:

The food of the Wood Pewee is almost exclusively derived from the animal kingdom, only a little more than one per cent of it being vegetable. This consists almost entirely of wild fruits such as those of elder, blackberry, dogwood and poke-berry. Spiders and millipeds are eaten regularly but in small quantities, comprising only a little over two per cent of the whole subsistence. Besides the items mentioned the remainder of the food of the Wood Pewee consists entirely of insects. The more important groups are flies (about 30 per cent of the total food), hymenoptera (28 per cent), beetles (14 per cent), lepidoptera (12 per cent), bugs (6 per cent), and grasshoppers (3 per cent). Among forest pests consumed by the Wood Pewee are carpenter ants, tussock and gipsy moths, and cankerworms, click beetles, leaf chafers, adults of both flat-headed and round-headed wood borers, leaf beetles, nut weevils, bark beetles, and tree hoppers * * * The Wood Pewee consumes also various useful insects, as parasitic wasps, ladybird beetles, and certain others, but on the whole is a very good friend of the woodlot.

F. E. L. Beal (1912), basing his conclusions "upon the examination of 359 stomachs taken in 20 States of the Union, the District of Columbia, Ontario, New Brunswick, and Nova Scotia," says in his summary:

The one point most open for criticism in the food of the wood pewee is that it eats too many parasitic Hymenoptera. There is no doubt that all birds which prey upon Hymenoptera at all destroy some of the useful species, but the proportion in the food of this bird is greater than in other birds whose food has been investigated. As these insects are for the most part smaller than the more common wasps and bees, it would seem natural that they should be preyed upon most by the smaller flycatchers, which very likely acounts for the fact that the wood pewee eats more of them than the kingbirds. But even so the bird does far more good than harm. The loss of the useful Hymenoptera can be condoned when it is remembered that with them the bird takes so many harmful or annoying species.

Walter Bradford Barrows (1912) says: "The food consists very largely of insects taken on the wing, yet it not infrequently hovers before a twig or leaf and snaps up small insects which appear to be stationary, sometimes descending to the grass for this purpose. * * * In Nebraska Professor Aughey found seven grasshoppers and many other insects in the single specimen which he examined."

As we watch a wood pewee feeding—dashing out from its perch repeatedly, often among the interstices of forest trees where the light is not over strong—we are impressed by the large number of very small insects it must capture. These are so small that we do not always catch sight of them in the air, but we may infer their number from the bird's actions, by hearing the click of its bill as it snaps them up, or attempts to do so, and sometimes by seeing more than one insect in the bird's beak after it alights. Forbush (1907) noticed this habit and remarks: "It usually perches on dead branches at some height from the ground, and flies out to some distance, taking one or many insects at each sally."

Dickey in his manuscript states that "the birds flit out from woodland margins to feed in clearings and over corn, wheat and oat fields. They are prone, too, to hover beside the webs of spiders and extract flies that have been snared, and they make repeated trips out over marsh-land and return to the woods, their beaks filled with appendages of insects."

That the food of the wood pewee is not restricted to small insects is shown by A. Dawes DuBois (MS.), who reports that he saw a parent bird come to a nest "with a good-sized butterfly, a red admiral, which the young bird swallowed, wings and all."

Bendire (1895) quotes George A. Seagle, superintendent of the Wytheville (Va.) Fish Commission station, who stated: "This little bird has frequently been seen to catch young trout from the ponds soon after they had been transferred from the hatching house."

Behavior.—The wood pewee is an obscurely marked, smallish flycatcher, only slightly larger than the little birds that make up the genus *Empidonax*. Wilson (1831) says: "It loves to sit on the high dead branches, amid the gloom of the woods." In such surroundings it is not easily seen, for its plumage appears in the field as brownish gray above and grayish white below, colors that harmonize with the filtered light of the forest. In fact, were it not for its voice, we should rarely notice the bird even when it is darting about, high overhead in its leafy retreat. It is a seclusive, apparently peace-loving little bird, quiet, although very quick in its motions, and seldom asserts itself, being wholly free from the aggressiveness that marks the behavior of some of the larger flycatchers. We meet it almost invariably alone, or in the company of its mate or its brood of young.

Here in eastern Massachusetts the wood pewee is not a common bird; it has diminished in numbers noticeably during the past 20 years. Both Wilson and Aubudon speak of it as more common than the phoebe. At the present time the reverse is true here, in the proportion, it seems, of ten to one.

Speaking of the wood pewee's relations with other species of birds, A. Dawes DuBois (MS.) says: "The pewees would not tolerate redwinged blackbirds or red-headed woodpeckers, although they were not agitated by the presence of flickers. With chipping sparrows they were on very friendly terms. The toleration of another species I once saw displayed even in the vicinity of the nest. A least flycatcher was building its deep nest about 60 feet from the ground in a tall slender tree, while a wood pewee sat unconcerned in her own flat nest about 5 feet away in the same tree."

Beatrice Sawyer Rossell (1921) points out an exception to the bird's usual peaceful behavior. She relates: "My attention was suddenly attracted by a Wood Pewee, which flew to a dead twig, not 3 feet above my head. I called my companion's attention to it, and as I spoke the bird darted at my head, coming so close that I instinctively swerved. He flew back to his perch, and in a minute made another dart, almost brushing me with his wings. * * * For a few seconds he fluttered around me, then made a dart and pecked my finger with his sharp little bill."

Ira N. Gabrielson (1922), who had lowered a nest containing three eggs to within 3 feet of the ground, says: "We were regarded with absolute indifference as we approached to within six feet to take a photograph. * * * The brooding bird was not disturbed by my entrance into the blind but as the camera lens appeared in the opening of the blind she left the nest and dashed repeatedly at the lens, snapping her mandibles vigorously. This continued for several minutes before she finally returned to the nest. At intervals during the morning she renewed her attack on the lens but aside from this paid no attention to either the blind or my movements."

Voice.—The wood pewee has a very attractive voice—a sweet, pure, tranquil whistle delivered calmly in short, slow phrases. The leisurely notes, sliding smoothly and evenly as they change in pitch, give the impression of restfulness and peace, almost of sadness. Bradford Torrey (1901) calls the song "an elegy." All day from dawn to dusk it goes languidly on, *pee-a-wee*, (a pause) *peea*, phrase after phrase, often with long pauses between them, never hurried, always serene. The song continues well into hot, parched August, when most birds are silent. Aretas A. Saunders (1924), speaking of the uniformity of the wood pewee's singings, says: "Of a number of records made from different individuals at the same season of the year, the majority are likely to be almost, if not exactly, identical."

Perhaps, in the case of the wood pewee, the term song should be applied only to the bird's singing in the half light before dawn and in the evening long after sunset. At these times of day the bird devotes about 40 minutes in the morning and a shorter, less regular period in the evening to singing a song quite different from its daylight notes, a song so charmingly rhythmical that it has attracted the attention of musicians and excited their admiration.

I noticed it for the first time on June 3, 1911, and wrote in my notes: "At 3.40 this morning (sun rose at 4.09) a wood pewee sang over and over with perfect regularity a song of five drawling notes—*pee-a-wee, pee-wee*—both phrases ending on a rising inflection. The syllables and the pauses between them were so regular that I could time by my breathing. *Pee-a-wee* corresponded exactly with an inspiration, then, after a short pause the *pee-wee* finished at the end of expiration. Then a longer pause—just as long as the rest between breaths—and after this he repeated his song with my next breath. I was breathing, I suppose, 16 times a minute, and the bird slowly fell behind, but he fell behind not from any irregularity, but because his rate was slightly slower than mine."

In listening to the twilight song in more recent years I have noted that, as the song goes on and on, a bird will occasionally introduce into it, among the phrases that rise in pitch, a phrase of falling inflection. This phrase brings to the song a restful effect. Indeed, Henry Oldys (1904), taking this infrequent phrase as the final theme of a four-line song, points out "that it is constructed in the form of the ballad of human music." He explains that "the arrangement of the ordinary ballad frequently consists of a musical theme for the first line, an answering theme for the second line that leaves the musical satisfaction suspended, a repetition of the first theme for the third line, and a repetition of the second theme, either exactly or in general character, but ending with the keynote, for the fourth line." Illustrating with a verse of "Way Down upon the S'wanee River," he shows that the wood pewee's song is governed by the same principles, and that the final keynote (of the falling phrase) completely satisfies the ear.

When the bird combines his phrases in this way, as he does from time to time, he converts his long soliloquy into a song of great beauty. But we must bear in mind that it is only through the fortuitous arrangement of its parts that the singing assumes for a moment the ballad form, and that the introduction of the key phrase is purely inadvertent.

Mr. Oldys slyly remarks at the end of his interesting exposition of the twilight song: "In closing this brief account I would call attention to the remarkable fact—perhaps a joke on us—that a bird which we have classed outside the ranks of the singers proper should deliver a song that judged by our own musical standards takes higher technical rank than any other known example of bird music."

The reader is referred also to two articles by Wallace Craig (1926, 1933) analyzing the twilight song.

Taverner and Swales (1907) write of the wood pewees at Point Pelee: "Their voices can be heard any hour of the day uttering their pathetically plaintive note; and often in the night, as we have lain awake in the tent, some Pewee has aroused itself and a long drawn 'pewee' has punctuated the darkness with its soft sweetness."

Harrison F. Lewis, in a letter to Mr. Bent, writes: "On July 12, 1920, I saw a wood pewee dash out of a tree at the height of about 40 feet from the ground and fly wildly and erratically about in a small area, crying rapidly and unceasingly, in a high-pitched squeaky voice, *whee-chuttle-chuttle*, *whee-chuttle*, etc., for about half a minute."

I have heard similar notes in midsummer from a pewee perched in a tree—seven or eight short whistled syllables given as a rapid twitter which suggested a goldfinch's voice, and wholly lacked the usual drawling quality of the pewee's.

The wood pewee's call note is a soft monosyllable, lower and less sharply enunciated than the explosive *chip* of the phoebe.

Francis Beach White, who has studied the voice of the wood pewee for more than 20 years, calls attention (MS.) to a seasonal variation in the notes. He says: "In the last week of May, prolonged singing analogous to the twilight song is heard, but this is not developed fully until June. In June the notes take on a somewhat richer tone. In July the *phee-ew* is heard in long series, especially at dawn and after sunset, and excited jumbles of song notes may also be heard occasionally, as well as antiphonal calling. In August, notes of more emotional tone are given, and toward the last of the month *pu-ee* is often heard with a strident element in the last syllable. The human reaction to the notes endues the last-mentioned call with a petulant anxiety, and the twilight song with a paradoxical mournful happiness."

Field marks.—The following excellent differential diagnosis is quoted from Ralph Hoffmann (1904): "The long-drawn song, when given, distinguishes the Wood Pewee from any of the other Flycatchers, but when the bird is silent it may be confused either with the Phoebe or with the Chebec. It may be distinguished from the former by its smaller size and by its *well-marked wing-bars;* moreover, it never flirts its tail after the manner of the Phoebe. It is considerably larger than the Chebec, and, when it faces an observer, the middle of its breast shows a light line separating the darker sides."

Enemies.—The wood pewee is subject only to the dangers that beset most of the small passerine birds. Notable among its enemies is the blue jay, which may rob it of its eggs or young.

Dr. Friedmann (1929) calls the wood pewee "a not uncommon victim of the Cowbird. * * * As many as four Cowbird eggs have been reported from a single nest of this species, but such cases are extremely rare. * * * About 3 dozen records have come to my notice but these probably represent a small percentage of the cases found."

Fall.—At Point Pelee, Taverner and Swales (1907) found wood pewees "very abundant in the early days of fall. It is evident that the first fall movement of this species begins early in the season. The 24th of August, 1907, we found the woods of the Point already in possession of innumerable hosts of Wood Pewees, and through early September we have always found them the most prominent bird in the landscape."

Alfred M. Bailey and Earl G. Wright (1931), writing of southern Louisiana, says: "In November it is fairly numerous, and one of the delights of still hunting for deer along the wooded regions, is to watch the small birds working through the tree tops. Small warblers are ever on the move, but the pewee often sits motionless on twigs over the water, and then comes suddenly to life long enough to flutter into space, seize an insect, and drop back to perch."

William Brewster (1937) gives an instance of a young wood pewee caught by inclement weather in northern Maine late in the season:

1899, October 1.—A bitter day for the season with harsh north-west wind, over-clouded sky and frequent flurries of snow melting as fast as it struck the ground in the lowlands, but whitening the mountain crests from morning to night. Visiting our boat-house by the river-bank in Upton this afternoon I was not a little surprised to find a Wood Pewee there, cowering under the lee of the building with ruffled and somewhat bedraggled plumage, looking benumbed and disheartened. Nevertheless its eyes shone brightly and it made occasional dashing forays among the thick falling snowflakes apparently mistaking them for white-winged insects or at least treating them as such. It spent most of its time, however, on the ground, or rather on piles of chips and pieces of boards, where it fluttered or hopped from place to place picking up food the nature of which I failed to ascertain.

DISTRIBUTION

Range.—Eastern United States and southern Canada; south in winter to northwestern South America.

Breeding range.—The eastern wood pewee breeds **north** to southern Manitoba (Fairford, probably Lake St. Martin, and Shoal Lake); southern Ontario (Indian Bay, Gargantua, North Bay, and Algonquin Park); southern Quebec (Hatley); northern Maine (Kineo, Presque Isle, and Easton); and Nova Scotia (Pictou). **South** along the Atlantic coast from Nova Scotia (Pictou and Halifax) to Florida (St. Augustine and Samsula). **South** to Florida (Samsula, Micanopy, and Waukeenah); Louisiana (New Orleans, Bains, and New Iberia); and southern Texas (Houston and San Antonio). **West** to central Texas (San Antonio and San Angelo); central Oklahoma

(Norman and Minco); central Kansas (Wichita and Manhattan); eastern Nebraska (Red Cloud, Lincoln, and Scribner); eastern South Dakota (Yankton, Sioux Falls, and Fort Sisseton); eastern North Dakota (Wahpeton, Jamestown, and Stump Lake); and Manitoba (Margaret, Aweme, and Fairford).

Winter range.—The winter range extends **north** to central Costa Rica (Zarcero and Jimenez); Panama (Gatun); and Colombia (Santa Marta, Bonda, and Cacagualito). **East** to Colombia (Cacagualito and Villavicencio); and Peru (Chanchamayo and Vista Alegra). **South** to central Peru (Vista Alegra) and southwestern Ecuador (Zamora). **West** to Ecuador (Zamora, Guayaquil, San Jose de Sumaco, and Rio Suno) and Costa Rica (San José and Zarcero).

Spring migration.—Early dates of spring arrival are: Florida—Whitfield, April 4. Alabama—Coosada, April 9. Georgia—Macon, April 16. South Carolina—Charleston, April 14. North Carolina—Raleigh, April 18. Virginia—Variety Mills, April 23. District of Columbia—Washington, April 29. Pennsylvania—Waynesburg, May 2. New Jersey—Englewood, May 10. Connecticut—Hartford, May 14. New York—Ballston Spa, May 15. Massachusetts—Amherst, May 11. Vermont—Randolph, May 19. Maine—Portland, May 21. Quebec—Montreal, May 18. New Brunswick—Scotch Lake, May 23. Mississippi—Bay St. Louis, March 25. Arkansas—Helena, April 12. Tennessee—Athens, April 23. Kentucky—Eubank, April 26. Missouri—St. Louis, April 28. Indiana—Brookville, May 4. Ohio—Oberlin, May 2. Michigan—Grand Rapids, May 4. Iowa—Keokuk, May 4. Illinois—Chicago, May 5. Minnesota—Minneapolis, May 10. Wisconsin—Sheboygan, May 12. Texas—San Antonio, April 9. Oklahoma—Norman, April 27. Kansas—Onaga, May 9. Nebraska—Red Cloud, May 10. South Dakota—Dell Rapids, May 8. North Dakota—Arnegard, May 15. Manitoba—Treesbank, May 16.

Fall migration.—Late dates of fall departure are: Manitoba—Treesbank, September 8. North Dakota—Argusville, September 24. South Dakota—Faulkton, October 1. Texas—Corpus Christi, October 30. Minnesota—Minneapolis, October 5. Wisconsin—Sheboygan, September 29. Illinois—Chicago, October 3. Ontario—Toronto, September 26. Michigan—Detroit, October 7. Ohio—Wauseon, October 4. Kentucky—Eubank, October 15. Tennessee—Athens, October 22. Louisiana—New Orleans, November 2. Quebec—Montreal, September 11. New Brunswick—Scotch Lake, September 13. Maine—Phillips, September 30. Massachusetts—North Amherst, October 3. New Jersey—Passaic, October 10. District of Columbia—Washington, October 12. North Carolina—Raleigh, October 13. South Carolina—Spartanburg, October 19. Georgia—Athens, October 26. Florida—Punta Rassa, November 23.

Casual records.—A specimen was taken at Bermuda on April 30, 1852, and one was obtained at Santiago de las Vegas, Cuba, on April 12, 1928. One was collected at Springfield, in southeastern Colorado, on May 12, 1905.

Egg dates.—Illinois: 21 records, May 20 to July 22; 11 records, June 18 to July 16, indicating the height of the season.

Massachusetts: 31 records, June 8 to July 12; 15 records, June 15 to 27.

New York: 37 records, May 22 to July 23; 19 records, June 15 to 20.

South Carolina: 6 records, May 22 to June 24.

MYIOCHANES RICHARDSONI RICHARDSONI (Swainson)

WESTERN WOOD PEWEE

PLATE 41

HABITS

A study of museum specimens would strongly suggest that *richardsoni* should be considered a subspecies of *virens*, but those who are familiar with the two birds in life recognize certain differences in voice and nesting habits that seem to warrant the separation of the western bird as a full species, distinct from our eastern wood pewee. In this connection the reader is referred to a discussion of the subject by no less an authority than Dr. Joseph Grinnell (1928b), who says, after "examination of large series of specimens," that "there is practically complete intergradation by way of individual variation between *richardsonii* and *virens*, in structural characters. Why should differences in voice or in nesting habits weigh against the use of the trinomial in this case any more validly than they do in other quite similar cases where the trinomial is in current undisputed employment?"

Another keen observer, A. J. van Rossem (Dickey and van Rossem, 1938) seems to concur in this view, for he lists the western wood pewee as *Myiochanes virens richardsonii* (Swainson). The western wood pewee enjoys a wide distribution, as a summer resident, over the western half of this country, as the eastern bird does over the eastern half. Whether and to what extent the two forms intergrade where their ranges meet does not seem to have been satisfactorily determined.

Samuel F. Rathbun writes to me of its haunts in Washington: "In the Puget Sound region the western wood pewee can be regarded a common summer resident. But, although its distribution is general, for it is found alike in the cities and towns as well as the outside country, I have observed that it is somewhat partial to certain localities and absent from others of the same general character. It shows a liking for some cultivated valley through which a stream flows, or the open deciduous tree growth along the borders of lakes and small bodies of fresh water; and in such localities one can expect to find the species year after year. Seldom are there more than a pair or two of these birds in any certain locality; usually they seem to be somewhat separated."

James B. Dixon writes to me of California haunts: "This is a common breeder from the Pacific Ocean to the tops of our highest coastal range in San Diego County. It is commoner in the sycamore groves of the stream beds but is found in the conifers of the higher elevations and also is common in the aspen thickets at the higher elevations in the Mono Basin in Mono County." In Arizona we found the western wood pewee in the lower, more open portions of the canyons where there was a heavy growth of large sycamores, cottonwoods, and other trees along the beds of the mountain streams, at elevations of 5,000 to 6,000 feet but not in the higher portions of the mountains or in the lower valleys.

Nesting.—In the choice of a nesting site the western wood pewee shows no special preference for any particular species of tree, provided it can find a suitable fork or horizontal branch on which to place its nest. And nests are often placed on dead branches or on wholly dead trees; dead aspens seem to be highly favored. J. Donald Daynes writes to me from Salt Lake City, Utah, that within an area of 40 acres of aspen trees he found five nests; three of these were in dead aspens and all were on dead limbs; they ranged in height from 10 to 50 feet above ground. Sycamores and cottonwoods seem to be popular trees, and nests have been recorded in walnuts, boxelders, ashes, birches, alders, various oaks, maples, hackberry, and eucalyptus trees, madrones, various pines, larches or tamaracks, cedars, firs, and spruces, as well as orchard trees. W. E. D. Scott (1879) mentions a nest that "was built where three branches crossed in a brushheap two feet from the ground." This, of course, is a very unusual location and a very low one; most of the nests are placed at heights ranging from 15 to 30 feet above ground; often they are as high as 40 feet and sometimes 50 or even 75 feet.

Mr. Rathbun writes to me of the nesting habits of this pewee in Washington: "The breeding period of this flycatcher, in this section at least, appears a little extended, or from quite early in June to about the middle of July. Our experience with its nesting habits has shown us that when the first nest is taken it is very quickly replaced by another. On one occasion when I took a nest, at the end of several days a second had been completed and held an egg. In another instance, on the eleventh day thereafter the new nest had three eggs; and in each case the second nest was placed very near where the first one had been.

"The nest of this flycatcher, though somewhat shallow, is a beautiful one. In the many I have found the materials used consisted of small pieces of plant fibers, often a downy substance from cottonwood blooms, bits of fine dry grasses, and at times a few bud scales of small size, all neatly interwoven. This represented the bottom of the nest, which then was decorated outwardly with lichens, bits of grayish moss, now and then a few little bud scales, all skillfully bound thereon by spider webs and filmy plant fibers, with occasionally a few horsehairs. Always the lining was of fine dry grasses. The nests were either saddled on an open branch or attached snugly on a small horizontal fork of a limb, at heights that varied from 15 to 40 feet; and all were placed in deciduous trees of not large size."

Baird, Brewer, and Ridgway (1905) described a nest from California that differs somewhat from that described above: "The base and sides of this nest are largely composed of the exuviae of chrysalides, intermingled with hemp-like fibres of plants, stems and fine dry grasses. The rim is firmly wrought of strong wiry stems, and a large portion of the inner nest is of the same material. The whole is warmly and thoroughly lined with the soft fine hair of small quadrupeds and with vegetable fibres."

All the nests I have seen differ from the nests of the eastern wood pewee in being somewhat larger, more compact, and more deeply hollowed; the usual lining of bright-yellow grasses is generally conspicuous; and, most important of all, the outer covering of lichens, so conspicuous in nests of the eastern bird, is usually lacking or replaced with some other material. Most other observers agree on these points, though a number of them have reported some use of lichens. Henshaw (1875), quoting C. E. Aiken, said: "No lichens at all are used in its construction, but instead the gray dead leaves of a minute plant that grows abundantly in the mountains is often found upon the outside." I am inclined to think that this may be the material that has been mistaken for lichens; also, bits of chrysalids or cocoons, frequently used, look much like pale gray lichens. However, most nests are well camouflaged with a great profusion of spider webs with which the nest is bound to the branch, or with other material that matches the branch. Another point of difference is that some nests of the western bird are more or less lined with various bird feathers, sometimes brightly colored ones; these, so far as I know, are never used by the eastern bird.

Many years ago Ridgway (1877) made the surprising statement that "the nest of this species, as is well known, differs very remarkably from that of *C*[*ontopus*] *virens*, being almost invariably placed in the crotch between nearly upright forks, like that of certain *Empidonaces*, as *E. minimus* and *E. obscurus*, instead of being saddled on a horizontal branch."

Major Bendire (1895) challenged this statement, as being quite at variance with his experience, and said: "If the Western Wood Pewee places its nest occasionally in a crotch, which I do not deny, it is exceptional and not the rule." There is, however, some more recent evidence that this pewee does occasionally build its nest in an upright crotch. Mr. Dawson (1923) says that "occasionally the nest is set in an upright crotch of a willow or some dead sapling." And J. A. Munro (1919) says of the nests found in British Columbia: "They are usually built saddle fashion on a rather large limb, generally at a crotch, but I have found two that were built in upright forks like a Yellow Warbler's nest. These two nests were in half-dead peach trees in an orchard."

Major Bendire (1895) quotes Charles E. Aiken as saying: "I have found several settled in the angle formed by the trunk of the tree and a horizontal branch, and in one instance, where a large limb had been torn from the tree by the wind, a nest was placed flatly upon a broad, board-like splinter."

John W. Mailliard (1921) had some favorable opportunities to watch some western wood pewees at their nest building activities, of which he writes:

One female gathered its building material by pecking small bits of bark from the branches of a dead willow, which was but a few yards from the large yellow pine in which the nest was placed. At times small bits of this material could be distinguished in the bill of the busy bird, while at other times nothing was discerned, the presence of such only being evidenced by the operations of the bird upon its return to the nest. Meantime the male perched in the near vicinity, or darted after its prey, sometimes perching in, or darting from, the very tree in which the nest was being constructed. * * *

Building operations seemed to consist solely in a constant pecking-weaving process, and the shaping of the nest was accomplished by the bird twisting its body while in the nest, and arching its neck so that its throat was over the rim and against the side of the nest. The head was then moved back and forth along the rim much as one sharpens a razor on a strop. With a similar effect the tail was often thrown down and compressed against the outside of the nest, but no lateral motion could ever be discerned.

Another pair built three nests in rapid succession. The first nest was destroyed by a storm, and a new nest was built and the first egg laid within eight days; this nest, with its three eggs, was collected. The third nest was constructed and the first egg laid in it within seven days. All the nests were within a few rods of the first site, and all in aspens.

The measurements of four nests before me show considerable variation in outside diameter, from 4 to 2½ inches, but the inside diameter is more constant at about 2 inches; the outside height varies from 1½ to 1¾ inches, and the depth of the inner cavity from 1 to 1¼ inches.

Eggs.—The western wood pewee usually lays three eggs, sometimes only two, and very rarely four. These are indistinguishable from those of the eastern wood pewee, which the reader will find described under that bird. The measurements of 50 eggs average 18.3 by 13.6 millimeters; the eggs showing the four extremes measure **19.5** by 13.2, 19.1 by **15.0** and **16.0** by **13.0** millimeters.

Young.—The incubation period is probably about 12 days. The female evidently does all the incubating, but the male assists in the care of the young. Probably only one brood is raised in a season. Mrs. Wheelock (1904) implies that the male does most of the feeding of the young, and says: "Small butterflies, gnats, all sorts of small winged insects are the orthodox food for infant flycatchers, and are swallowed at the rate one every two minutes. Nor does the supply ever quite equal the demand, for every visit of the devoted father is welcomed with wide-open mouths and quivering wings. At first all this feeding must be by regurgitation, the adult swallowing the insect first and partially digesting it in some cases, and in others merely moistening it with the saliva. After four or five days most of the food is given to the young in a fresh state."

Plumages.—I can find nothing in the molts and plumages of the western wood pewee that are in any way different from those of the eastern wood pewee; all that has been said about the eastern bird will apply equally well to the western.

Food.—Professor Beal (1912) reports on the contents of 174 stomachs of the western wood pewee, in which 99.93 percent of the food was animal matter and only 0.07 percent vegetable. Beetles of 19 species amount to 5.44 percent, of which only 0.95 percent are useful beetles, ladybird beetles, and predaceous ground beetles. Hymenoptera, wasps, bees, and ants amount to 39.81 percent of the food and were found in 107 stomachs, 17 of which contained no other food. Parasitic species were noted in 8 stomachs and ants in 10. No trace of a honeybee was found, and he never heard any complaints against the bird on this score. Diptera (flies) seem to be the largest item of the food, amounting to 44.25 percent. They were found in 162 stomachs, 30 of which were entirely filled with them. They included horse flies, snipe flies, crane flies, robber flies, and house flies. Hemiptera amount to only 1.79 percent, and no trace of grasshoppers or crickets was found. Moths were found in 24 stomachs and caterpillars in 5, making an average of 5.17 percent for the season. Dragonflies, lace-winged flies, Mayflies, white ants, and spiders together make up 3.47 percent, the remainder of the animal food.

Vegetable matter was found in only four stomachs. "In one of these it consisted of 3 seeds of elderberries (*Sambucus*); in another, of a bit of fruit skin, with a trifle of rubbish; in another, of one seed of wild oats; and in the fourth, of rotten wood."

He once watched one of these pewees flying out from its perch for insects "and noted the number caught in three minutes." He says:

In the first minute 7 were taken, in the second 5, and in the third 6, or 18 in three minutes. These observations were made at 10 A. M., when the air was warm and many insects were on the wing. At 9 A. M. the next day the same perch was again watched, and 17 captures were noted in 8 minutes. This morning was much cooler than the previous one and fewer insects were abroad. The mean of these two observations is 4 insects per minute, and if this rate is kept up for even 10 hours a day, the total is 2,400 insects. It seems hardly possible that one bird can eat so many unless they are very small, but this bird is rarely seen when it is not hunting. When there are young in the nest to feed, the havoc among the insects of that immediate vicinity must be something enormous.

In spite of the fact that Professor Beal found no traces of honey-bees in the 174 stomachs that he examined, some do occasionally eat bees, for Frank Stephens wrote to Major Bendire (1895) : "I have known apiarists to be compelled to shoot a great many to protect their bees; one in San Diego County told me that he shot several hundred in a season. They capture both workers and drones, and I have examined many stomachs which had stings sticking in them."

Dr. A. K. Fisher (1893) says that, in Death Valley, "one day, when the wind was very high, a number were seen sitting on the bare alkaline flats near the [Owens] Lake, where they were picking from the ground the flies which swarmed there, as grain-eating birds do seeds."

Behavior.—In a general way the habits and actions of the western wood pewee are similar to those of its eastern representative. It sits in a vertical attitude for long periods on the top of some dead tree, or oftener on a dead branch on the edge of the woods or under the shadow of the open forest, whence it makes frequent sallies into the air after insects and returns to its lookout perch. Fruit orchards and shade trees about houses are also often used as hunting grounds. It is very lively in its movements, darting about in the air after its prey, which it seizes with a click of the mandibles. It is a most industrious and persistent flycatcher, spending most of its time in the pursuit of these tiny insects, of which an enormous number seems to be needed to satisfy its appetite.

As a rule, it seems to be quite tolerant and peaceful toward its avian neighbors, but it knows how to discriminate between the harmless and the dangerous ones. Jays and other nest robbers are immediately attacked by the guarding male, if they venture too near the nest; with cries of protest and much snapping of mandibles, the unwelcome visitor is set upon, driven away, and pursued relentlessly until forced to leave the territory. Mr. Rathbun writes to me: "If their nest is disturbed the birds show much excitement, not only by giving their notes often and continually shifting from one perch to another, but at times one of the pair will make a feint to strike the intruder; and these actions continue as long as they are molested."

Voice.—Mr. Ridgway (1877) states that it seems "to be more crepuscular than the eastern species, for while it remains quiet most of the day, no sooner does the sun set than it begins to utter its weird, lisping notes, which increase in loudness and frequency as the evening shades deepen. At Sacramento we frequently heard these notes about our camp at all times of the night. This common note of Richardson's Pewee is a harsh, abrupt lisping utterance, more resembling the ordinary rasping note of the Night-Hawk (*Chordeiles popetue*) than any other we can compare it with, though it is of course weaker, or in strength proportioned to the size of the bird. Being most frequently heard during the close of the day, when most other animals become silent and Nature presents its most gloomy aspect, the voice of this bird sounds lonely, or even weird."

Mr. Rathbun (MS.) says that "the note of this flycatcher is similar to that of its eastern relative, although it is more abbreviated; but it has the same plaintive cadence so suggestive of the drowsy summer days."

Dr. Loye Miller (1939) in his study of the song of the western wood pewee writes it *tswee-tee-teet*, *tswee-tee-teet*, *bzew*, and says: "The *tswee-tee-teet* is designated as a triad, it is once repeated and is then followed by *bzew*, a downward slur, which completes the pattern of three equal measures. This pattern is then repeated without interruption of rhythm for an extended and metronomic performance. Each measure lasts for about one and a half seconds. The triad is a rising sequence with the strong accent on the first note. The slur is a downward slide equivalent in length to the three rising notes of the triad. The tone quality of the triad notes is entirely different from that of the slur, the latter being a roughened buzz, whereas the former are clear and sweet."

What is apparently the same song has been expressed in various syllables by different observers, but all seem to give the same impression. The doleful "dear me" written by Dawson (1923), appeals to me as expressing the tinge of melancholy that the song of the wood pewee always suggests to me; it is not a particularly joyful song.

Ralph Hoffmann (1927) describes another song as follows: "When a pair are together in the mating season they utter a hoarse, gurgling note, *ahée-up chée-up*, and the male encourages the female during the nest-building by a musical *pip, pip, pip, pip, pée-a*, or at times mounts into the air and flies about calling *pit, pit, pit.*"

Field marks.—The wood pewee is only slightly larger than the other small flycatchers, and it has no very conspicuous field marks. The dark sides of the breast are divided by a narrow, light-colored line, there are no very conspicuous wing bars, except in the young bird, and it lacks the white eye-ring, so prominent in the Empidonaces.

Its habitat is different from those of the other small western fly-catchers, and its notes are quite distinctive.

Enemies.—Dr. Friedmann (1929) says that "this species seems to be a rather uncommon host of the small, southwestern race of the Cowbird—the subspecies *obscurus.*"

DISTRIBUTION

Range.—Western North America, south to northwestern South America.

Breeding range.—The western wood pewee breeds **north** to central Alaska (Fairbanks and near Circle); southern Yukon (Little Salmon River); southwestern Mackenzie (Fort Simpson, Fort Providence, and Fort Smith); northeastern Alberta (Athabaska delta); central Saskatchewan (Big River and Cumberland House); and southeastern Manitoba (Winnipeg). **East** to eastern Manitoba (Winnipeg and Treesbank); northeastern North Dakota (Grafton); southwestern South Dakota (Elm Mountains and Hot Springs); western Nebraska (Henry); central Colorado (Fort Collins, Denver, and Beulah); eastern New Mexico (Montoya and Roswell); western Texas (Fort Hancock); Veracruz (Perote and Presidio); Guatemala (Progreso); and Costa Rica (La Hondura and Boruca). **South** to Costa Rica (Boruca); Chiapas (Tonala); Guerrero (Chilpancingo); and southern Baja California (San Jose de Rancho and La Laguna Mountains). The **western** limit of the breeding range extends northward along the Pacific coast from Baja California (La Laguna Mountains and the San Pedro Martir Mountains) to Alaska (Chickamin River and Fairbanks).

Winter range.—The winter range of this species is imperfectly known, but at this season it appears to be concentrated in Central America and northwestern South America. It has been recorded **north** to Oaxaca (Topanatepic) and **south** to Bolivia (Nairapi) and Peru (La Gloria and La Merced).

As outlined the range includes three currently recognized subspecies. The typical form (*Myiochanes richardsoni richardsoni*) is found in summer from Alaska and Mackenzie south to northern Mexico and in winter south to Ecuador, Bolivia, and Peru; the large-billed wood pewee (*M. r. peninsulae*) breeds in southern Baja California and is found in winter on the mainland of Mexico and south to Guatemala; the third form, the Mexican wood pewee (*M. r. sordidulus*), breeds in the highlands of southern Mexico and winters south to Peru.

Spring migration.—Early dates of spring arrival are: Texas—San Antonio, April 15. New Mexico—State College, April 25. Colorado—Boulder, May 3. Wyoming—Cheyenne, May 13. Montana—Columbia Falls, May 20. Saskatchewan—Eastend, May 20. Mani-

toba—Aweme, May 17.   Arizona—Tombstone, April 21.   Utah—
Salt Lake City, May 8.   Idaho—Rupert, May 6.   Alberta—Edmonton, May 8.   California—Buena Park, April 14.   Oregon—Coos Bay,
April 28.   Washington—Yakima, May 3.   British Columbia—Chilliwack, May 9.   Alaska—Fairbanks, May 12.

Fall migration.—Late dates of fall departure are: Alaska—Taku
River, September 8.   British Columbia—Okanagan, September 13.
Washington—Pullman, September 15.   Oregon—Coos Bay, October
1.   California—Buena Park, October 2.   Alberta—Camrose, August
28.   Idaho—Preston, September 8.   Arizona—Huachuca Mountains,
October 29.   Saskatchewan—Eastend, September 8.   Montana—Missoula, September 10.   Wyoming—Laramie, September 30.   Colorado—Fort Morgan, October 10.   New Mexico—Mesilla Park, October 25.   Manitoba—Aweme, September 13.   Oklahoma—Kenton,
September 26.   Texas—Runge, September 21.

Casual records.—According to Kumlien and Hollister (1903) several typical western wood pewees have been taken at Lake Koshkonong, Wis., one pair of which with their nest and eggs, were identified by Dr. Elliott Coues.   Dr. A. K. Fisher also has so identified a
specimen taken on July 31, 1890, at Alden, Wis.   There are no known
recent records.   One was collected on July 1, 1898, at Point Barrow,
Alaska.

Egg dates.—California: 100 records, May 9 to August 1; 50 records,
June 10 to 27, indicating the height of the season.

Colorado: 9 records, May 6 to July 7.

Oregon: 14 records, June 4 to July 18; 8 records, June 18 to July 3.

Washington: 7 records, June 3 to July 22.

MYIOCHANES RICHARDSONI PENINSULAE (Brewster)

LARGE-BILLED WOOD PEWEE

HABITS

When William Brewster (1891) described and named this Lower
California subspecies of the western wood pewee he characterized
it as "much smaller than the northern race, the color of the upper
parts slightly grayer, the yellowish of the throat and abdomen
clearer or less brownish and more extended, the pectoral band narrower and grayer, the light edging of the inner secondaries and
greater wing-coverts broader and whiter."

He also said: "In the coloring of the under parts this form resembles *C. virens*, the yellowish of the throat and abdomen being
of about the same shade and fully as extended as in that species.
The breast and sides, however, are less olivaceous and more as in
richardsonii, but grayer, with the pectoral band almost invariably

narrower. The coloring of the upper parts is essentially similar to that of *richardsonii*, but perhaps a trifle paler. The wings and tail are much shorter or about as in *virens*. The bill averages considerably larger (both longer and broader) than in either *virens* or *richardsonii*."

About all we know of its distribution and habits is contained in the following statement by Mr. Brewster (1902): "This near ally of *C. richardsonii* was discovered by Mr. Frazar on the Sierra de la Laguna, where it appeared about the middle of May, the males arriving two weeks in advance of the females. It soon became very common, frequenting open places in the woods, and usually taking its station at the extremity of some dead branch. Its note is 'a sharp cutting *pee-ee-e*, the second syllable with a falling, the last with a rising, inflection.' On June 9 while descending the mountain Mr. Frazar found these Flycatchers common to its base as well as afterwards at Triunfo and San José del Rancho. An adult female killed on June 20 at Triunfo was incubating, but no nests were found."

The measurements of three eggs, all that I have been able to secure, are 19.0 by 15.6, 18.7 by 15.6, and 18.5 by 15.3 millimeters.

NUTTALLORNIS MESOLEUCUS (Lichtenstein)

OLIVE-SIDED FLYCATCHER

PLATES 42–44

HABITS

As the nomenclature of the 1931 Check-List is being followed in this work, the above name heads this chapter. "More recently, however, Van Rossem has located Lichtenstein's type in the Berlin Museum to find that it is a species of South American flycatcher. This circumstance allows return again to the familiar name *borealis* as the specific term for this attractive flycatcher" (Wetmore, 1939). This illustrates the instability of the supposedly stable scientific names; fortunately, the good old common name has proved more permanent. Dr. Wetmore also states, in the same paper, that in his opinion the recognition of a western race of this species is not warranted.

The generic name *Nuttallornis* seems to have been well chosen, for, although the first specimen was taken by Richardson on the banks of the Saskatchewan and described by Swainson (Swainson and Richardson, 1831) as *Tyrannus borealis*, only one specimen was taken and nothing was learned about its habits, and it was Thomas Nuttall who secured the next specimens and gave us the first account of the habits of the species. Supposing it to be a new species, Nuttall

(1832) named it *Muscicapa cooperi* and says: "This undescribed species, which appertains to the group of Pewees, was obtained in the woods of Sweet Auburn, in this vicinity, by Mr. John Bethune, of Cambridge, on the 7th of June, 1830. This, and a second specimen, acquired soon afterwards, were females on the point of incubation. A third individual of the same sex was killed on the 21st of June, 1831." Nuttall showed Audubon his first specimen of this species in Brookline, Mass., and the latter (1840) gives an amusing account of the capture, on August 8, 1832, of the bird from which his plate was drawn.

The olive-sided flycatcher is now known to enjoy a wide distribution as a summer resident in the coniferous forests of Canada, central Alaska, and Newfoundland and to extend its range southward in the Canadian and Transition Zones into some of the Northern States, and, mainly along the mountain ranges, in the east as far as North Carolina and in the west nearly to the Mexican border. It migrates through Mexico and Central America, to spend the winter in northern South America.

The characteristic and favorite haunts of the olive-sided flycatcher are the opener coniferous forests of the northern woods, where tall spruces, firs, or balsams lift their towering spires along the edges of clearings, or "deadenings," about the shores of wilderness lakes, along the banks of wooded streams, or around the borders of northern bogs and muskegs. In such places the loud, ringing, strongly accented notes of the male may be heard long before his upright form is discovered perched on some dead branch or the top of some tall dead tree, whence he can sally forth in pursuit of his insect prey or warn his mate of approaching intruders. His territory is well guarded, and no others of his own kind are allowed to nest in his vicinity.

Dr. Samuel S. Dickey tells me that, "although ordinarily closely associated with evergreen trees," these birds do, even in Canada, occasionally "depart from such cover and are observed to feed and disport themselves among growths of aspens, poplars, birches, maples, and mountain-ash trees."

On several occasions the olive-sided flycatcher has remained to breed for a few seasons in parts of eastern Massachusetts, far south of its normal breeding range, establishing itself temporarily in situations quite different from those mentioned above. The birds discovered by Nuttall (1832) were evidently established for at least three years, 1830 to 1832, "in the solitude of a barren and sandy piece of forest, adjoining Sweet Auburn," in an area "circumscribed by the tops of a cluster of tall Virginia junipers or red cedars, and an adjoining elm, and decayed cherry tree." Whether they continued to nest there does not seem to be known; at least it is not recorded. But William Brewster (1906) found them breeding there in 1867 and has published several more nesting records for that general region up to 1877 or 1878. He says: "Since then no one, so far as I am aware, has noted the Olive-sided Flycatcher in the Mount Auburn region, even during migration."

In 1900 and 1901 we located two pairs of olive-sided flycatchers breeding in the vicinity of Triangle Pond in Plymouth, Mass. This is rather high, rolling, sandy country, covered with a scattering, open forest of pitch pine (*Pinus rigida*) and a rather dense undergrowth of scrub oaks and other underbrush, with several ponds in the hollows. These birds continued to breed here until 1905, but since then, though I have hunted the region repeatedly (and such noisy birds could hardly be overlooked), I have never seen them there. Outram Bangs found two sets of eggs in the adjoining town of Wareham, where the forest growth is similar, in 1892 and 1894, but, though I have also hunted this country quite thoroughly, I have found no evidence that olive-sided flycatchers have nested there since that time. Apparently this species no longer breeds in eastern Massachusetts.

Dr. Dickey writes to me: "In the Canadian red-spruce (*Picea rubens*) belts of West Virginia and western Maryland, where a few pairs of olive-sided flycatchers annually summer and breed, I have observed them among those singular bog formations locally known as 'cranberry marshes,' or 'The Glades,' at altitudes above 2,700 feet." They also inhabit "fringes of evergreens" along some of the rivers, and "certain taller groves of spruces in Garrett County, Maryland." He also found a few pairs spending the summer and breeding, for several seasons, "in the white pine and hemlock forests along the upper reaches of the Allegheny River in Pennsylvania."

S. F. Rathbun says in his notes: "When I came to Seattle more than 40 years ago, the shoreline of Lake Washington presented an almost unbroken line of forest. At that time the species was abundant, and every mile or so of the woods along the lake appeared to have its pair of these birds. This flycatcher is more or less common in the Puget Sound region and is well distributed. I have found it from as high as 4,000 feet (and no doubt it ranges higher) to the very border of the Pacific Ocean; where one walks along the beach, the calls of this bird often mingle with the long, slow wash of the waves on the sands."

Major Bendire (1895) says: "In Colorado the Olive-sided Flycatcher reaches an altitude of 9,000 or 10,000 feet in summer. In the San Pedro Martir Mountains, in Lower California, Mr. A. W.

Anthony informs me that this Flycatcher was occasionally observed by him up to 11,000 feet, and evidently nesting."

The olive-sided flycatcher is a typically boreal bird. It would seem from the above statements as to its haunts that its chief requirements for a summer home are coolness, the presence of coniferous trees, and the proximity of water; all these are combined in its more northern haunts, and in its mountain resorts it enjoys the coolness of the higher altitudes. But it seems rather strange to find it breeding contentedly almost down to sea level in the San Francisco Bay region and even in the city of Berkeley, as will be referred to later. Here the cool, often foggy, climate seems to offer a congenial summer home among the pine and eucalyptus groves, with suitable nesting sites in the Monterey cypresses.

Courtship.—Dr. Dickey sends me the following brief note on this subject: "It arrives from its winter quarters from May 15 to June 1 and is pretty sure to be found year after year in favorable habitats. Then it is that the males are brimful of vivacity. They combat for the possession of territory, sometimes dashing at each other and actually causing the feathers to fly. At this season males are seen to pursue females across the canopies of evergreen forests. Throughout a period of at least two weeks such immoderate activities prevail among these birds. They then have finished their mating and have chosen their breeding sites."

Nesting.—The first nests were found by Nuttall (1832) at Mount Auburn, near Boston, Mass., in 1831 and 1832. The first was "discovered, in the horizontal branch of a tall red cedar 40 or 50 feet from the ground. It was formed much in the manner of the Kingbird, externally made of interlaced dead twigs of the cedar, internally of the wiry stolons of the common cinquefoil, dry grass, and some fragments of branching *Lichen* or *Usnea.*" The nest that he found in 1832 was similar, but only 14 or 15 feet up in a "small juniper."

In the same locality, in 1867, Mr. Brewster (1906) found two nests "on the horizontal branches of isolated pitch pines." And in 1868, in the same locality, he again found two nests; "one, with three fresh eggs, was in an apple tree (an unusual situation), near the extremity of a long, drooping branch and about twelve feet above the surface of a little pond"; the other was "in the same small pitch pine in which one of the nests of the preceding year had been placed. On both occasions the birds built so very near the ground that I could look into their nest by pulling down the slender branch on which it rested."

The seven nests that we found near Triangle Pond in Plymouth and the two found by Mr. Bangs in Wareham, Mass., were all similarly located and made of similar materials. They were all in pitch-pine woods, where the trees were small or of moderate size, seldom over 30 feet in height, and more or less scattered, with an undergrowth of scrubby oaks and other underbrush. The nests were usually not far from the shores of small ponds, and almost invariably on the lowest limb of the tree, often on a branch overhanging a woodland path. The heights from the ground varied from 7 to 20 feet, and they were placed from 3 to 5 feet out from the main trunk, usually not far from the end of the limb. If it had not been for the noisy activities of the birds in advertising the fact that we were near their nests, we should have had great difficulty in finding them, for they were very well concealed. They were all on horizontal branches, usually in a cluster of upright twigs, raised an inch or two above the branch and well hidden among the pine needles; it was sometimes necessary to climb a tree and examine a suspicious-looking bunch before we could recognize it as a nest. The nests were mainly made of *Usnea barbata*, the familiar hanging moss or lichen of that region, on a foundation of dead twigs, mixed with straws, rootlets, and dead pine needles; there was usually a smooth lining of the usnea, but sometimes a few fine rootlets, grasses, or pine needles appeared in the lining. The birds were very noisy and aggressive in the defense of their nests, often snapping their bills and dashing at the intruder, but oftener only indulging in loud vocal protests. We found that after a nest had been robbed a new nest was built and a second set of eggs laid within about three weeks or less. The three beautiful eggs, lying in a shallow cup of pale green usnea and surrounded by a framework of dark green pine needles make a most attractive picture.

The only other nest that I have seen was found near Bay of Islands, Newfoundland, on July 1, 1912. It was 12 feet from the ground on a drooping limb of a small balsam fir in a scattered open growth of coniferous trees; it was 4 feet out from the trunk, contained 3 fresh eggs, and was similar in construction to the Plymouth nests. Mr. Brewster (1937) mentions several nests found near Lake Umbagog in Maine that were 25 to 50 feet above ground in red spruces. Dr. Dickey (MS.) says that in West Virginia and Maryland these flycatchers nest in the red spruces (*Picea rubens*) and the intermediate fir (*Abies intermedia*) and that in Pennsylvania they are partial to the hemlock trees (*Tsuga canadensis*). In Canada, he says, they are notably partial to the spruce timber, the white spruce (*Picea canadensis*), and the black spruce (*Picea mariana*).

"The nest itself, which is completed in mid-June and even up into July, is a rather loose weave of dead spruce or other evergreen twigs, masses of cones, mats of the gray-green *Usnea barbata*, or old-man's-beard lichens, rootlets, and pads of moss, such as knight's-crest (*Hypnum*) and hair moss (*Polytrichum*). It is shallow inside and lined with lichens and pine needles or rootlets." A nest that he measured was 6 inches in outside diameter and 2½ inches in inside diameter; the outside depth was 2⅛ inches and the inner depth 1 inch. This seems rather large, as nests that I have measured have been 5 inches or less in outside diameter; and Bendire's figures agree very closely with mine.

Robie W. Tufts tells me that nests of this species have been found in spruce, fir, hemlock, apple, elm, and locust trees, but a fair estimate for Nova Scotia would be 24 out of 25 in conifers.

The above data all refer to eastern nestings; some western nesting sites and nests are quite different. Chester Barlow (1901) says that in the Sierra Nevadas "the average heights of nests of this species seems to be from 60 to 70 feet, firs being the favorite tree." He mentions a nest, found by Mr. Carriger, that was 72 feet up in a Douglas spruce, and one that he collected himself at about the same height in a silver fir. "The nest was composed of rootlets with which was mixed a quantity of bright yellow dry moss (*Evernia vulpina*) so common in the Sierras." At the other extreme, as to height, is a nest in the Thayer collection, taken by C. I. Clay near Eureka, Calif., that was only 5 feet from the ground in a redwood sapling; this nest is made of small, dry twigs, coarse weed stems, mixed with a small quantity of a mosslike lichen; the lining of the nest is smooth and firm, but the material is only slightly finer. J. E. Patterson sent me a photograph of a nest at Buck Lake, Oreg., that was on the tip of a horizontal branch of a lodgepole pine, 30 feet above ground. Another nest in the Thayer collection, taken by F. M. Dille in Estes Park, Colo., is made of coarse and fine dead twigs, weed stems, and grasses and is lined with very fine grasses and rootlets, but with no moss or lichens.

The presence of the olive-sided flycatcher in summer in the San Francisco Bay region, notably in Berkeley, had been noted by a number of observers before Donald D. McLean and Joseph Dixon actually discovered a nest there. Dixon (1920) reports that on June 12, 1920, a set of four slightly incubated eggs was secured by them "from a slender Monterey cypress that stands on the south-facing hillside just north of the Claremont Hotel, in Berkeley." "The nest," he says, "was placed fifty-seven feet above the ground, by actual measurement, and thirty inches from the tip of a long slender upper branch of a broken-topped cypress. The situation was exposed, but the brooding bird was partially screened from above by an overhanging branch. The nest was firmly ensconced on top of a cluster of twelve cypress cones, the main limb itself at this point being insufficient, as it was only one-half inch in diameter. The foundation of the nest consists of dead bare cypress twigs and a few dry grass stems. It is lined with fine dry pine needles, stiff fibrous rootlets, and horsehair. The outside dimensions of the nest are 6×6½×2½ inches * * * and the inside dimensions, 3½×3¾×1½ inches."

Charles Piper Smith (1927) reports finding two California nests in deciduous trees; one was on "a drooping limb of an alder, *Alnus rhombifolia*, the distance from the ground about fifty feet, the locality a canyon floor at the eastern base of Black Mountain (Monte Bello Ridge), Santa Clara County, elevation probably about 200 feet. * * * Coast redwoods were available in the canyon, but these birds evidently used the alder by pure choice." In another canyon at an elevation of about 1,800 feet he found a nest about 60 feet up in the top of a canyon live oak (*Quercus chrysolepis*). And Allan R. Phillips (1937) tells of finding a nest, near Flagstaff, Ariz., "10 feet up in a small crotch against the trunk of a small Gambel oak. The tree was too slender to be climbed. * * * The nest, placed on the east side of the tree, was composed of twigs and yellow-pine needles, with a grass lining and rim decorations of mossy, green rootlets."

Mr. Rathbun, in his notes from Seattle, Wash., says: "The nests we have taken were outwardly made of small dry conifer twigs, closely interlaced with coarse, dark rootlets, and all had a lining of fine, black rootlets. Once, from a distance of less than 20 feet, I watched one of these flycatchers at work on its nest. It lightly settled upon it, then used its bill to press into the inner wall of the nest the material it had brought, after which the bird settled in the nest and by a circular motion of its body smoothed the material into place; and while this was taking place, it continued to give its note in a soft key."

Eggs—The olive-sided flycatcher lays almost invariably three eggs, occasionally as many as four; Bendire (1895) says that about one set in 20 contains four eggs; I have never heard of more than four. The eggs are usually ovate, rarely somewhat short-ovate or nearly elongate-ovate. They are practically lusterless. They are among the most beautiful of birds' eggs, especially when fresh enough for the yolk to show through the delicate shell and enrich its coloring. The ground color varies from creamy white to "light buff" or "pale ochraceous-salmon." This is rather lightly, but conspicuously, marked with small blotches or spots, usually concentrated in a ring or loose wreath about the larger end, of various bright browns, such as "hazel," "cinnamon," "russet," or "chestnut," and with different shades of "vinaceous-drab," "cinnamon-drab," or "drab-gray," in underlying spots or small blotches.

Mr. Dixon (1920) describes his oddly marked set as follows: "The ground color of the eggs is normal for the species, being light ochraceous salmon, but the markings of the eggs are odd. Instead of being wreathed about the larger end with clusters of fairly well defined spots, all four eggs have a single heavy splotch or smudge, six by ten millimeters in extent in one case, on one side or surface of the egg, while the opposite surface is practically unmarked. These splotches are light vinaceous drab, fading about the edges to cinnamon-rufous."

The measurements of 50 eggs average 21.7 by 16.1 millimeters; the eggs showing the four extremes measure 23.7 by 16.9, 23.1 by 17.3, 20.1 by 15.5, and 21.2 by 15.2 millimeters.

Young.—Burns (1915) gives the period of incubation as 14 days. Bendire (1895) says "not over fourteen days. The young are said to remain in the nest about three weeks." In the nest that Nuttall (1832) watched, "the young remained in the nest no less than 23 days, and were fed from the first on beetles and perfect insects, which appeared to have been wholly digested without any regurgitation. Towards the close of this protracted period the young could fly with all the celerity of the parents; and they probably went to and from the nest repeatedly before abandoning it. The male was at this time extremely watchful, and frequently followed me from his usual residence, after my paying him a visit, nearly half a mile."

Anna Head (1903) watched a brood of fully fledged young, of which she writes:

They were a pretty sight as they stretched their little wings, craned their necks, and tip-toed along the fir-twigs. They were rather more brightly colored than their parents, whose plumage was somewhat worn at that season. * * * Only the Yellow Gape showed immaturity, and they spent a great deal of time preening their glossy feathers. The parents visited them often, catching insects and delivering them on the wing, with a light, swallow-like action. * * *

The next morning the young took their first flight, already seeming quite expert, and choosing bare twigs to perch on, like all their race. They gave the characteristic, three-syllabled call clearly the first day, though more softly than their parents. For more than a week the family kept together near the nest. The last part of the time there seemed to be a good deal of flutter and scolding going on. I think the old ones were trying to induce the young to catch their own game.

Dr. Mearns (1890) says: "Like many other mountain species [in Arizona] it ranges down hill with its young after the breeding season. On Oak Creek, in the cypress belt below the pines, it appears in families during the first half of August." Probably before the end of that month, both old and young have started on their southward migration.

Plumages.—I have not seen the natal down or any but fully fledged young. In full juvenal plumage the young bird is essentially like the adult in color pattern, but darker above and brighter below. The pileum, wings, and tail are deep "clove brown," and the rest of the upper parts are deep "olive-brown"; the under parts are "primrose yellow," narrowed to a median line on the breast by "olive-brown" streaking on the throat and sides; the median and greater wing coverts are edged with "ochraceous-buff," and the secondaries and tertiaries are tipped with brownish white.

Dr. Dwight (1900) says: "First winter plumage acquired by a late postjuvenal moult beginning in September which possibly is complete. I have seen no extra-limital specimens but I should expect to find them retaining the brown wing edgings. Pale wing bands are probably acquired at this moult when young birds become practically indistinguishable from adults. First nuptial plumage acquired apparently by wear. Birds return from the south in fresh little worn plumage, the young birds with a dull clay-colored lower mandible. Old worn feathers may be found mixed with the new in some specimens, very strongly suggestive of a recent limited prenuptial moult."

Older birds have a complete postnuptial molt, which may begin in August, but probably oftener in September, or after they have migrated southward. They may also have a partial prenuptial molt, as suggested above. The sexes are alike in all plumages. Fall birds, in fresh plumage, are more brightly colored than spring birds.

Food.—Professor Beal (1912) examined the contents of 69 stomachs of the olive-sided flycatcher, collected in 12 States and in 3 Provinces of Canada. In his summary he writes:

The food of this bird is interesting, as it represents the food of a typical flycatcher. With the exception of the vegetable matter in 1 stomach, everything it eats could be taken on the wing. Caterpillars, spiders, and millepeds, although found in the stomachs of most flycatchers, are entirely absent. * * *

The most prominent fact in the food habits of the olive-sided flycatcher is its consumption of honeybees. As it eats no vegetable matter worth mentioning, its record must rest on its insect food, and honeybees constitute entirely too large a quota for the best economic interests. Were the bird as abundant and as domestic as either of the phoebes, there is no doubt that it would be a pest to bee keepers. * * *

Hymenoptera are the staff of life of the olive-sided flycatcher and form a large percentage of the food of each month. The fewest were taken in May, when they amounted to 74.50 per cent. The average consumption for the season from April to September was 82.56 per cent. They were found in 61 of the 63 stomachs, and 26, that is, over 41 per cent of the whole, contained no other food. Of all the birds examined by the Biological Survey, not one subsists so nearly exclusively upon one order of insects. Winged ants were found in 10 stomachs and entirely filled 2 of them. A few useful parasitic species were identified, but more interesting than these were 63 honeybees (*Apis mellifera*), found in 16 stomachs, or 25 per cent of the whole number. Of these, 36 were workers and 27 were males or drones. Thus the bird shows a very decided fondness for hive bees, but not the special preference for drones manifested by kingbirds.

Other animal food included useful beetles, 0.45 percent; harmful beetles, 5.79 percent; Diptera (flies), 0.88 percent; Hemiptera (bugs), 3.25 percent; grasshoppers, 3.36 percent of the food for two months, or 1.12 percent for the season; moths, 4.13 percent; and dragonflies, 1.77 percent. Animal food made up 99.95 percent of the whole. The remainder consisted of some unidentified fruit pulp, spruce foliage, and rubbish.

To partially offset the impression created above by Beal, that the olive-sided flycatcher might prove to be a menace to beekeepers, W. L. McAtee (1926) aptly suggests that "there can be no doubt that most of the bees eaten come from the wild colonies that are so frequent in woodlands, and not from hives on bee farms." As to the economic relation of this flycatcher to woodlots, he says: "Insects injurious to woodlands which were identified in the diet of the Olive-side were carpenter ants, click beetles, adults of both flat-headed and round-headed wood borers, leaf chafers, nut weevils, bark beetles, and cicadas. One of the flat-headed borers (*Melanophila fulvoguttata*) is destructive to hemlock and spruce, while another (*Asemum moestum*) attacks pine, spruce and other trees."

Behavior.—The olive-sided flycatcher seems to prefer the solitudes of the forests to the vicinity of human habitations; the experience with it in Berkeley, Calif., was, however, a notable exception to the rule. In its wilderness home each pair establishes a definite territory from which it drives away any other individuals of its own species and often shows hostility to some birds of other species. Verdi Burtch tells me that Clarence F. Stone saw a scarlet tanager drive one of these flycatchers from one of its perches; but the flycatcher returned later and drove the tanager away. It would probably attack any hawk, crow, or jay that came too near its nest, though no such case seems to be recorded. On the approach of a human intruder, it starts its alarm note, *quip, quip, quip*, repeating it constantly as it flies nervously about, alighting first on one tree and then on another within its chosen territory. From such actions the collector realizes that the nest is near and begins to hunt for it. If he climbs the tree to examine or rob the nest, the bird's activities are intensified; both birds may now attack the intruder, flying about excitedly, snapping their beaks, screaming incessantly, and even darting down at and almost striking the man's head. Some less bold birds are content to perch on nearby trees and scold, with crests erected, bills clicking, and tails wagging.

The olive-sided flycatcher is oftenest seen perched erect on the topmost spire of some tall tree in its chosen haunts, where its presence is first detected by its loud and striking notes long before it is seen.

From this lookout perch it can warn its mate of approaching danger, or oftener sally forth in any direction to snatch its insect prey. Often it returns to the same perch or, if its flight has been a long one, it may select any one of its favorite observation posts. Anna Head (1903) watched one "as it whirled and tumbled in the air in frantic pursuit of a moth, it almost seemed to be coming to pieces, so loosely was it jointed, till a loud *click* of the beak announced success, and in an instant it was back on its perch, looking as if it had always sat there."

W. F. Henninger (1916) gives the following account of an unusual flight behavior that he noted near New Bremen, Ohio, on October 11, 1911: "In the dry tops of two large trees about 17 meters apart from one another, there were two specimens of this species. While the one sat perfectly motionless preening its feathers occasionally, the other one began to fly upward in very short spirals and then to descend in a number of jerky drops with quickly expanded and closed wings. After doing this a number of times it finally flew so high that it disappeared from sight altogether and it did not return at all."

Voice.—Its voice is the most prominent and characteristic feature of the olive-sided flycatcher, by which its presence is oftenest detected. On its breeding grounds its loud, ringing, somewhat musical calls may be heard from the first crack of dawn to the deepening twilight of evening; its notes are so constantly uttered at times that they become fairly tiresome. Mr. Rathbun says, in his notes: "On one occasion, when I was at a mountain lake resort, the owner asked me to 'shoot up' a pair of the olive-sided, which he said made so much noise that they annoyed some of the guests." Grinnell and Storer (1924) say that, in the Yosemite, "the Olive-sided Flycatcher is one of the earliest birds to call in the morning and one of the last to be heard in the evening. This is probably due in some degree to its choice of surroundings, for in the tree tops it is apprized of the coming of dawn long before that news reaches the earthward dwelling species, and in the same places it enjoys the lingering daylight for some time after the glades and thickets below are lost in the shadows of the evening."

Many different descriptions of the voice of the noisy olive-sided flycatcher have been published, far too many to be quoted here, and I have also a number of contributed notes on the subject. The most distinctive note is the 3-syllabled call note, which might be considered the song of the male on the breeding grounds. The first of the three syllables is short and sharp, but not so loud as the others and not audible at so great a distance, so that the note may then appear to have only two syllables. The other two syllables are very loud, emphatic notes and are strongly accented. Perhaps the best rendering of this call is that given me by Francis H. Allen, as *whip-whée-péeoo*, "the *whip* not carrying very far and perhaps sometimes omitted—a loud and striking note." Various interpretations of the same "song" given by others are: *gluck, phe-béa; whit, pray-téar; quip, gree-déal; whet, we-wéa; pit, perwhéer*, etc. The "song" has also been well expressed in words, which convey a very good impression of it, such as "look three deer"; "its me here"; "who be you"; "quick three beers"; "come right here"; and "whew take care."

Another very common note is the warning cry, or the angry note of protest, uttered when its territory is invaded or its nest threatened. I recorded this in my notes as *wheup, wheup, wheup*, or *weap, weap, weap*. But it is oftener written by others as *quip, quip, quip*, or *pip, pip, pip*. This note is short, rather loud, and decidedly emphatic, almost a scream under intense excitement.

Anna Head (1903) says that "a third note was more like a twitter, and was uttered during excitement, chiefly when the young were learning to fly. It sounded like 'why, why, why,' repeated very rapidly a number of times. Sometimes this note was given as a prelude to the real song." A. Dawes DuBois, in his notes, mentions a 2-syllabled "plaintive song having a quality similar to the notes of the wood pewee. The notes may be whistled as *too-wee* in ascending pitch, slightly suggesting the towhee's notes."

Verdi Burtch tells me that the *pip-pip-pip* notes, always in series of threes, seemed to be made mostly by the female in the vicinity of the nest and that all the nests he found were located after hearing these notes. The 3-syllabled call, which he writes as *kuk, pu-wheu*, was given mainly by the male from his high perch. Dr. S. S. Dickey writes to me that the outcries of this flycatcher are "interpreted with variable exclamations, especially when the bird is excited by the combativeness of other males, or is approached on its nesting grounds. Then it becomes actually pugnacious, and its notes are a series of piercing *quoits* and *pee-quoiks*. On clear days these notes may be heard three-quarters of a mile, or even a mile, away."

Field marks.—The erect pose of the olive-sided flycatcher, as it sits on the top of some tall forest tree, its stout figure, dark color, short neck, and large head are all distinctive. If facing the observer, it may be recognized by the white abdominal region narrowing to a point between the dark olive of the sides of the breast and throat. If seen from one side or from the rear, two glaring white patches of silky plumage may be seen, on each side of the lower back, often showing above the wings; these conspicuous patches are excellent field marks and may often be seen in flight. On migrations, when it is mainly silent, the above marks must be looked for, but on its breeding grounds its voice is its most distinctive character.

Fall.—As summer wanes this flycatcher becomes more silent and its note is heard only occasionally. Beginning in August and prolonged through much of September, the leisurely southward migration is in progress from Canada to northern South America. Alexander F. Skutch has contributed the following notes: "Although the olive-sided flycatcher is generally considered to be merely a bird of passage through Central America, I have a few midwinter records from the valley of El General in southern Costa Rica, and Carriger cites a January record for the Cerro de Santa Maria, somewhat farther to the north. Whether on migration or settled in its winter home, this flycatcher is always solitary while in Central America and never associates with others of its kind or joins in the mixed flocks of small birds that attract each a single individual of certain species that, during the winter months, are unsocial with respect to their own kind. It is never abundant; the greatest number of individuals that I have recorded in one day is six, seen on a 6-mile journey from San Isidro del General to Rivas on September 13, 1936, when they had just arrived from the north. If it utters any call while in the Tropics, I have so far failed to recognize it. The olive-sided flycatcher arrives in Central America early in September and sometimes delays its northward migration until the end of the first week of May."

DISTRIBUTION

Range.—North and South America, from Alaska and Newfoundland to Peru.

Breeding range.—The olive-sided flycatcher breeds **north** to Alaska (head of the North Fork of the Kuskokwim River and Circle); western Mackenzie (probably Fort Norman, probably Fort Resolution, and Fort Smith); northern Alberta (Smith Lodge and Fort Chipewyan); northern Saskatchewan (Lake Île à la Crosse, Sandy Lake, and Cumberland House); Manitoba (Trout River); Ontario (English River Post and North Bay); Quebec (St. Margaret and Mont Louis); and Newfoundland (Nicholsville and Base Camp). From this point the range extends southward through the coniferous forests of the **East** to eastern Pennsylvania (Hazelton and Pottsville) and western North Carolina (Black Mountain and Highlands). **South** to southwestern North Carolina (Highlands); eastern Tennessee (Mount LeConte); northern Wisconsin (Kelley Brook); northern Minnesota (Gull Lake); central Colorado (St. Peters Dome and Crested Butte); central Arizona (White Mountains and Flagstaff); and northern Baja California (San Pedro Martir Mountains). The **western** boundary extends northward along the coast from northern Baja California (San Pedro Martir Mountains and Juarez Mountains) to Alaska (Nizina River and the head of the north fork of the Kuskokwim River).

Winter range.—Present information indicates that in winter this species is concentrated in northwestern South America, from Colombia (Cincinnati, Bogota, Minco, and La Concepcion) south through Ecuador (Gualea) to Peru (Chaupe, Huambo, and Chancha). A specimen taken on October 28, 1913, at Galipan, near Caracas, Venezuela, seems to be the most easterly record in the winter range.

Spring migration.—A late date of spring departure from the wintering grounds is March 22, from Cincinnati, Colombia, while it has been observed in migration at San Salvador, El Salvador, on April 24.

Early dates of spring arrival in North America are: Florida—Fort Myers, March 31. Virginia—Wise County, May 5. District of Columbia—Washington, May 9. New Jersey—Englewood, May 12. New York—Rochester, May 14. Massachusetts—Harvard, May 16. New Hampshire—Monadnock, May 18. Vermont—St. Johnsbury, May 19. New Brunswick—Scotch Lake, May 22. Nova Scotia—Halifax, May 26. Quebec—Montreal, May 31. Missouri—St. Louis, May 8. Illinois—Urbana, May 12. Michigan—Ann Arbor, May 1 (exceptional). Ontario—London, May 13. Iowa, Hillsboro, May 15. Minnesota—Minneapolis, May 9. Manitoba—Aweme, May 9. Arizona—Huachuca Mountains, April 20. Colorado—Loveland, May 11. Wyoming—Laramie, May 16. Montana—Columbia Falls, May 21. Saskatchewan—Regina, May 20. Alberta—Belvedere, May 21. California—Pasadena, April 24. Oregon—Corvallis, May 4. Washington—Tacoma, May 15. Alaska—Kenai, May 26.

Fall migration.—Late dates of fall departure are: Alaska—Port Snettisham, August 29. British Columbia—Atlin, September 2. Washington—Prescott, September 5. Oregon—Prospect, September 25. California—Pasadena, September 26. Montana—Fortine, September 10. Wyoming—Laramie, September 18. Manitoba—Aweme, September 4. Minnesota—Minneapolis, September 16. Illinois—Chicago, September 15. Michigan—Sault Ste. Marie, September 15. Ohio—New Bremen, October 11. Quebec—Anticosti Island, September 3. Nova Scotia—Halifax, September 3. New Brunswick—St. John, September 10. Maine—Winthrop, September 13. New Hampshire—Dublin, September 14. Pennsylvania—Erie, September 18. New York—Jamaica, September 26; Alabama (Genesee County), October 17. It has been observed in southward migration at San Salvador, El Salvador, on August 28.

Casual records.—On the north shore of the Gulf of St. Lawrence the olive-sided flycatcher has been detected on several occasions indicating that it may breed in that region. It was recorded from Godbout on June 6, 1883; a pair was seen during several successive days between July 25 and August 10, 1912, on the Natashkwan River; one was seen repeatedly on June 11, 12, and 13, 1921, at Mingan, and another was noted by the same observer at Piashte Bay, on June 16, 1921; while two were seen on June 10, 1927, at the mouth of the Kegaska River.

A specimen was taken at Nenortalik, Greenland, on August 29, 1840.

Egg dates.—California: 48 records, May 20 to July 4; 24 records, June 9 to 25, indicating the height of the season.

Colorado: 16 records, June 16 to July 20; 8 records, June 23 to July 3.

Maine: 13 records, June 12 to July 26.

Massachusetts: 20 records, June 8 to July 10; 10 records, June 15 to 22.

Nova Scotia: 56 records, June 11 to July 8; 28 records, June 19 to 28.

PYROCEPHALUS RUBINUS MEXICANUS Sclater

VERMILION FLYCATCHER

PLATE 45

HABITS

The bird lover, traveling in the hot, semiarid, or desert regions in our extreme Southwestern States or Mexico, will receive one of the thrills of those interesting regions when he sees for the first time this brilliant, flaming gem, with his prominent crest of fiery scarlet and his equally bright scarlet-red breast. *Pyrocephalus*, firehead, is a good name for it. One looks for somber colors in the denizens of the desert and is, indeed, surprised to see this outburst of gleaming color, which seems to outshine even the most brilliant scarlet flowers of the springtime desert. We found it to be an abundant and con-spicuous species in all the lower valleys of southern Arizona, especially so in the valley of the San Pedro River, where willows and cottonwoods with thickets of smaller trees and underbrush grew along irrigating ditches, separating the fertile areas from the more arid surroundings; we also found them near the wooded banks of other streams and along the dry washes on the plains where these extend outward from the mountains; but we seldom saw them even in the mouths of the canyons and never at higher altitudes.

Spring.—Although a few individuals may remain throughout the winter, the vermilion flycatcher is mainly a summer resident in southern Arizona. Major Bendire (1895) says that the "first mi-grants usually return about March 1, the males preceding the females about a week, and by the 10th of the month both sexes are common."

Courtship.—When we arrived in its haunts in April, the males were busy with their courtship flight songs, a curious and most brilliant display. Starting from his perch on the top of some tall weed stalk, or low dead branch, he mounts upward 20, 30, or even 50 feet, in an ecstasy of excitement, his fiery crest erected, his glowing breast expanded, his tail lifted and spread, and his wings vibrating rapidly, as he hovers like a sparrow hawk in rising circles; and at frequent intervals he pours forth a delightful, soft, twittering, tinkling love song, all for the delectation of his chosen mate, clad in somber colors and hidden in the foliage below; then slowly he flutters down to claim her, and the two fly off together, unless perhaps some other more fortunate suitor has won her. William Beebe (1905) has expressed it very well as follows:

Up shoots one from a mesquite tree, with full, rounded crest, and breast puffed out until it seems a floating ball of vermilion—buoyed up on vibrating wings. Slowly, by successive upward throbs, the bird ascends, at each point of vibrating rest uttering his little love song—a cheerful *ching-tink-a-le-tink! ching-tink-a-le-tink!* which is the utmost he can do. When at the limit of his flight, fifty or seventy-five feet above our heads, he redoubles his efforts, and the *chings* and the *tinks* rapidly succeed each other. Suddenly, his little strength exhausted, the suitor drops to earth almost vertically in a series of downward swoops, and alights near the wee gray form for which he at present exists. He watches eagerly for some sign of favor, but a rival is already climbing skyward, whose feathers seem no brighter than his, whose simple lay of love is no more eager, no more tender, yet some subtle fate, with workings too fine for our senses, decides against the first suitor, and, before the second bird has regained his perch, the female flies low over the cactus-pads, followed by the breathless performer.

Nesting.—The nests that we found in Arizona were in willows or sycamores, in horizontal forks, at heights ranging from 8 to 20 feet from the ground, generally not far from an irrigation ditch, stream, or other body of water. Other nests have been found there in cottonwoods, oaks, mesquites, paloverde, and hackberry trees. Mr. Dawson (1923) mentions nests as high as 40 and even 60 feet above ground. Dr. Joseph Grinnell (1914b) found a nest in the Colorado Valley that "was fifty-four inches above the ground, saddled on the bare forking branch of a dead mesquite standing in an open area thirty-five yards from the river bank." And Dr. J. C. Merrill (1878) says that in southern Texas "the nests are usually placed upon horizontal forks of ratama-trees, growing upon the edge of a prairie, and rarely more than six feet from the ground."

With the exception of one nest, in the Thayer collection, which was evidently built on a drooping limb of a sycamore tree and was supported by upward-slanting twigs, all the nests that I have seen or read about were placed in horizontal crotches or forks, generally on rather small limbs. The nest is a flat, well-made structure, usually sunken well down into the fork, so that its rim projects very little above the supporting branch and is very inconspicuous; as viewed from below or from one side it appears like a slight enlargement on the branch, and the materials of which it is made add to the deception.

The foundation and much of the body of the nest consist of short pieces of dead twigs, 1 to 3 inches in length; mixed with these are finer twigs, shreds and pieces of weed stalks, fine grasses, rootlets, and other plant fibers, bits of dry leaves and lichens, fine strips of inner bark, cocoons, and spider webs; the exterior is often, but by no means always, profusely decorated with small bits of lichens; the whole structure is firmly bound together and anchored to the branch with spider webs. The lining consists of finer pieces of similar material, plant down, horsehair, cow's hair, fur, and feathers, with occasional bits of thread or string. Among the feathers used are those of doves, the desert quail, and the yellow feathers of the Arkansas kingbird; some nests have very few small feathers in the lining, and others are profusely lined or decorated with larger feathers.

The outer diameter of the nest usually varies from 3 to 2½ inches, rarely 3½, and its height from 2 inches to 1 inch; the inner cavity is 2 to 1¾ inches in diameter, and is only 1 inch or less in depth.

Eggs.—Three eggs seem to constitute the usual set for the vermilion flycatcher, though sometimes only two are laid and rarely four. The eggs are quite distinctive, being very boldly marked in striking contrasts. They are ovate or short-ovate, sometimes rounded ovate, and are lusterless. The ground color varies from pure white to creamy white, or rarely "cream color." They are usually heavily marked, chiefly near the larger end, with very dark browns, "sepia," "bone brown," or "clove brown," and with underlying, smaller blotches or spots of pale shades of drab or lavender. Occasional eggs are more evenly marked with small spots all over the surface, and in some the markings are concentrated in a well-defined wreath. The measurements of 50 eggs average 17.4 by 13.1 millimeters; the eggs showing the four extremes measure **19.0** by 12.7, 18.0 by **14.2, 15.0** by 12.7, and 16.3 by **12.5** millimeters.

Young.—The incubation period is said to be about 12 days. This duty is performed mainly by the female, although Bendire (1895) says that "the male assists to some extent, as I have on two occasions seen one sitting on the eggs." He also says: "I believe two broods are occasionally raised in a season. On June 6 I found a nest of the Vermilion Flycatcher in a small grove of cottonwood trees, with no other shrubbery nearer than 600 yards; it was placed on a horizontal fork of one of these trees, about 20 feet from the ground, and contained three fresh eggs; close by the male was feeding a full-grown young bird; no other pair appeared to occupy this grove, and it seems very probable that it belonged to these birds. The fact that I also found fresh eggs as late as July 16 further strengthens this supposition." Mrs. Wheelock (1904) writes:

On April 24 the first egg was laid, and one each day thereafter until there were three. Twelve and a half days were required for incubation, and during this time I never saw the male nearer to the nest than six feet. The almost naked nestlings were salmon-pinkish; and, as in the case of most newly hatched birds, the eyes were covered with a membrane. On the fourth day this parted in a slit, giving them a comical, half-awake look, while grayish down stood out thickly on the crown and along the back. On the tenth day they were fairly feathered, but remained in the nest until the fourteenth and sixteenth days, when one and two, respectively, fluttered out on untried wings. The father took charge of the one that left home first, while the patient mother fed and coaxed the lazy ones.

Plumages.—Mrs. Wheelock (1904) describes the almost naked nestlings as salmon-pinkish, with grayish natal down standing out thickly on the crown and along the back. But Mr. Dawson (1923) says that "the chicks are black for a few days after hatching, with some outcropping of white down."

The young bird in juvenal, first plumage differs considerably from the adult female; the upper parts are grayish brown, more golden-brown on the rump, and the feathers are margined with pale buff or whitish, giving a scaled appearance; the median and greater wing coverts, the secondaries, and the tertials are margined with pale buff; the outer web of the outer tail feather is pale buff or whitish; the breast and sides of the abdomen are thickly marked with *rounded* spots of brownish gray, instead of being streaked as in the adult female; the under parts are otherwise white, tinged with pale yellow posteriorly. This plumage is apparently worn for only a short time, and is evidently replaced by a complete postjuvenal molt early in the fall. I have seen a young male completing this molt on October 27; it has completed the body molt, but is still molting the wings and tail.

In this, the first winter plumage, the sexes begin to differentiate; young males are like the females at first, but they soon begin to acquire more or less red on the under parts and in the crown; I have seen a series of young males, taken in October, November, December, February, and March, showing the progress of this change. Probably, however, the fully adult plumage is not assumed until the next post-nuptial molt the following summer or fall. Some say that two years are required to assume the fully adult plumage, but I have been unable to find any young males in spring that still wore the adult female plumage, suffused with salmon-pink or orange-red on the posterior under parts. Probably, however, the most brilliantly colored males are the older birds, as there is much individual variation in the intensity of the scarlet.

I have seen one, apparently adult, male that has a patch of yellow feathers on one side of the breast; and another has the whole pileum "cadmium orange" to "cadmium yellow", the throat "deep chrome", and the under parts "salmon orange", mixed with "deep chrome".

Mrs. Bailey (1902b) mentions a rare melanistic phase of plumage that "is uniform dark brown tinged in male with wine purple on crown and lower parts."

Adults have a complete postnuptial molt in August and September and at least a partial prenuptial molt early in spring.

Food.—No very comprehensive study of the food of the vermilion flycatcher seems to have been made. Bendire (1895) says: "Its food consists of insects which are mostly caught on the wing; but I have also seen it alight on the ground to pick up a grasshopper or small beetle, returning to its perch afterwards, beating its prey against it, and devouring it at leisure." We often saw it darting out into the air, after the manner of other flycatchers, in pursuit of insects, as well as picking them up from the ground. Both large and small grasshoppers are captured, as well as small beetles, flies, and other small flying insects. When living near apiaries, it has been known to kill many honeybees; but otherwise its food habits are probably more beneficial than harmful.

Behavior.—We found this flycatcher to be rather tame and unconcerned about our presence, flitting nervously from one perch to another, from some low tree or bush to a tall weed stalk and then back again, making frequent sallies after insects, or executing his spectacular nuptial flights. The male is a bold and fearless fighter in defense of the nest and rather aggressive against intruders. Mr. Dawson (1923) witnessed the following rather peculiar behavior: "In watching the antics of a certain Vermilion dandy, I saw him resort twice to a tiny fork on a horizontal branch, remote from any possible proximity of a mate, and indulge in a very peculiar set of motions, bowing and turning, and lying supine with outstretched wings and dangling feet. Careful reflection showed the act to be an outcropping, through suggestion, of what we call a secondary sex character, viz., a demonstration of the nest-building instinct, excited by the presence of an especially attractive site."

Voice.—The flight song, given in courtship, is well rendered by William Beebe (1905) as "*ching-tink-a-le-tink.*" Mr. Dawson (1923) writes it "*tutty tutty tutty zzüngh.*" Ralph Hoffmann (1927) calls it "a slight call of two or three notes, *pitt-a-see, pitt-a-see*, jerking his head upward at each utterance." However interpreted, it is a striking song, given in an outburst of ecstasy and with considerable energetic effort; it is given by the male alone.

Both sexes have a short call note that sounds like *pisk*. Mrs. Wheelock (1904) says that the call "is a characteristic loud and constantly repeated 'peet, peet', or 'peet-ter-weet'."

Field marks.—The male is unmistakable with his bright scarlet crown and breast, in marked contrast with his brown back. The demure female is brown above, with indistinct whitish wing bars, a white breast streaked with dusky and with a wash of salmon or yellow on the lower belly. Both have the mannerisms of flycatchers.

Enemies.—Mrs. Wheelock (1904) witnessed a fight between a vermilion flycatcher and an Arkansas kingbird. The flycatcher happened to fly too near to a tree where the kingbird was perched and was immediately attacked by the latter. "The result was a kaleidoscopic mingling of yellow, red, and brown tumbling earthward, the birds fighting as they fell. The Vermilion had been taken by surprise, and was no match for his antagonist, but he fought gallantly. As he landed on his back on the ground, with feet and bill still eager to finish, the kingbird rose a few feet above him, poised over him as a hawk over a field mouse lair, hesitated, and for some occult reason flew back to his own perch."

Dr. Beebe (1905) writes: "This beautiful creature must have had some talisman which guarded him from the fate which overhangs brilliantly coloured birds, for he seemed to have no fear of showing his beauty. There was no attempt at skulking or concealment. * * * Although we watched long and carefully, we never saw a Vermilion Flycatcher assailed or threatened by shrike or hawk. Sometimes a Ground Squirrel rushed at one in a rage, but the bark of a Ground Squirrel is much worse than its bite, so this sham threatening meant little and the flycatcher acted as if he knew it."

Dr. Friedmann (1929) lists this flycatcher as "an uncommon victim of the Dwarf Cowbird." Dr. J. C. Merrill (1878) reported one such case, and Major Bendire (1895) reported two.

Fall.—Mr. Swarth (1904) says: "During August, families of young with the parents in attendance, were frequently seen, and at this time I found them more shy and difficult to approach than at any other. The males are, in my experience, singularly tame and unsuspicious for such bright, gaudy plumaged birds." During September the fall migration begins, southward and westward, and by the first of October very few are left in their summer homes in Arizona and New Mexico.

Winter.—A few individuals remain all winter in southern Arizona, and a few are scattered westward into southern California, even to the coastal counties as far north as Santa Barbara. Strangely enough, the species is not known as summer resident in California, except in the extreme southwestern corner in the vicinity of the Colorado Valley. James B. Dixon tells me that he has seen the vermilion flycatcher in San Diego County only in winter.

DISTRIBUTION

Range.—Southwestern United States; Central and South America; accidental in Louisiana and Florida; not regularly migratory.

The range of the vermilion flycatcher extends **north** to southern California (Cushenbury Ranch); southern Nevada (probably Alamo and Pahrump); southern Utah (probably St. George and the Virgin River Valley); New Mexico (Alma and Mesilla); and southern Texas (San Antonio). **East** to Texas (San Antonio, Somerset, Corpus Christi, and Brownsville); Tamaulipas (Altamira); Quintana Roo (Camp Mengel); British Honduras (Belize and Sibun River); eastern Honduras (Segovia River); northeastern Colombia (Valencia); Venezuela (Altagracia and Ciudad Bolivar); Brazil (Caviana Island, Canuman, Forte de Rio Branco, Cuyaba, and Taquara); and Uruguay (San Vincente). **South** to Uruguay (San Vincente and Paysandu); and central Argentina (Cape San Antonio, Lomas de Zamora, and Victorica). **West** to western Argentina (Victorica, Mendoza, and Concepcion); southwestern Bolivia (Tolomosa); Peru (Lima and Pacasmayo); Galapagos Archipelago (Charles, Narborough, Bindloe, and Abingdon Islands); Guerrero (Chilpancingo); Sinaloa (Escuinapa); Baja California (Santiago, San Joaquin, and probably San Ramon); and California (Mecca Indian Wells, and Cushenbury Ranch).

The range as outlined is for the entire species, which has been separated into several races. Only one, *Pyrocephalus rubinus mexicanus*, occurs in the United States. This subspecies ranges south to southern Mexico.

Casual records.—A specimen was collected at Avery Island, La., on December 22, 1934, and another at Baton Rouge, in that State, on February 7, 1938. One was taken on March 25, 1901, near Tallahassee, Fla.

Egg dates.—Arizona: 70 records, March 4 to July 4; 36 records, April 30 to May 28, indicating the height of the season.

California: 4 records, March 20 to May 9.

Lower California: 36 records, April 7 to June 3; 18 records, May 4 to 18.

Texas: 16 records, March 25 to June 23; 8 records, April 14 to May 3.

CAMPTOSTOMA IMBERBE Sclater

BEARDLESS FLYCATCHER

HABITS

This curious little flycatcher finds the northern limit of its range in southern Arizona and the valley of the lower Rio Grande in Texas; it is widely distributed in Mexico and Central America but is rare and not well known at the northern limits of its range. William Brewster (1882b) attempted to separate the Arizona bird under the subspecific name *ridgwayi;* A. J. van Rossem (1930) concurred at least partially in this view; but Robert Ridgway (1907) did not recognize the northern race; and Ludlow Griscom (1934) confirms Mr. Ridgway's judgment. The 1931 Check-list agrees with the views of the last two authors. As its name implies, the rictal bristles, so prominent in other flycatchers and so useful in capturing their prey, are exceedingly small or nearly lacking in this species.

Dickey and van Rossem (1938) found this flycatcher "a fairly common resident of the coastal plain and locally, where swampy conditions prevail, a short distance inland [in El Salvador]." They say further:

The thinly foliaged, low, deciduous forest along the peninsula of San Juan de Goso was the only locality in which beardless flycatchers were at all common. There, in January, 1927, the sparse scrub along the lagoon contained a pair or more for every hundred yards of beach, and one was seldom out of sound of their sharp, piping call-notes.

Although the very densest jungle is avoided, still, many were heard in the open, middle heights of the swamp forests about Lake Olomega, Puerto del Triunfo, and Rio San Miguel. In such places it is usually difficult to take specimens of this tiny, rather sedentary, and very inconspicuously colored flycatcher. Were it not for the sharp and unmistakable call-notes which draw one's attention, the species could be very easily overlooked.

In Arizona we failed to find, or to recognize, the beardless flycatcher, though we spent much time in the valley of the Santa Cruz River, which Mr. van Rossem (1936) says "is evidently the center of its range in Arizona." And my companion, Frank C. Willard, evidently had never seen it, though he had collected in this region for several years. Probably we both overlooked it. However, Mr. van Rossem (1936) says:

I believe this species to be common in southern Arizona, and that the chief reason why it has not been detected more often is its close resemblance in color, size, and call notes to the Verdin.

I first heard the "verdin" notes of this species at Continental on April 24, 1931, and caught occasional flashes of a bird in a dense patch of mesquite and second growth cottonwoods along the nearly dry stream bed. On the 26th, Dr. Miller and I went back to the same spot and succeeded in taking both members of a pair which was nearly ready to breed. We next met the Beardless Flycatcher in a grove of oaks, alders, and sycamores near the mouth of Madera Cañon in the Santa Ritas, at an altitude of about 4000 feet and just at the juncture of Lower and Upper Sonoran Zones. * * * At Tumacacori we found Beardless Flycatchers to be common in the groves of cottonwoods and willows along the dry river bed. I estimated the population to average a pair to every quater-mile for at least two miles either way from our camp. However, pairs were by no means regularly spaced.

Austin Paul Smith (1916) obtained specimens of this flycatcher in January and February near a salt-water estuary 30 miles north of the Rio Grande, in the vicinity of Harlingen, Tex. He writes: "This diminutive bird showed a persistent partiality for the low bushes that constitute the greater portion of the chaparral of the region, never being observed in arborescent growth, although trees grew rather plentifully along the Arroyo, and some large mesquites were scattered through the chaparral proper. This was at variance with my previous experience with the species in Mexico; and it is quite likely that I would have overlooked the bird entirely had its notes not given me the clew. In size, color, and movements, the Beardless Flycatcher bears a superficial resemblance to several other small birds thronging the chaparral during the winter, such as the Verdin, Orange-crowned Warbler, and Ruby-crowned Kinglet."

Nesting.—What were probably the first two sets of eggs, with the nests, of this rare little flycatcher to be collected, were taken by Gerald B. Thomas in British Honduras, near the Manatee River, on May 7 and 16, 1906, and were purchased by Col. John E. Thayer (1906). Mr. Thomas wrote to Col. Thayer as follows:

The first set of Beardless Flycatchers was taken from a nest in a small palmetto about 4½ feet from the ground. The palmetto was on the edge of quite a clump of its kind and was situated in a flat sandy stretch of low land about five miles from the coast. The nearest fresh water was about two miles away.

The other nest was in a similar location about two miles from where the type was found. This nest was about 7 feet from the ground and only a few rods from a freshwater creek. Two other nests—old ones—were found and both were built in palmettos, one about 12 feet from the ground and the other about 6 feet.

The parent birds were very bold and perched within two feet of the nest while I was examining it, continually uttering their clear piping call and ruffling the feathers on their heads into a small crest. The female sat very close and almost allowed herself to be touched before flying.

Both parents were collected with the first set. I have examined these two nests, which are very much alike in situation and in composition. They are nearly globular, with the entrance partially on the side away from the stem of the palmetto, which came with the nest. They are rather firmly lodged between the bases of the thorny stems of the palmetto fans. They are well made and deeply hollowed in a downward-slanting direction and are composed mainly of the reddish-brown fibers from the stem of the palmetto, strips of inner bark, shredded weed stems, bits of lichens, and other vegetable fibers. The first nest is well lined with what the collector calls "cottony seed fibers of orchids"; the other has two feathers of some dove and a green feather of a parrot in the lining. The outside diameter is about 2 by 2½ inches, and the inner cavity is fully 2 inches deep.

Three Arizona nests, the only ones for which we have the data, are all quite different from the Honduras nests in situation and composition, but all seem to be much like each other in these respects.

Mr. van Rossem (1936) says: "A nest was found on May 13. It was about twenty-five feet above the ground and was tucked into the pendant stems of a clump of mistletoe which grew at the tip of a cottonwood branch. * * * It was a thick-walled, rather loosely packed, four-inch globe of grasses and fine weeds with the interior well padded with plant-down, feathers and a small amount of rabbit fur. The entrance was in the side, slightly above center."

H. H. Kimball took a set of eggs of the beardless flycatcher near Tucson, Ariz., on July 10, 1922, that is now in the P. B. Philipp collection in the American Museum of Natural History, in New York. Mr. Amadon tells me that, according to the data furnished with it, the nest was "in a bunch of mistletoe in a large ash tree on a high point of land between two forks of a creek, 40 feet from the ground, 12 feet from the tree trunk."

There is another set of eggs in the Thayer collection, taken by Mr. Kimball near Tucson on June 25, 1923; the nest was 50 feet from the ground and 30 feet from the trunk of a large cottonwood; the nest came with the set and appears to have been built in the midst of a clump of mistletoe, as it is surrounded with the twigs, leaves, and blossom stalks of the mistletoe, now faded to a warm buff color; the nest is similar to the one described by Mr. van Rossem, with the mistletoe material added to the exterior.

Eggs.—Perhaps the normal set of eggs for the beardless flycatcher consists of three, but incubated sets of two and even one have been collected. What few eggs I have seen, only six, are ovate and without gloss. They are pure white and are finely sprinkled, chiefly about the larger end, more sparingly elsewhere, with small spots or minute dots of dark brown, light brown, or reddish brown and shades of drab. The measurements of 9 eggs average 16.5 by 12.2 millimeters; the eggs showing the four extremes measure 17.0 by 12.7, 17.0 by 13.1, and 14.8 by 11.4 millimeters.

Young.—William Brewster (1882b) says: "On May 28 Mr. Stephens met with a young bird which had but just left the nest. It was accompanied by the female parent, who showed much solicitude and frequently uttered her shrill cries, to which the offspring responded in nearly similar tones. Both individuals were secured, but neither the nest nor the remainder of the brood—if indeed there were any more—could he find. On the following day this episode was repeated, a second female being found in attendance on another young bird of nearly the same age as that obtained on the previous occasion."

Plumages.—According to Ridgway (1907), young birds in juvenal plumage are "essentially like adults, but general color of upper parts nearly hair brown, the pileum concolor with back, etc., and the

wing-bands and edges of remiges wood brown or cinnamon." In
the young birds that I have seen, the wing bands are decidedly
"cinnamon," and the edgings on the secondaries and tertials are
merely "wood brown" or pale buff. How long this juvenal plumage
is worn we do not seem to know.

Adults evidently have a complete molt late in summer or in fall,
perhaps before they go south. I have seen adults in badly worn
plumage in July; and Mr. Swarth (1929) reports a female, shot on
September 13, that had nearly finished the annual molt and was in
fresh fall plumage. I have also seen adults, taken in April, that are
in worn plumage, which may indicate a partial prenuptial molt.
Adults in fall plumage are more decidedly olivaceous above and
more yellowish below than spring birds.

Mr. Ridgway (1907) says that "some males have the feathers of
the pileum much longer, especially on the occiput, forming a very
distinct crest, and these elongated feathers are sooty in color, margined with grayish olive." But Dickey and van Rossem (1938)
write: "The variation in the color of the crown seems to be individual
and not due to age or sex, as supposed by Ridgway. Some fully
adult males which have completed the fall molt and are, therefore,
more than one year old, have the pileum absolutely concolor with
the back, and others (as certainly adult) have very dark, almost
sooty crowns. The same is true of females. An immature male
in first winter plumage is average in color, that is, with the crown
slightly, but obviously, darker than the rest of the upperparts."

Food.—Allan R. Phillips wrote to me that he sent two stomachs
of the beardless flycatcher, collected near Tucson, Ariz., to the Biological Survey for analysis. These have recently been examined by
L. W. Saylor, and Dr. Clarence Cottam has sent me the results of
the analysis. Both stomachs contained 100 percent animal matter.
The contents of one were: Coccidae, 55 percent, five or more scales
and fragments thereof; Membracidae, 15 percent, fragments of head
and eye; one Lepidoptera larva, 22 percent; pupal fragments of two
Diptera pupae, 8 percent; and a trace of insect chitin. The other
contained: Coccidae, 48 percent, at least 14 scales and fragments;
two Lepidoptera larvae, 12 percent; two Coccinellidae larvae, 20
percent; three or more *Crematogaster* species, 5 percent; two Membracidae, 5 percent; and undetermined animal debris, 10 percent.

Probably much of the above food was gleaned from the foliage
and the branches of trees and shrubs, for Mr. Phillips tells me that
it often gleans after the manner of kinglets, vireos, and warblers.
And Austin Paul Smith (1909) says: "In many instances have I
watched this mite simulate the Vireo's habit of branch inspection,
in the same time-careless manner." Evidently insects are not the
sole food of this flycatcher, for he "found it the premier seed-eater

of the family." And Dr. Beebe (1905) says: "All of this species which we saw later were feeding on small berries and not on insects."

Behavior.—Most observers seem to agree that this little flycatcher likes to spend most of its time well up in the tops of the trees, where its small size, inconspicuous coloring, and superficial resemblance to some other small birds make it very difficult to detect. But in southern Texas Austin Paul Smith (1916) found that it showed "a persistent partiality for the low bushes" in the chaparral. Others have noticed that it often resorts to the mesquite thickets and other small trees, probably for feeding purposes. Mr. Smith (1909) writes of its habits in southern Mexico:

You cannot be in these parts long before you detect a very peculiar bird note, the author of which may perhaps be detected in the nearest tree; for the Beardless Flycatcher (*Camptostoma imberbe*) is of a friendly disposition at times. Impressions of early acquaintance would class him as a Flycatcher, Vireo, or Titmouse, dependent upon his action at the time of your observation. The flycatcher nature is less in evidence than the other two. In many instances I have watched this mite simulate the Vireo's habit of branch inspection, in the same time-careless manner. And again, I might be startled by a titmouse-like note from the brush near at hand, only to discover a chickadee-mimic in Camptostoma. Where observed following the Tyrannidae instincts, it was from the tops of the tallest trees, when it remained very quiet.

Mr. Stephens told Mr. Brewster (1882b) that "they were very shy, and specimens were obtained only at the expense of much trouble and perseverance. Their loud calls were frequently heard, but when the spot was approached the bird either relapsed into silence or took a long flight to resume its calling in another direction. In their motions they resembled other small Flycatchers, but their tail was less frequently jerked."

Voice.—All who are familiar with the beardless flycatcher in life have remarked on its loud, sharp, far-reaching notes, which are sure to attract attention and help in locating the inconspicuous little bird. Mr. Brewster (1882b) writes: "The males had a habit of perching on the tops of the tallest trees in the vicinity of their haunts, and at sunrise occasionally uttered a singular song which Mr. Stephens transcribes as '*yoop-yoop-yoope-deedledeeè*, the first half given very deliberately, the remainder rapidly.' A commoner cry, used by both sexes in calling to one another, was a shrill '*pièr pièr pièr pièr*,' beginning in a high key and falling a note each time."

Mr. van Rossem (1936) refers to "verdin-like notes"; but Mr. Phillips tells me that "the call notes have none of the sharpness of the verdin's, and the usual one is more musical than the verdin's. Possibly, of course, the species has some verdin-like *alarm* note near the nest which I have never heard." A. P. Smith (1909) writes one note as "seetee-tee-tee-tee, often kept up continuously for five minutes"; and again, he (1916) sets down what is probably a call note, as "*whee-e-oop*, often repeated, and distinguishable at a considerable distance."

Field marks.—The beardless flycatcher is evidently not easy to recognize in life, as it has no very striking color characters. Its resemblance in size, general color, and behavior to the verdin, small vireos, warblers, and kinglets has been mentioned, and it might easily be overlooked on this account. But, Mr. Phillips suggests (MS.) that there is no close resemblance in color to the verdin, the tail is much shorter and less active and the bill shape is different. Although tiny, it is after all a flycatcher and shows many of the characteristic traits and colors of that group. Perhaps its characteristic voice is the best field mark.

DISTRIBUTION

Range.—Southern Arizona south to Costa Rica.

Breeding range.—The beardless flycatcher nests **north** to southern Arizona (Tucson) and southern Texas (Lomita and Hidalgo). **East** to Texas (Hidalgo); Quintana Roo (Cozumel Island); British Honduras (Manatee River); and Costa Rica (Tenorio). **South** to Costa Rica (Tenorio, Bebedero, and Carillo); Oaxaca (Tapanatepec and Chivela); Morelos (Cuernavaca); and Nayarit (San Blas). **West** to Nayarit (San Blas); Sinaloa (Mazatlan); Sonora (San Javier) and Saric); and Arizona (Papago Indian Reservation and Tucson). A specimen taken March 1, 1911, 60 miles north of Tucson, is the northernmost record.

Winter range.—The species appears to be nonmigratory in the southern part of the range, although in winter it withdraws entirely from the United States. At this season it has been detected **north** to Sonora (Alamos); southern Tamaulipas (Altamira and Tampico); and Quintana Roo (Cozumel Island).

Migration.—The species has been noted at Lomita, Tex., as early as March 2 and at Tucson, Ariz., on April 22. No data are available for the fall migration.

Egg dates.—Arizona: 3 records, May 13 to July 10.

British Honduras: 2 records, May 7 and 16.

Family ALAUDIDAE: Larks

ALAUDA ARVENSIS ARVENSIS Linnaeus

SKYLARK

CONTRIBUTED BY FRANCIS CHARLES ROBERT JOURDAIN

HABITS

The skylark has a double claim to a place in the American list, for it is said to have occurred, as a genuinely wild bird, accidentally in Greenland and in Bermuda. The Greenland records are, however, not very satisfactory; Herluf Winge (1898) relegates them to the hypothetical list at the end of his book, as unsubstantiated. The Bermuda record rests on better evidence; H. B. Tristram, writing to Dresser, says that he secured one, a storm-driven waif, in Bermuda; and this has been generally accepted. The other ground for recognition is the fact that it has been introduced and is resident on Vancouver Island, British Columbia, and was introduced at several localities in the United States, but failed to establish itself permanently, though for several years it was resident and bred on Long Island, N. Y.

If we include the *gulgula* forms as subspecies of *Alauda arvensis*, we shall find that some race or other occurs everywhere not only in the whole of Palearctic region, but also in a great part of the Indo-Malayan subregion. Altogether, nearly 30 forms have been separated, but here we are concerned only with the typical race, a common bird throughout the British Isles and the greater part of Europe, with the exception of the Mediterranean region, where it is replaced by other races.

Skylarks are birds of the open country. The long hind claw shows at a glance that we are dealing with a terrestrial rather than an arboreal species. True, I have occasionally seen one perch for a few moments on the flat, closely cropped top of a quickset hedge, but it was obviously uncomfortable, and, though the word "never" is a dangerous one for the ornithologist to utter, one may assert with some confidence that it does not perch on trees. Even otherwise suitable country, divided up into small plots, with high hedges and tall hedgerow timber, is avoided by them. Grassy downs, wide spreading cornlands, broad meadows, marshy flats, sandy coasts, barren heaths, and rough mountain pastures are the chosen home of the skylark, though it does not penetrate far into the heather-clad hilltops, which are the chosen haunt of the meadow pipit (*Anthus pratensis*).

Spring.—In the British Isles the skylark is a resident and a partial migrant. Severe weather, especially deep snow, will drive them off their territory and force them to work southward and westward in search of uncovered ground. In mild winters, especially in the south, they may be seen (and frequently heard) in every month of the year, except for a short break in August and September when they are passing through the molt. A fine day, even in January, on the south coast of England will bring many a cock skylark up into the air to cheer us with his tinkling song; and the wanderers, driven south by stress of weather, lose no time in working their way northward again as soon as conditions improve and the green fields again become visible. As a partial vegetarian the skylark can always pick up a living when he can see the ground.

Courtship.—An early breeder, the skylark begins courtship as soon as winter shows signs of relaxing its grip. When two males are in pursuit of the same female the trio may come hurtling over the hedgerow even along the highway, oblivious of the passersby. One gets a momentary glimpse of a scuffle on the ground when they have passed by with broken song notes audible as they fly. Then one bird drops out, and after a time we hear the clear, sweet notes of the song as the cock rises higher and higher in the air. It is true that there is not a great deal of range in the notes or variety in the song, yet, as the notes come to us from the sky overhead there is an exhilarating tone about them that is in keeping with the season. Up he goes, higher and higher, sometimes swinging round in a circle, and then begins gradually to drop again to earth, finally by a steep descent before "flattening out" to alight.

In the actual display the cock runs round the hen with raised crest and drooping wings; the tail also is expanded. The hen is less demonstrative but is said to quiver her wings before mating.

Nesting.—The usual nesting site is either in meadow grass or among crops; it may also be found exposed on sandy ground or occasionally among stones on a shingle bed. The nest itself is an artless affair—merely a hollow in the ground with a lining of roots and grasses and finer material, sometimes, but not always, with horsehair added. Except when among crops that are regularly worked over, only a small proportion of the nests is ever found, for, as a rule, the nesting pairs are well spaced out and in grass or growing crops finding the nest is largely a matter of chance.

Eggs.—The normal clutch is three or four; in some districts five are quite a rarity, but in others occasional. Six and even seven in a nest have occurred, and certainly in some cases the extraordinary resemblance left no doubt that they were produced by one hen. An instance of eight eggs was, however, due to two females. Normally the eggs are neither beautiful nor interesting, often with a yellow cast, thickly and uniformly covered with pale spots of olive or hair brown, so that little of the the grayish white ground is visible. Yet in a selected series there is remarkable range of variation. Some clutches are pure glossy white without a single mark; others have a few fine brown and gray spots on a white ground; some of the zoned types, in which the dark brown markings are concentrated in a wide zone round the big end, are decidedly handsome; one set, with almost black zones and a white ground, is very striking. There are also warm red-brown sets, and a few cases are on record in which the markings and even the tinge of ground color are purely erythristic.

Some eggs have a very green tinge; others show dark blotches on every egg in the clutch.

The shape also varies a good deal; besides the usual ovate, almost spherical eggs have occurred, as well as extremely elongated ones. One hundred British eggs, measured by the writer, averaged 23.77 by 17.05; the four extremes measuring 26.6 by 17.3 and 22.8 by 18.4; 21.8 by 16.2 and 25.1 by 15.3 millimeters.

Young.—The eggs are laid on consecutive days and incubation begins on the completion of the set. All the evidence goes to show that incubation is performed by the hen only. As to the period, I have before me a number of field observations, and in every case the results vary between 11 and 12 days.

Yet incubator tests by W. Evans (1891) showed 13 to 14 days, and one would have expected them to be slightly less, as the absences of the hen to feed must result in frequent cooling of the eggs and consequent prolongation of the period. Continental observers also usually give 14 days, but whether from original observation or not is not stated. The young usually hatch out on the same day and, as soon as their down has dried, present a curious appearance. They are covered with long, very pale straw-colored down, so that when one glances down at the nest nothing is visible but a mass of hairy down. The only sign of life is the breathing of the young birds. When one or the other of the parents arrives at the nest (for both sexes share in feeding the young) all is changed, for three or four yellow mouths, widely opened, spring into sight. The cock and hen do not fly directly to the nest but alight at some distance and work their way toward it on foot, adding to the store of insects carried in the bill on the way. Generally each bird has a favorite track by which it approaches the nest.

The young develop rapidly, and shortly before they leave the nest they acquire the habit of running out to meet the parents. It is not uncommon to find the nest deserted after 9 or 10 days, but at that age the young are quite unable to fly and are skulking in the neighboring herbage; the actual fledgling period is about three weeks. Two broods are regularly reared and occasionally three.

Plumages.—The juvenile plumage is not unlike that of the adult, though less distinctly marked; and it is completely molted in August–September. For full description of the plumages, see "The Handbook of British Birds," by H. F. Witherby, F. C. R. Jourdain, N. F. Ticehurst, and B. W. Tucker, vol. 1, 1938.

Food.—This has been studied volumetrically by W. E. Collinge (1924–27) on the basis of stomach contents of 69 adults and 36 nestlings. There is a slight preponderance of vegetable food, 54 percent, over animal matter, 46 percent. An analysis of the vegetable matter showed that weed seeds figured largely, 43.5 percent; grain, 9.5 percent; and leaves of crops, 1 percent. Of animal matter, 35.5 percent was injurious insects, 3.5 percent neutral, and 2.5 percent beneficial insects; worms, 2 percent; slugs, 1 percent; and miscellaneous matter, 1.5 percent.

The food of nestlings was entirely beneficial, consisting of insects and larvae. It is evident that on the whole this species is beneficial, but in hard weather, when large flocks are present, considerable damage may be done to leaves of autumn-sown young corn, and an appreciable amount of grain is taken. During a brief spell of Arctic weather on the south coast of England in December 1939, when the ground was covered with a few inches of snow down to the seashore, skylarks were migrating along the coast westward in countless thousands, and where green crops (cabbages and sprouts) stood out above the snow, the ground was brown with larks, even in small gardens in suburban districts, where normally the birds are never seen. The leaves of the cabbages were rapidly reduced to mere skeletons, and apparently the whole crop was ruined, but, as the hearts were seldom reached and usually only the side leaves destroyed, the plants subsequently recovered. Some of the dead birds picked up were mere skeletons, obviously starved to death; and, from observations in Kent and southeast England, it became evident that many of these birds were immigrants from the Continent.

Among insects taken we have records of Collembola, Orthoptera, Hemiptera (Aphides), Lepidoptera (including *Hepialis*, also various larvae), Coleoptera (*Phyllotreta*, larvae, etc.), Hymenoptera (*Ichneumon*, Formicidae), and Diptera (larvae). Spiders and millipeds also are taken.

Voice.—The song of this species has been described already, as it forms an integral part of the courtship, and serves as an indication of territory already occupied. One remarkable point about it, which has not been mentioned, is the remarkable length of time during which it may be continued. Continuous song by one male for at least 7 to 10 minutes, and possibly longer, is not infrequent. O. G. Pike's statement that it has been known to remain in the air for half an hour is probably due to observations on more than one bird. The way in which it rises higher and higher, singing all the time, till it can only just be distinguished as a mere speck in the sky, is unparalleled by any other bird. The duration of the period is also long. H. G. Alexander (1935) records full song from February to June (inclusive) and again through October, and partial song from November to January and in July; so that August and September are the only months when it is silent, and then only for most of the time. The other note, which one frequently hears, is described by B. W. Tucker (1938) as a "liquid, rippling 'chirrup'" and is uttered at intervals during the low, undulating flight.

Field marks.—In size the skylark is about 7 inches long. It is readily distinguished from the woodlark by its longer tail and by having only a faint light streak through the eye. Its greater size and more deliberate movements distinguish it from the pipits, while it has also a distinct crest, which is not nearly so long as the crested lark but which is frequently raised during times of excitement. The two outer tail feathers are white, but the rest of the coloring is not very distinctive, as the general effect is various shades of earthy brown, with light under parts.

Fall.—The abnormal migration witnessed during hard weather in winter has been described already under "Food", but besides this there is a normal southward trend of birds from the more exposed parts of northern Scotland; some of these birds apparently cross the Channel or migrate west to Ireland. There is also a great immigration from September to November on our east Britain coast from the Baltic countries, but the presence of these visitors is not normally noticeable on our south coasts, except under conditions already described. Migration takes place both by day and night, but preferably by day.

DISTRIBUTION

Breeding range.—This includes the whole of the British Isles, the Faeroes, and on the Continent **north** to the shores of the Arctic Ocean, latitude 71° N. in Norway, 68½° in Sweden, Finland (scarce in east and north), and northern Russia. Its southern limits extend through middle Europe **south** to southern France, northern Italy, Austria, northern Hungary, and central Russia. It is replaced by local races in the Iberian Peninsula, northwest Africa, most of Italy, and southeast Europe, as well as in Asia.

Winter range.—This includes the greater part of the British Isles, while typical birds occur in the Iberian Peninsula, northwest Africa east to Cirenaica, also in Italy, most of the Mediterranean islands (Balearic Island, Corsica, Sardinia, Sicily, and Malta), the Balkan Peninsula, and the Caucasus in southern Russia. It has also been recorded in winter from Madeira and the Canary Islands.

Egg dates.—In the British Isles the first eggs may be found quite exceptionally late in March, but most eggs are laid from late in April or in May onward to June and July. Late nests have been recorded in August, September, and even October (young in nest in Orkney Isles on September 19, and 3 eggs nearly fresh on October 17 in Lancashire).

England: 5 records, March 26 to April 26; 34 records, May 1 to 27 (22 after May 16); 16 records, June 1 to 17; and 4 records, June 30 to July 20.

In northern Europe the dates are rather later, from the beginning of May to July in Finland.

OTOCORIS ALPESTRIS ARCTICOLA Oberholser

PALLID HORNED LARK

HABITS

This large, pale race of the horned larks is about the size of *alpestris alpestris*, but is decidedly paler. Dr. Oberholser (1902) says of it: "This form is one of the best marked of all of the races of *Otocoris alpestris*, differing from the typical subspecies in its very much paler upper surface, more pinkish nape, upper tail-coverts and bend of wing, as well as in the pure white of throat and eyebrow. * * * This is the race to which, through misapprehension of the identity of Dr. Coues' type, the name *leucolaema* has, by common consent, been applied. Examination of the rediscovered type, however, proves it to belong to another race, as fully explained under its proper heading, and leaves the present subspecies without a name."

At the time Dr. Oberholser described it, its breeding range was supposed to be confined to Alaska, "(chiefly the interior), with the valley of the Upper Yukon River." Since then it has been found, apparently breeding, above timberline on several of the interior mountain ranges of Alaska, and in Alpine-Arctic regions in the mountains as far south as Washington.

In the Stikine River region of southeastern Alaska and northern British Columbia, Harry S. Swarth (1922) reports it as "seen in small numbers on the mountain tops above Doch-da-on Creek. There, on July 11 and again on July 23, they were found on the open, moss-covered slopes above timber line, associated with rosy finches and pipits."

It remained for Dr. Walter P. Taylor (1925) to extend the known breeding range of this lark to a point much farther south than any of the localities in British Columbia from which it had previously been reported. He writes:

During field work in the State of Washington in 1919 we found the subspecies breeding well south of the international boundary line. On August 5 of that year a small Horned Lark, as yet unable to fly, was captured at an altitude of 7,300 feet on Panhandle Gap, Mount Rainier. The locality is well above timberline, in the Alpine-Arctic Zone, and a favorite resort for Ptarmigan and mountain goat. The so-called Gap is in reality a broad ridge, to the north dropping off abruptly to the Sarvent Glaciers, on the south sloping gently to Ohanapecosh River. Although the date of capture of the young bird (August 5) seems a little late, at least for localities at lesser altitudes, the season on Panhandle Gap was at its height, and the ground, only recently uncovered by the snow, was blanketed with grass and flowers. On being picked up the young Horned Lark disgorged three locust-like insects and a small green worm.

The mother remained close at hand while we watched the young bird, uttering a solicitous call-note resembling *chipew, chipew.*

Specimens taken in this and neighboring localities proved to be typical of *arcticola*. He sums up the status of the pallid horned lark in Washington as follows: "The subspecies occurs as a common migrant and breeding bird at least from July to September in the Alpine-Arctic Zone of the Cascade Mountains south at least to Mount Rainier (Taylor); east to Chopaka Mountain (Taylor); and west to Mount Baker (J. M. Edson); in winter, as early as November and probably to March, it is found in the lowlands of eastern Washington, north and east to Cheney, south to Walla Walla (Lyman); and west to Moses Lake (Cantwell) and Benton County (Decker); it is of accidental occurrence, during migration, in western Washington (A. K. Fisher)."

I cannot find any information on the nesting of this subspecies except the following brief statement by Maj. Allan Brooks (1909): "Mr. [C. de B.] Green this year took the eggs of the Pallid Horned Lark on the high mountains above timber line, between the Okanagan and Similkameen valleys and collected the female, which is now in my collection. This is the breeding form on all the high mountains of the Province, *Otocoris a. merrilli* being restricted to the arid lower levels; nowhere do their breeding ranges impinge on each other."

I do not know what became of the eggs collected by Mr. Green, who is not now living.

Probably the nesting habits of this race and its eggs are not materially different from those of *alpestris* or *hoyti*.

The measurements of three eggs from the Pearson Mountains, British Columbia, are 23.7 by 16.5, 23.2 by 16.1, and 23.3 by 15.5 millimeters; three eggs from the Ashuola Mountains, Wash., measure 24.1 by 16.5, 22.8 by 16.5, and 22.8 by 16.0 millimeters; these are well within the limiting measurements of eggs of the northern horned lark and, in fact, average about the same in size.

The plumage changes of the pallid horned lark are apparently similar to those of the other races of the species. Mr. Swarth (1922) writes: "Two adult males, taken July 23, are beginning the annual molt, shown mostly in the wing coverts. The young bird, taken July 23, is in juvenal plumage throughout. Compared with the young of various southwestern subspecies of *Otocoris alpestris*, it is extremely dark colored. Ground color of the upper parts is blackish, throat and lower belly are white, and there is hardly a trace of rufous or vinaceous anywhere."

The food and general habits of this race are probably similar to those of the other northern subspecies, with due allowance for the difference in its alpine habitat.

HOYT'S HORNED LARK

HABITS

Dr. Louis B. Bishop (1896) named this bird in honor of his friend, William H. Hoyt, of Stamford, Conn., and described it as "similar to *Otocoris alpestris* but with the upper parts generally paler and more gray, the posterior auriculars gray rather than brown, and the yellow of the head and neck replaced by white, excepting the forehead, which is dirty yellowish-white, and the throat, which is distinctly yellow, most pronounced toward the center."

The breeding range is generally understood to extend from the west shore of Hudson Bay westward to the valley of the Mackenzie River, northward to the Arctic coast, and southward to at least Lake Athabaska. Just where it intergrades with *arcticola* on the west, or with *leucolaema* and *praticola* on the south, does not seem to be definitely known. It is the central one of the three northern races, occupying most of central-northern Canada. This is just where we might expect to find an intermediate form. Reference to the comparative descriptions of the three races by Oberholser (1902) and Ridgway (1907) will convince the reader that it is strictly intermediate in all of its characters between the dark *alpestris* and the pale *arcticola*. The wisdom of describing and naming an intermediate race seems open to question, as it immediately produces two more sets of intermediates. Such intermediate forms seem to occur in regions north of Hudson Bay, such as Baffin Island and Southampton Island. J. D. Soper (1928) collected 36 specimens on Baffin Island, of which he says:

The great majority are typical *O. a. hoyti*, are as large as *alpestris*, but have white eyebrows and much white on face and sides of neck. Five specimens from Nettilling lake represent birds found associating with the typical *O. a. hoyti*. These five, if not typical breeding *alpestris*, are much nearer to that race than to *hoyti*. Amongst the specimens are several that are intermediate between these extremes. A male specimen taken at Nettilling lake, June 25, has a pure white instead of well-marked yellow, throat and seems indistinguishable from typical *O. a. arcticola*. A few other birds have white feathers in mosaic pattern, over the yellow throat, suggesting a mixture of bloods rather than a fortuitous development of white feathers.

Referring to the above remarks, Dr. George M. Sutton (1932) says: "I think Mr. Soper is wrong in calling 'the great majority' of this series *hoyti*, or in referring to them as having white eyebrows and white faces. In looking at these birds in an off-hand way, their faces *do* appear to be rather pale; but when compared with the Southampton breeding birds, the decidedly *yellow* face and *heavy, stubbed bill* show up immediately; and furthermore, when compared with a series of seven breeding birds from the Labrador coast (supposedly *alpestris*) they do not appear to be in any way *greatly* dissimilar from them." Referring to Southampton Island, he says:

Hoyt's Horned Lark is a common summer resident in the region of South Bay. It is not so common farther west, at Capes Low and Kendall, and I do not know whether it occurs at all in the extreme eastern, higher part, where its place may be altogether taken by *alpestris*. Unfortunately no summer collecting was carried on about East Bay and Seahorse Point, so I do not know which race breeds there. *Alpestris*, apparently, is the only form which occurs there as a migrant.

Hoyt's Horned Lark arrives in the spring a little later than the Snow Bunting and Lapland Longspur, and departs somewhat earlier and more definitely in the fall. It has never, to the best of my knowledge, been recorded in winter.

Frank L. Farley tells me that both *alpestris* and *hoyti* are common summer residents at Churchill, Manitoba.

Spring.—Referring to the spring passage of the pallid and Hoyt's horned larks at Aweme, Manitoba, Stuart and Norman Criddle (1917) write:

They usually arrive within a few days of each other and with the Lapland Longspurs in large flocks about April 6. Soon the ploughed fields are swarming with them and their value as destroyers of noxious weed seeds must be considerable. * * *

It is an interesting sight to see these birds, in company with thousands of Longspurs, circling for miles around some large hawk, though their object in doing so is a mystery and seems to be almost ignored by the hawk. Their music, as they fly around in millions, fills the air, producing an effect which is long remembered. Both Horned Larks and Lapland Longspurs may also be seen to rise some 30 feet, uttering as they drop a short song. It is evident, however, that this is only a prelude to what is to come when the birds reach their true homes.

Taverner and Sutton (1934) found these larks very numerous when they arrived at Churchill, Manitoba, on May 28. "They were everywhere, feeding confidingly with Snow Buntings even about the doorsteps of the offices and workshops of the townsite. They sang more persistently and finely than we had ever heard them before. The male of a pair nesting close to our Churchill camp habitually perched on the ridge-pole of the tent and sang continuously for many minutes, deserting his post only for momentary feedings or when he flew to the adjoining tennis-court, where he continued to sing. About June 10 the species became less noticeable about the door-yards, but continued abundant all over the tundra." Both races, *alpestris* and *hoyti*, were present on their arrival, but "most of the yellow-faced birds left with the transients."

Nesting.—Dr. Sutton (1932) writes: "The female alone builds the nest, and performs all the duties of incubation. While the male occasionally brings her food, there are regular periods of the day when she leaves the nest to preen, bathe, and feed. The nest is usually built in the open quite a way from water, often on a sloping plain or plateau, and not often on the highest part of a ridge. It is placed in a cup-like depression in the tundra. It is made of stalks of weeds, grasses, and small leaves, lined with soft vegetable material, especially the tassels of 'bog cotton,' and the plumous pappi of some of the flowering plants. I did not note any 'pavements' near nests."

A nest found by Mr. Soper (1928) "was located in low, tundra-like ground, though fairly dry; was built into a small depression on the side of a grassy hummock; fashioned with a thin layer of dead grasses for the walls, and lined on the bottom with white down from the dwarf Arctic willow." This was discovered on June 16 in a small upland valley near Nettilling Lake, Baffin Island; it held five eggs.

Frank L. Farley tells me that at Churchill, Manitoba, the "nests are usually set deep into the tundra and well protected with the last season's growth of grass. They are made of grasses and liberally lined with ptarmigan feathers."

Eggs.—Hoyt's horned lark lays ordinarily four or five eggs. The eggs described by Bendire (1895), as quoted under my account of the northern horned lark, are, of course, referable to this race, as they were taken by MacFarlane near the Anderson River, which is supposed to be within the range of Hoyt's horned lark. The reader is referred to this account, which will apply equally well to most of the races of *Otocoris alpestris*. The measurements of 33 eggs average 23.2 by 16.5 millimeters; the eggs showing the four extremes measure 26.7 by 18.8, 31.1 by 19.0, 21.6 by 16.3, and 23.6 by 15.5 millimeters.

Young.—Dr. Sutton (1932) says: "The first newly hatched young were found on July 3. The period of incubation therefore is probably about thirteen or fourteen days. The young are fed by both parents on insect-food, which is abundant at that season. The fully fledged young go about with their parents during the rest of the season. Unless the spring is unusually early, but one brood of young is raised. The young at the time of leaving the nest are in a much spotted and very pretty juvenal plumage, which is completely moulted in late August, apparently at about the same time the adults perform the post-nuptial moult."

The plumage changes, food, and general behavior of Hoyt's horned lark are apparently similar to those of the other northern races of the species.

Dr. Sutton (1932) says that, on Southampton Island, "its principal enemies are the Parasitic Jaeger, which eats the eggs and captures both young and adults; the weasel, which searches the ground carefully for the nests of small birds; the Arctic Fox; the Snowy Owl, which chiefly captures the young birds at the time its own young have to be fed; and the Duck Hawk. The Herring Gull does not greatly disturb this species, for it hunts chiefly along the lake-shores and coast and not in the high country of the interior.

"Hoyt's Horned Lark leaves Southampton for the south among the earlier fall migrants. It has entirely disappeared before the last of the Snow Buntings, redpolls, pipits, and Lapland Longspurs have gone."

Winter.—According to the 1931 Check-list, Hoyt's horned lark wanders southward in fall and winter to Nevada, Utah, Kansas, Michigan, Ohio, New York, and Connecticut, thus spreading out over a wide winter range and apparently thinly distributed and mixed with some other subspecies. Walker and Trautman (1936) say that in central Ohio, for example—

Hoyt's Horned Lark (*Otocoris alpestris hoyti*) is by far the rarest of the three races that occur in this region. Most of our records are of one or two individuals associated with large flocks of *O. a. alpestris*. These birds,. with the white superciliary line and pale dorsal coloration of *praticola*, but fully as large as the *alpestris* with which they associate, are not difficult to identify in the field. The greatest number recorded, on December 29, 1928, at Buckeye Lake, was five in a flock estimated to contain 100 individuals of *alpestris*. Many large winter flocks of larks which we have carefully examined contained no *hoyti* nor have we found any flocks composed entirely of *hoyti*. * * * The available central Ohio records for this race range from November 26 to March 17. Upon a few occasions we have heard a short song from individuals of this race, and twice our attention was first attracted to the birds by a peculiar quality of the voice which seemed distinctly different from that of *alpestris*.

OTOCORIS ALPESTRIS ALPESTRIS (Linnaeus)

NORTHERN HORNED LARK

PLATES 46, 47

HABITS

The horned larks form a most plastic species that has been split into a large number of subspecies, more or less recognizable, scattered over much of the northern portions of North America, Europe, and Asia. Our northern horned lark, *O. a. alpestris*, stands as the type of the widely distributed species, because the name given by Linnaeus was based on Catesby's bird that was supposed to have come from somewhere in the Carolinas. But the European race, *O. a. flava*, is closely related to it and was once supposed to be identical with it. The northern horned lark is one of the largest and one of the darker-colored races of the North American subspecies. It might well have been called the northeastern horned lark, for Hoyt's horned lark ranges fully as far north and the pallid horned lark ranges much farther north than *alpestris*.

Most of us know the northern horned lark only as a winter visitor, for few of us have enjoyed the privilege of seeing this hardy bird in its summer home on the northern barrens. To my late friend and companion, Dr. Charles W. Townsend (1923), belongs the credit for the discovery of the southernmost breeding station now known of this lark on the barren summits of the Shickshock Mountains, near the northern coast of the Gaspé Peninsula, Quebec, and about 200 miles south of Canadian Labrador. On the tableland above tree limit, on the summit of Mount Albert, 3,640 feet, he found a breeding colony of northern horned larks and pipits. He secured two specimens of the bird for identification, and says:

It was breeding in considerable numbers, for, at a very moderate estimate, I concluded there were twenty pairs. I saw several full-fledged young, and the old birds flew about with insects in their bills, scolding me anxiously. Occasionally I heard the flight song and saw the bird high in the air. * * *

The summit of Mount Albert consists of a table-land some fifteen miles in extent, rising a little at the edges to plunge down in chasms and precipices. Protected by the northern rim of hard schists is a straggling forest of black spruce and fir, rising to a height of five or six feet, with tops blasted by the arctic gales, and, on its southern edge, a little lake imbedded in the mossy and grassy tundra. Beyond are great plains of brown serpentine rock masses, riven' and heaved about by the frost, and beyond are other plains that appear almost as green and smooth as a lawn.

The flora is arctic in character, and comprises many species common to Labrador, such as curlew-berry, Labrador tea, pale-leafed laurel, moss campion and creeping birch and willows.

Late in the spring and early in the summer of 1909 Dr. Townsend and I cruised along the south coast of the Labrador Peninsula, the north shore of the Gulf of St. Lawrence, traveling from Quebec to Esquimaux Point by steamer and from there to Natashquan, about 85 miles, in a small sailboat. On our arrival at Esquimaux Point, on May 24, we saw small flocks of northern horned larks; these were evidently migrating birds, for on our return there on June 2 these birds had all left. At Natashquan, on June 1, we collected a pair on the open, dry tundra near a small pond a short distance inland; they were evidently breeding there, for the female showed the well-known signs of incubation. This is probably the western limit of the regular breeding range of the northern horned lark on this coast; from this point eastward the coast becomes more progressively Arctic in character. Audubon's breeding record was much farther east, near Bras d'Or.

In 1912 I spent the month of June in Newfoundland, where I found the northern horned lark living and probably breeding on the treeless and tundra-like plains about Gafftopsail near the center of the island. During that same season I cruised down the northeast coast of Newfoundland Labrador with Capt. Donald B. Mac-

Millan as far north as Okak. We found a few pairs of these larks scattered all along the coast, but more commonly from Battle Harbor to Nain, wherever they could find the open and exposed situations that they like on the treeless coastal strip or on the rocky, moss-covered tops of the numerous islands. They seem to prefer the barren hilltops, where large beds of reindeer moss or other lichens of various colors partially cover the rocks and tundra and where there are no trees except the diminutive dwarf willows and birches, which grow only a few inches high and spread out over the ground, prostrated by the Arctic gales.

Spring.—Most of the wintering flocks of horned larks leave New England before the end of April, though a few may linger well into May. We found a few small migrating flocks at Esquimaux Point, Quebec, on May 28, but by June 2 they had left. Lucien M. Turner says in his unpublished notes that in the vicinity of Fort Chimo, Ungava, "these birds are common in the spring migration only, arriving just after the middle of May"; but he found them breeding later on near the mouth of the Koksoak River. Taverner and Sutton (1934) found *alpestris* migrating in company with *hoyti* on May 28 at Churchill on the west coast of Hudson Bay. Dr. Sutton (1932) collected two from a flock of five males on South-ampton Island on May 19, which were evidently migrating; he thought that *alpestris* occurs on this island only as a migrant, *hoyti* being the breeding form there. Some of J. D. Soper's birds, collected on Baffin Island, seemed to him to be referable to *alpestris*, though Mr. Soper (1928) says that the "great majority" of the breeding birds are typical of *hoyti*. Evidently these two races mingle on migration and probably intergrade in the regions north of Hudson Bay.

Courtship.—I saw the courtship flight song of the male when I was in Labrador, but, as I could not hear it very well, I prefer to quote Dr. Townsend's (Townsend and Allen, 1907) excellent account of it:

The bird suddenly mounts high into the air, going up silently in irregular circles, at times climbing nearly vertically, to such a height that he appears but a little speck in the sky, several hundred feet up. Arrived at this eminence he spreads his wings and soars, emitting meanwhile his song, such as it is— one or two preliminary notes and then a series of squeaks and high notes with a bit of a fine trill. The whole has a jingling metallic sound like distant sleigh bells, although the squeaks remind one strongly of an old gate. The whole effect, however, is not unpleasant,—even melodious. Having finished one bar of his song, he flaps his wings a few times, closes them and sails again, repeating the song. One bird repeated his song twenty-four times and remained in the air one and a half minutes; another remained in the air three minutes, during which he repeated his song thirty-two times. During all this time the bird is flying in curves or irregular circles, sometimes in straight lines, or if the wind be strong, he heads up into it and remains in the same place. The performance ended, he plunges head foremost down to earth, reaching it in a marvelously short space of time. The descent is as silent as the ascent.

Nesting.—Audubon (1841) gives the first account of the nesting of this species, as follows:

The Shore Lark breeds on the high and desolate tracts of Labrador, in the vicinity of the sea. The face of the country appears as if formed of one un-dulating expanse of dark granite, covered with mosses and lichens, varying in size and colour, some green, others as white as snow, and others again of every tint, and disposed in large patches or tufts. It is on the latter that the Lark places her nest, which is disposed with so much care, while the moss so resembles the bird in hue, that unless you almost tread upon her as she sits, she seems to feel secure, and remains unmoved. * * *

The nest is imbedded in the moss to its edges, which are composed of fine grasses, circularly disposed, and forming a bed about two inches thick, with a lining of Grouse's feathers, and those of other birds.

Townsend and Allen (1907) say: "At Frenchman's Isle on July 16th, we found the nest of a Horned Lark composed of dry grass and a few large feathers, deeply sunk into the reindeer lichen and moss in a level piece of ground. There was no shelter or covering of any sort. It contained three dark-skinned young, clothed sparingly in sulphur-yellow down. Their eyes were not yet open." Bendire (1895) quotes, from some notes sent to him by E. A. McIlhenny, an account of a nest found by him on an island near Cape Charles Harbor on July 18, 1894, as follows: "The nest was embedded in a slightly inclining bank of moss and entirely below the surface of the moss; it contained five richly marked eggs, slightly incubated. When I found the nest it gave me the impression of being very small for the bird; but this was due to the fact that the entrance was small, and the hollow was enlarged under the moss. The nest was deeply cupped, having a thickness of about 1 inch of fine dry grass; it was lined with the down from reindeer moss and the white feathers of Ptarmigans." C. W. G. Eifrig (1905) reports a nest found near Cape Chidley: "The nest, placed on the ground, partly sunk in the moss, is made of moss, plant stems, grasses, finer toward the cup; this is lined with feathers and caribou hair. The outside diameter is 5 in., of cup 2 in., depth of cup, 1.75–2 in., outside depth, 2–2.50 in."

Eggs.—The northern horned lark lays three to five eggs, probably oftener four than any other number. Major Bendire (1895) gives a very good description of the eggs of the "pallid" horned lark (=*hoyti*), which would apply equally well to those of this and the other races. He says that they—

are mostly ovate in shape, less often elongate ovate. The shell is close grained, rather strong, and shows little or no gloss. The ground color is mostly drab gray, sometimes grayish white; in an occasional specimen a faint greenish tint is perceptible, which fades out in time. The entire surface of the egg is profusely blotched and sprinkled with different shades of pale brown. In some specimens the markings are bold and well defined; in others they are minute, giving the egg a pepper-and-salt appearance; and again they are almost confluent, causing a uniform neutral brownish appearance. In some specimens the markings are heavier and become confluent about the larger axis of the egg, forming a wreath and leaving the ground color on the smaller end of the egg plainly visible; in fact, there appears to be an endless variation in color and markings as well as in size among these eggs and scarcely any two sets are exactly alike.

The measurements of 29 eggs of the northern horned lark average 22.6 by 16.5 millimeters; the eggs showing the four extremes measure **24.1** by 16.5, 24.0 by **17.4**, **21.3** by 16.8, and 23.5 by **15.5** millimeters.

Young.—Audubon (1840) writes:

The young leave the nest before they are able to fly, and follow their parents over the moss, where they are fed about a week. They run nimbly, emit a soft *peep*, and squat closely at the first appearance of danger. If observed and pursued, they open their wings to aid them in their escape, and separating, make off with great celerity. On such occasions it is difficult to secure more than one of them, unless several persons be present, when each can pursue a bird. * * * By the first of August many of the young are fully fledged, and the different broods are seen associating together, to the number of forty, fifty, or more. They now gradually remove to the islands of the coast, where they remain until their departure, which takes place in the beginning of September. They start at the dawn of day, proceed on their way south at a small elevation above the water, and fly in so straggling a manner, that they can scarcely be said to move in flocks.

Plumages.—The nest found by Townsend and Allen (1907) "contained three dark-skinned young, clothed sparingly in sulphur-yellow down." The juvenal plumage, which is acquired before the young bird leaves the nest and is alike in both sexes, is a fine example of concealing coloration, for it blends in so well with the surrounding lichens and mosses as to make the bird almost invisible in its open and unprotected nest. The crown is dark brown, almost black, and spotted with brownish white; the back is slightly lighter brown, mixed with dusky, and each feather is tipped with a spot of yellowish white, giving the whole upper surface a conspicuously spotted effect; there is a subterminal black bar on each of the scapulars with a broad terminal margin of yellowish white; the lesser and median wing coverts have large terminal spots of yellowish white; the greater wing coverts and remiges are margined with brownish buff; a superciliary stripe and a spot below the eye are pale yellow, as are the chin and throat, this color extending up the sides of the neck almost to the superciliary stripe; the chest is pale brownish buff, spotted with dusky; the rest of the under parts are very pale yellow or yellowish white.

A complete postjuvenal molt late in summer, mainly in August, produces a first winter plumage, in which the sexes are different. Dr.

Dwight (1900) describes the first winter male as "above vinaceous buff, brightest on nape, vinaceous cinnamon on rump flanks and wing coverts streaked on head and back with sepia. Forehead, lateral 'horns', lores, auriculars and triangular breast patch black, veiled by overlapping pale buff or pinkish feather tips. Wings deep sepia, primaries much darker, edged with whitish, the rest of the wing feathers edged with vinaceous cinnamon. Tail brownish black, the outer rectrices edged with white, the middle pair paler, broadly edged with pinkish Isabella color. Below, dull white, the chin, sides of the head and forehead strongly suffused with lemon or canary yellow, a buffy band across breast below the black patch, flecked with dusky spots."

The first winter female is similar but lacks the black forehead, which is streaked instead, the breast patch is smaller, the back is more streaked, and the colors are duller. The first nuptial plumage is acquired by wear, with possibly some slight evidences of a prenuptial molt; the wearing away of the light-colored feather tips brings the black areas into clear prominence; and old and young birds look much alike, though young birds usually show more dusky streaking on the chest and flanks than adults.

Adults have one complete postnuptial molt on their breeding grounds, about August, and the nuptial plumage is acquired by wear, as in the young birds. Adults in fall are quite similar to the young at that season, the black areas being obscured by brownish tips, the yellow areas deeper yellow, and the white of the under parts more or less streaked with grayish brown.

Food.—W. L. McAtee (1905), in his paper on the relation of horned larks to agriculture, publishes a long list of the vegetable food, mainly seeds, and the animal food, mainly insects, eaten by these birds, most of which does not apply to the northern horned lark. He has much to say about the injurious effect of weeds on agriculture and the cost to farmers in their control. Horned larks feed largely on seeds, perhaps mainly weed seeds, and so do many other birds, but I have always felt that the good that birds do in destroying weed seeds is a myth. Nature is so prolific in the production and so effective in the distribution of the seeds of plants, that only an infinitesimal percentage of those distributed can possibly find room to germinate; and no matter how many the birds pick up, there are always many times more than enough to cover the ground with verdure in a remarkably short time. Has any one ever known of a case where birds have kept even one square yard of ground free from weeds by eating the seeds? I certainly have not. Therefore, it seems to me that the eating of weed seeds is a neutral rather than a beneficial factor in the economic status of birds.

Cottam and Hanson (1938) have published a paper on the food of some Arctic birds, in which they give the results of the examination of three stomachs of this lark, taken at Indian Harbor, Labrador, in July. This shows what a large percentage of the summer food consisted of insects. They say:

Adult and larval Lepidoptera were consumed in numbers by each bird and averaged more than a fourth (27.33 per cent) of the entire amount consumed. These included the genus *Agrotis* sp. and undetermined Geometridae and Noctuidae. Large ants (*Camponotus* sp.) were next in the order of importance of the animal foods, averaging 7 per cent of the total. Other hymenopterous material, including ichneumon wasps, added another 5.67 per cent. A number of dipterous forms were next in order with 4.33 per cent, followed by spiders with 3.33 per cent. Leaf-hoppers, aphids, and other true bugs supplied 2.67 per cent, while mollusks, mostly a small *Mytilus edulis*, made up the remaining 1.67 per cent animal food, making an aggregate of 52 per cent.

Of the 48 per cent vegetable material, 31.67 per cent consisted of fruits and seeds of the bog bilberry (*Vaccinium uliginosum*), while the remaining consisted of cyperaceous seeds and undetermined vegetable debris.

While with us in winter the food must consist almost wholly of seeds, such as waste grain in the stubblefields, the seeds of forage plants in the fields, the seeds of eel grass, sedges, wild oats and mallows along our coasts, various grass seeds, and the seeds of ragweed and a host of other weeds. Probably some dried fruits or berries are eaten, and perhaps some insects in their dormant winter stages.

Most of its food is apparently picked up from the ground, where it spends most of its time walking nimbly along among the stubble, in the short grass or over the salt marshes. Dr. Townsend (1905) says: "It picks at the grass-stalks from the ground, never alighting on them as do the snow buntings and longspurs. It sometimes flies up from the ground, seizing the seeds on the tall grass or weed-stalks, at the same time shaking many off onto the ground, which it picks up before flying up to repeat the process. Horned larks are frequently found in roads picking at the horse-droppings, especially when much snow has covered the grasses and weeds. They also come into the farm-yards for scraps of food."

Behavior.—As we see them in winter northern horned larks are decidedly gregarious, occurring in flocks that range in size from half a dozen to a hundred or more birds; they are seldom seen singly or in pairs as in their summer haunts. As we walk across some flat salt marsh near the shore, or some bare stubblefield farther inland, we may be surprised to see a flock of these birds arise from the ground, where their quiet movements and concealing coloration had rendered them almost invisible. They rise all together, and we hear their faint sibilant twittering as they circle about, now high in the air in scattered formation, now close to the ground in more compact order, showing a bright glimmer of white breasts as they wheel away from us, then suddenly disappearing from our view against the dark background as they turn their backs toward us, and finally vanishing entirely as they all alight on the ground not far from where they started. Their flight is light and easy, with a somewhat undulating motion; and the flocks are rather loose and irregular, yet they are apparently all in touch with each other and guided by a common impulse. As they alight on the ground they scatter out and walk about rapidly on their short legs, taking rather long steps, as shown by the marks of the long hind claw in the soft mud or sand. Horned larks are essentially ground birds; I have never seen one alight in a tree, and, so far as I know, no one else has. The top of a rock, stone wall, or low stump, not over 3 or 4 feet above ground, is about as high as they care to perch, and that not very often. They prefer open ground, especially bare ground or where the grass is short, and they are almost never seen where the vegetation grows rank and high. Among the stubble or tufts of short grass, they walk or run in a crouching attitude, reminding one more of mice than of birds; often they squat and hide until too closely approached. They are not particularly shy, if carefully approached, and seem to feel aware of their ability to conceal themselves in scanty cover. If we remain motionless while the bird is hiding, it will soon lift its head and look about, but at the slightest movement on our part it squats again or runs or flies away.

Dr. Townsend (1905) says: "It is a persistent fighter or extremely playful, whichever you will, and is constantly engaged in chasing its fellows. I have seen two face each other for a moment, with heads down like fighting cocks, the next instant twisting and turning in the air, one in hot pursuit of the other. When in flocks with the other winter birds, they more frequently chase them, especially the smaller Longspurs. I have also seen them chase Snow Buntings, and often Ipswich Sparrows that were feeding with them, and once, what appeared to be a Prairie Horned Lark."

Voice.—Aside from the courtship flight song, described above, the vocal performances of the horned lark do not amount to much. Townsend and Allen (1907) say of other notes, heard in Labrador: "The familiar sibilant squeaking call note was commonly used, and also a note which we do not remember to have heard during the migration in Massachusetts. This sounded like *zzurrit* and was often preceded by another note thus, *whit-zzurrit*. These notes were occasionally so soft and sweet that they recalled the trilling whistle of the Least Sandpiper."

Ralph Hoffmann (1927) writes: "The common note of the Horned Lark is a shrill *tsee*, or *tsée-de-ree*, and a still sharper double-syllabled *ti-sick*. The song is thin and unmusical, suggesting the syllables, *tsip, tsip, tsée-di-di.*" Dr. Townsend (1905) writes it "*tssswee it, tsswt,* the sibilant being marked." This is the note commonly heard in winter, usually uttered in flight, but often given from the ground or from some slight eminence. O. J. Murie writes to me that one of these birds appeared to answer his imitation of its notes.

Field marks.—If the bird is facing the observer, the conspicuous head markings are unmistakable, though in young birds and females, especially in fall and winter, these markings are much obscured. While walking away from the observer on the ground, no conspicuous field marks appear, but a horned lark can generally be recognized by its thick-set appearance, by its habit of walking instead of hopping, and by its mouselike movements. It is larger than the pipit or the Lapland longspur or any of the sparrows with which it is likely to be associated. The pipit has more white in the tail than the lark; and the lark does not wag its tail as the pipit does. When the lark is flying overhead its black tail shows in sharp contrast with the white under parts.

Winter.—The old name, shore lark, seems very appropriate for this bird, for while with us in winter in New England it is far more abundant along the coast than elsewhere. Here it is often associated with the snow buntings and the Ipswich sparrows in the sand dunes and on the beaches, or with the pipits and Lapland longspurs in the salt marshes and meadows. Flat, open, brackish meadows along our tidal rivers are favorite resorts, and the birds are often seen about the shores of lakes and even in stubble fields and plowed lands farther inland.

In Massachusetts the northern horned lark is more abundant as a migrant than as a winter resident, though it is here in some numbers all winter. Dr. Townsend (1905) says of its seasons in Essex County: "During the first half of October, Horned Larks are found in small numbers, but they become abundant in the latter half of the month, increase through November, and reach their height in December. During most of January they are common but in the latter part of that month and in February and early March comparatively few are to be found, while in the latter half of March they again increase in numbers but are never as common as in the fall, and a few may occasionally be found early in April."

Walker and Trautman (1936), referring to its status in central Ohio, write:

The Northern Horned Lark (*O. a. alpestris*) is unquestionably the dominant race during the winter months. * * * Flocks of from twenty to one hundred individuals are usually present by early November. The peak of abundance occurs during December, January, and February when flocks of 200 or more are frequently encountered. The largest flock noted by us was estimated to contain 600 individuals and was seen in the cornfields of the Scioto River bottom-lands a few miles south of Columbus on February 18, 1928. The largest number recorded in a single day was that of an estimated 2000 individuals, the combined number of several flocks which were encountered along a three mile stretch of road immediately south of Buckeye Lake on February 14, 1929. During the month of March there is a rapid decline in numbers.

Severe winter weather or cold storms, especially snowstorms, sometimes drive these northern larks as far south as North Carolina or even South Carolina, where they seek their food in the shelter of bare furrows or in the lee of tufts of grass in the fields. When the ground is covered with snow they manage to find food by scratching little hollows in the snow, or they resort to the barnyards to pick up hayseed and waste grain. Probably most of the horned larks seen in the Southern States in winter are of the prairie subspecies.

DISTRIBUTION

Range.—Circumpolar; from the Arctic coast of both hemispheres south to northern Africa and South America.

Breeding range.—The North American breeding range extends **north** to Alaska (St. Michael and Fort Yukon); Yukon (Herschel Island); and the Northwest Territories (Liverpool Bay, Horton River, Kent Peninsula, Cape Fullerton, Bowman Bay, and Resolution Island). **East** to the eastern part of the Northwest Territories (Resolution Island); extreme northeastern Quebec (Button Islands and Cape Chidley); Labrador (Okak, Davis Inlet, and Rigolet); eastern Quebec (Battle Harbor, Cape Charles, and Loup Bay); Newfoundland (Canada Bay and Cape St. Mary); New Brunswick (Scotch Lake); Maine (Eustis and Waterville); rarely eastern Massachusetts (Essex County, Plymouth, Barnstable, and Nantucket Island); rarely western New Jersey (Mount Holly and probably Gloucester and Salem Counties); rarely the District of Columbia (Washington); central Virginia (Lynchburg and Naruna); western Tennessee (Nashville); eastern Arkansas (Helena); eastern Texas (Galveston, Corpus Christi, and Brownsville); Tamaulipas (Miquihuana); Hidalgo (Real del Monte); Veracruz (Mirador and Perote); and eastern Oaxaca (San Mateo). **South** to southern Oaxaca (San Mateo, Oaxaca, and Mitla); Mexico (Valley of Mexico); Guanajuato (Silao); Durango (Durango); and central Baja California (Santa Rosalia Bay and San Ignacio Lagoon). **West** to Baja California (San Ignacio Lagoon and San Quintin); California (San Clemente Island, San Miguel Island, Santa Cruz, and Red Bluff); Oregon (Fort Klamath and Wapinitia); western Washington (Seattle and Tacoma); British Columbia (Spence Bridge, Chilcotin, 150–mile House, and Wilson Creek); and Alaska (Kenai Mountains and St. Michael).

The range as outlined is for the entire species, which has been separated into no less than 16 currently recognized subspecies or geographic races in North America. The typical form, the northern horned lark (*Otocoris alpestris alpestris*), breeds in the Ungava Peninsula, Labrador, and Newfoundland; Hoyt's horned lark (*O. a. hoyti*) occupies northern Canada from Hudson Bay west to the mouth of the Mackenzie River and south to the northern parts of the Prairie Provinces; the pallid horned lark (*O. a. arcticola*) breeds in Alaska (except the Pacific coastal region) and south in the mountains through British Columbia to Washington; the desert horned lark (*O. a. leucolaema*) occupies the region from southern Alberta to New Mexico and Texas, east on the Great Plains to South Dakota and Kansas, and west to Nevada; the prairie horned lark (*O. a. praticola*) is found from southern Manitoba and Quebec south to eastern Kansas, Arkansas, Kentucky, and Virginia; the Texas horned lark (*O. a. giraudi*) occupies the coast of Texas and northeastern Tamaulipas; the streaked horned lark (*O. a. strigata*) inhabits the Pacific coast region of Washington, Oregon, and northern California; the dusky horned lark (*O. a. merrilli*) is found in southeastern British Columbia, eastern Washington and Oregon, northeastern California, and northwestern Idaho; the island horned lark (*O. a. insularis*) is confined during the breeding season to the Santa Barbara Islands of California; the California horned lark (*O. a. actia*) is found in California south of San Francisco Bay, east to the San Joaquin Valley and south to northern Baja California; the Magdalena horned lark (*O. a. enertera*) is confined to the central part of Baja California between Santa Rosalia and Magdalena Bays; the ruddy horned lark (*O. a. rubea*) is the race of the Sacramento Valley, California; the Montezuma horned lark (*O. a. occidentalis*) breeds in central Arizona and New Mexico; the scorched horned lark (*O. a. adusta*) appears to be confined to a relatively small area in southeastern Arizona; the Mohave horned lark (*O. a. ammophila*) nests in the Mohave Desert and Owens Valley, Calif.; and the Sonora horned lark (*O. a. leucansiptila*) occupies a region extending along the Colorado River from southern Nevada and western Arizona south to northeastern Baja California. Additional races of this species are found in Europe, Asia, and Mexico, as well as one (*O. a. peregrina*) that appears to be localized in the vicinity of Bogota, Colombia.

Winter range.—The species is found throughout the year over most of the breeding range, although it withdraws during the winter from the northern regions. At this season it is found **north** to southern British Columbia (Arrow Lake); southern Alberta (Warner and Sullivan Lake); southern Saskatchewan (Eastend and Skull Creek); northern North Dakota (Charlson and Grafton); northern Minnesota (Fosston and Iron Range); southern Wisconsin (Westfield and Greenbush); southern Ontario (Guelph, Toronto, and Ottawa);

central Maine (Bangor and Calais) ; and Nova Scotia (Grand Manan and Kings County). Occasionally they are recorded in winter from points farther north in Alberta (Battle River) and Saskatchewan (Dinsmore) ; as well as from southern Manitoba (Winnipeg), southern Quebec (Montreal and Quebec), and southern New Brunswick (Scotch Lake). In the eastern part of the country the **southern** limits of the winter range extend south of the breeding range to Florida (St. Augustine and Apalachicola, rarely Daytona Beach and Miami) ; and northern Alabama (Decatur and Leighton).

Spring migration.—Because of the fact that in the East the horned lark winters regularly to the northern parts of the United States and southern Canada, and in the West is resident south to and beyond the Mexican border, dates of arrival and departure are not numerous for the seasonally unoccupied regions.

In the southeastern part of the country late dates of spring departure are: Florida—Hastings, March 18. Georgia—Athens, February 25. North Carolina—Pinehurst, February 22.

Early dates of spring arrival at points north of the winter range are: Quebec—Montreal, February 23. Labrador—Gready, April 15. Newfoundland—Raleigh, April 24. Mackenzie—Fort Simpson, April 28. Yukon—Forty-mile, May 10. Alaska, Demarcation Point, May 6.

Fall migration.—Late dates of fall departure from the North are: Alaska—Swan Lake, August 19. Yukon—Russell Mountains, September 4. Mackenzie—Fort Simpson, October 3. Labrador—Ticoralak, October 12. Newfoundland—October 14. Quebec—Montreal, November 15.

Early dates of fall arrival south of the breeding grounds in the eastern part of the country are: North Carolina—Raleigh, November 21. South Carolina—Chester, December 3. Georgia—Clayton County, November 16. Florida—Daytona Beach, November 20.

Egg dates.—Arctic Canada: 7 records, June 14 to July 9.

Arizona: 15 records, April 8 to July 28; 8 records, April 29 to June 10, indicating the height of the season.

California: 106 records, March 20 to June 23; 52 records, April 17 to May 20.

Colorado: 26 records, April 9 to July 16; 14 records, May 10 to June 9.

Illinois: 32 records, March 3 to July 6; 16 records, April 25 to June 17.

Iowa: 19 records, March 27 to July 16; 10 records, May 14 to June 19.

Labrador: 7 records, June 1 to 30.

Lower California: 2 records, April 6 and 23.

Montana: 46 records, April 10 to July 7; 23 records, May 7 to June 4.

New Mexico: 11 records, May 20 to July 6.

New York: 14 records, March 19 to May 30; 8 records, April 1 to 22.

Northwest Territories: 5 records, June 1 to July 18.

Santa Barbara Islands: 5 records, April 4 to May 14.

Saskatchewan: 4 records, May 15 to June 9.

Texas: 22 records, February 20 to June 20; 12 records, April 24 to May 23.

Washington: 43 records, May 20 to June 25; 21 records, April 25 to May 30.

Wisconsin: 11 records, March 21 to June 25; 7 records, May 15 to June 18.

OTOCORIS ALPESTRIS LEUCOLAEMA (Coues)

DESERT HORNED LARK

PLATES 47, 48

HABITS

On the more barren plains of the far West, we find this horned lark replacing the familiar prairie horned lark of the more fertile prairie regions of the Middle West. It is about the same size as and only slightly paler than *praticola*. In comparing it with neighboring races, Dr. Oberholser (1902) says: "This form may be distinguished from *praticola* by the markedly more cinnamomeous tint of cervix, upper tail-coverts and bend of wing, as well as by the paler color of the back, where the blackish of *praticola* is replaced by sandy brown. From *arcticola* it differs in reduced size, usually yellow throat, nape more tinged with cinnamomeous, lighter and brownish instead of blackish back; from *giraudi* in larger size, generally paler throat, together with paler, much more brownish upper surface; from *merrilli* in larger size and lighter, more brownish coloration."

In southwestern Saskatchewan, in 1905 and 1906, we found horned larks very common on the prairies, on the barren hills north of Maple Creek, and on the alkaline plains. As we drove along the narrow wagon trails over the rolling plains, the monotony of the scenery was often relieved by seeing one of these little brown-backed birds running along in the wagon ruts ahead of the horses, perched on a clod of earth beside the road, or springing into the air to pour out its quaint little ditty, not quite equal to the rapturous flight songs of the chestnut-collared longspurs but, nevertheless, pleasing.

We collected a series of horned larks here and in Alberta, most of which, particularly those taken on the prairies in the more eastern portions of the region, were referable to the new form described by Dr. Oberholser (1902), *O. a. enthymia*, while those collected on the alkaline plains and sagebrush plains farther west were more clearly referable to *leucolaema*.

Courtship.—Although the habits of the various races of the horned lark are all very much alike, different observers have described them somewhat differently, illustrating certain phases of behavior more clearly than has been done by others. For this reason, I shall, at the risk of some duplication, quote freely from two excellent papers by A. Dawes DuBois (1935 and 1936) on the habits of the desert horned lark. He noted that the male, in his upward nuptial flight, usually ascends at an oblique angle, but that against a strong wind he rises almost vertically. He then goes on to say (1936) : "After remaining aloft for a time, singing his best song, which comes to the human ear but faintly from so great a height, the bird suddenly folds his wings and drops like a bullet. With ever increasing velocity he descends until one might fear for his life; but he spreads his wings just in time to avert a violent end, skillfully turning his course into a glide which carries him horizontally, near the ground, until his momentum has been spent. He then alights quite easily, as though nothing important had happened. * * * I doubt that the bird world holds a more awe-inspiring event than this headlong drop from the sky."

He thinks that this spectacular dive "far surpasses the performance of the nighthawk" but does not compare it with the thrilling dives performed by some of the hummingbirds, which seem equally inspiring. He says:

Another ceremony of the season is the fighting exhibition, which takes place in the air a few feet above the ground. The two males engaging in it begin their advance and attack while on the ground but immediately rise together in a whirl and flutter of gallantry. * * * There seems never to be an injury, nor even a victory. I have never seen a drop of blood drawn or a feather lost in the encounters. * * *

The third sort of maneuver is an exciting chase. Two, three, or four birds usually take part in it. They fly in close formation with great swiftness and remarkable skill. It looks like a game of follow the leader, with instant response to every change of the leader's course—a course of rapidly changing, meandering curves.

Nesting.—Mr. DuBois (1935) has given us a very full account of the nesting habits of this lark in Teton County, Mont., based on four years of study of 58 nests. His data are given in far too much detail to admit of more than a few extracts from them, as follows:

Two peaks of nesting activity occur, one about the end of April, the other early in June, indicating two broods in a year. * * * The Desert Horned Larks avoid the more luxuriant growths which are to be found in moist situations

They prefer the dry bench lands. There is no special preference as to surface contour so long as the situation is a dry one. Nests occur on knolls or slopes, or in the dry depressions of the benches. * * * The number of nests which one finds near old dried droppings of horses, and sometimes of cattle, seems much greater than the laws of chance would account for. * * * Only one nest was found in a cultivated field. * * * It was in a field of young spring wheat which stood in drills, about two inches tall, the ground being otherwise bare. * * *

The nest is invariably built in a rounded hollow in the ground, which is evidently scratched out by the birds, the excavated dirt in the form of fine scratchings being thrown out to one side of the nest. The dirt is almost always on the east side, which is also the side least protected by vegetation. Usually the top of the nest structure is flush with the ground surface. * * * The materials used for the body of the nest are dead grasses, including both stems and blades, usually without any other materials.

The nest built on cultivated land contained rootlets and old dead grass. The linings were more varied, including a bit of rag, some tiny bits of rabbit fur, soft, silky, white plant down, and seed pods, heads, tips, or leaves of yarrow, which when dried are gray or white and of soft, velvety texture.

"All nests examined, with only one exception, were provided with pellets of dried mud at the entrance or elsewhere around the nest. These are little cakes or broken pieces of the cracked crust which forms on the surface of mud when baked by the sun. * * * The pellets are used chiefly to cover the loose dirt thrown out in excavating the hole for the nest."

The inside diameter of nine nests averaged 2.49 inches, and their inside depth averaged 1.92 inches. The ground hollow for one nest was 4 inches in diameter and 2 inches deep. The time required to build the nest, after the hollow was dug, varied from 2 to 10 days. As to concealment of the nests, he says: "The prevailing short grasses of the bench lands do not afford much cover. The concealment of nests in general, so far as the surrounding grass is concerned, is very incomplete, sometimes quite meager. Nevertheless, the nests are not easy to see. In most cases there is some protection from grass on the west side; sometimes it slants over the nest, owing to the prevailing winds."

Eggs.—The desert horned larks usually lay three or four eggs, perhaps very rarely five, though none of Mr. DuBois's 58 nests contained five. These are practically indistinguishable from those of the other races of similar size. Some sets are somewhat paler and less heavily marked than those of the darker races. The measurements of 50 eggs average 22.1 by 15.6 millimeters; the eggs showing the four extremes measure 23.9 by 15.5, 23.4 by 17.0, 20.6 by 15.2, and 19.6 by 15.2 millimeters.

Young.—In his second paper (1936) Mr. DuBois gives a full account of the developments of young desert horned larks, to which the reader is referred for details, as only portions can be quoted here. He writes: "Apparently the male never assists in the duty of incubation, but both parents take part in feeding the young as soon as the eggs have hatched. The food for the young is evidently of a solid nature from the beginning. A parent was observed carrying a smooth green caterpillar in the afternoon of the day of hatching. Large larval insects are fed to the young birds after they are strong of flight. Parent birds were seen carrying excreta away from the nest when the nestlings were two and three days old." He tells of the hatching of a chick from an egg held in his hand:

When it was ten hours old, nearly all of its natal down was dry, fully three-eighths of an inch long, and very fluffy—a marvelous transformation! * * * At the age of seven or eight days the nestlings are fairly well feathered and the natal down is confined to the feather tips. * * * When ten days old the young have left or are, in most cases, leaving the nest. They are not able to fly but can run very well. It appears that they usually leave in the fore part of the day. * * * It is easy to identify the horned lark nestlings, at any stage of their development, by looking into their mouths. The mouth lining is orange, and there are distinct black marks in the mouth and on the tongue. This distinguishes them at once from the nestlings of McCown and Chestnut-collared longspurs, which have plain pink mouths and throat linings. When the young larks have grown up, the orange color fades and the black marks disappear.

Leon Kelso (1931) also gives a detailed account of the development of the young in four nests of this lark, to which the reader is referred. He states: "It is evident that the rate of growth and length of time spent in the nest by nestling Desert Horned Larks varies according to the time of the year. The young of nests 1 and 4, in the months of April and May, respectively, remained in the nest at least ten days, while those of 2 and 3, in July, stayed in the nest not more than 7 days. The size attained appeared to be comparable in all instances. * * * The first nest has a lining of thistle-down, contained 1505 pieces of material and weighed 16.75 grams; the second had no lining, was built of 805 pieces, and weighed 7.7 grams."

Enemies.—Mr. DuBois (1936) writes:

On their nesting grounds the Desert Horned Larks have to contend with their share of enemies and sources of accident. Among the natural enemies, weasels, skunks and ground squirrels came to my attention, not to mention man, whose poisoned baits set out for ground squirrels apparently kill more birds than spermophiles. One day, by quick action, I intercepted a weasel on his way to a nest to get the last nestling. The birds, of course, are powerless to defend their young against weasels and skunks. It is believed that the abundant ground squirrels often destroy eggs, and possibly sometimes take a nestling; but the adult larks are not afraid of them. It is common to see the larks driving a trespassing squirrel away from their premises. They go after him from the air, in a series of dashes; and quite often the two birds attack together.

The barbed wire fence, new in part of the region when these notes were made, was a source of unexpected danger. Several carcasses of horned larks were found at different times beneath wire fences. * * *

Storms cause the greatest destruction of nestlings. Eggs and young are kept dry during all ordinary rains. But in some years the destruction of nests by severe and protracted storms is doubtless, over an extensive region, practically total. A continuous rainstorm of three days duration, coming first from the east, then from the north and finally from the northwest, killed all the nestlings that were known to me.

William G. Smith wrote to Major Bendire (1895): "While I lived in the Platte River Canyon, 40 miles west of Denver, Colorado, a terrible snowstorm set in suddenly in April, and with it came thousands of these birds, which tried to shelter themselves under projecting banks. The majority were soon so chilled by the intensely cold wind which was blowing at the same time, that they could not move, and were quickly smothered by the drifting snow; and after this melted bushels of their dead bodies could be picked up everywhere."

Probably this horned lark is imposed upon occasionally by the cowbird, but Dr. Friedmann (1929) cites only one authentic record; it would seem as if horned larks might be frequently victimized where cowbirds are common.

Winter.—After the second brood of young is strong on the wing these horned larks gather into immense flocks and roam about the country preparing to move southward. Many extend their winter range into the more arid regions of our southern border States and even into Mexico. But through a large portion of its range this race is largely resident and is found less commonly as far north as Montana in winter. It is common in winter at least as far north as Wyoming. In Colorado, it is very abundant in winter, traveling about in enormous flocks in company with some of the other subspecies. When the ground is bare the flocks spread out over the plains and fields, but when the snow covers their feeding grounds they congregate about the ranches and farmyards to feed on the waste grain or come into the towns and cities to be fed by the residents; sometimes in severe weather thousands of the birds come into the towns, where people feed them regularly on millet and other seeds, scattered on bare spaces, where the birds often gather so thickly as to almost cover the ground.

Claude T. Barnes writes to me from Utah: "Though the day be cold and drear, and all vegetation well nigh covered with snow, in the fields west of Salt Lake City, flocks of pretty horned larks are daily seen feeding on the seeds of the pigweed, saltbush, ragweed, amaranth, and other noxious weeds, which here and there protrude through the snow. If the snow becomes too deep for them, they even venture into the city."

OTOCORIS ALPESTRIS PRATICOLA Henshaw

PRAIRIE HORNED LARK

PLATES 49–52

HABITS

CONTRIBUTED BY GAYLE PICKWELL [1]

Out at the bleak end of the ecological series of bird habitats, which begins with the heavy forests and ends with the barrens, lives America's only true lark, *Otocoris alpestris* (Linnaeus).   In that region extending from Missouri to the Atlantic and from Kansas to Ontario the particular form of this lark is *Otocoris alpestris praticola* Henshaw, the prairie horned lark.   Far from the treeless Arctics, far from the deserts, this lark finds as its barrens the plowed fields of the Midwest, the tree-denuded, wind-swept hilltops of the Northeastern States, and those peculiarly unnatural and artificial barrens, the hazards of these modern-day golf courses.

If for no other reason than that here is a bird nesting where no bird has a right to nest, a bird in a niche that demands not vegetation but lack of it, a bird alone and unique in its nesting site without a competitor and far out at the end of the series—if for no other reason than this purely ecological one—the prairie horned lark invites close study.   But if we add to this the fact that it is a lark, a representative of our only lark, with the song of a lark, the ways of a lark, and many a habit and idiosyncrasy peculiarly its own, and that it is an intriguing bird of the open field, then the bird becomes even more interesting.

The prairie horned lark, because of its tendency to occupy the most barren regions as its home, interested me very early, for desultory observations of this bird were begun while still a boy in eastern Nebraska.   The lark nests were found on the ridges of listed corn and an observation of a song still remains clear and trenchant.   We were shocking wheat, hence it was mid-July, when a lark was seen climbing the air for his song.   We watched him against the vivid sky during his long minutes aloft; were amazed by that final headlong drop to earth.

Study of the prairie horned lark was initiated in eastern Nebraska, continued intensively in northern Illinois (Evanston) for two years, then transferred to Ithaca, N. Y., and concluded.   Since that time I have lived in California, where the prairie horned lark is replaced by the California horned lark, a closely related subspecies.

---

[1] Derived largely from Pickwell, "The Prairie Horned Lark" (1931).

Henshaw (1884) erected the subspecies *Otocoris alpestris praticola*, splitting it from *O. a. alpestris*. Prior to this records of a new form of lark and new lark breeding records were published from lower Ontario and New York. These were variously interpreted as a "paler form" or as a southward extension of *O. a. alpestris*. Following 1884 a consistent and progressive series of records demonstrated that the prairie horned lark, coming up probably from Michigan through Ontario, invaded successively New York, Vermont, Massachusetts, New Hampshire, Maine, and Connecticut. From New York or Vermont it seems to have invaded Quebec much later; and lastly on the north, probably from the New England States, New Brunswick, Prince Edward Island, and Nova Scotia. Shortly after its entry into New York the lark appeared in western Pennsylvania, then farther east in that State, and south into Maryland and West Virginia. Recently the prairie horned lark has been recorded as breeding in the District of Columbia and Virginia.

Less complete evidence seems to show that Indiana, Michigan, Ohio, probably northern Kentucky, and southern Missouri have been occupied by this lark since the white man has entered and altered those regions. The regular advance of the bird, always consistent with geographic conditions, is suggested as an irrefutable evidence that such an extension is bona fide. It is suggested that this extension of range has resulted from changes that civilized man has made by deforestation and cultivation; thus creating permanent or seasonal semibarren conditions that the prairie horned lark requires.

The drier portions of the prairies of Illinois have probably long been occupied by this lark. The studies of Forbes and Gross (1922) seem to indicate that the lark, though it probably breeds in Lower Austral, Upper Austral, and Transition Zones, seems to prefer the Transition in that State. It is suggested that the prairies of northern Missouri, Iowa, Wisconsin, Minnesota, eastern portions of Kansas, Nebraska, North and South Dakota, and southern Manitoba probably formed the ancestral home of the subspecies. Nearly all this vast region would have been suitable for two broods in March, April, and early May, though the bird would have been forced to the more barren regions as the grasses became vigorous in late May, June, and July. That this lark species is versatile in the matter of occupation of new territory seems to be further demonstrated by the observation of Gätke and Saunders in Europe with regard to *O. a. flava*.

Courtship.—Prior to the establishment of well-defined territories, fighting between males is promiscuous; after that fighting takes place only on territory boundaries, where two lark areas juxtapose. The males, at the boundary line, frequently strut before each other and often peck the ground furiously, like barnyard cocks, but all fighting is in the air. On a boundary this fighting often results in a curious game of "tit for tat," as the male larks chase one another back and forth. Every adventitious lark, wandering into established territories, is promptly evicted by the male. Such a bird will leave without protest. So far as noted, the female is never the direct cause of fighting; in fact fighting is most frequently noted when the female is brooding and the male is no longer attending her. Only once was a female noted driving out another lark, a male. She was defending a recent nestling.

The female has no courting maneuvers and was never observed to sing. Only once was she seen to importune sexual attention and then by a crouch and flutter similar to the actions of the English sparrow. The male struts frequently before the female with wings dropped, tail spread, and horns up. He will assume this attitude before another male at the territory boundary.

Nesting.—The literature shows a surprisingly large range of habitats in which the prairie horned lark has been known to nest. These habitats, resulting for the most part from agricultural activity or other human agencies, are those that most nearly result in barren conditions. It does not matter that these barrens may be seasonal or otherwise very temporary, if they are suitable for the initiation of nesting. That bare ground is the determinant is shown by the fact that variations of moisture, soil, elevations, and temperature will all be tolerated in the selection of nest sites. The prairie horned lark thus, it seems, does not differ greatly in the ecological condition of breeding habitats from other horned lark subspecies.

Some typical Chicago marsh in the Evanston region was drained for a golf course. The course was later cut up into real-estate subdivisions; sewers were laid exposing a wide area of bare soil in the streetways; and old sand hazards remained here and there. This series of activities provided nesting sites for many larks. More than a score of nests were located on this area (about 90 acres) in 1926. A plot of vegetable gardens bordering this region on the west, where larks had probably nested for some years, was also subdivided and the vegetation subsequently neglected. Here several larks also nested.

The advent of vegetation in both areas and the demand of the lark for bare ground forced a seasonal succession of horned-lark breeding sites first from lot surface, to streetway, to sand hazard, to vegetable garden, in the order that each was successively occupied by verdure.

At Ithaca, N. Y., one nest was located on the overturned sod of a former hay meadow. Most of the observations there made, however, were on a tract of ground that was largely fall wheat, partly fall rye, and the remainder devoted to experimental vegetable gardens. The growth of the wheat forced the larks from its surface by late May. The gardens and portions of the fall rye area that were turned under as green manure remained suitable throughout. Clean vegetable gardens will always present a considerable amount of bare soil, and the prairie horned lark is usually able to occupy such gardens until late in June.

A breeding territory was delimited by a male lark on February 7, 1926, at Evanston, Ill. From his selected territory he could not be driven. This territory was about 100 yards square. Late March snows disrupted all territories, and it was not learned whether the original sites were ultimately resumed or whether the same territory was maintained through more than one nesting. The pressure of vegetation in late May and June greatly modified the territories at Evanston and caused, eventually, the abandonment of most of those on the erstwhile golf course.

At Ithaca, N. Y., a male lark was forced to mark territory for the first time on March 13, 1927, though it had undoubtedly been established some time before this. Territories voluntarily marked were somewhat larger than those indicated when the birds were forcibly driven about. The regions of a breeding territory most frequently occupied were those boundaries that joined the territories of a neighboring lark.

The territores at Ithaca were much larger than those at Evanston, possibly because fewer larks attempted to occupy them. At Evanston they were seldom over 100 yards square, whereas at Ithaca they ran out to lengths of 300 yards and widths of 200 yards, in March and April. In general all suitable territory was occupied at Ithaca and most boundaries were established by the margins of unsuitable areas, though a large amount of suitable territory, extending beyond, was used only in part by the bird. Boundaries between males were often definitely established on ground that had no natural marker whatsoever.

The territory history of three pairs of larks was followed from March to June at Ithaca. One influence only modified the territories, namely, the growth of vegetation. One territory, completely on fall wheat, was abandoned by the close of the second nesting in May. Another territory, in part on fall wheat and in part on the gardens, was gradually reduced to the gardens, from an area once 300 by 200 yards to an ultimate area about 100 by 50 yards. A third territory, almost entirely on the gardens, suffered no major reduction. But the owner of this third territory, which abutted that of the second, gave no ground to the latter.

Though most of the feeding was done on the nesting territories, a neutral feeding territory was discovered, and others were indicated because, now and then, the larks would go off on purposeful flights entirely out of their areas.

The female would mark the same territory as that marked by the male, and if anything she was more closely restricted to it than the male. She selected the nest site with little or no regard to the center of the area.

The literature contains four February records of nests and many records of March nests in many States, and two or three records of nests in July. I have records of nests from about March 21 to July 12, in 1926, at Evanston, Ill.; from about March 11 to June 28, in 1927, at Ithaca, N. Y.

It is suggested that such a strange phenomenon as that of a passerine bird nesting in March in eastern United States cannot be easily explained. The bird has too long a nesting season to explain it on the conditions that might exist in early spring alone; and then, in the range where the prairie horned lark was studied, nests are frequently destroyed by inclement weather and many young die of starvation at this season. Since this bird demands barren conditions, and not verdure, for a nest site, the conditions are suitable very early, and it is suggested that an early-nesting physiological cycle may have been acquired in a more propitious climate and subsequently carried north and east. It is further noted that *O. a. actia* of California nests in March where conditions are quite ideal.

With one exception all of the 14 observed nests of March and April were not begun until the mean temperature rose above 40° F. for two or more days in succession. The exception was the initiation of a nest on the *first* day that the temperature rose above a mean of 40° F. Once weather conditions suitable for the initiation of nesting activities prevailed, no subsequent weather, no matter how severe, except deep snow only, would inhibit these activities. Even birds that had nested in March and whose nests were destroyed by late March and early April snows, would not renest until weather conditions were as given, though this necessitated a delay of nearly three weeks in two cases at Ithaca, N. Y. That this was a delay caused by the weather is easily demonstrated by the fact that an exceptional case, as noted above, began renesting on a single suitable day, but two other larks waited two weeks longer for renesting or until weather again was suitable and for a longer period. It is known that two of these birds, and probably all, had former nests.

On the basis of this known weather control it was possible to calculate the frequency, over a period of years, with which nestings would occur in March, by a study of weather summaries for the month. The results showed one year when nesting was impossible and 16 years of possible nestings, at Evanston, Ill., for the years 1910 to 1927, inclusive, except 1924. During 10 of the 16 years nestings could have been successful; during 2 they would have been destroyed by snow; and during 4 weather and snow would have made success problematical. At Ithaca, for the years 1916 to 1927, inclusive, the summaries showed one year when nests were impossible and 11 years of possible March nests. During 5 of the 11 years, nestings could have been successful; during 4 they would have been problematical. Summaries could not be obtained for years previous to the earliest here noted. On the basis of those obtained it is shown that northern Illinois has more favorable weather in March than southern New York. New York, it will be noted, is a State recently occupied by the lark. It is concluded that 3 or more inches of snow, lasting two or more days, would destroy a nest.

It is suggested that the discovery of nests during nest building is possible by locating first the calling or singing male. At this period the male will be attending the female closely and she will be discovered shortly. The status of nesting can always be determined by the actions of the female. During nest building she is very restless, runs here and there, flies up and away, but shortly returns. Eventually she may disclose the site of the nest excavation. These reactions are instinctive responses to the desire for nest concealment. All nest building seems to be done by the female.

During egg laying the discovery of a nest is at best accidental. Neither male nor female has been noted to approach a nest during this period. They express no solicitude beyond that of nest concealment, thus displaying a remarkable nonchalance, especially on the part of the female. This reaction is so marked that an observer can nearly always be assured of the status of nesting whenever it is noted.

When incubation has begun the behavior is very different, as is also the behavior after the eggs have hatched. These reactions will be noted later. During these periods nests may be located by a systematic search that involves driving the male about until the female is noted. She will flush from the nest and the male will go to her. Then a patient watch of the female will, after a variable length of time, disclose the nest. When young are being fed the male will, at times, disclose the nest much more quickly than the female, for he assists in feeding and has nest-concealing instincts that are very poorly developed. Though the nest of the prairie horned lark is never concealed from above, it fits its semibarren environment so closely that a promiscuous search over a breeding territory is nearly always tiresome and unavailing. An incubating or brooding lark, as will be discussed later, often remains close to her nest on a chilly day or very early in the morning or toward evening. Nests can be found under these circumstances by a systematic search of likely habitats and so flushing the bird from the nest.

No evidence of the use of a natural depression was noted either at Evanston or at Ithaca; all were dug by the female. According to Sutton (1927) and my own observations of *O. a. strigata* in western Oregon, this excavation is dug with both beak and feet. The nest is constructed usually at the edge or partially under a grass tuft or clod, which, in the case of the prairie horned lark, lies most frequently on the west, northwest, or north, possibly because the cold and violent winds of the early nesting season come from this direction. The body of the nest consists of coarse stems and leaves with a finer lining within. The time spent in nest construction varies from two to four days.

The majority of the nests of the prairie horned lark showed a variable amount of clods, pebbles, or similar items laid about the margin usually on the side away from the protective tuft or clod. These so-called "pavings" were always composed of the material most easily obtained regardless of its permanency. It is suggested that the purpose of "pavings," if there is a purpose, arises from the method of nest construction and from the desire of the larks to have a bare-ground nest approach.

Eggs.—The egg has a background of gray with an occasional greenish tinge, which background is almost completely concealed with a fine speckling of cinnamon-brown. The cinnamon-brown often forms a denser ring about the larger end. The average size was found to be 2.25 cm. by 1.55 cm. The eggs of natural second sets seemed to be a trifle larger than first sets of the same individual. The number of eggs per set varied from two to five; the average was about four, the sets of fewer numbers occurred early, those of larger number, later. [AUTHOR'S NOTE: The measurements of 50 eggs average 21.6 by 15.7 millimeters; the eggs showing the four extremes measure 24.6 by 15.5, 23.1 by 17.3, 18.3 by 15.0, and 21.6 by 14.5 millimeters.]

Incubation.—The incubation period was determined to be 11 days. Only the female incubates.

The male shows little or no solicitude during the incubation period. The female has a highly developed series of automatic instincts of solicitude, which are modified by time of day, condition of weather, and frequency of disturbance. The most highly developed and probably the most recently acquired of these has been given the name *nest concealment by abandonment*, or *casual abandonment*. The female leaves the nest, in this reaction, when an intruder is at a long distance, and flies quietly away, low against the ground, and does not show other solicitude for a very considerable period. The distances of the intruder from the nest during this reaction vary from 25 to 100 yards or often farther, a greater distance, it will be noted, than would disturb even a timid lark under other circumstances. A reaction that in many ways is the reverse of this, but still a marked exhibit of solicitude, is that called *distress simulation*. This consists of a precipitate flushing and rapid flutter over the ground after the nest has been approached closely. This reaction would be given most frequently on very cold days, in the dusk of very early morning or evening, and when the bird was flushed very shortly after she had returned to the nest. It is certainly more primitive than the first reaction here described and is probably a culmination of the more frequent distraction display that most birds present when their nests are disturbed. Between concealment by abandonment and distress simulation there occurs a complete gradation, which, since the reactions are exact opposites in expression, involves a curve that drops from the first to the zero point and then rapidly ascends to the expression of the latter. Thus, between the two, lessened expressions of either reaction would result, with a curious hiatus midway in which the incubating bird would allow an intruder to approach closely and then leave without an expression of either type of solicitude. Experimental flushing of an incubating bird from a blind showed that the bird, in one case, would give distress simulation if flushed in an interval that was less than two minutes from the time of her return; but would give casual abandonment if flushed after an interval of five minutes. A female lark, shortly after being forced from a nest, would express her agitation by aimless ground pecking, and, to be sure, would eventually be driven by the incubation urge to return to the nest even though an intruder might be much nearer than he had been when the nest was originally abandoned. This complex of instincts involved both the urge to incubate and the urge to protect. The instinct to protect, by whatever method, would all be overshadowed in time by the instinct to incubate.

Young.—With the exception of those nests of early April, in which incubation began before the set was complete, all young hatched within an hour or two of each other. The young are fed within an hour or two following hatching. In most cases the male assisted the female in feeding the young. In carefully observed cases he visited the nest less often but brought greater burdens and fed more young at a visit than did the female. A total number of observed feedings during one day (April 30, 1926) was 108. The male fed 39 times and the female fed 69 times.

Observations of the adults and dissection of a few nestlings showed that some vegetable matter (weed seeds) is fed early in spring, but that even in March most of the food is animal matter. Later in the season grasshoppers become conspicuous in the diet. The adults dig up both cutworms and earthworms.

The male shows solicitude for the nest and its contents for the first time after the hatching of the eggs. His solicitude is restricted to calls. The female will leave her brooding in typical concealment by abandonment when conditions are appropriate as when incubating; likewise she will go from the young in distress simulation under conditions as noted previously. Proportionately the number of concealments by abandonment decreases and distress simulations increase slightly with young in the nest. Other reactions, which are various primitive expressions of solicitude, or intermediates of the two just mentioned, increase proportionately. Perhaps a return of more primitive instincts indicates a sum total of greater solicitude. Since the female is frequently absent from the nest in food foraging, she will come in, as an intruder approaches, with calls and cries. One or two references in the literature show that the reactions to dogs is the same as to man, but that hens are driven off by entirely different methods.

The larks removed all excreta throughout the full extent of nest occupancy. Early in the season much of the excreta was eaten by the adults; later it was dropped to the ground 50 or more feet away. This seasonal change of habit may have been related to the available food supply. The instinct compelling excreta removal proved itself very powerful, at times overcoming strong solicitude for nestlings and even fear.

The young showed a psychic development closely related to their rate of growth and not to their age. Young of the same age or but one day younger than their nest mates often presented a psychic development two to four days behind them. This was due to uneven feeding, which occurred frequently early in spring because of uneven hatching or an inadequate food supply.

Normal nestlings give a food response indiscriminately up to the fifth or sixth day. Just prior to this time their eyes open. Following this they respond not at all or momentarily only. They withdraw at a touch from the hand on the sixth day and sink back quietly into the nest in crouch-concealment between the seventh and ninth days. Upon being removed from the nest at this age they sit quietly upon any object upon which they are placed; prior to this time they wriggle about when taken from the nest. They leave the nest on the tenth day and then express fear by hopping and calling wildly when disturbed. An expression of this type of fear, prior to the tenth day, would take them from the nest prematurely.

Weight-growth curves show a gradual increase over the first three days, a very precipitate rise (except for April nestlings) for the next three or four days, a marked leveling during the seventh and eighth (in one case the sixth), and a gradual rise during the ninth and tenth. Nestlings in May grew slightly more in the same period than a nestling in June and much more than a nestling in April. This discrepancy of growth seems closely (though perhaps indirectly) correlated with the temperatures of these seasons.

A lessening in weight growth occurs, normally, between the seventh and eighth days. This is brought about by the simultaneous unsheathing and drying of most of the feathers. On the other hand, growth in length shows, if anything, an acceleration at this period due to the extension of the tail. Marked variations in growth occurred in the various broods measured and in the different young of the same brood. This was brought about by two things: The fact that a slight difference in age gave the older larks a great advantage in securing food from the parents; and the fact that food was more plentiful later in the season than at the beginning.

Length-growth curves show a precipitate rise during the first three days, a slight leveling during the next three days, and a precipitate rise during the sixth, seventh, eighth, ninth, and tenth days. The cause for the intermediate leveling is not understood, but the rise toward the end of the nesting period is brought about by the growth of the tail.

In the early breeding season the enemies of the young are weather and a scanty food supply. The weather may result in snows that bury them from sight. The scanty food supply may result in the starvation of one or more of the nestlings. Starvation results from the automatic feeding reaction of the adults wherein the nestling nearest that part of the nest habitually approached by the adults will receive the first feeding; if the food is scanty this bird will receive all or nearly all the food. Only when food is so abundant that the first nestlings fed do not swallow promptly will the remainder of the brood be fed. In this case food is withdrawn from the mouth and put in the next and so on. The female lark rarely brings more than will go into one mouth; the male may feed two or more, but never four or five, at a time. Those young larks that have a few hours' advantage in hatching—a full day in several cases in the early spring—will have the advantage in size that will allow them to push to that side of the nest over which the food always comes. They survive; the others may perish. Such occurred in many observed nestings in April.

Predacious enemies cause a greater and greater loss as the season advances into June and July. The optimum season for the welfare of the young is shown to be May.

One case of cowbird parasitism was observed and followed. A lark, which hatched before the cowbird, came to maturity. The cowbird probably did not. It is suggested that the early nesting season and the exposed habitat may mitigate against such parasitism as may also the early departure of the young larks from the nest. However, since the adult lark will tolerate the parasitism and the food of June and July is suitable, other reasons prevent more extensive parasitism at this later season.

The young leave the nest, normally, on the tenth day, some three to four days before they can fly. Their protection during this interval is silence and a very effective "freeze" or crouch-concealment. Their plumage is remarkably adapted for this. The actions of the parents, especially the female, with her abandonment concealment, are calculated to take advantage of the protective color of nest and young at all ages.

Young leave the nest usually by following a parent that has brought food. One case was noted wherein a female enticed a belated nestling from the nest with a food morsel. The young fly in about five days after leaving the nest. They hop for some days after nest leaving, whereas the adults walk. This hopping may be anatomical or an atavism.

Plumages.—The recently hatched nestlings are rather heavily covered with down, a necessary protection against sun and cold in their exposed location. The down is cream-buff in color. At nest-leaving age the young lark is in full juvenal plumage but presents an appearance quite unlike that of the adults; each feather of the upper surface has a triangle of brown at its tip, the under surface is white except the throat, which is gray. [AUTHOR'S NOTE: Subsequent molts and plumages are described under the northern horned lark.]

Food.—McAtee (1905), in his extensive account of the food of the horned larks, writes that in August and September many grasshoppers are taken (7.1 and 8.9 percent of the total food, respectively) and that weevils constitute 18 percent of the food in August. He says further that spiders are taken in every month. The conspicuous weed seeds that he lists (foxtail grasses, smartweeds, bindweeds, amaranth, pigweeds, purslane, ragweed, crab and barn grasses) are probably largely consumed in fall, winter, and early spring. The total of 79.4 percent of vetetable matter taken in the year, as given by McAtee, is made up largely of these weed seeds. He found about 40 percent of food taken in August to be animal matter, 20 percent animal matter in September, between 10 and 20 percent in October, 5 percent or less in November, about 2 percent in December, 1.73 percent in January, and 3.11 percent in February. The animal matter of January and February consisted principally of weevils and cocoons of tineid moths.

Grain, chiefly waste oats, corn, and wheat, formed 12.2 percent of the food of larks, exclusive of California forms, and much of this would have been taken in the period under consideration.

The Main Subdivision at Evanston, where the most extensive observations were made by the writer, had, in the winter of 1925–26, great quantities of *Agropyron repens* (quack grass), *Setaria* (foxtail), and *Amaranthus* (pigweed), all of which had been allowed to mature seeds. Of this the seeds of the quack grass were eaten first and wherever their long stems had fallen over the sidewalks the larks would invariably be found in January and February. When quack grass failed, foxtail was eaten, and lastly *Amaranthus* was substituted when no other seeds were available. Once or twice larks were noted along the roads feeding on the oats of horse droppings, when snow covered all the weed seed of the subdivision. And again at Ithaca the compost heaps, put out for fertilizer along the garden margins, supplied some food when snow lay deeply over the ground.

At Ithaca, during the spring of 1928, prairie horned larks were observed feeding on *Setaria* (March 1), on *Ambrosia artemisiaefolia* (April 1). A pair of larks were frightened away from an arctiid moth larva (*Apantesis arge*), which I observed the female dig up (March 3). Finally a few adults were collected in March at Ithaca (Connecticut Hill), and examination of the stomachs of six individuals showed that the vegetation consisted of oats, *Setaria, Ambrosia artemisiaefolia,* and waste buckwheat.

In summary of the feeding habits of the prairie horned lark in autumn, winter, and early spring, all that need be said is that the bulk of food taken is that of weed seeds, and the animal food, a much smaller proportion, is almost entirely of those forms harmful to the agriculturist. The lark, in feeding habits, finds for his food those things that appertain to the waste lands he inhabits.

Behavior.—Breeding birds, such as females in abandonment concealment of the nest, or males in flight song, exhibit several distinct flights, but at other seasons the flight is of but one definite character. This is a choppy undulation brought about by three or four rapid, even strokes of the wings interrupted by the space of about two beats during which the wings are closed. A note is uttered on the climb of each undulation. Or again, on prolonged flights, the character of the wing beat is as follows: Long strokes are made, one, two, three (or one, two), with a pause of about one wing beat between each stroke wherein the wings are folded. Then come four to six rapid and successive strokes, which cause a climb. At this time the note *zeet-it* is uttered. Then comes a pause of the length of one or two beats, with wings folded, causing a drop in elevation. These repeat. The bird goes thus: jump, jump, jump, climb (call also), drop, jump, jump, jump.

Voice.—The horned lark, like the goldfinch, usually advertises itself in flight by a definite, unmistakable note. Except for an occasional song, this is about the only sound from the birds in fall and winter. The flight and call notes are several in number, some of them appertaining more especially to the breeding season than to wintering birds, and in that connection they will be considered again. The chief stock in trade of the lark and the one most commonly heard is *p-seet* or merely *zeet*. It is uttered casually on the climb of the ordinary undulating flight, especially on long journeys and in flights of young birds. Adults frequently make low flights over the ground without uttering a note. This *p-seet* is occasionally, sometimes frequently, lengthened to *p-seet-it* during the flight. When flushed the note is *zu-weet* or *zur-reet* (long drawn), *zeet-eet-it*, or *zeet-it-a-weet*, which is so high-pitched and mournful in character that it makes the birds indeed a part of the winter's gale that whips them away.

The season of song extended, in the Evanston region, from mid-January until early in July; in the Ithaca region from mid-February to late in June. With flight songs used for a criterion, it was found that May was the optimum month. The lark sings both from the ground and in the air, under all conditions of weather, though flight songs are most numerous on quiet, mild days, perhaps a little more numerous when the sky is overcast than when it is clear.

The most vigorous period of song extends through nest building, egg laying, and incubation. Perhaps of this period that portion of it when the female incubates allows most song from the male, since he attends the female carefully during nest building and egg laying. The period of least song occurs when the young are in the nest, for the male assists in feeding. Ground songs are regularly distributed throughout the entire day; flight songs seem to be most numerous toward noon and near sundown.

For three months the prairie horned lark is the only singing bird in the open field; but with the coming and establishment of other migrants late in May and in June many other songs will be heard in that region. On June 16, 1926, at Evanston, the prairie horned lark, the last to begin singing that morning, went into flight song at 4:00 A. M. However, the lark almost always closes the singing at night with a long period of recitative which in mid-June, would not close until after 8:00 P. M.

The literature contains several descriptions of the flight song of the prairie horned lark, that of Langille (1884) seeming to be most accurate. He describes the flight. The song he describes as "*quit, quit, quit, you silly rig and get away.*" This is the intermittent type; nowhere in the literature has a description of the recitative been found.

Songs are sung from the ground, from a clod or any other slight elevation, the greatest elevation noted being the roof of a sample apartment put up on the Evanston area; and from the air. The ground songs are similar to the flight songs though rarely as long or as systematically presented. The urge to flight song may come at any time or after an invading male lark has been evicted from occupied territory. Larks will also go into flight song upon approach of a human being or they can be forced to go up by driving them for a time about their territory.

The climb to flight song is distinctive and usually executed without a sound from the bird. The songs, in the air, are of two types: A recitative or rapid monotony of notes usually uttered at the beginning of the flight song, though occasionally at other periods, never over a few seconds in duration, accompanied by a steady beat of the wings; and an intermittent song uttered while the lark sails, about two seconds in duration, followed by a somewhat longer silent period during which the lark flutters up. The recitative can be transcribed as *pit-wit, wee-pit, pit-wee, wee-pit;* the intermittent as *pit-wit, pit-wit, pittle wittle, little, litle, leeeeee.* Large circles are described overhead during the flight song, or the bird heads into the wind if it is strong. The lark closes flight song by a headlong drop to earth with wings tightly folded.

Female larks seem to be unaware of the males in flight song, though other males note the bird overhead. The territory a bird may occupy in flight song is very extensive. Never were two visible birds noted in such a performance simultaneously. The one in the air is left undisturbed though his performance may carry him over many other lark breeding grounds below. Breeding territories are not vertical for a distance above a few feet; the flight song territory is something quite different.

Of several methods employed to determine the heights of larks in flight song, the most accurate was found to be the use of a binocular with an ocular scale. It was determined thus, through measurement of 25 songs, that the lark sings from elevations that vary from 270 to 810 feet. The average was 464.4 feet. Differences in height seemed to be individual variations or due to weather. Thirty times flight songs varied from one minute to five; the average was 2.34 minutes. Intermittents, regularly given, averaged 11.9 a minute.

An Evanston bird sang from song posts on the ground, which, during one entire day, varied a few feet from the incubating female out to 100 yards. The average was 38.66 yards. Ithaca birds, with bigger territories, sang frequently as far as 150 yards from the nest.

Fall.—Young larks flock shortly after nest leaving. If the breeding ground has become untenable owing to vegetation, they seek other regions. Flocks grow larger through addition of adults in

August and September and then smaller as migration begins. In flight the flocks are comparatively compact, but they spread widely when the birds alight to feed or pass the night. During autumn and winter they occupy regions essentially like those in which they breed in March and April, that is, semibarren or almost denuded areas, which may be natural or due to some seasonal condition of agriculture. The Lapland longspurs and the shore larks (*Otocoris alpestris alpestris*) are the only other birds that occupy a habitat with conditions just like those in which the prairie horned lark occurs in fall and winter.

Subspecies of the horned lark vary from such highly migratory forms as the northern horned lark to such strictly sedentary subspecies as the California horned lark. The prairie horned lark is intermediate between these extremes and leaves its breeding grounds for a period of one to two months during winter. This bird breeds north to the southern edge of Canada and migrates south to South Carolina, Kentucky, and Texas. From the northern part of this range it is absent during the month of December and part or all of January. Throughout the remainder of the breeding range some individuals are always present.

OTOCORIS ALPESTRIS GIRAUDI Henshaw

TEXAS HORNED LARK

HABITS

Along the coast of Texas from Galveston Bay to the mouth of the Rio Grande, and for a short distance into Tamaulipas, Mexico, we find this race of the horned lark, rather widely separated from its nearest relatives and usually on the salt marshes or not far from the seacoast. It seems to be most closely related to the prairie horned lark. Dr. Oberholser (1902), in comparing it, says: "This race is quite similar to *praticola*, though considerably more grayish, rather smaller, and with the yellow of throat usually deeper and suffusing also the superciliary stripe. In winter plumage the dark streaking on the breast is frequently heavier. It is fully as gray above as *arcticola*, but is of course easily distinguishable by its reduced size and yellow of throat and eyebrow. It is so much smaller and more grayish than either *hoyti* or *alpestris* that it does not need special comparison."

W. E. Grover wrote to Major Bendire (1895): "The Texan Horned Lark is locally here known as 'Chippie' and 'Road Chippie', as it is essentially a ground bird. It frequents the level, grassy prairies along the Gulf shore, and may frequently be observed in the wagon roads; hence its local name. I do not know how early it arrives in this vicinity [Galveston]; I noticed a few on April 1, and by May they are abundant. The nest is built in a saucer-shaped hole scratched out by the birds, and here it is nearly always placed alongside of bunches of wild chamomile (*Matricaria coronata*) growing close to the road; it is constructed of dry prairie grass and lined with thistle down. The top of the nest is even with the surrounding ground."

Bendire goes on to say: "All the nests of the Texan Horned Lark I have seen are much more substantially built than any of the balance of the subspecies breeding within the United States." He says of an unusually bulky one: "It was placed in a pile of dry cow droppings near the shore of Aransas Bay. The outer walls are chiefly constructed of salt-cedar twigs (*Monanthochloe littoralis*), and the lining consists of dry sea moss picked up on the shore. It measures 6 inches in outer diameter by 2½ inches in height. The inner cup is 3 inches in width by 2 inches in depth." An average nest is only about 4 inches in outside diameter, "and it is sparingly lined with blades of dry grass, and a few feathers."

The Texas horned lark lays ordinarily three or four eggs, which are practically indistinguishable from those of other horned larks of similar size. The measurements of 40 eggs average 21.5 by 15.7 millimeters; the eggs showing the four extremes measure 23.4 by 16.0, 22.6 by 16.6, 19.7 by 15.3, and 21.2 by 14.9 millimeters.

OTOCORIS ALPESTRIS STRIGATA Henshaw

STREAKED HORNED LARK

HABITS

The above name applies to the horned larks that breed in the Pacific coast belt of Washington and Oregon, west of the Cascade Mountains, and southward to Siskiyou County, Calif.

Dr. Oberholser (1902) writes: "This race differs from *merrilli* in much smaller size, deeper and more extended yellow suffusion below, and in the decidedly more brownish color of the upper parts. In autumn and winter, when *merrilli* is often brownish above and shows sometimes as much yellow below as *strigata*, size is the best means of identification. In color it much resembles *alpestris*, but in summer the back is more blackish, in winter the yellow suffusion is more extensive, while its smaller size will of course distinguish it at all seasons. It differs from *hoyti* as from *alpestris*, with the additional character of a deep yellow eyebrow."

While I was in Seattle, in 1911, Samuel F. Rathbun showed me the breeding grounds of the streaked horned lark, of which he says in his notes: "Almost invariably this lark frequents what are known as gravelly prairies. The gravelly prairies are distinctive. They are of rather limited extent, and have a soil that is mostly of fine, smoothly worn gravel, which at times represents the greater part of the soil. In spring and early summer these prairies are strewn with wild flowers, but when the flowers disappear the prairies look more like sterile areas."

J. H. Bowles (1900) says that "around Tacoma these little larks are extremely local in their distribution, large areas of prairie being altogether untenanted, while an exactly similar piece of land will be swarming with them." They nest commonly on the grounds of the Tacoma Golf Club. "The surrounding prairie extends for miles where hardly a dozen pairs of the birds can be found in a day's walk, while on the links last summer I estimated that fully one hundred pairs must have nested. Indeed, so sociable are they that only an occasional nest is placed more than a few feet from the 'putting green' or the 'tee-off'."

Nesting.—Mr. Bowles (1900) says: "The location of the nest may be almost anywhere on the ground, but the soil must be extremely dry. As a rule the birds scratch out a hole for themselves about two and one-half or three inches deep, both birds working, but I have found nests in the hoof prints of cattle, in cart ruts, holes made by dislodged stones and one that was placed in an unused golf hole." One very large nest was "well lined with grass, fir needles and feathers."

Mr. Rathbun (MS.) says of several nests that he has found: "Each time it was in a slight depression of the ground on the dry, open prairie, oftener close to the edge of a dried cow dung, and just within or against the growth of grass, which would be more dense in such a spot. A nest found in very early June was made of dry, gray mosses, bits of dead weed stalks, dead grasses, with various soft substances of plant kind, and it was neatly lined with fine, dry grasses. Its dimensions were: Outside diameter, 4 inches; inside diameter, 2½ inches; height, 2 inches; depth, 1¼ inches."

Eggs.—Three or four eggs constitute the usual set for the streaked horned lark. These are practically indistinguishable from those of other horned larks of similar size, with the usual variations. The measurements of 32 eggs average 21.0 by 15.8 millimeters; the eggs showing the four extremes measure **23.4** by **16.9, 19.0** by 15.7, and 20.0 by **14.9** millimeters.

Enemies.—Mr. Rathbun writes to me that the "crows used to pester the breeding larks, for the grass was so short when and where these were nesting that many a nest was broken up by the crows. Several times I watched the black rascals carefully working over the nesting areas, and I found the torn nests and now and then an eggshell."

OTOCORIS ALPESTRIS MERRILLI Dwight

DUSKY HORNED LARK

HABITS

In naming this race in honor of Dr. J. C. Merrill, who collected the type near Fort Klamath, Oreg., Dr. Dwight (1890) described it as "larger, more broadly streaked above, and blacker than *strigata*, with less yellow about the head and throat, the nape pinker.* * * This is the blackest-backed of all the races, the dark brown of *strigata* having a decidedly yellowish shade, particularly in autumn specimens, whereas *merrilli* is black-brown in spring and strikingly grayish and streaked in autumn."

Its breeding range lies chiefly in the Transition Zone from southern British Columbia, through Washington and Oregon east of the Cascades, and from extreme northern Idaho to northeastern California and northwestern Nevada.

Major Bendire (1895) says of its haunts: "This subspecies is essentially a bird of the foothills (the so-called 'bunch grass country,' *Festuca* sp.?) as well as of the more open and grass-covered valleys and plains occasionally found in the mountains, while it is either rare or entirely absent in the more arid sagebrush plains found interspersed through the same regions."

Dr. Walter P. Taylor (1912) says that, in northern Nevada, these "horned larks exhibit a very marked preference for the vicinity of the fields and dry meadows, as along Quinn River. The birds were frequently encountered, however, on the most inhospitable deserts, although they were more numerous in pleasanter surroundings. We did not observe them at a greater altitude than 7000 feet, although Ridgway noted them as high as 11,000 feet."

Referring to the Lassen Peak region in California, Grinnell, Dixon, and Linsdale (1930) write: "Horned larks in the eastern part of the section were seen on ground where the cover of vegetation was slight. The birds frequented both cultivated land and unbroken land. Areas of sandy as well as hard ground were foraged over. * * * Single birds were sometimes seen perched in the tops of sage bushes. On July 25, 1928, * * * many horned larks were seen among small and scattered sage bushes. During the hot mid-day hours the birds sought shade beneath these small bushes."

Nesting.—C. E. McBee (1931) has found the dusky horned lark nesting very abundantly on the grassy plateau and on the cultivated lands about his ranch in Benton County, Wash. He gives an excellent description of the nests, which I quote:

The nests of the larks are built in a cup-shaped hollow scratched out by the birds, with the top of the nest even with the level of the ground. They are composed of various kinds of vegetable material; sometimes small rootlets, pliable twigs, or pieces of a soft kind of plant which grows on the plateau, but more often a good share of the nest is made up of the thin coverings of wheat stubble and pieces of grass. In only two instances has the writer found any evidence of animal material; these were both small pieces of rabbit fur used as lining. The birds place the nests in the shelter of a clod, small plant, or bunch of stubble. The prevailing spring winds are from the southwest and the nests are always placed directly opposite, so that they are open to view from the northeast. This is an important point to remember when looking for nests; it is much easier to find them while traveling in a southwesterly direction. Although the birds seem to prefer wheat stubble fields for nesting sites, the writer has found nests in plowed ground, in freshly seeded fields, and in grassy sod of abandoned ranches.

Eggs.—He says that "the number of eggs to a set is either two or three—very seldom four." In five years he found "only three sets of four eggs, and the markings on them lead to the belief that one of them was deposited by another female. This was true in each set. Three is the common number, although late sets sometimes contain only two eggs when complete."

"The eggs," he continues, "vary a great deal in size, shape and coloration. The shape varies from ovate to elongated ovate, as a rule, and an average egg measures about .82 by .62 inches. The ground color varies from light gray to, rarely, almost white, sometimes faintly tinged with greenish. The markings are usually dots and small spots that vary in color from dark gray to light brown, sometimes sprinkled rather evenly over the entire egg, in others with a distinct ring around the larger end."

Some eggs of this race that I have seen have a yellowish-white ground color, with a wreath of "buffy brown" and "dark olive-buff" spots around the center of the egg, otherwise finely sprinkled with the same colors. The measurements of 35 eggs average 21.5 by 15.4 millimeters; the eggs showing the four extremes measure 24.4 by 15.8, 21.8 by 16.5, 20.1 by 15.2, and 20.2 by 14.0 millimeters.

Enemies.—Mr. McBee (1931) says that "natural enemies of the birds are probably many. Ravens undoubtedly account for some; snakes for many young birds of the hopping age, and the very plentiful 'Sagerat,' which travels far and is believed to eat eggs, destroys some." But, as the birds prefer to nest in stubblefields, spring plowing "leads to the destruction of untold hundreds of nests, eggs and young birds." He continues:

Allowing a minimum of one nest to every two acres, which the writer believes is too conservative, various operations during the nesting season on the McBee ranch alone would account for the destruction of some three thousand nests. There is no possible way of preventing the destruction, except by rescuing those nests which happen to be noticed by the tractor or team driver. Most of the farmers are acquainted with the fact that the larks are valuable insect eaters, so will not knowingly destroy them. It is a sight, indeed, to see an expensive tractor outfit, or a twenty-horse team, stopped while the driver leaves his place to set a tiny nestful of eggs or young birds to a place of safety. Observations by the writer tend to show that in most cases the parents will return to their young, even though they may be moved several feet; sometimes to heavily incubated eggs, but very seldom to fresh or incomplete sets of eggs. They will also immediately start rebuilding their destroyed nests.

Winter.—The dusky horned lark is present all the year round in eastern Washington and Oregon. In winter immense flocks of this and the pallid horned lark (*arcticola*), with perhaps some other forms of the species, form one of the most conspicuous features in the bird life of the open plains. Mr. McBee (1931) writes:

Weed seeds and kernels of wheat are the main articles of food for the larks, and while there is no snow, these are easily obtained. However, when the ground is covered to some depth, the birds are forced to pick off the seeds from the tumbleweeds which protrude. Soon after the snow falls it is stained in the vicinity of protruding weeds by dirt and pieces of stalk which are thrown off by the birds in the process of obtaining seeds. When all visible seeds have been eaten, hunger forces the larks to the strawstacks, pig-pens and barns of the ranches. During the heavy snow of the winter of 1929–30, the writer fed the birds cracked wheat and succeeded in tolling them to within a few feet of a large bay window in the ranch-house, where for two days they fed in great numbers. Many interesting observations were made during that time of the feeding habits of the birds. Especially amusing was the occasional display of a terrible temper by the larks. A few members of the flocks were bullies of the first water and absolutely refused to allow any other birds to feed on the square patches of ground which were cleared for them. One especially bad tempered bird succeeded in keeping for its sole use an entire feeding plot, almost three feet square, for upwards of an hour. At times fifty or sixty birds would be sitting around in the deep snow surrounding the feed, hungry almost to the point of starvation, feet heavy with droplets of ice, but more content to sit and suffer rather than brave the wrath of the terrible tempered one. Newly arriving individuals, upon alighting in the cleared space, soon found the lay of the land and joined their companions in the snow.

Professor Beal (1910) makes the following tribute to the hardiness of horned larks:

The writer has met them on an open prairie when the temperature was nearly 30 degrees below zero, and though a fierce gale was blowing from the northwest they did not exhibit the least sign of discomfort, but rose and flew against the wind, then circled around and alighted on the highest and most windswept place they could find. Probably they remain through the night in these bleak spots, for they may frequently be seen there after sunset. Most animals seek shelter from wind and cold, even though it be nothing but the leeward side of a ridge or hummock, but the horned lark refuses to do even this, and by preference alights on the top of the knoll where the wind cuts the worst. It seems strange that in so small a body the vital heat can be maintained under such adverse conditions, but if one of these birds be examined, its body will be found completely covered with a thick layer of fat, like the blubber on certain marine animals. This indicates that horned larks have plenty to eat, and that their food is largely carbonaceous.

OTOCORIS ALPESTRIS INSULARIS Townsend

ISLAND HORNED LARK

HABITS

The island horned lark is the small, dark race of the species that inhabits practically all the Santa Barbara Islands off the coast of southern California, though it is more abundant on some of the islands than on others. It was formerly supposed to be identical with the streaked horned lark and was so listed by some of the earlier writers. Dr. Oberholser (1902) describes it as follows:

Like *Otocoris a. strigata*, but darker, somewhat less ochraceous above, less yellowish on breast and abdomen. * * * Notwithstanding Dr. Dwight's statement that he could not distinguish the Santa Barbara Islands birds from *strigata*, they constitute an easily recognizable race, which, though curiously enough most closely allied to *strigata*, yet differs in the darker color above, particularly on cervix and bend of the wing; in the more grayish tone of the back and scapulars; the absence of yellow on the breast; and the much more conspicuous streaking on this part. * * * From *merrilli* the island bird differs in smaller and much more reddish coloration; while from *actia* of the adjacent mainland it may be separated by its conspicuously darker coloration throughout.

I made my acquaintance with the island horned lark on San Nicolas and Santa Barbara Islands, far off shore in the southern group of islands, which I visited as a guest of J. R. Pemberton in his power boat. San Nicolas is the outermost island of the group, over 60 miles from the mainland; it is a barren, windswept isle, about 7 miles long and 3 miles wide. It is mainly formed of soft sandstone, and the high cliffs that tower above the wide sandy beaches are broken by deep rocky canyons and are sculptured into all sorts of fantastic shapes by the erosion of sand and wind. The upper part of the island, about 800 feet above sea level, is a nearly level, or rolling, grassy plateau, which serves as grazing land at times for sheep and horses. Here we found the larks very abundant on February 24, 1929. They were mostly in pairs, or trios, flying about as if mating or preparing to nest. As we found no nests we were probably too early for eggs. We saw the larks also on the benches below the sandstone hills, where scanty grass and other low vegetation was growing. On the sloping sand dunes near the beaches there were many large clumps of iceplant in which the larks have been known to conceal their nests.

We landed the next day on Santa Barbara Island, about 35 miles from the mainland, much smaller and more precipitous than San Nicolas. From the only available landing place, a flat rock near a colony of barking sea lions, we climbed up a steep and narrow trail to the top of the island, which at its highest point is over 500 feet above the sea. The top of the island is a rolling grassy plateau on which herds of sheep were grazing. It is broken by numerous deep gullies, which were overgrown with a curious plant called the "tree dahlia"; it looks like a small tree with a thick woody stem and rough surface, often branching like candelabra, with a growth of leaves and a cluster of bright yellow, daisylike flowers at the end of each branch. The Santa Barbara song sparrows, which were very abundant here, are said to build their nests in these plants. On the hills and the open grassy parts of the plateau, the island horned larks were very common but apparently not yet nesting; at least we found no nests.

Nesting.—There are two interesting nests of the island horned lark, preserved with the surrounding vegetation, in the Thayer collection in Cambridge. They were collected by C. B. Linton on San Nicolas Island on May 12 and 14, 1910, and are accompanied by photographs. The first nest contained three eggs and was sunk into the sand under a thick patch of iceplant that was growing in a dense mass on the sloping side of a ravine and only about 50 feet from the ocean. The nest was made of fine grasses and weeds and was profusely lined with white feathers, apparently gull feathers. The pale eggs in the white nest, framed by the pale gray foliage of the iceplant, make a harmonious picture. The second nest, containing four eggs, was placed among a more scattered growth of iceplant on a sandy slope; it was made of very fine grass and bits of iceplant but contained no feathers.

Another nest in the same collection was taken by O. W. Howard on San Clemente Island on April 4, 1905; this nest, containing four eggs, was located near the summit of the island near a cattle trail; it was compactly made of various coarse and fine weed stalks, fine grasses, lichens, bits of wool, and plant down and was lined with finer bits of the same materials.

Dr. Joseph Grinnell (1897) found a nest on San Clemente Island on June 3, 1897, that "was on the ground in a depression under the broad, obliquely-inclined leaf of a cactus. It was thus well-protected, as no fox could reach the contents without encountering the stiff spines."

Major Bendire (1895) states that H. W. Henshaw found a nest on Santa Cruz Island on June 4, 1875, that was "placed within the cavity of an abalone shell, one of a large heap lying half overgrown with herbage. The whole cavity of the shell was filled by the material, and the eggs looked very pretty as they lay contrasted with the shiny pearly shells clustered about them."

Eggs.—The island horned lark lays three or four eggs, which are practically indistinguishable from those of the other western horned larks of similar size. What few I have seen are of the finely speckled pale type, but they probably show all the variations seen in eggs of the streaked horned lark and the California races. The measurements of 28 eggs average 20.7 by 15.7 millimeters; the eggs showing the four extremes measure 22.3 by 16.0, 20.9 by **16.2, 18.8** by 15.2, and 19.8 by **15.1** millimeters.

The sequence of molts and plumages, feeding habits, and general behavior are apparently similar to those of the neighboring mainland form, the California horned lark, from which this race probably originated, developing independently certain characters similar to those of the streaked horned lark, with which it was formerly confused.

OTOCORIS ALPESTRIS ACTIA Oberholser

CALIFORNIA HORNED LARK

PLATE 53

HABITS

The California horned lark was formerly considered to be identical with the Mexican horned lark (*chrysolaema*), but Dr. Oberholser (1902) has shown that it differs from that race and has given it the above name. He describes it as "similar to *Otocoris a. chrysolaema*, but upper surface paler, more rufescent; yellow of throat and head of not so deep a shade. * * * From *rubea*, with which it intergrades in central California, *actia* differs in the much more pinkish tint of cervix, rump and bend of wing, as well as in its more grayish back which is usually in more or less abrupt contrast to the color of the nape." It is somewhat smaller than *chrysolaema*. Although there are other races of the horned lark in California, *actia* seems to enjoy the widest range, which extends from San Francisco Bay south to northern Lower California and east to the San Joaquin Valley.

James B. Dixon (MS.) says of its haunts: "This bird is a common resident in the salt-grass pastures and drier barren areas, from 2,600 feet elevation to sea level. Its favorite nesting locations are the dry, high humps in salt-grass cow pastures that are well grazed and do not have very much cover. They also nest in vineyards at the bases of the vines and in sparsely growing grainfields of all kinds. Their liking for closely cropped barren areas is well demonstrated by their nesting commonly on golf courses."

John G. Tyler (1913) says of its haunts in the Fresno District: "In former years, when large tracts of land north and east of Fresno were devoted to grain farming, the California Horned Lark was one of the most abundant birds to be found in the district; but it has not responded favorably to the settlement of the country and is now rare in many parts of the valley. * * * It seems that for feeding and nesting these birds must have dry, barren ground almost free from shrubbery."

Nesting.—My experience with the nesting of the California horned lark has been limited to the finding of two nests on March 20, 1929, while driving over the open plains in southern Kern County with

J. R. Pemberton. These rolling plains lie south of Maricopa and east of the Wheeler Ridge; they are covered with a scanty growth of short grass and are much grazed over in places by sheep, of which we saw some large flocks. The larks were abundant here and were mostly running about in pairs. We saw a female running about actively and feeding in a hurried, nervous manner. Mr. Pemberton said that she had just come off her nest to feed and would return to it within 10 minutes. We stopped to watch her, and in about 5 minutes we saw her settle down on some higher ground where there were a number of stones scattered about. We walked up and flushed her off her nest in a little hollow; it was a typical horned lark's nest, made of coarse grasses and weed stems, mixed and lined with finer grass and a few bits of soft, cottony material; it held three well-incubated eggs. The other nest was found by flushing the bird, as we drove within 5 feet of it; it was sunk into the ground between two small pieces of dry cow manure; it also held three eggs.

Mr. Tyler (1913) says: "Nests of this species are built most often in summer-fallow fields, but sometimes in very young vineyards, hay fields from which the crop has been cut, and on the uncultivated plains. Sometimes they are found at the base of a clod or a small accumulation of trash, but in the majority of cases that have come under my observation a small weed or plant, frequently the California poppy, is chosen, probably more for the shade it affords than with any thought of concealment."

Judged from the dates at which eggs have been found, this lark must raise at least two broods in a season and perhaps sometimes three.

Eggs.—The California horned lark lays two to five eggs to a set, usually three. In 35 sets recorded by Mr. Tyler (1913), there were four sets of two, eight sets of four, two sets of five, and the others sets of three. The eggs are similar to those of other horned larks of similar size.

The measurements of 40 eggs average 20.2 by 15.3 millimeters; the eggs showing the four extremes measure 21.5 by 16.1, 21.4 by 16.2, 19.2 by 13.5, and 20.0 by 13.8 millimeters.

Food.—According to W. L. McAtee (1905), "the food habits of the California subspecies (*Otocoris alpestris actia*) were found to differ so remarkably from those of the other horned larks as to merit separate notice. Briefly stated, the difference consists in the high percentage of vegetable—as compared to the animal—food consumed by the California birds." Based on a study of 267 stomachs, collected in every month but May, he says that the "vegetable food composes 91.44 percent of the diet of the California horned larks, while the

larks in the remainder of the country take less than 80 percent of
the same class of food." McAtee continues:

Of the vegetable matter, weed seed, which is 51.1 percent, is less than the
amount of the same kind of food taken by the other horned larks. The rest
of the vegetable food, 40.2 percent, is grain, including that from wild as well
as from cultivated plants. * * *

Of the grain eaten by the horned larks of California, 31.1 percent consists
of oats and 9.1 percent of wheat, corn having been eaten by but one bird.
Oats, then, are the favorite food, and on this account the horned larks are
liable to damage the crop. However, a great part of the oats consumed
probably comes from the wild plants so abundant in all parts of the State,
and the destruction of these is a benefit. * * *

The California horned larks consume only 8.56 percent of animal food [ants,
grasshoppers, and other insects], while the other forms collectively eat 20.61
percent. * * * It appears that the highest percentage of animal matter
is taken in June. This, however, is only 27.7 percent, not much more than
half the highest monthly average for the other members of the species.

Professor Beal (1910) remarks in his summary: "In the final analysis of the food habits of the horned lark there is but one tenable
ground of complaint, namely, that it does some damage to newly
sown grain. This can be largely remedied by harrowing in immediately after sowing, and can be wholly prevented by drilling. The
bird's insect diet is practically all in its favor, and in eating weed seed
it confers a decided benefit on the farmer. It should be ranked as
one of our useful species, and protected by law and by public
opinion."

OTOCORIS ALPESTRIS ENERTERA Oberholser

MAGDALENA HORNED LARK

HABITS

The horned larks of central Lower California, from Santa Rosalia
Bay to Magdalena Bay, have been given the above name. Dr. Oberholser (1907) describes it as "similar to *Otocoris alpestris ammophila*,
but smaller, the upper parts paler and more grayish, the cinnamomeous
of nape, upper tail-coverts, and bend of wing more pinkish. * * *
This new race is in color very similar to *Otocoris alpestris leucolaema*,
but is more grayish above, at least when in good plumage; and has the
eyebrow usually more yellowish; furthermore, the greatly inferior
size of *Otocoris a. enertera* distinguishes it at once. From *Otocoris
alpestris actia*, whose range it approaches most closely, it differs very
much more than from either *Otocoris a. ammophila* or *O. a. leucolaema*, being strikingly paler and more grayish throughout, as well
as somewhat smaller."

It seems to be a resident form on open ground, locally, in the
Lower Sonoran Zone. Nothing seems to have been published on its

habits, which probably do not differ from those of other southwestern subspecies of the horned lark. There is a set of three eggs in the Doe collection, taken by J. S. Rowley near Rosalia Bay, Lower California, on April 23, 1933; these measure 22.8 by 16.5, 22.3 by 15.8, and 21.9 by 15.6 millimeters.

OTOCORIS ALPESTRIS RUBEA Henshaw

RUDDY HORNED LARK

HABITS

Dr. H. C. Oberholser (1902) says of the restricted range of the ruddy horned lark: "The present race seems to be strictly resident, occupying a comparatively circumscribed area in the region drained by the Sacramento River, passing south into *actia* at about the latitude of San Francisco, and northeastward into *merrilli*."

This subspecies is described by Ridgway (1907) as "most like *O. a. actia*, but much more rufescent, the occiput, hindneck, shorter upper tail-coverts, lesser wing-coverts, and sides of breast deep chestnut-vinaceous or dark vinaceous-rufous, the back decidedly brown, broadly streaked with darker. Adult female most like that of *O. a. oaxacae*, but darker and browner above, with hindneck distinctly rufescent; similar also to that of *O. a. alpestris*, but much smaller, back browner with spots less dark, and hindneck and sides of breast more rufescent."

Dr. Oberholser (1902) says of it: "This form is easily distinguishable from all the other horned larks by the peculiar color of the occiput and cervix, which is bright brick red with very little tinge of pinkish, particularly in summer; the remainder of the upper surface is much suffused with the same shade, further differentiating *rubea* from both *insularis* and *strigata*, which races in other respects it closely resembles."

Grinnell, Dixon, and Linsdale (1930) say of the haunts of the ruddy horned lark in the Lassen Peak region: "The main habitat chosen by this race of horned lark within our section was the rocky mesa just above and bordering the Sacramento Valley on the east side [their cut shows a level grassy plain, with numerous rocks scattered over it]. Whenever visits were made to this 'waste' mesa land, in April, May, June, or December, this bird was conspicuous for its large numbers and for being almost the only one to live on that rocky type of ground. In winter the compact flocks would start up when disturbed and would fly so near the ground, below the horizon line, as to be nearly indistinguishable against the background of ruddy-hued earth and rocks."

I can find nothing peculiar mentioned on the nesting or other habits of the ruddy horned lark, which are apparently similar to those of the closely related California horned lark. The eggs are also similar to those of the neighboring race, though what few I have seen are rather more richly colored, more yellowish. Bendire (1895) mentions a peculiar set of eggs collected by Dr. Charles H. Townsend, that are "so suffused with rich reddish brown as to be unrecognizable"; this is evidently a case of erythrism, which occurs occasionally in the eggs of other species. The measurements of five eggs, all I have been able to gather, are **20.8** by 14.7, 19.8 by **15.2, 18.5** by **14.5,** 20.5 by 14.8 millimeters.

OTOCORIS ALPESTRIS OCCIDENTALIS McCall

MONTEZUMA HORNED LARK

HABITS

Although this race of the horned lark was described and named as long ago as 1851, it was not recognized in the first two editions of our Check-list, published in 1886 and 1895, mainly because several good ornithologists confused it with other races. It was, however, included in the third edition in 1910, as Dr. Oberholser (1902) demonstrated that it is a recognizable race. He says of it:

The geographical variation exhibited by this race has been obscured, since Dr. Coues included *occidentalis* in his *leucolaema;* Mr. Henshaw referred it to *arenicola,* and Dr. Dwight to *adusta;* but the form is well worthy of recognition. * * *

From *adusta,* to which it is most closely allied, *occidentalis* differs in its much larger size and decidedly less ruddy colors above, the nape being more pinkish, the back more dusky. It is distinguished from *oaxacae* by much paler, less rufescent colors above, and by decidedly larger size; from both *actia* and *chrysolaema* by greater size, together with paler, more brownish coloration. Although of the same dimensions as *leucolaema,* this form may be separated by the darker, more cinnamomeous or rufescent shade of the entire upper surface, this in summer being particularly noticeable on the cervix; and these characters will serve to determine even doubtful specimens at all seasons.

In its nesting habits, molts, food, behavior, and other habits it probably does not differ materially from other southwestern races.

The measurements of 6 eggs, two sets of three, average 20.7 by 15.6 millimeters; the eggs showing the four extremes measure **21.3** by 14.8, 20.4 by **16.6, 19.9** by 16.1, and 20.7 by **14.8** millimeters.

OTOCORIS ALPESTRIS ADUSTA Dwight

SCORCHED HORNED LARK

PLATE 53

HABITS

In southeastern Arizona, from the eastern slopes of the Huachuca and Santa Rita Mountains, vast grassy plains extend eastward toward the valley of the San Pedro River; these plains are much like virgin prairies, pure grass lands, entirely devoid of other vegetation except for the straggling lines of cottonwoods, sycamores, and willows that mark the underground courses of the mountain streams that flow out from the canyons onto the plains at a few points, but they are hardly in sufficient numbers to make any appreciable break in the continuity of the prairies. These grassy plains form the best grazing lands in that region, but the cattle need vast areas of land to support them even here, and water must be provided for them from driven wells, windmills, and watering tanks, to which the cattle resort from miles around.

Here was the typical home of the scorched horned lark, as we found it abundant and apparently breeding in 1922, though we did not succeed in finding a nest. Similar plains extend westward for a long distance from the Chiricahua Mountains to the eastward, and as far north as Wilcox in Cochise County, where J. S. Rowley tells me that he found it breeding "rather abundantly." A. J. van Rossem (1936) says that it "does *not* reach its western limit at the Santa Ritas, but continues west along the mesa lands to, and all along, the east base of the Baboquivaris." Between this latter range, in Pima County, and the valley of the Colorado River there seem to be no suitable breeding grounds for horned larks, as they shun even a scanty growth of bushes or the ordinary desert vegetation.

In naming this race, Dr. Dwight (1890) describes it as "similar to *chrysolaema* [which he then included with what we now call *actia*], but of a uniform scorched pink or vinaceous-cinnamon above."

Nesting.—Mr. Rowley writes to me that, on the plains near Wilcox on May 24, 1936, he found these birds flying about and mating and located a nest in process of construction, "the female doing a lot of fussing around with a blade of grass in her bill." Five days later, the nest contained three fresh eggs, and the female was sitting very closely. "The bird did her best to use what shelter was available by excavating under the side of a dried cow dung, to get even a little relief from the hot sun."

There is a nest of this horned lark in the Thayer collection in Cambridge that was collected by O. W. Howard in Cochise County, Ariz., on June 10, 1902, and held three eggs. It was placed at the edge of a tuft of grass and was made entirely of very fine grasses, mixed with pappus down; in its somewhat flattened condition, it measures nearly 4 inches in outside diameter.

Eggs.—Apparently the usual set consists of three eggs. The three eggs in the Thayer collection are elliptical-oval and have very little gloss. The ground color is dull white, and the eggs are finely and lightly dotted and spotted over the entire surface with very pale "olive-brown"; on two of the eggs there is a concentration of spots of darker brown and drab in a wreath about the larger end. Probably a larger series of eggs would show the usual variations in colors and markings seen in the southwestern races of the species. The measurements of 31 eggs average 21.4 by 15.7 millimeters; the eggs showing the four extremes measure 23.2 by 15.5, 20.9 by 16.6, 19.4 by 14.8, and 19.8 by 14.4 millimeters.

Fall.—Referring to the vicinity of the Huachuca Mountains, Harry S. Swarth (1904) writes: "Toward the end of July and early in August, young and old gathered together in immense flocks, and were at this time very restless and difficult to approach, flying a long distance when disturbed. They seemed to depart for the south soon after, for on September 5, 1902, on a drive of over twenty miles over country in which they had bred in the greatest abundance, not a single Horned Lark was seen." Later he says (1929): "In the fall there proved to be but a small proportion of *adusta* among the enormous flocks of horned larks that frequented the plains." The predominant number of *occidentalis* at that season indicated "a deserting of the breeding grounds by *adusta* during the winter months."

OTOCORIS ALPESTRIS AMMOPHILA Oberholser

MOHAVE HORNED LARK

HABITS

In describing and naming this subspecies, Dr. Oberholser (1902) states that it is like *O. a. actia* but "may be easily distinguished by its very much paler color above, while its decidedly smaller size, conspicuously more cinnamomeous shade of nape, upper tail-coverts and bend of wing render it readily separable from *leucolaema*. Compared with *occidentalis* it is paler, of smaller size, with the cervix, upper tail-coverts, and bend of wing more cinnamomeous, the upper surface less uniform. It is somewhat smaller than *adusta* and paler, particularly in winter, with the back scarcely or not at all reddish, the demarcation line between cervix and back usually well marked. * * * The young of *ammophila* differ markedly from the young of *actia* in their paler, much more grayish upper parts; being practically indistinguishable from *leucolaema* of the same age."

This subspecies occupies, in the breeding season, a rather limited range in the desert regions of southeastern California, from the Mohave Desert northward to Owens Valley. Dr. Oberholser (1902) says: "This desert race seems to be most typical in the region immediately southwest of Death Valley, California, whence a good series of specimens was brought back by the Death Valley expedition of 1891. The breeding birds in this series were identified as *arenicola* (=*leucolaema*), the winter specimens as *chrysolaema* (=*actia*)."

I can find nothing in the report of this expedition, or elsewhere, to indicate that the habits of this subspecies are in any way different from those of the other desert-loving races.

OTOCORIS ALPESTRIS LEUCANSIPTILA Oberholser

SONORA HORNED LARK

HABITS

From extreme southern Nevada southward along the Colorado River in western Arizona and extreme southeastern California and into northeastern Lower California, we find this extremely pale desert race. In describing it, Dr. Oberholser (1902) says: "This new race is the palest of all the American horned larks, not excepting *pallida* itself, from which form it further differs in lacking much of the cinnamomeous tinge of the upper parts, particularly on the cervix and bend of the wing. Other characters distinguishing *leucansiptila* from *actia* and *ammophila* are the more uniform upper surface and the much more pinkish shade of the cervix, upper tail-coverts and bend of wing; from *occidentalis*, the decidedly smaller size; from *adusta*, the conspicuously less reddish upper surface; from *leucolaema*, reduced size and more uniform upper parts."

I can find nothing published on its habits, which probably are similar to those of the other desert races in neighboring regions.

Family HIRUNDINIDAE: Swallows

CALLICHELIDON CYANEOVIRIDIS Bryant

BAHAMA SWALLOW

HABITS

This lovely bird, with its velvety green crown and back, its steely blue wings, pure white under parts, and long forked tail, is one of our most beautiful and graceful swallows. Except for occasional wanderings it seems to be confined mainly to the Bahama Islands, but recent collectors do not seem to find it very common even there. Dr. Henry Bryant (1859), who described and named this swallow, says: "I saw them during the whole of my stay at Nassau [Jan. 20 to May 14], but only on the first mile of the road leading to the west end of the island. They were so abundant there that thirty or forty could be seen at almost all times."

According to C. J. Maynard (1896), it was not so abundant at the time of his visit as Dr. Bryant's account would indicate, for he says: "I do not think that the entire number inhabiting the island of New Providence, even in June, amounted to over fifty individuals."

He seemed to think that this swallow is given to wandering about among the islands, or elsewhere, disappearing from some sections at certain times, its movements probably influenced by the abundance or scarcity of its insect food; he says that it disappeared from New Providence in March and returned again in June.

The Bahama swallow was added to our fauna by W. E. D. Scott (1890), who collected a specimen, an adult male, on Garden Key, Dry Tortugas, Fla., on April 7, 1890. William Brewster (1897) purchased from Mr. Scott "a young bird in practically unmixed *first plumage* but with fully developed wings," which was collected by W. S. Dickenson at Tarpon Springs, Fla., on September 3, 1890; this bird was supposed by the collector to be a young tree swallow. Both of these specimens are now in the Museum of Comparative Zoology in Cambridge.

Mr. Maynard (1896) reports two sight records of the Bahama swallow in Florida in winter: "Once I saw quite a flock of them sailing high in air over Key West, and once a single specimen passed within a few feet of my head as I stood on the banks of Indian River. This last bird was in company with White-bellied Swallows, but on both occasions the deeply-forked tail rendered the bird at once distinguishable."

Outram Bangs (1914) reports that W. Cameron Forbes collected a pair of Bahama swallows at Nipe Bay, Cuba, on March 8, 1914; these specimens are also in the Museum of Comparative Zoology. "Mr. Forbes says that this swallow was exceedingly abundant and generally distributed at Nipe Bay, feeding in the manner of its kind or resting on the telegraph wires. * * * Whether the Bahama swallow is resident in northeastern Cuba or only occurs there as an abundant winter visitor we cannot say."

Nesting.—We do not know much about the nesting habits of this swallow, and, so far as I know, there are no eggs in American collections. Dr. Bryant (1859) says: "I did not succeed in finding their nests, and could not ascertain whether it bred on the island or not. I killed no specimen after the 28th of April; up to this date the genital organs exhibited no appearance of excitement." Mr. Maynard (1896) says that, at Nassau, "they did not become common until June 4th, and on the 10th they were flying about the streets moving quite slowly and heavily as all Swallows move when about to breed. A few days later I saw the Swallows alighting on the ground about the house late in the day picking up strings, feathers, etc., for the nests, and when on the schooner, 'Isle of June,' just before sailing for New York I saw one enter a hole under the eaves of a building which stood on the wharf, so it is highly probable that in breeding habit they resemble the White-bellied."

Plumages.—Ridgway (1904) describes the young bird, presumably in juvenal plumage, as "brown above with a strong luster of oily green, a little more pronounced on the back and wing-coverts; head and upper tail-coverts more sooty brown, as also the upper margin of the ear-coverts; cheeks, ear-coverts, and under surface of body white, with a patch of sooty brown on the sides of the upper breast."

The sex of the specimen from which the above description was taken is not stated; but if Maynard (1896) is correct in stating that the sexes are unlike in the immature plumage the description fits the young female rather than the young male. He says that the young male is "quite similar to the adult female", which he says is similar to the adult male, "but duller. * * * Beneath the white is much less pure, being inclined to slaty beneath the eye, on the ear coverts, across breast and on sides of breast."

We have no information on the molts, but there is probably a post-juvenal molt, more or less complete, and subsequently one complete postnuptial molt each summer or fall.

Food.—The stomachs of those dissected by Dr. Bryant (1859) "contained almost entirely small dipterous insects, some of them extremely minute." Maynard (1896) says that "they feed upon diptera and coleoptera".

Behavior.—Dr. Bryant (1859) thought the flight of the Bahama swallow more like that of the barn swallow than that of the tree swallow, but Mr. Maynard (1896) considered it halfway between the two, the Bahama swallow being swifter on the wing than the tree swallow, but not so swift or so graceful as the barn swallow. The latter observer states:

They have the habit of pausing, or seemingly pausing in the air, observed in the White-bellied, especially when flying against the wind. Their movements are, I think, influenced by the heat as they fly more slowly on a warm, still day. * * * No note whatever has yet been heard. * * *

On March 8th we came upon a flock of some twenty or more in a cup-like hollow in the piney woods far to the west of the town of Nassau where the trees had been cut away for the purpose of making charcoal, leaving an open space of some five or six acres. The birds were darting gracefully and rapidly about, much lower than I had ever seen them keep before, often nearing the tops of the low herbiage. Here we shot four fine males, the first of which fell only wounded. As I approached him he uttered a clucking note, which attracted the attention of his hurrying comrades and they darted down at me as they dashed past, some answering the cries of the prostrate bird with a low, musical chirping. Upon shooting two or three times at them, however, they speedily dispersed, and in a few minutes disappeared.

Field marks.—The adult male or female should be easily recognized on the wing, as it is the only one of our swallows that has a white breast and a deeply forked tail. The white-breasted tree swallow has a nearly square tail; and the barn swallow, with a forked tail, has a brown breast. The immature Bahama swallow closely resembles the young tree swallow and might easily be mistaken for it, though it has a grayer head and back.

Range.—This is a nonmigratory species confined to the Bahama Islands, chiefly the northern part of this archipelago (Great Bahama, Abaco, Eleuthera, New Providence, Andros, and the Anguilla Islands).

Casual records.—A specimen was collected on Garden Key, Dry Tortugas, Fla., on April 7, 1890; and another was taken on September 3, 1890, at Tarpon Springs, Fla.

TACHYCINETA THALASSINA LEPIDA Mearns

VIOLET-GREEN SWALLOW

PLATE 54

HABITS

This beautiful swallow is well named; the soft, velvety plumage in subtle hues of violet and green on the upper surface, the conspicuous white patches on the sides of the rump, and the pure white lower surface combine to make a charming whole, a dainty feathered gem. It enjoys a wide distribution west of the Great Plains and from Alaska to Mexico, and in some places it is one of the most abundant species. A. E. Shirling (1935) writes: "The violet-green Swallow * * * is to the Colorado mountains what the English Sparrow is to eastern and central states. It is the most common bird about cottages and towns. In respect to relative abundance, it exceeds the English Sparrow for the sparrow's range is confined to human surroundings of houses, barns, and picnic grounds. The violet-green Swallow, while most abundant in the neighborhood of human dwellings, ranges widely up the mountain slopes and unfrequented forest lands."

Near the coast in southern Alaska, as far north as Ketchikan, in British Columbia and in the vicinity of Seattle, Wash., we recorded the violet-green swallow almost daily, where it seemed to be a common summer bird at low altitudes; here it frequented the towns and villages, much as our tree swallow does in the East, apparently trustful of human society and ready to accept such nesting accommodations as human structures offered. It was also seen commonly in clearings in the woods where there were dead trees for nesting sites, especially in the vicinity of lakes and streams. In California, also, its haunts and habits are similar, strongly reminding us of our familiar tree swallow, which it largely replaces in the West, though sometimes associated with it.

In the mountain ranges of Arizona we found it at the higher elevations, nesting mainly above 7,500 feet and among the pines near the summits; however, a few pairs could generally be found at lower levels in the wooded canyons. In New Mexico Mrs. Bailey (1928) found it breeding in the mountains at elevations varying from 11,300 to 6,500 feet. James B. Dixon tells me that it is a bird of the higher elevations in California; he found it nesting near Mono Lake at 6,500 feet and at Escondido at 1,400 feet. Evidently it prefers to nest in the mountains throughout the southern portion of its range.

Spring.—The violet-green swallow is an early migrant and has a long journey before it to reach its northernmost summer resort. Theed Pearse writes to me that he has seen it passing through British Columbia in great numbers in spring, during the last few days in March and through April.

Mr. Rathbun says, in his notes from Seattle: "The first arrivals in spring will generally be seen hawking about over some low meadow. Then, for a short period following, they are likely to be observed flying about the localities where they have been in the habit of nesting in previous years. Then they disappear, not to be seen again during the day. As the season progresses, the birds begin to linger longer about these particular spots, and, when early April comes, the swallows become fairly established in localities where it is their intention to breed. Their twittering notes are a most welcome sound in the spring."

He tells me that in 1936 bad weather, low temperature, and several inches of snow on March 27 caused the death of many of these swallows. One of his friends found 20 or 30 dead violet-green swallows within the space of a week along a short stretch of highway, and he had reports of others that had succumbed.

Courtship.—Mr. Rathbun (MS.) records in his notes his experience with early-morning flight songs, heard long before sunrise, of the violet-green swallow, which appear to be a part of the nuptial performance. I quote directly from his notes as follows: "June 5, 1923. This morning I arose at 1.45 to make some notes. It is mild, the stars shine brilliantly, and at this hour there is no moon. 2.26 A. M. The first notes are heard coming from a violet-green swallow. Within the next two minutes, notes from a number of the swallows were heard, the birds flying rapidly around. At this time, faint traces of the dawn show in the sky at the northeast, but the stars are very bright as it is clear. At 2.33 A. M., the violet-greens are heard on every side, their notes increasing within the next five minutes, the birds seeming to be at quite a low height. As the morning light slowly grows, the calls of the swallows are even stronger; interjected at times were some musical notes heard only occasionally at the time of the nuptial period. 3.08 A. M. The notes given by the swallows are now becoming less. At this time, the stars are faint. 3.11 A. M. The swallows have ceased calling, having flown about for 45 minutes."

In recording a similar experience on June 3, 1929, he writes: "The twittering notes, given by the birds in the very early morning hours, are somewhat like their ordinary ones, but there is this difference: the notes are of a shrill, chattering character, more rapidly given, and almost continuous, as if the bird was excited, though sometimes there are very short pauses between the notes. The flight of the swallow is also very erratic, more of a dashing or twisting kind, and each bird seems to somewhat confine itself to a limited territory in the vicinity of where it is nesting. From a long observation of the violet-green swallow in the dim morning hours before sunrise, I incline to the opinion that those in flight are the male birds, and their unusual actions at this time can be attributed to the procreative impulse which is at its height at this time, for the actions are at an end in the latter part of June, when young swallows are in the nest."

Nesting.—The violet-green swallow nests in holes, cavities, and crevices in a variety of situations. In suitable localities, where the birds are often very abundant, the demand is sometimes greater than the supply, competition for the available cavities is keen, and the birds cannot be too particular in the choice of a nesting site. Where nesting cavities are numerous, the swallows often form colonies, with many nests in a suitable tree or cliff. Charles F. Morrison (1888), in Colorado, has "seen as many as twenty pair in a single dead pine, and four or five pair in one limb which had been used first by the woodpeckers."

Throughout much of its range this swallow still continues to nest in localities more or less remote from human habitations and under primitive conditions, such as in deserted woodpecker holes and natural cavities in trees, or cracks, crevices, or holes in various kinds of rocky cliffs. Dawson and Bowles (1909) say that, in eastern Washington, they still nest "to a large extent upon the granite or lava cliffs. In the last-named situations they utilize the rocky clefts and inaccessible crannies, and are especially fond of the smaller vapor holes which characterize the basaltic formations. Favorable circumstances may attract a considerable colony, to the number of a hundred pairs or more."

In Arizona they nest mainly high up on the mountains in old woodpecker holes in the pine belt, but a few pairs nest in the limestone cliffs that form the walls of the upper canyons. In other places they have been reported as using the old nests of cliff swallows and even the burrows of bank swallows. An unusual nesting hole was discovered by Edward R. Warren in Colorado; he says in his notes: "It was in a dead cottonwood about 5 feet above ground, in what appeared to be an abandoned flicker hole. The entrance was rather curiously situated. The bark had been split at the above height and thence down to the ground, the split widening to a foot or more. The bark had kept on growing and the edges were rolled in. The entrance to the nest was under the roll, and was not visible until one stooped down and looked up."

In the Yukon Valley, Alaska, Dr. Louis B. Bishop (1900b) "frequently saw colonies of from six to ten birds of this species, and one near White River that must have contained over fifty. They were nesting about the cliffs as a rule, but several times we saw them enter holes in banks similar to those of *Clivicola riparia*, while at Fort Selkirk they were nesting in the interstices between the logs of the cabins."

In the vicinity of Seattle and Tacoma, Wash., the violet-green swallow, according to Thomas D. Burleigh (1930), "has readily accepted the benefits to be derived from the proximity of man, for during the breeding season they were rarely seen far from houses. * * * A nest found June 9 held four slightly incubated eggs, and was on a beam in a corner inside the attic of an old unused house. It was a large mass of weed stems, grasses and feathers, the middle being neatly cupped and well lined with large chicken feathers. Another nest found June 13 held six half incubated eggs, and was in a cavity between two logs in the side of an old log cabin."

S. F. Rathbun writes to me from Seattle: "For 16 consecutive years a pair of these swallows nested in a box placed under the eaves of our house in the city." Some time was usually consumed in the selection of nesting sites among the four boxes that he had set up, as much rivalry took place between these swallows, a pair of tree swallows and a pair of English sparrows. The sparrows were allowed to occupy the least desirable box, and he kept them busy all summer by removing the eggs at intervals. He says that the violet-green swallows construct their nest in a very leisurely manner, sometimes requiring as much as three weeks to complete it, and then a day or two may elapse before the first egg is laid. "The nesting material consists of an abundance of straws and dry, dead grasses often with bits of string, and is warmly lined with an abundance of feathers, at times a few horsehairs being woven into the lining."

In the Kootenay Valley, British Columbia, Joseph Mailliard (1932) found this swallow "commonly nesting, in the height of the flood, in the dead trees and stumps of the river bottom in company with the preceding species [tree swallow]. A number of pairs of the Northern Violet-green Swallow were found in possession of a lumber yard in town and were nesting inside large, square piles of board lumber that was loosely cross-laid so as to leave space for circulation of air for drying purposes."

Eggs.—The violet-green swallow generally lays four or five eggs to a set, but sets of six are not very rare, and as many as seven have been recorded. The eggs are usually ovate and pure dead white, without markings. The measurements of 50 eggs average 18.7 by 13.1 millimeters; the eggs showing the four extremes measure 20.8 by 12.7, 17.3 by 13.7, 16.3 by 12.7, and 17.3 by 12.2 millimeters.

Young.—Mr. Rathbun tells me that the incubation period is 13 or 14 days, and that the young are on the wing within 23 days; they remain near the nest for a few days and then disappear from the locality, to be seen only occasionally thereafter. During the first few days after they leave the nest they make only short flights, but as the birds become stronger these aerial excursions become more extended. When at rest on some favorite perch, such as a telephone wire, the young are constantly calling to their parents and are fed by them from time to time. Frequently one launches itself into the air to meet its parent and take its food while on the wing, a graceful performance.

Leon Kelso (1939) made some observations in Colorado on the feeding of young violet-green swallows. "The female secured most of the food for the young; the male brought something only occasionally. When returning with food they would sweep about the nest in wide arcs, then, coming to a point about 50 ft. or more in front of it, would fly directly to and in the entrance." He noted that food was brought at frequent but widely varying intervals; during the first afternoon the "visits came at 1 to 30 second intervals." On another day, August 1, the intervals varied from 20 seconds to four minutes, considerably over a minute intervening in most cases. Periods spent in the nest between feedings varied from one second to over two minutes.

Mr. Shirling (1935) noted on two occasions that more than one female attended a brood of young, entered the nest and apparently fed them:

One male bird stood guard and rarely entered the nest. He was kept busy chasing other male swallows away, but did not seem to object to either of the females coming. [At another nest] there was more than one female swallow interested in the nest. At one time three female birds and one male were at the nest. The male, keeping guard, paid little attention to the other birds unless it was a bird of some other species that arrived, or a male Violet-green Swallow. * * *

Female No. 1 seemed to have priority of claim to the nest. She often remained with her head at the doorway and pecked at intruders. She was also on very good terms with the male. The other female birds seemed to be merely meddlesome busybodies who had no home of their own nor young to care for, and, like a cat that has lost her kittens, just had to have some one to mother.

Apparently only one brood is reared in a season; Dawson (1923) says so, and I can find no evidence of a second brood. Probably a second attempt would be made if the first attempt failed. Harry S. Swarth (1904) says that in the Huachuca Mountains, where these swallows breed at the higher elevations, "toward the end of July, 1902, after the young were out of the nest, they moved down into the lower parts of the mountains, where young and old were seen together in large flocks; the young birds being, in many cases, still fed by their parents."

Plumages.—I have seen no unfledged young, which are probably like those of the closely related tree swallow. Ridgway (1904) describes the young bird in juvenal plumage as "above plain sooty grayish brown, darker on back, where faintly glossed with purple, violet, or bronze; a white patch on each side of rump, as in adults; lores dusky gray; auricular region and postocular spot mottled sooty brown and grayish white, or uniformly of the former color; under parts grayish white anteriorly, pure white posteriorly, the chest usually tinged with sooty brown, especially laterally, where sometimes with a distinct narrow transverse patch of brown."

This plumage is worn through the summer and perhaps early fall; apparently a complete postjuvenal molt occurs in September and October, producing a first winter plumage, which is practically the same as the winter plumage of the adult. Both adults and young birds, in fresh fall plumage, have the tertials conspicuously margined and tipped with white; these white edgings disappear by wear during the winter. Adults have one complete annual molt in late summer and in the fall. Wear produces only a slight effect on the spring plumage. Females, after the postjuvenal molt, are always much duller in color than the males, with gray mottling on the sides of the head, making them readily distinguishable.

Food.—The violet-green swallow seems to live entirely on insect food, taken on the wing. It does not differ materially from other swallows in this respect. Prof. F. E. L. Beal (1907) examined 67 stomachs and found that bugs (Hemiptera), mostly leafhoppers and leafbugs, constituted 36 percent of the food. Diptera (flies) came next, 29 percent. Hymenoptera amounted to 23 percent and in July were mostly made up of ants; six stomachs taken on one day were entirely filled with ants, and another, taken the next day, was half full of them; the ants were evidently swarming on the wing at that time and were easily caught, as very few were taken at other times. The remainder of the Hymenoptera eaten were wasps and wild bees. Beetles made up over 11 percent of the food, 3 percent of which were useful species and 8 percent harmful beetles. "Three stomachs, collected at the same time in Carmel Valley, are of interest. They contained respectively 42, 45, and 40 percent of scolytid or engraver-beetles. This was in the region of the Monterey pine (*Pinus radiata*), and there is no doubt that these insects prey upon those trees, and probably were taken when migrating in a swarm to fresh foraging grounds. A few moths, with some unidentified insects, make up the remainder of the animal food, a little more than 1 percent."

Behavior.—All swallows are swift and graceful in flight, but the violet-green combines these attractive qualities with exquisite beauty of form and color and with fearless confidence in human friendliness. It is a pleasure to watch these dainty birds, as they sweep by in loose flocks or smaller groups low over the herbage in the open fields, over the surface of some small pond, or up and down the bed of some canyon stream, winnowing the air for insect prey. It is delightful to have one show its confidence in human nature by gliding by us almost within arm's reach, showing alternately its snow-white breast and the metallic colors of its back, with an occasional flash of golden sheen. But they do not always course low over the ground or water; on clear, warm days, when the insects are flying high, they are often seen circling at a great height, when they seem to be just white-breasted swallows. They are often seen perched in long rows on the telegraph wires, like other swallows, or sunning themselves in the tops of leafless trees. They have endeared themselves to many bird lovers by the almost friendly way they have accepted the artificial homes erected for them in our towns and cities, returning each spring to greet their human friends.

Voice.—The early-morning flight notes are referred to under courtship. Ralph Hoffmann (1927) mentions these as follows: "Before dawn when the Robin chorus is in full swing, Violet-green Swallows fly about in the darkness repeating over and over two or three slight notes, *tsip tseet tsip*. Their ordinary notes are a rapid twitter."

Field marks.—The violet-green swallow might easily be mistaken for a tree swallow at a distance, though its wing strokes are more rapid and it sails less; it appears to me somewhat chunkier in form, though it is actually very little shorter. When reasonably near, the white patches on the sides of the rump are very conspicuous from either above or below. At short range, the white of the throat may be seen to extend well up on the sides of the neck and over the eyes, whereas in the tree swallow the eyes are entirely surrounded in the black of the crown. The colors of the back may be distinguished only in good light.

Enemies.—The English sparrow is one of the worst enemies of this and other swallows, as it preempts the nesting boxes or attempts to oust the swallows from them. The sparrows begin nesting operations earlier than the swallows and thus have the advantage, but, when once well established, the swallows are not easily driven out. Bird lovers that want to have swallows in their bird boxes must keep the sparrows under control.

Violet-green swallows are sensitive to weather conditions; migrants in early spring are often confronted by a sudden cold spell, which forces them to retreat. At such times, when insects are scarce or driven to cover, it is impossible for the swallows to obtain sufficient nourishment and many of them perish and are picked up in an emaciated condition.

Mr. Rathbun tells me of one that was building a nest in one of his boxes, using some horsehair in the lining; its wing became entangled in one of the hairs, and it might have perished if it had not been released by Mrs. Rathbun.

Fall.—Mr. Rathbun (MS.) says: "After the young are on the wing these birds disappear from the localities where they have nested unless such happen to be in the vicinity of water, for the species seems to be partial to such surroundings in the vicinity of streams or lakes or even along tidewater. Here the birds will be seen hawking above the low-lying fields or along the littoral line and, on occasions, perched in numbers on the telegraph wires in the vicinity. These localities appear to be the resorts of the violet-greens up to the time when they take their departure in the autumn."

Theed Pearse tells me that this is the first of the swallows to leave the vicinity of Vancouver Island in summer, very few being left by the end of July. Farther south they gather in large flocks during August and depart from Colorado and New Mexico early in September.

Dr. Joseph Grinnell (1908) writes: "Many adults and full-grown young were found congregated about the shore of Bear lake [San Bernardino Mountains, Calif.] July 30 to August 2, 1905. On the bare branches of one dead pine on the north shore of the lake, July 31, hundreds (without exaggeration) of violet-green swallows were perching, mostly young-of-the-year. Individuals were constantly coming and going, and occasionally nearly the entire flock would launch out with loud twitterings, only to gather again within a few minutes. It made me dizzy to watch the restless throng. A similar gathering, though on a smaller scale, was witnessed near the South Fork of the Santa Ana, July 24, 1906."

Winter.—A few violet-green swallows spend the winter in the relatively warm Imperial Valley in southern California, in the lower Colorado Valley, and in northern Lower California. But the main

winter range is in Mexico and Central America, as far south as
Guatemala and Costa Rica. Dickey and van Rossem (1938) refer
to its status in El Salvador as an "abundant, though extremely local,
midwinter visitant to the seacoast and mountains." They say
further:

The abundance of this species at tidewater and in the interior mountains,
with a vertical gap of some 6,000 feet in distribution, cannot be explained at
the present writing. In the winter of 1925–1926, violet-green swallows were
to be seen in large numbers over the tidal flats at Puerto del Triunfo, where
they fed from about an hour before sundown to dusk, in company with the
even more common rough-wings. During the day they spread out over the
coastal plain a short distance inland where their actual numbers were not
so apparent. This was the only locality in which the species was detected
until, in 1927, it was found to be extremely common above 6,000 feet on Los
Esesmiles. Here, flocks were seen daily from February 1 to March 10. On
warm days they were usually circling back and forth over the cloud forest at
8,000 to 9,000 feet, but when, as was usual, that hunting ground was blanketed
in wind-driven fog and clouds, they worked over the pines at 6,000 to 7,000
feet on the sunny, southern slope of the mountain. All of the specimens taken
are typical *lepida*.

DISTRIBUTION

Range.—Western North America, and Mexico.

Breeding range.—The breeding range of the violet-green swallow
extends **north** to Alaska (Iliamna Lake, Lake Clark, Fairbanks, and
Circle); Yukon (Dawson, Fort Selkirk, and Fifty-mile River); and
central Alberta (probably Henry House and Red Deer). **East** to
Alberta (Red Deer and Canmore); central Montana (Great Falls,
Little Belt Mountains, and Billings); western South Dakota (Spear-
fish, Elk Mountain, and Indian Creek); northwestern Nebraska
(Squaw Canyon); central Colorado (Fort Collins, Boulder, Little-
ton, and Beulah); central New Mexico (Taos Mountain, Glorieta, and
San Mateo Mountains); western Chihuahua (Pinos Altos); Veracruz
(Orizaba); and Oaxaca (Mitla). **South** to Oaxaca (Mitla); State
of Mexico (Valley of Mexico); and southern Baja California (Cape
San Lucas). The western limits of the range extend northward
along the Pacific coast from southern Baja California (Cape San
Lucas, San Bernardo Mountain, and La Paz); to Alaska (Thomas
Bay, Cordova, and Iliamna Lake).

Winter range.—While the species has been recorded at this season
north to central California (Point Reyes, Point Lobos, and Hayward), the normal northern winter limits appear to be southern California (Salton Sea) on the west and northern Veracruz (mouth of the
Tuxpan River) on the east. The winter range extends **south** to western Guatemala (San Geronimo and Chichicostenango) and El Salvador (Los Esesmiles and Puerto del Triunfo). This species has
been recovered twice from Costa Rica (Matina River and Bebedero),
but this country seems to be south of the regular winter range.

The range as outlined is for the entire species, which has been separated into three subspecies. The typical race, known as the Mexican violet-green swallow (*Tachycineta thalassina thalassina*), is confined to the Mexican tableland; the northern violet-green swallow (*T. t. lepida*) occupies all the balance of the range except the Cape district of Baja California, which is occupied by the San Lucas swallow (*T. t. brachyptera*).

Spring migration.—Early dates of spring arrival are: New Mexico—Silver City, April 17. Colorado—Boulder, April 22. Wyoming—Yellowstone Park, May 11. Montana—Anaconda, May 6. Alberta—Red Deer, May 7. Arizona—Tombstone, February 18. California—Los Angeles, February 19. Oregon—Portland, March 7. Washington—Tacoma, March 9. British Columbia—Okanagan Landing, March 6. Alaska—Fairbanks, May 8.

Fall migration.—Late dates of fall departure are: Alaska—Thomas Bay, August 22. British Columbia—Okanagan Landing, October 15. Washington—Seattle, October 19. Oregon—Portland, October 7. California—Kernville, October 28. Montana—St. Marys Lake, August 10. Wyoming—Yellowstone Park, August 18. Colorado—Boulder, September 9. New Mexico—Chloride, September 23. Arizona—San Francisco Mountain, September 28.

Casual records.—A specimen was taken at South Kenwood in the Calumet region, near Chicago, Ill., on May 4, 1897. In Alaska, two specimens were obtained on St. Paul Island on August 22, 1914, and one in immature plumage was taken at Point Barrow on August 26, 1929.

Egg dates.—California: 31 records, May 1 to July 1; 15 records, May 25 to June 11, indicating the height of the season.

Baja California: 3 records, May 1 to June 16.

Oregon: 19 records, May 27 to July 5; 9 records, June 4 to 17.

TACHYCINETA THALASSINA BRACHYPTERA Brewster

SAN LUCAS SWALLOW

HABITS

The violet-green swallow of the southern half of Lower California was named and described by William Brewster (1902) as follows:

Similar to *T. lepida* Mearns, but with the wing decidedly and apparently constantly shorter. * * * The Violet-green Swallow of the Cape Region furnishes an interesting illustration of the recognized fact that isolated, non-migratory birds are given to having shorter wings than those which regularly perform extended journeys, for in respect to the length of the wing it is almost if not quite as much smaller than the form which breeds in the regions further to the northward (i. e. California, Oregon, Washington, and British Columbia) as the latter is smaller than true *thalassina* of the Mexican table-land still further to the southward. * * *

This is the characteristic Swallow of the Cape Region, if not the only representative of the Hirundinidae, excepting the Western Martin, which breeds there regularly and plentifully. About La Paz and other places on or near the coast it perhaps occurs only in winter, as Mr. Belding indicates, but Mr. Frazar found it common on the Sierra de la Laguna in May and early June, and at Triunfo and San José del Rancho in late June and July. On the summit of La Laguna it was nesting late in May, and one was seen flying over the highest peak of this mountain on December 2.

J. S. Rowley writes to me: "Comondú was the only place we encountered these little swallows at all abundantly, and here they were rather common because of the creek running nearby and the only surface water for miles." He found several nests here on May 1, 1933, "containing eggs in various stages of incubation. Sets of two and three seemed more common, but one set of five was taken. All the nests found were made of feathers and hair, and placed well up in old woodpecker holes in the cardons [saguaros]."

There is an interesting nest, with two eggs, of the San Lucas swallow in the Thayer collection in Cambridge, taken by W. W. Brown, Jr., at La Paz on May 31, 1908. It is reported to have been placed "in a depression on the face of a cliff among the rocks." It must have been in a rather large cavity that sloped downward at the outer edge, for the base of the nest measures over 5 inches in diameter at its widest part, while at the inner end of the cavity the nest proper is only 2 inches high. The bulky base of the nest is made of coarse weed stems and long, fine grasses, circularly and firmly woven to hold the nest cup in place; the nest cup is neatly and compactly woven of the finest grasses and fibers, and is lined with black and white hair; there are no feathers anywhere in its composition. The inner cup measures about 2 inches in diameter and about 1 inch in depth. It is not only the most peculiar swallow's nest that I have ever seen, but decidedly the neatest.

The eggs are ovate, pure white, and have hardly any gloss. The measurements of 9 eggs average 17.3 by 12.9 millimeters; the eggs showing the four extremes measure 18.3 by 13.0, 16.8 by 13.5, 16.8 by 13.0, and 17.0 by 12.0 millimeters.

IRIDOPROCNE BICOLOR (Vieillot)

TREE SWALLOW

PLATES 55-57

HABITS

CONTRIBUTED BY WINSOR MARRETT TYLER

Spring.—The tree swallow is the earliest of the six swallows that move up the Atlantic coast in spring on the way to their breeding grounds in the Northern States. It is a hardy species; many individuals spend the winter in the southern part of the United States and hence have a comparatively short migration to their summer homes. As they press northward, arriving in New England often late in March, they may encounter frost and snow, for at this season the advance of spring is very variable here. Under such conditions they often disappear for a time, and it is supposed that they retreat until spring steps northward again.

It is interesting to watch these early arrivals when caught by unfavorable weather. As an example of their resourcefulness, Richard J. Eaton (MS.) relates the following account: "On the morning of March 19, 1939, G. W. Cotterell and I closely observed three tree swallows feeding on the ice on Heard's Pond, Wayland, Mass. The fields were blanketed with hard, crusty snow. The day was sunny and calm, the temperature slightly below freezing in the shade. We first saw the birds from a distance, flying back and forth over a restricted portion of the old white ice. The entire pond and the flooded meadow between it and the Sudbury River were frozen tight. We hitched ourselves (in a sitting position) to within 10 feet of the feeding swallows. Watching at this close range, we could see that they were feeding on what looked like seeds frozen to the surface of the ice in long windrows or irregular bands about 20 yards from shore. Maneuvering within a few inches of the ice in a zone about 30 feet long, they picked up the food while they were *on the wing*, making an audible clicking sound with their bills when pecking at the seeds. The swallows had discovered that a vigorous dig with the bill was necessary to dislodge the seeds. They accomplished this with an emphatic downthrust of their heads without interrupting their somewhat deliberate flight. Frequently the birds rested on the ice for a few seconds and occasionally pecked at the food in this position, like very young chickens.

"These swallows seemed to be plump, vigorous, and in good condition. After they left we carefully inspected the ice where they had been feeding and found an abundance of seeds, chiefly *Scirpus* and *Carex*. We discovered no living animal matter—not even snow fleas."

Walter Faxon and Ralph Hoffmann (1900) speak of conspicuous flights of tree swallows in Berkshire County, Mass. They say: "Vast numbers collect at the head of Pontoosuc Lake during the vernal migration, where we have seen them take their departure for the north as late as the 22d of May. After sitting toward the close of day upon the low bushes that protrude above the surface of the lake, at half past seven o'clock, myriads at once soared into the air, parting to form two flocks, one of which took a course due north, while the second struck off to the W.N.W."

As we see tree swallows arriving normally in "spongy April," when the weather is mild and insects are in the air, they often follow the course of a river, either low down over the water, feeding as they go, flying in long curves, often sailing on set wings over the greater part of a wide circle, or well up in the air in loose flocks, 20 to 30 perhaps, exchanging in their progress high, delicate salutations.

In Florida, late in February, there is a conspicuous migration from farther south. I have often seen large numbers passing over the Kissimmee marshes, moving steadily toward the north in wide array on a broad front, hundreds flying past within a few minutes. At such times they make use of a monosyllablic call that I have rarely heard on the breeding grounds. At a little distance it resembles the rough note of the bank swallow.

Courtship.—The tree swallow's courtship apparently consists in a pursuit that enables the male to display his proficiency and expertness in flying. Dr. Samuel S. Dickey (MS.) describes thus "the maneuver of mating" of a pair just before the eggs were laid: "They gyrated rapidly to and fro, up and down; then mounting still higher than their previous level they dallied in midair. The male drew up to the female and grasped her breast feathers with his feet, and the two birds tumbled downward together, not parting until they were near the ground. The female then flew to the vicinity of the nest and settled down on a perch, lifted her wings slightly, and expanded her tail. The male glided above her and, dropping his wings, alighted on her back. I have never seen these birds have sexual contact in midair."

Francis H. Allen (MS.) states that in the courtship flight the wings are never raised above the horizontal.

Observations on banded tree swallows show that a pair may breed together in successive seasons. Mrs. Kenneth B. Wetherbee (1932), Oscar McKinley Bryens (1932), and Lewis O. Shelley (1934a) speak of this habit, but Laurence B. Fletcher (1926) reports a case in which two birds, mated in 1925, returned the next year and paired off with different mates.

Shelley (1935) has shown that male birds not only may change mates from year to year but may have two mates at the same time. He explains irregular matings thus: "The first Tree Swallows of the season arrived at the station on April 3d, when the migration of the species was nearly over. About May 1st brown females commenced to appear, and they were common throughout their migration, which lasted to June 16th. * * * Since these unmated birds arrive to such an extent after nesting by older pairs is well under way, they are susceptible to mating, and they do mate with paired males, who often desert their former mates for the new-comers."

Observers have sometimes seen more than a single pair of adult birds feeding young in a nest, and Forbush (1929) says "occasionally three birds, usually two males and one female, engage in preparing a nest, incubating the eggs and feeding the young."

Harrison F. Lewis writes to Mr. Bent: "A pair of tree swallows nested on the grounds about my residence near Quebec, Canada, in 1920. The young birds left the nest on July 13, a few days later than did many other broods in the general vicinity. On July 11 I noticed that the young in my box were being fed by two adult males and one adult female. Later in the day I saw one adult tree swallow on the box, one on the wire nearby, and at the same time four more circling close overhead. It would appear that the young swallows in the box were being cared for by four to six adults, of which at least three were males and at least one a female. The males seemed to work together very harmoniously, but the female sometimes acted as if she objected to having so many males about her home. Whether the assistants to the actual parents of the brood were birds that had not bred successfully or were the parents of other broods already able to feed themselves is a matter of speculation."

Seth H. Low (1934) accounts for some instances of such behavior. He studied a colony of tree swallows whose broods had been decimated by a summer hail-storm, after which he says: "Three and even four adults were caught actually bringing food to the same brood. In each of three different boxes a female whose young had previously died was caught along with the rightful parents feeding the brood. Apparently this is a carrying-over of the maternal instinct."

Nesting.—Before North America was settled the tree swallow built its nest in hollow trees, but with the advent of civilized man the bird quickly appropriated as nesting sites the houses that our ancestors put up to accommodate the purple martins—always a popular bird with the early settlers. Both Wilson and Audubon mention this habit.

At the present time tree swallows still build in old apple orchards and in holes in trees, especially when they stand in open ground near meadows or bodies of water, but they seem to prefer wooden boxes, even ramshackle affairs affording incomplete shelter.

The Austin Ornithological Research Station recently made an interesting experiment with tree swallows on Cape Cod, Mass. They put up a large number of breeding boxes and obtained astonishing results that indicate that the birds sometimes have difficulty in finding places to nest. Seth H. Low (1933), summarizing the results, says: "With 98 wooden boxes, mostly in the open, the population jumped in 1931 from 4 pairs to 60 pairs. In 1932, with over four hundred boxes available in favorable sites, there were 113 pairs of breeding birds." Austin and Low state (1932) : "They had a choice of boxes in open fields, in partial cover, on the salt marsh, and in dense woods. They showed a preference for those in the open fields, most of which were occupied, and used none of the ones that were sheltered by vegetation."

Forced, perhaps, by a scarcity of breeding sites, tree swallows have been found nesting in several unusual places. For example, Dr. Dickey (MS.) reports that they build their nests in the "eaves and cracks of log shelters erected by campers and foresters in lumber camps in Ontario and in old excavations of woodpeckers, notably those of the yellow-bellied sapsucker and arctic woodpecker." Milton S. Ray (1903) found a nest "in a hole of a pile of an old wharf, over the water." Henry Mousley (1916) says: "In my experience the nesting site here [Hatley, Quebec] is generally some small cavity in the eaves or cornices of farm buildings." Hartley H. T. Jackson (1923) says: "A nest containing five eggs was found in a fence post by the roadside, June 7, the entrance to the cavity being in the top of the post." John Treadwell Nichols (1920) speaks of a nest in an abandoned hydrant, the opening to the nest only a couple of feet from the ground.

The nest of the tree swallow consists of an accumulation of dry grass and straw, hollowed out, and lined with feathers. The birds show a marked preference for white feathers and often arrange them so that the tips curl upward over the eggs.

Oliver L. Austin, Jr., and Seth H. Low (1932), speaking of nests on Cape Cod, say:

> The time occupied by the nest-building varied individually from a few days to two weeks. The peak of these activities occurred during the last week in April and the first week in May.
>
> In general the foundation and bulk of the nests were of upland or marsh grasses, pine-needles, or a combination of these materials woven together. A hollow was formed in this foundation, sometimes in the center, but as often in one side or corner, and profusely lined with gray and white feathers packed in tightly with the quills buried in the grass or pointing away from the central hollow. In forty-six boxes were found over 3,300 feathers, ninety-nine per cent of which were those of the Herring Gull, though a few feathers of domestic fowl, Black Duck, Scaup Duck, Wood Duck, Canada Goose, Great Horned Owl, and Red-tailed Hawk were mixed in with them. There was an average of 72 feathers to a nest, but over a hundred were found in each of ten boxes; one contained 132, one 134, and another a maximum of 147.

A. Dawes DuBois (MS.) describes a nest "profusely lined with soft, pure white chicken feathers—a beautiful nest. After three eggs had been laid, I saw the male bird catch a large downy white feather that was floating on the breeze and carry it into the box." Thomas D.

Burleigh (1930) found a nest "ten feet from the ground in a cavity in an old rotten willow stub in underbrush bordering a stream, * * * built entirely of large chicken feathers."

The tree swallow commonly nests in isolated pairs, showing none of the strictly communal habits of the purple martin and the cliff swallow, but the birds are frequently found nesting in groups, their nests scattered about, not far apart, in favorable feeding localities. Ralph Works Chaney (1910) speaks of such a case in Michigan. He says: "Large colonies of these swallows nested in cavities of dead stumps which projected out of the lake." Charles L. Whittle (1926) sums up an investigation on the distribution of nests by concluding that "the determining factor is the adequacy of a nearby feeding area, or areas, be they meadows (old, filled lakes), marshes or water, to furnish the necessary quantity of food for the young at the requisite period."

The female bird builds the nest, aided little or none at all by her mate. Winton Weydemeyer (1934b) says: "Although the male often makes a pretense at gathering straws, and occasionally carries feathers to the nest, his principal job is that of overseer. * * * Only occasionally do the male birds share the task of incubating the eggs; frequently—especially during days when few insects are in the air—they carry food to their mates in the nests. Generally, however, the females during the day leave their eggs long enough to secure food for themselves." Both parents feed the young and remove excreta from the nest, dropping the sacs from the air a few yards away.

Winton Weydemeyer (1934b) states: "As a rule, at no time during the entire season do the males share the houses at night. * * * Frequently the males perch on the houses for an hour or more after their mates have retired, not leaving for their own sleeping places until after darkness has settled." The same observer (1935) reports from Montana: "In the case of sixty nests a full record has been obtained of the percentage of hatch and survival of the nestlings." The latter shows 94.7 per cent for the first brood and 84.6 for the second brood. Austin and Low (1932) calculate a reproductive efficiency of 56.5 percent from 278 eggs.

Eggs.—[AUTHOR's NOTE: The tree swallow usually lays four to six eggs; as many as 10 have been found in a nest, but any over five or six eggs may be the product of two females. The eggs vary from ovate, the usual shape, to elliptical-ovate or, rarely, elongate-ovate. They are pure white, unmarked, and usually without gloss. The measurements of 50 eggs average 18.7 by 13.2 millimeters; the eggs showing the four extremes measure 20.3 by 13.5, 19.3 by 13.7, 16.8 by 13.2, and 17.8 by 12.7 millimeters.]

Young.—Austin and Low (1932) from a study of a large number of nests "found the length of the [incubation] period, estimated from the day the last egg was laid to the day the first one hatched, to vary from a minimum of 13 days to a maximum of 16." According to their records the young birds remained in the nest 16 to 24 days, and to account for this variation they point out "that the most food per young will be delivered in those nests containing the fewest nestlings, and hence the rate of growth will be slowest where the broods are largest." They state that "at no time were the young birds observed to return to the boxes once they had flown." Winton Weydemeyer (1934b), however, says: "For a few days after taking to the air, the young birds enter and leave the houses frequently, and remain in them all night."

The nestling tree swallow is an attractive little bird when, well grown, it comes to the doorway and peers about, watching for its parents to come through the air with food. As it waits at the entrance its low forehead and immaculate throat call to mind a little frog sitting there in the box. Its eyes shine eagerly, and when the parents come near it stretches out toward them, its throat gleaming white against the dark interior.

A. Dawes DuBois (MS.) says: "The young are strong of wing when they leave the nest." He speaks of one young bird which "took to the air like a veteran, both parents accompanying it." Austin and Low (1932) state that "usually they showed remarkable ability on their first flight, often remaining in the air well over a minute, and flying a quarter of a mile." George Nelson tells me that he has often watched the young birds leave the boxes in his garden. They launch out, then fall, fluttering, nearly to the ground sometimes, when, of a sudden, the power of flight comes to them, and they rise into the air and fly off, seemingly as ably as their parents.

Plumages.—[AUTHOR'S NOTE: The juvenal plumage of the tree swallow is quite unlike the plumage of either adult. The upper parts, including the sides of the head and neck, are dark sooty brown or very dark brownish gray, "dark mouse gray", without any trace of the iridescent bluish green of the adult plumage but with a fine silky gloss. The feathers of the interscapular region are at first faintly edged with pale fawn color, but these edgings soon wear away. The wings and tail are slaty brown, with slight greenish reflections, the secondaries and tertials with faint grayish edges and tips. The under parts are duller white, less silky, than in the adult; and there is a very faint and incomplete pectoral band of ashy brown.

A complete postjuvenal molt takes place, beginning late in August and continuing into October. This produces a first winter plumage, which is practically indistinguishable from the winter plumage of the adult. Dr. Dwight (1900) says of this plumage: "Above iridescent green, sometimes with steely blue reflections. Wings and tail deep bottle-green slightly iridescent, the tertiaries broadly tipped with white. Below, pure white slightly smoky gray on the sides."

The white tips on the tertials are characteristic of the winter plumages of both adults and young birds; but these tips wear away before spring. There appears to be no spring molt, but a complete postnuptial molt begins about the middle of August and is usually completed before the birds go south. This is the only one of our swallows that completes its molt before migrating; it breeds early, molts early, and migrates late.

The sexes are much alike, except in the breeding plumage, when the female is duller, the upper parts often being largely dark grayish brown with only the tips of the feathers glossy greenish.]

Food.—F. E. L. Beal (1918) points out that "in its food habits this species differs somewhat from other American swallows in that it eats an appreciable quantity of vegetable food, frequently filling its stomach completely with berries or seeds." In an examination of 343 stomachs, collected in every month of the year over a wide range, Professor Beal found that "the food divided into 80.54 per cent animal matter to 19.46 per cent vegetable," and he states: "The vegetable food is made up of a few varieties of seeds and berries, but more than nine-tenths of it consists of the fruit of a single shrub, the bayberry, or waxberry (*Myrica carolinensis*)."

Of the animal food he says beetles make up 14.39 percent, ants 6.37 percent, and that Diptera form the largest item of the tree swallow's food (40.54 percent). Minor items are grasshoppers, dragonflies, and spiders. He summarizes his findings thus: "In the food of the tree, or white-bellied, swallow one point is prominent— in its vegetable food it has no relation to man. Every item is wild and of no use. In its insect diet it destroys some parasitic Hymenoptera, some carnivorous Diptera, and a few other useful insects, but this fault it has in common with most other insectivorous birds, and in common with them it is engaged in reducing the great flood of insect life to a lower level. Let it be protected and encouraged."

When we watch feeding tree swallows we see them chiefly in the rôle of flycatchers. They tour over meadows, ponds, and rivers, veering from side to side, doubling back with marvelous quickness, snatching up insects as they overtake them or meet them in the air, coursing low down over the meadow grass where flies abound, or, mounting, crisscross through the swarms of higher-flying insects, gorging their throats with the tiny bodies. Ever on the move, they pass back and forth across their feeding grounds, their quick turns evincing success in capture after capture.

Arthur C. Bent speaks in his notes of seeing a flock of tree swallows alight on a marshy shore and feed from the ground on what appeared to be minute insects almost too small to be seen; also of their catching insects on the wing under the lea of a hillside.

John J. Elliott (1939) during his study of a company of tree swallows that wintered on Long Island, N. Y., collected their excrement. He reports that "analysis of the fecal material by the U. S. Biological Survey showed that crustacean material (*Orchestia platensis*) formed the bulk of the food, along with water-boatmen, spiders, bulrush, sedge, bayberry and smartweed seeds, and fragments of rose thorns."

George Nelson had an interesting experience with tree swallows in Florida on November 24, 1938. He was crossing a hammock about 10 o'clock in the morning when he noticed, rising straight up from the path in front of him, a thin pillar of what he took to be a wisp of smoke no bigger round than a pencil. Following the column upward with his eyes, he noted that, at a height of 10 to 12 feet above ground, it swayed in the light breeze, broke up, and became dispersed. When he stepped up close to the wisp of smoke, he saw that it was not smoke at all, but a closely packed column of winged red ants issuing in countless numbers from a board walk. They mounted straight toward the zenith and then spread out like a funnel and scattered over the space of an acre. Soon a tree swallow flew past, up in the air where the insects had separated. It turned back and flew over and over again through the swarm, snapping up the insects. Before long more tree swallows appeared, until within 15 minutes many hundreds had collected, all swooping back and forth at great speed where the insects were flying. In less than half an hour the insects had disappeared and not a swallow was in sight.

Mr. Nelson was surprised to see so many tree swallows together, for at this season of the year the bird is not abundant in Sebastian where the incident occurred. They had assembled, apparently, in the way sea birds collect at points in the ocean where food is plentiful, bird after bird being drawn to the spot by seeing from a distance others feeding.

On another occasion, when Mayflies were rising from an extensive marsh, Mr. Nelson saw tree swallows gathered in a great cloud, so thick and dense that in the bright sunlight the flock cast a dark shadow on the marsh.

Behavior.—As we watch swallows in flight we notice that they do not all fly in the same fashion, and after long watching we become able to tell them apart when they are far away, or at least to suspect which is which. For example, perhaps the most distinctive in its manner of flying is the barn swallow. It is characteristic of this species to drive along through the air, seemingly with a strong push.

At the end of each stroke, the tips of the wings are brought backward until the primaries are nearly parallel with the long axis of the body. A robin also shows this peculiarity but to a less degree. The bird swings to the right and left, to be sure, but there are periods of straight flying or sailing, and always there is the impression of a steady drive through the air, with a good deal of power for so small a bird.

The tree swallow, compared with the barn swallow, appears to be less steady in the air, although doubtless it possesses complete mastery over it. There is a suggestion of flickering in its flight, due perhaps to the quicker, less forceful motions of its wings. Flying at a distance, it sometimes resembles a starling—another quick-moving bird—but most characteristic is the habit of hunching up its back, or seeming to do so, and lowering its wing tips as it sails, like an inverted saucer in the sky. Francis H. Allen (MS.) speaks of their flight as "largely a succession of reaches and runs with periods between them when the bird seems to hang in stays for a while—to speak in nautical terms."

In the air the tree swallow resembles somewhat the purple martin, the similarity being due probably to the triangular shape of the wing in both birds—a triangle with a sharp apex and a fairly broad base. The bank swallow is readily distinguished from the tree swallow by its habit of hugging its wings close to the side of its body when it sails and by the suggestion of soft fluttering in the motion of its wings.

Tree swallows do not linger long about their nests once the young are on the wing. Both the young and the adult birds apparently retire to broader feeding grounds—the meadows bordering river valleys or marshes near the seacoast—where, gathering in increasing numbers, they form the nucleus of the autumn flocking.

Often because of a scarcity of nesting-holes or boxes, but sometimes because of a preference for a certain site, tree swallows come into conflict with other species of hole-nesting birds as well as individuals of their own species. F. Seymour Hersey (1933) relates a remarkable example of such an encounter. A pair of bluebirds were breeding in a box in his garden, and the female was incubating a set of eggs. He continues the story:

Then, one day, a pair of Tree Swallows arrived and decided they wanted that particular box. I hurriedly put up boxes No. 2 and No. 3 but the Swallows paid no attention to these new nests and after a day of constant bullying the Bluebirds surrendered their nest and eggs and retired to box No. 2. The Swallows remodelled the Bluebird's nest, incidentally disposing of the eggs in some way, and the Bluebirds started another nest in box No. 2. For awhile peace and quiet reigned and both pairs of birds had young a few days old when a second pair of Swallows put in an appearance. Once more there was fighting of a rather general nature among all three pairs of birds, but soon the Swallows in box No. 1 managed in some way to make good their title and the scene of battle centered about box No. 2. For two or three days this second pair of Swallows constantly harassed the Bluebirds, so that it was difficult for them to bring any food to their young, and at the end of this time they abandoned their nest and left the garden and vicinity.

The following year the swallows had a pair of starlings to deal with. Mr. Hersey says: "Their method of getting rid of the Starlings was interesting. While either of the Swallows was away from the nest the other was on guard, perched on the roof of the box. When the Starling appeared she was either attacked and driven away, or the Swallow immediately entered the box and sat looking out the hole, effectively blocking the entrance. For several days I did not see the nest left unguarded for a moment and the Starlings soon went elsewhere."

The swallows are not always able to oust a bluebird; a bluebird may even drive them away from the vicinity of its nest. Helen J. Robinson (1927) tells of such a case. She says: "The Bluebirds kept a vigilant lookout from their own tree, a young oak, and watched the Swallows circle about the other boxes. When the Swallows seemed about to enter, one or both Bluebirds charged them, straight as an arrow. The mere sight of the enemy was usually enough to put the Swallows to flight, circling and screaming as they retreated a short distance, but returning as soon as the coast seemed clear. Sometimes Swallows flew bravely to attack the Bluebird, but such birds were always borne to earth by the larger bird, which then tweaked the victim's crown feathers without mercy." The same author speaks also of the discord that arises among the swallows. She says: "The usual internal warfare among the Swallows themselves proceeded briskly, beginning the day of arrival and continuing a full calendar month."

There are several reports in the literature of strange behavior on the part of the tree swallow. Lewis O. Shelley (1934b) describes the actions of a female bird with a lust for killing the nestlings of other pairs of tree swallows; he (1936b) tells of another female that reared a nestling cliff swallow, which he introduced into her nest just as her own brood was about to fly, and J. A. Munro (1929) gives an account of a male that fed the 8-day-old nestlings of a pair of western robins whose nest was built on the top of the swallow's bird-house.

Voice.—The voice of the tree swallow has a pleasing, gentle quality. *Silip*, he seems to say, a quick, rapidly pronounced note, sometimes rippled into three or more syllables. It may run into a chatter, but it is never jarring; it always retains its delicacy. A little excitement brings more emphasis to the voice and introduces a long doubled *e*, so that the note suggests our word, cheery. Real alarm calls out greater power; the tone now rises to a squeal, almost a shrill whistle.

Francis H. Allen (1913) speaks of the tree swallow as "one of the very earliest singers in the morning concert." Of the song he says: "It is really a remarkable performance regarded as an exhibition of endurance." He describes thus the song of a bird singing at 2.53 A. M.: "He sang continuously, apparently without interruption, from the time I first heard him till 3.40. The song came and went, as the Swallow flew about over the pond, now nearer, now farther away, now to the right, now to the left, but never stopping,—a constant *tsip-prrup*, *tsip-prrup-prrup*, *tsip-prrup*, *tsip-prrup-prrup-prrup*, *tsip-prrup-prrup*, *tsip-prrup-pruup-prrup-pruup*, varied only by the varying number of bubbling notes following each *tsip*. The ending of the performance seemed to come gradually."

Winton Weydemeyer (1934a) says: "Singing is done both in flight and from perches near the nests. A series of phrases, repeated over and over in slightly varying order, at the rate of 125 to 140 a minute, is given for several minutes or as much as an hour without pause."

Field marks.—The underparts of the tree swallow are pure white. It is the only swallow that shows this character except the violet-green, which may be distinguished by an opalescent coloring of the back. Jonathan Dwight (1900) speaks of "a very faint incomplete sooty collar on the jugulum" in the juvenal plumage of the tree swallow, but this mark is lost by the time the first winter plumage is acquired in October and should never be confused with the broad pectoral band that the bank swallow possesses in all plumages.

John Treadwell Nichols (1920) points out two excellent diagnostic points. He says: "When one gets a good view of them, our different Swallows are well marked and easy to identify. They also present differences in size, flight and call-notes which one learns to recognize. However, it may aid in the determination of a bird darting by at a difficult angle, to call attention to the white on the Tree Swallows' flanks, which encroaches on the dark upper parts in front of the tail so as to be conspicuous. The Tree Swallow also has an angle in the posterior outline of the wing unlike the other species, as though the primary feathers projected more abruptly beyond the secondaries."

Enemies.—The speed and agility of the tree swallow render it comparatively safe from attacks by birds of prey. Tree swallows come into competition for nesting sites with other hole-nesting birds, but as shown under "behavior," often hold their own.

A nest sent to the Bureau of Entomology, Washington, D. C., by Mrs. Kenneth B. Wetherbee (1932) "contained fifty-two *Protocalliphora splendida*, var. *sialia*, thirty *Mormoniella vitripennis*, and one hundred and sixty fleas."

The chief danger in the life of the bird is the inclement weather that it may be subjected to during the winter in the Southern States or that it may meet after its arrival on its breeding grounds, as noted under "Spring."

Fall.—Fall is a season of drama in the tree swallows' yearly cycle. A single idea, or an urge, seems to grip every swallow in the land. The nesting season with its quarrels over, the swallows draw together with a common interest in preparation for their next step, the long migration they will take in companies of hundreds or thousands. In August and September we see them gathering in the great marshes by the sea, where they linger for many days in ever-increasing numbers, young and old, sometimes associated with other species of swallows, notably the barn swallow.

The following quotations show the tree swallows gathering in autumn: John Lewis Childs (1900) gives an idea of the great number that collected at Barnegat Bay, N. J., in September. He describes the birds as he saw them flying overhead between their feeding grounds and their night quarters. He says: "Not a Swallow was seen until the solid column of the flight appeared, and it was at once apparent that where there were hundreds two weeks previous there were now thousands. The flight was compact like a swarm of bees and at times almost darkened the sky. Most of the time there were two distinct columns, one flying low just over the water, and the other high up in the air. I watched the flight for hours, and the air in both directions seemed alive with them as far as the eye could reach." Of his observation on the following day, he adds: "After watching the birds nearly all of the forenoon we made a careful estimate of the number that had passed and we calculated that it was not to be reckoned by tens of thousands or hundreds of thousands, but by millions."

Lynds Jones (1910), speaking of tree swallows at Cedar Point on the shore of Lake Erie, in Ohio, says: "After the breeding season, during late July, I have seen great companies gathering to roost in the swamp vegetation east of the mouth of Black Channel. They formed the characteristic funnel group before finally settling into the vegetation for the night."

Bradford Torrey (1893) describes tree swallows as seen at Ipswich, Mass., in fall. "At eight o'clock," he says, "when we took the straggling road out of the hills, a good many—there might be a thousand, I guessed—sat upon the fence wires, as if resting. We walked inland, and on our return, at noon, found, as my notes of the day express it, 'an innumerable host, thousands upon thousands,' about the landward side of the dunes. Fences and haycocks were covered. Multitudes were on the ground,—in the bed of the road, about the bare spots in the marsh, and on the gray faces of the hills. Other multitudes were in the bushes and low trees, literally loading them. Every few minutes a detachment would rise into the air like a cloud, and anon settle down again."

We often see tree swallows using the sea beach as a path of migration, flying either over the sand itself or over the ocean a short distance from the shore. As they course along they sometimes swoop playfully at a shorebird, put it to flight, and chase it, twisting and dodging, rising and darting down in unison with it, at the same time continuing their southerly course.

Years ago Walter Faxon and I watched about 500 tree swallows circling around the steeple of a church at Ipswich, Mass. It was in the latter part of an afternoon in mid-September at the height of the swallow migration. Suddenly their haphazard flight changed to an orderly procession in which about half the birds wheeled in a great spiral and, mounting high in the air, sailed away due south.

Winter.—Alfred M. Bailey (1928) says of the tree swallow on the gulf coast of Louisiana in the winter: "Very common over the marsh, where thousands were seen at once."

John J. Elliott (1939), who watched a little company of 28 tree swallows through the fall and winter of 1937–38 on Long Island, N. Y., found that the majority of the birds survived, although the locality is far to the north of the normal winter range. He says:

I decided to give this flock, which to all appearances was going to winter, as much of my time as possible, and to learn of their winter habits and peculiarities. To this end I made 78 trips and spent 131½ hours. The frequent trips had the advantage of permitting me to observe the actions of the birds in many types of weather. I found cold rains, windy and snowy weather, extremely discouraging for flying birds, and after hours of watching, saw not a bird in the air. During these periods they remained in their sheltered situations, chiefly north of the pond, subsisting principally on bayberries. Cloudy days, especially if cold, also discouraged them to a certain extent from taking the air. In fact, I found that, generally, the brighter the sun, the more I saw of flying birds, and the warmer the sun and surrounding air, the higher their flight, both proportionately, except during windy days. At no time did they use telephone wires or any high exposed perch, as in summer; no long flights high in the upper air. * * *

They are given to much resting in sunny sheltered places, in little huddled groups of four or five birds each, and resort to low broken stubs of bayberry bushes, strong blackberry canes, squatting on the sand, or on boards imbedded in the sand. They are usually silent but on occasion utter a cheery, reedy, double-noted twitter when sunning themselves and in moderate weather.

[AUTHOR'S NOTE: Since the above was written, accounts have been published of the great mortality among tree swallows and other birds during the "big freeze" in southern Florida during January 1940. Bayard H. Christy (1940) writes:

On the morning of January 28, 1940, after ten days of north wind and continued cold, the temperature at Coconut Grove, Florida, fell below the freezing point. There was widespread destruction of cultivated plants, and small fishes in countless numbers lay dead on the tidal flats. * * * As in the preceding days the cold grew more intense, these birds were seen to hover more closely above the water surfaces, and to leeward of the walls of houses. On the morning of January 28, about fifty of them were found, densely packed in the cavity of a Pileated Woodpecker, formed long since in the stub of a palmetto, standing among mangroves a few yards from the margin of the bay. Twenty-eight were already dead or died soon after removal. A few flew away at once; others revived in the sunshine. On a sea-wall near by, about fifty more were resting in the sun, clustered like bees, some on the level top of the wall, others on its rough and sheltered face. Later in the day a half-dozen more were picked up dead on the lawn of the adjacent property."

A few of the bodies of the dead swallows found in the wood-pecker cavity were opened and their stomachs were found to be empty. "No doubt there had been considerable shortage of food; no doubt the massing within the cavity had increased the destruction, but the fundamental cause of death was the cold."

The next morning he found only about a dozen swallows in the cavity, three of which were dead. Driving across the Tamiami Trail he found further evidence of disaster to the swallows, many of which had been run over by passing cars. "Even on wing the living birds seemed to have lost their usual agility, and twice, to my regret, my moving car struck birds in the air. * * * I found them thronged in an outhouse on the brink of the canal—one dead, two or three others fluttering to lie widespread upon the ground. At another place the graveled parking area was strewn with bodies—perhaps fifty of them."]

DISTRIBUTION

Range.—North and Central America; occasional in Cuba and casual on the coast of British Guiana.

Breeding range.—The tree swallow breeds **north** to northern Alaska (Point Barrow and Fort Yukon); Mackenzie (Fort Goodhope, Fort Norman, Fort Providence, and Fort Resolution); northern Manitoba (Lac du Brochet and probably Fort Churchill); northern Quebec (Chimo); and northern Newfoundland (Bard Harbor). **East** to Newfoundland (Bard Harbor, Nicholsville, and St. George Bay); and south along the Atlantic coast to Virginia (Wallops Island). **South** to Virginia (Wallops Island); Maryland (Fairhaven, Hagerstown, and Crellin); central Ohio (Columbus); Indiana (Terre Haute); Missouri (St. Louis and Bolivar); Kansas (Wichita); Colorado (Salida and Durango); Utah (La Sal Mountains and Provo Bay); central Nevada (Carson); and southwestern California (San Onofre). The **western** limits of the breeding range extend northward along the Pacific coast from southern California (San Onofre); to Alaska (Lake Aleknagik, St. Michael, Cape Prince of Wales, and Point Barrow).

Winter range.—The normal winter range appears to extend **north** to southern California (Santa Barbara, Buena Vista Lake, and Salton Sea); northern Chihuahua (San Diego); central Nuevo Leon (Monterrey); southern Texas (Brownsville and Houston); Louisiana (Cameron, Avery Island, and New Orleans); Mississippi (Biloxi); and coastal North Carolina (Lake Mattamuskeet). The **eastern** limits of the range at this season extend southward along the Atlantic coast from North Carolina (Lake Mattamuskeet and probably Fort Macon), to southern Florida (Miami and Royal Palm Hammock); and occasionally eastern Cuba (Nipe Bay). **South** to occasionally Cuba (Nipe Bay and San Francisco); Yucatan (Chunyache and Camp Mengel); British Honduras (Belize); Honduras (Tolva Lagoon); and southern Guatemala (Quiriqua, San Geronimo, and Volcan de Fuego). **West** to western Guatemala (Volcan de Fuego and Lake Atitlan); State of Mexico (Tlalpam and Mexico City); Baja California (Cape San Lucas, La Paz, and Calexico); and California (San Onofre and Santa Barbara).

Occasionally tree swallows are noted in winter on the Pacific coast north to central California (Point Reyes and Sonoma) and on the Atlantic coast to southern Connecticut (Saybrook) and New York (Gardiners Island).

Spring migration.—Early dates of spring arrival are: District of Columbia—Washington, March 24. New Jersey—Englewood, March 24. Pennsylvania—Philadelphia, March 30. New York—Canandaigua, March 28. Connecticut—Hartford, March 29. Massachusetts—Dennis, March 13. New Hampshire—Monadnock, March 24. Maine—Ellsworth, March 27. New Brunswick—Scotch Lake, April 17. Quebec—Montreal, April 7. Missouri—St. Louis, March 14. Illinois—Canton, March 11. Indiana—Hobart, March 20. Ohio—Oberlin, March 13. Michigan—Ann Arbor, March 18. Ontario—London, March 25. Iowa—Davenport, March 18. Wisconsin—Madison, March 25. Minnesota—Minneapolis, March 29. Manitoba—Winnipeg, April 13. Oklahoma—Norman, March 30. Nebraska—Nebraska City, April 11. North Dakota—Larimore, April 24. Saskatchewan—Indian Head, April 29. Arizona—Tucson, March 10. Colorado—Durango, April 9. Wyoming—Yellowstone Park, May 2. Montana—Fortine, March 21. Alberta—Camrose, April 25. Mackenzie—Fort Resolution, May 12. Oregon—Mercer, February 21. Washington—Tacoma, February 22. British Columbia—Okanagan Landing, March 6. Alaska—Craig, April 10.

Fall migration.—Late dates of fall departure are: Alaska—Craig, August 15. British Columbia—Okanagan Landing, September 15. Washington—Seattle, October 19. Oregon—Benton County, September 24. Alberta—Birch Lake, August 25. Montana—Fortine,

September 3. Wyoming—Yellowstone Park, September 8. Colorado—Durango, September 26. Saskatchewan—Eastend, September 4. North Dakota—Jamestown, September 20. Nebraska—Badger, September 18. Kansas—Lawrence, September 30. Oklahoma—Norman, October 2. Manitoba—Aweme, October 7. Minnesota—Minneapolis, October 13. Wisconsin—New London, October 17. Iowa—Keokuk, October 31. Ontario—Toronto, October 25. Michigan—Vicksburg, October 26. Ohio—Huron, November 2. Illinois—Chicago, October 22. Missouri—St. Louis, October 19. Arkansas—Hot Springs, November 15. New Brunswick—St. John, October 2. Maine—Lewiston, October 20. Massachusetts—Dennis, October 22. Connecticut—Middletown, October 31. New York—New York City, November 5. New Jersey—Demarest, October 30. Maryland—Cambridge, November 2.

Casual records.—In British Guiana tree swallows have been reported (date?) from Georgetown, and they were reported as "numerous" on Abary Island between April 12 and 15, 1909. On April 27, 1909, one flew on board a vessel a few miles north of the island of Trinidad. A flock of five was seen at Corinto, Nicaragua, on March 7, 1917. A large flight was reported and a specimen was taken on Bermuda on September 22, 1849. The species was noted in the spring of 1925 at the Post at Coral Inlet, on Southampton Island, Northwest Territories; a specimen was killed by a native at a point 10 miles east of Demarcation Point in Yukon Territory, on June 7, 1914; and one was taken on St. Paul Island, Alaska, on May 25, 1922.

Egg dates.—California: 27 records, April 28 to June 29; 13 records, May 13 to June 2, indicating the height of the season.

Illinois: 13 records, May 10 to June 9; 7 records, May 19 to 28.

Massachusetts: 20 records, May 4 to June 25; 10 records, May 20 to 28.

Ontario: 12 records, May 17 to June 17; 6 records, May 31 to June 7.

Virginia: 3 records, May 24 to June 17.

RIPARIA RIPARIA RIPARIA (Linnaeus)

BANK SWALLOW

PLATES 58, 59

HABITS

CONTRIBUTED BY ALFRED OTTO GROSS

The name swallow immediately brings to our mind a bird with a peculiar charm and grace and possessing a dexterity of flight that is seldom excelled by birds of other groups. The swallows are insectivorous and most of the insects are winged types that are captured during the course of their flying. All the species share these qualities, which serve as a strong bond of comradeship among the members of the family. Especially is this true during autumn, when the individuals flock together at the common roosting places and later share with each other the vicissitudes of the long migration.

The bank swallow is distinguished from the other swallows by its unique habit of nesting in burrows, which it cleverly excavates well into the vertical sides of a bank of clay, sand, or gravel. This characteristic habit has given origin to both its scientific and series of common names. The scientific name *Riparia* (Latin *riparia*, riparian; *ripa*, bank of stream) refers to its living in a bank of a stream. Likewise the common names are obviously suggestive of this mode of life. The name bank swallow is the one generally accepted in this country, but others, such as sand swallow, ground swallow, bank martin, and sand martin, also suggest the characteristic nesting habit of this swallow.

The 1931 A. O. U. Check-list does not recognize a subspecific difference between the American bank swallow and the bank swallow of Europe. Oberholser (1938), however, points out that "the American bank swallow differs from the European Bank Swallow, *Riparia riparia riparia*, in shorter wing, relatively larger feet and bill, darker and more sooty (less rufescent) brown upperparts. * * * It was long ago distinguished from the European bird by Leonhard Stejneger, but his diagnosis has subsequently been overlooked or ignored, although the American bird is, however, readily separable as above indicated. Stejneger named the 'American variety' of bank swallow *Clivicola riparia maximiliani;* and his type, subsequently designated, is an adult male, No. 8325 of the United States National Museum collection, taken at Ipswich, Mass., May 20, 1870, by C. J. Maynard." Oberholser, therefore, proposes the name *Riparia riparia maximiliani* (Stejneger) for *Riparia riparia riparia* of the 1931 A. O. U. Check-list. Whether or not the proposal to differentiate, subspecifically, the American and European representatives of the bank swallow is accepted by the A. O. U. Committe on Nomenclature, it seems advisable to limit the present account to the bank swallow in America. Wetmore (1939) does not agree with Oberholser that there is a line of demarcation between the American and European birds.

Courtship.—The activities of courtship begin after the birds have arrived at their nesting sites in the spring. At such times, Stoner (1936b) says that "several individuals congregate in a particular part of a sand pit or on a given section of a creek bank, with much chattering and fluttering and occasional mating." More rarely one may see a pair of swallows emerge from the nesting holes and tussle with each other in the air, sometimes falling together to the base of the nesting bank and there apparently going through the act of copulation. Harold B. Wood mentions a similar observation in correspondence we have received from him. Dr. Wood also points out that an observation made by Beyer (1938) is more properly interpreted as a part of the courtship than of fighting. Beyer's statement is as follows: "In the early stages of burrow excavation some fighting occurs among the occupants of a nesting site, apparently in settling territorial claims to burrow locations. The contestants peck each other vigorously and sometimes fall together to the earth in front of the bank in the intensity of their struggles."

Beyer (1938) writes: "One day early in June I saw a white feather floating high in the air just above a bank where a large colony of Bank Swallows was located. Suddenly a swallow darted at the feather, caught it and carried it for a short distance and then released it. Another bird caught the feather and released it and then another and another. * * * They seem to like to poise on beating wings before the face of the bank where the nests are located, holding their position for a few seconds and then wheeling away out over the nearby fields, only to return soon again to repeat the performance. This they do in companies of eight to a dozen or more." Beyer considers this as mere play but it is conceivable that these activities are a part of the intricate courtship performances.

William Brewster (1898) made an interesting observation on August 22, 1896, at Lake Umbagog, N. H., of a group of immature swallows, including bank swallows, which he interprets as a premature courtship:

There had been a heavy rain during the night and the road was very muddy. The birds alighted about the edges of one of the larger puddles in great numbers and walked slowly about fluttering or quivering their half-opened wings like so many big butterflies. At first I supposed that they were drinking or picking up insects, but what was my astonishment to find that the Eave Swallows were filling their bills with mud and the White-bellied and Bank Swallows gathering pieces of hay or straw. * * * Each bird, on obtaining a satisfactory load of mud or grass, flew with it to the fence and after shifting it about in its bill for a few moments, finally dropped it and at once returned to the road for a fresh supply. * * *

While the birds were clustered about the mud puddle, scarce a minute passed when one or two pairs were not engaged in copulation. Perhaps I should say in attempted, rather than actual, copulation, for as nearly as I could see, the sexual commerce was in no instance fully or successfully accomplished. The females (or at least the birds that acted that part) submitted willingly enough to, and in some instances as I thought, actually solicited, the attention of the males. * * * Every one of the swallows which visited the mudpuddle and engaged in collecting mud and straw or in attempted copulation, *was a young bird.* * * * The remarkable behavior of the birds which alighted in the road was simply an expression of premature development, in the young, of the instincts and passions of nest-building and procreation.

Frank Chapman (1900), in his interesting account of the swallows observed at the Hackensack marshes in northern New Jersey, records a similar behavior of flocks of young swallows. The birds were engaged in picking up bits of dried grass and flying up in the air with it, sometimes carrying it for fifty yards or more. Here also he observed the immature birds attempting the act of copulation. Chapman agrees with Brewster's conclusion that it represents a "premature exhibition of the procreative and nest-building instincts."

Spring.—There are few ornithological experiences that provide a greater thrill than the arrival of the birds in spring. Swallows, like other gregarious birds, attract unusual attention because of their numbers, and this fact coupled with their extraordinary exhibitions of flight makes their appearance an event of unusual significance.

A vivid account of the bank-swallow migrations in the Province of Alberta, Canada, is given in correspondence received from Frank L. Farley. "In company with a companion," he says, "I witnessed a remarkable migration of bank swallows on May 11, 1930, at a point on the Athabaska river, about 125 miles northwest of Edmonton, Alberta. The river here is about 700 feet in width and flows in a northerly direction. Reaching the stream shortly before noon we camped for lunch. At once our attention was drawn to large numbers of bank swallows, flying in irregular, zigzag manner, only a foot or two above the surface of the water. Evidently they were feeding on flies and other winged insects, and so intent were they in covering all likely places where food would be found that their progress toward the north was estimated to be at a rate of not more than 5 miles an hour. With the aid of binoculars we found the birds to be in just as great numbers for a mile, both up and down the stream. None were ever noted to reverse their course, and we came to the conclusion that they were migrating, many of them, no doubt to their summer homes on the lower stretches of the Mackenzie River and its many tributaries. A count of this procession was attempted as the birds passed a point in front of our camp, and it was found that about 50 birds was the average for every minute of the three hours we spent at the river, a figure approaching the 9,000 mark during our sojourn. Such a movement, should it continue uninterrupted even for one day, would reach enormous proportions."

The forerunners of the spring migration reach New England about the middle of April, but it is not until the latter part of the month that the bank swallows appear in appreciable numbers. It is then that they may be seen in flocks in company with other species of swallows in the vicinity of human habitations but more frequently along the streams and lakes, where at this season more flying insects are there to attract them. During the first two weeks after their arrival they are very erratic in their behavior, and their chief objective is to obtain an adequate food supply.

After the flocks of swallows break up for their nesting activities the bank swallow is not so much in evidence as the barn, cliff, and tree swallows, which favor nesting sites in or about the barns and houses of man. At this time the bank swallow resorts to the more isolated sections in which its nesting colonies are located. The first preliminary excavations I have noted in Maine were May 12 but it is not until about 10 days later that nest building starts in earnest. By the first of June the colonies are well established.

Nesting.—The bank swallow with very few exceptions builds its nest at the end of a burrow that it excavates usually near the top of a nearly vertical bank of a lake or stream or that is provided by the excavations of a railway or gravel and sand pits. They exhibit little inclination for human society and have not departed from their primitive nesting habit to accept new environmental conditions provided by man, as have the barn, cliff, and tree swallows. So tenacious are they of their method of nest building that large colonies are restricted in their distribution to places where banks or artificial pits of proper material structure for their excavations exist. As would be expected, the bank swallows are most numerous in the glaciated sections of the country where glacial deposits of sand and gravel abound.

A departure from their nesting in banks of natural deposits is the use of sawdust piles left by lumbering operations. On June 6, 1902, Barrows (1912) saw from a train large numbers of bank swallows about sawdust piles at Ostego Lake, Mich. There were numerous holes in the vertical sides of the sawdust heaps. He was unable to determine definitely whether the holes were occupied by nesting birds but presumed that they were. Bradford Torrey (1903) reports observations of Mrs. Annie Trumbull Slosson made at Franconia, N. H., in the summer of 1902. She saw no less than 20 holes that had been excavated in a sawdust pile, and apparently all were occupied by the swallows, which were carrying on their usual activities entering and leaving the holes as at any other colony. Such a nesting site, however, is likely to prove precarious.

E. S. Rolfe (Barrows, 1912) found bank swallows nesting abundantly in the walls of an abandoned dry well about 15 feet deep. The perpendicular walls were honeycombed with the nesting holes.

C. R. Stockard (1905) states: "On May 2 of this year I found a Bank Swallow's nest placed in a Kingfisher's deserted tunnel. The tunnel was six feet long, and three feet from the entrance it made a bend of 45 degrees, and at this place the swallows had placed their nest." We have no reason to suppose that Stockard was confused in his identification, although it is well known that rough-winged swallows nest in such situations, and furthermore this species is more numerous as a nesting bird in Mississippi, where his observations were made, than is the bank swallow.

The depth of the nesting burrow varies somewhat, depending on the texture of the material in which it is excavated. The average depth of 20 nests located in a sand bank near Brunswick, Maine, was 34 inches, the longest being 48 inches, whereas the average depth of nine nests excavated in a clay bank where there were numerous pebbles and larger stones was 19 inches, the shortest being only 14 inches. Stoner (1936b), in his study of the nests in the Oneida Lake region of New York states: "Occasionally the birds constructed a nest in a burrow not more than 16 or 18 inches in depth, but usually the completed burrow was from 22 to 36 inches deep." For 89 occupied burrows the average depth was 28 inches with a minimum of 15 inches and a maximum of 47 inches. Burrows 40 to 50 inches in depth were not uncommon, the deepest burrow which he measured was 65 inches. Stoner noted that burrows which are begun early in the season, in soft sandy soil, are deeper as a rule than those begun later in the season after ovulation has started.

A burrow excavated in a colony near Topsham, Maine, had an external opening measuring 1 by 2 inches. The burrow had a length of 33 inches, and the cavity containing the nest measured roughly 3 by 4½ by 6 inches. The nesting cavity was about 4 inches higher than the entrance of the tunnel, which took an upward course. This was true of many of the nests examined. Whether this is done for a purpose one cannot be sure, but it does prevent water from running into the nest during a heavy downpour of rain. Some of the nests were built on a level, and a very few were even directed downward. Stoner states that in general the burrows are excavated directly into the bank with the long axis approximately at right angles to the face of the bank. An occasional one, however, may form an acute angle with the base of the bank and thus comes to intersect other burrows in its course. Sometimes also the sandy partitions between the adjacent burrows are so thin that they give way, two or more burrows thus ending in a common chamber lying side by side. Merrill (1881) found three burrows in a colony examined on the Cranberry Islands, Maine, each of which contained two nests and all nests contained fresh eggs.

Stoner (1936b) gives an excellent account of the excavating activities of the bank swallow:

With the onset of the breeding season the first excavating efforts of the bank swallow are intermittent and seemingly aimless * * * not until about May 10 did the excavation of burrows begin in earnest. * * *

When starting a burrow it appears that the birds first cling to some slight projection on the face of the bank from which they can reach the point of attack, either with the claws or bill or both. After a slight concavity is formed, its sides are creased with the marks of claws and bill. As the work proceeds, the ceiling takes on a distinctly arched form, while the floor is practically flat.

Observations indicate that the more deeply scarred appearance of the inner or apical part of the cavity is due to the use of the bill. The bird clings to the walls and dislodges the particles of sand or gravel by pecking with a rapid side-to-side movement of the head. As soon as the shelf has resulted the feet with their long sharp claws are brought into action. Both bill and claws of captured individuals were caked with the moist earth. In digging, the tail is frequently used as a support, as in the woodpeckers.

Dissection of a few specimens taken showed that both male and female take active part in the task of digging. Often one individual of the (apparently mated) pair remains clinging to the face of the bank immediately beside the one that is working. When the latter hesitates or flies away its place is immediately taken by the other.

Until about May 15, the bank swallow seems to have a strong impulse to dig, but many more burrows are begun than are later occupied. However, as the season advances the impulse evidently becomes stronger, and excavating is confined to single burrows which increase in depth rapidly and steadily. At this time, too, the birds exhibit sensitiveness to disturbance.

With the deepening of the excavation the dislodged materials fall to the floor of the shallow burrow whence they are ejected by frequent vigorous kicks accompanied by a kind of wriggling movement of the body. In this action the wings also are "shuffled" rapidly in a backward and forward direction as well as from side to side, thus aiding in whipping the sand out behind the bird. Ordinarily, excavating is interrupted by frequent flights, probably at times to feed, but at other times for no very evident reason.

When a swallow returns to a burrow under construction it often enters immediately, and at once sends forth a shower of sand from the burrow by the rapid backward kicking action of its feet. After a few seconds, digging evidently is resumed and a new pile of loose material results; this is ejected by the same individual or its mate. In colonies where the burrows are close together the considerable amount of sand which is thus dislodged from them often accumulates in distinct windrows on the bank below.

When old burrows are renovated, the labor involved, in many if not most instances, is probably less arduous, the task then being mainly that of enlarging or extending the tunnel, and clearing out the remains of the old nest as well as other debris.

Generalizing the results of his study of the bank-swallow excavations at colonies in the Oneida Lake region, N. Y., Stoner (1936b) states that the excavating activities were most marked between May 7 and 21 and little was accomplished after May 26. In general, burrow excavation progressed at the rate of 3 to 4 inches a day. The greatest rate of excavation was a burrow that was increased 65 inches in the course of 7 days; this was also the deepest burrow noted by Stoner.

Leonard K. Beyer (1938) writes as follows concerning his observations of burrow excavating at a colony near Milton, Pa.:

Both birds of a pair took part in the work. A bird would begin by clinging to the vertical face of the bank with feet and tail and pecking at the dirt with a side-to-side motion of the head. When the opening was deep enough for it to get partly inside it would use its feet also, kicking the loosened sand backward in vigorous little spurts. As the tunnel became deeper the bird disappeared from sight, but still the sand came spurting out as evidence of the work of the little miner inside.

Bank swallows seem to take the work of excavating their burrows very lightly, more like play than work. Indeed, an eager holiday spirit seems to pervade the flock. A swallow will work vigorously for a few minutes, the while many of its comrades are circling about over the bank talking to each other in their reedy, buzzing twitter. Soon it can no longer resist the temptation and it flies out for a ride through the air with them. But usually not for long, and after a few minutes it returns to its job. These activities continue throughout the day, though at intervals the entire flock may leave the bank for a time. As evening comes on they fly away to some favorite roosting place in a nearby marsh.

The birds begin carrying in nesting materials as soon as the burrows are completed. At first the larger and longer grass stalks comprising the foundations of the nest are transported, and later come finer materials to which may be added a few feathers for the lining of the structure. Generally large numbers of feathers are not added until after the set of eggs is completed and incubation under way. According to Stoner the nest is prepared by the combined efforts of the male and female. The birds obtain much of the nesting material while they are on the wing, but frequently they may be seen on the ground near the nesting colony picking up dried weed and grass stalks.

The following description is based on a bank swallow's nest in a large colony located in a gravel pit 5 miles northwest of Topsham, Maine. The entrance to the burrow was 1½ inches high and 2¼ inches wide and was 39 inches deep. The extreme outer extent of the nesting materials included an area roughly 6½ by 5¾ inches, but the diameter of the central depression occupied by the incubating bird was only 2 inches in diameter and about half an inch deep. The nest itself was a very flimsy structure and would scarcely hold together without the support of the underlying sand. The bulk of the materials was made up of dry grass stems ranging from a half inch to four inches in length. Interwoven with the grass stems were rootlets, white pine needles, two match sticks, and several tufts of sheep's wool. In the interior of the nesting bowl were four well-worn scraggly feathers and small parts of others. This nest, containing 5 eggs, was taken on June 10, 1932. Nests examined later in the season were of a similar structure but in general were lined with many more fluffy feathers. A nest found by Knight (1908) on June 16, 1900, at Veazie, Maine, was composed of dried grass and lined with white feathers. It contained five eggs and was located in a burrow 3 feet deep. The entrance opening was 3 inches in diameter, and the nest chamber was 5 inches across. Stoner (1936b) says:

> The type of material used depends more or less upon what is available in the near vicinity. If a freshly mown hayfield is at hand, grass stalks will be gathered from this source, both green and dried blades being acceptable. Long thin blades or stems to be preferred. If a farmyard is near, a considerable amount of straw is likely to be found in the nest. Weed stalks are also used and some of these are of surprising length and size for so small a bird to manage within the limited confines of the burrow. For example, a dried grass stem 25½ inches long was found in a burrow. * * * Rootlets, horse hair and pledgets of sheep's wool often form a part of the nest. Some nests, however, are composed almost entirely of blades of grass. In July, particularly, the dried, black rootlets of the field horsetail, which dangle from the sandy banks in so many places along Fish Creek, frequently comprise a large part of the nest material.

The first eggs, and sometimes all in the clutch, are usually layed in the unlined nest. However, as egg-laying and incubation proceed, feathers are added as a lining, so that by hatching time a warm cozy bed awaits the young.

Eggs.—The eggs are pure white, but Stoner noted a departure of this coloration in which two of a set of five eggs that he found May 27, 1931, had numerous, minute, rounded elongate or irregular brownish marks. In about 50 nests examined in the vicinity of Brunswick, Maine, four and five eggs constituted the complete set in the large majority of cases. Only one set of six eggs was found, but sets of six have been frequently reported from other localities. Sets as large as seven (Knight, 1908) and even eight eggs (Forbush, 1929) have been reported, but these larger sets are very rare.

The measurements of 50 eggs average 17.9 by 12.7 millimeters; the eggs showing the four extremes measure **19.8** by 12.7, 19.3 by **13.7**, **15.2** by 11.4, and 15.7 by **10.7** millimeters.

Young.—Both male and female take part in the incubation of the eggs and the brooding of the young. Stoner definitely ascertained this fact by determining the sex by dissection. The period of incubation varies somewhat, depending on a number of controlling factors. In studies made by Stoner it was determined to be about 14 to 16 days, counting from the date on which the last egg was laid to that of the first hatching. The following account of the young is taken from Stoner's (1936b) exhaustive studies of the bank swallow.

During the first few days of life the young bank swallow is brooded almost constantly by one or the other parent. General observations indicated that the parents probably share about equally in the task of carrying food and in removing pellets of excrement from the nest. * * *

Usually the nestling expels a pellet [1] of excrement immediately after being fed, and this pellet is then carried away by the parent upon its next visit. The pellet may be dropped just outside the mouth of the burrow or carried, farther— * * * The parent may swallow some of the pellets, especially those appearing during the first days of the nestling's life.

The young usually discharge the pellets on the edge of the nest, and seldom in it. With young approximately 7 to 12 days old the pellets are often deposited in a little group a few inches from the nest. By aid of a beam of light, young were observed to crawl out of the nest to defecate, and then to shuffle back again. The returning parents begin at once to remove the pellets, and keep at the task until it is finished.

Beyer (1938) writes "Sometimes while the female was brooding the young the male would squeeze himself in beside her on the nest and then actually push her off. She would leave reluctantly and return in a few seconds. Sometimes while the male was foraging he would come back to the entrance of the burrow, twittering cheerfully, look in for a second or two as if to see that everything was all right, then fly away."

Stoner (1936b) noted that the earliest date in which young of the year were able to fly was on June 18 (1931), when three individuals were flushed from a nest. The following day a few young swallows flew from their nests, apparently in their first aerial excursion. These birds launched themselves boldly into the air, with a sharp twitter and vigorous flapping of wings. At first their progress was wavering but steadied as they gained momentum. They soon tired and usually circled back to the home bank, to enter any convenient burrow. One bird was seen to fall into the creek after a short flight. It flapped its way over the surface for a distance of 40 yards and landed safely on the bank. Another swallow that had dropped into a pool of water flapped along for a few feet and after gaining strength and steadiness rose into the air and continued its flight until lost in the distance. After leaving the home burrow for the first time the young bank swallows often return to it to rest or to receive food, at more or less frequent intervals. Later they return only for the night. Then, as the season advances, less and less time is spent at the burrows until, by July 25, approximately, practically all the first brood young have permanently left the locality in which they were reared. For a few days after their first flights the birds of the year often congregate about the home bank, twittering and flying in and out of the burrows, dusting themselves in the sand and disporting themselves in apparently carefree fashion. It is while perched on the sandy ledges below or near the burrows that these young of the season first engage in excavating activities. A row of shallow cavities along a ledge apparently represents their first instinctive efforts at burrow digging. But the cavities are seldom more than a few inches deep and are used by the birds as resting places.

The following extracts concerning the food of the young are taken from Leonard K. Beyer's paper on the "Nest Life of the Bank Swallow" (1938), in which he records his observations from a pit blind excavated back of the nests. When the young were freshly hatched his observation was as follows:

During the hour which I spent in the observation pit the young were fed several times on small flies, the parent placing the food far down the mouths of the nestlings. * * *

On the next day, the fourth for our observations, feeding went on just as it did the day before, the male foraging while the female remained in the nest to brood the three young. Feeding occurred on an average of twice every five minutes, the diet consisting mostly of small flies and caddice flies. * * *

As the days passed the character of the food brought to the young gradually changed, probably due mainly to changes in the relative abundance of various kinds of insects in the vicinity. On the sixth day a brood of may-flies emerged along the lake shore and these insects began to appear in the diet [the birds were now five days old]. * * *

On the day they were six days old their food consisted mostly of may-flies. * * *

On July 3 the young, then ten days old, were very active and hungry, stretching their necks, opening their mouths, and calling eagerly for food when an old bird appeared at the entrance. They were fed thirteen times in an hour, the food being practically all may-flies.

On July 6 Beyer and his assistant spent the entire day in the blind alternating turns. Some of his observations on food and feeding are as follows:

During the fifteen hours that the nest was under observation food was brought one hundred and fifteen times * * * the length of time between feedings ranged from one to fifteen minutes, averaging somewhat less than five minutes for the day. Several times both parents came at the same time with food. May-flies again seemed to be the main article of diet. Occasionally one would be dropped while being passed from parent to young and it would struggle, in a more or less mutilated condition, along the floor of the burrow toward the entrance. Neither the young nor the old birds would pay any attention to it. During the period from 5:00 P. M. to 7:00 P. M. feeding occurred more often, many times at only one-minute intervals. But after 7:00 P. M. no more visits were made by the parents that day. * * *

From the data secured in the daily observations at the nest can be made certain general statements concerning the life of the Bank Swallow. When the young are first hatched, * * * they are brooded almost continuously, apparently by the female, while the other parent, apparently the male, forages for food. When the male comes with food the female often flies out for a brief time, usually returning by the time he is through feeding the young. If she has not come back by that time he remains to brood them until she arrives. Occasionally she brings food with her.

The food the first few days consists of small soft-bodied insects. In the nest under observation it was mostly of small Diptera. Feeding occurs quite often, averaging twice every five minutes in this nest of three young. Only one young is fed at a visit—that is, all the food brought by the parent on one trip is given to one young. When entering the burrow with food the parent calls in a series of sweet high-pitched notes to the young. If the young are not hungry the parent calls more insistently, at the same time trampling them gently to arouse them.

As the young get older they are brooded less and less, by the sixth day scarcely at all. When brooding is no longer necessary both birds seem to share about equally in the feeding. Many times they both return at the same time with food.

In the nest under observation the food during the middle and latter part of the time spent in the nest seemed to consist entirely of may-flies which were abundant along the shore of the lake at this time.

Young bank swallows that are fully fledged but not ready to fly often rush to the mouth of the burrow, in full cry and with wide-open mouths, as soon as they are aware of the approach of a parent. A beam of light cast into a burrow sometimes brings a similar reaction. Occasionally such eager young birds fall out of the burrow and perish.

Fear in young bank swallows seems to appear only after they are well fledged. Young 8 to 15 days of age when removed from the nest display a marked tendency to crawl under ledges of earth or other objects. When removed a little distance from the home burrow they almost invariably shuffle toward it or in the direction of its location, not away from it. Young birds at this stage of growth, too, have a tendency to shuffle backward when placed on the ground outside the burrow; but this manifests itself also before they have been removed from the burrow.

So far as could be determined, the average time spent by the young in the nest varies from 18 to about 22 days. The young in a nest studied by Beyer left in 18 days after hatching.

After the young birds are fairly well grown and no longer require brooding, the parents leave them for the night, joining other adult birds from their own and other colonies in the vicinity to roost together in the marshes. Beyer did not see any of the adult birds fly into the nest after 7:30 P. M., and by 8:00 P. M. on July 6, the day he made continuous observations from a pit blind behind the nest, all the adult swallows had gone to their roosting place.

In regard to the number of broods reared during a single season there seems to be a difference of opinion. Forbush (1929) states that there is just one brood yearly. Beyer (1938) found no evidence of second broods, though he looked persistently for them throughout July and August in the Sodus Bay region. He concludes that though it is possible two broods are reared it is of rare occurrence in the region where his studies were made. Stoner (1936b), on the other hand, presents concrete evidence that two broods are the usual thing in the colonies studied intensively by him. Stoner found that four or five eggs form the usual complement of the eggs of the first brood clutch and three or four in the second.

Plumages.—The following notes on the plumage of the young are from Leonard K. Beyer's (1938) paper:

The newly-hatched birds * * * were pink in color, with a scanty covering of gray down on the back of the head and neck, base of wings, and top of back. The eyes were very large and showed black through the closed lids. The inside of the mouth and the flanges on the bill were lemon-yellow, the bill yellowish-gray, the feet pinkish-gray. The tiny nestlings appeared quite weak and it seemed to be only with the greatest effort that they were able to raise their immense wobbly heads for the food their parents brought them. * * *

At this age [5 days] the feather sheaths have appeared in the feather tracts but as yet none of them have burst. The nestlings are much stronger and noticeably larger, and their eyes are open though they keep them closed most of the time. * * *

At this age of the young [7 days] many feathers in the dorsal tract have burst their sheaths, and also a few wing and tail feathers. * * * Now nine days old, they had become rather well feathered, most of the body feathers and many of the wing and tail feathers having burst their sheaths. However, their appearance was rough and unkempt, for the feathers were only partly grown and the scanty natal down still clung to the tips of many of the feathers. * * *

The remaining nestlings were thirteen days old * * * They were quite well feathered, with a considerable amount of natal down still clinging to them. * * *

On July 8 there was only one young bird in the nest. It was fifteen days old, fully feathered, and practically all the natal down was gone.

The natal down has disappeared by the sixteenth day and at this time they are able to fly. The young remain in the nest until about the eighteenth day but sometimes remain two or three days longer.

The plumages and molts of the bank swallow have been described by Dwight (1900), Brewster (1878), and Chapman (1917). I have drawn freely from their papers in the descriptions that follow.

The juvenal plumage is acquired by a complete postnatal molt. In this plumage the upper parts, including sides of the head, are brownish mouse gray, feathers edged with ferruginous, edging broadest on the rump and secondaries, narrowest on the crown and nape. The under parts are similar to the plumage of the adult but with the broad pectoral band strongly washed with clove brown, the feathers edged with cinnamon. The chin and throat are tinged with cinnamon and spotted with faint dusky dots. Lores dull black, feet sepia becoming black.

Some of the cinnamon edgings of the juvenal plumage are lost before the young start the autumn migration. Some specimens taken at this time have their secondaries tipped with white. The post-juvenal molt takes place after the birds arrive at their winter quarters. The first winter plumage, according to Dwight (1900) has the tail more deeply forked and is indistinctly barred, the chin is pure white without spots and the collar is darker. The young and adults become indistinguishable.

The first nuptial plumage is acquired by wear. The wings and tail are darker than those of the juvenal plumage. The upper parts are a lusterless grayish brown, darker on the pileum. Feathers of the rump, scapulars, tertials, and upper tail coverts have paler margins except in midsummer. Wings and tail are fuscous, underparts white interrupted by a broad band of grayish brown across the breast, which is continued along the sides; bill brownish black; iris brown, legs and feet dusky horn color. There is a small tuft of feathers on the tarsus near the base of the hind toe.

The sexes are alike in their plumages and molts, and there is but little variation with age. The adult winter plumage, according to Dwight (1900), is acquired by a complete postnuptial molt, which is assumed after the birds have moved southward in the winter. The adult nuptial plumage is acquired by wear.

Abnormal plumages such as albinism, a condition where there is an absence of pigment in the feathers, occurs in many groups of birds. While this condition is not common it may be expected in any species. On June 2, 1932, I noted a pure-white albinistic bank swallow among the members of a large colony near Topsham, Maine. It was seen three days later and then disappeared. As far as I was able to discover this individual was not nesting.

Food.—The bank swallow is primarily insectivorous. Vegetable matter such as seeds occurring in the stomach contents is purely accidental.

F. E. L. Beal (1918) found in the examination of 394 stomachs of bank swallows collected in 21 States and Canada that 17.9 percent of the food consisted of Coleoptera. Of these, May beetles, flea beetles, and the various species of weevils that have proved destructive to many crops were most numerous. Arthur H. Howell (1924) reports that 25 bank swallows taken over the cottonfields of Texas in September had eaten 68 boll weevils, one bird having eaten 14 of them.

Ants, chiefly winged forms, composed 13.39 percent of the food in the stomachs examined by Beal. The ants were found in 121 stomachs and formed the total contents of 11 of them. Various other species of Hymenoptera were found in 207 stomachs. The Hemiptera, chiefly leafhoppers, treehoppers, and plant lice, comprised nearly 8 percent of the food, but this amount of Hemiptera is small as compared to that eaten by the other species of swallows.

The Diptera, represented chiefly by house flies and crane flies, made up 26.63 percent of the food. The Lepidoptera, comprising only 1.21 percent, was made up of moths and caterpillars. One stomach was entirely filled with caterpillars. The Odonata (dragonflies), which are usually found about water, fall prey to bank swallows that fly along or that nest in the banks of streams. These large swift-flying insects comprised 2.11 percent of the food. Other miscellaneous insects eaten comprise 10.53 percent of the food. A few spiders were included in the stomach contents examined by Beal.

Dayton Stoner (1936b) reported on the food of 21 adult and 43 young bank swallows collected in the Oneida Lake region, New York, between May 14 and July 22, 1931. His determinations confirm Beal's results in that they clearly indicate that the food of the bank swallow consists primarily of insects that are captured during flight. While a few of the insects eaten are beneficial to man, the vast majority are injurious, placing the bank swallow high in the column of species of birds beneficial to man.

The food of the young is similar to that of adults, but that of the younger nestlings contained an excess of soft-bodied forms of insects. Such large, coarse-bodied insects as stoneflies are absent from the food of the young, and there are a smaller number of beetles and a greater number of flies and Homoptera in the food of the young.

Voice.—The bank swallow does not excel in its ability to sing; indeed its simple twittering notes at their best can scarcely be called a pleasing song. Forbush (1929) describes the notes as "usually rather silent, except when danger threatens; call notes, more harsh and 'gritty' than those of other swallows; 'song,' a mere twitter."

In correspondence, S. S. Dickey writes: "Their notes approach those of the rough-winged swallow, but they are more subdued and gentler. They sound not unlike the syllables *speedz-sweet; speedz-sweet*, oft repeated and notably audible around the nesting sites."

The calls and notes of the adults and young at the nest are described in detail by Leonard K. Beyer (1938):

When entering the burrow with food the male calls in a series of peculiarly sweet, fine notes much higher in pitch than the usual Bank Swallow call. This seems to be the food call to the young, for upon hearing it they raise their heads with mouths wide open. When the young were four days old, I heard for the first time the call notes of the nestlings, weak and rather frequent and resembling somewhat the call notes of the old birds. When the young were seven days old, their voices had now begun to resemble those of the adult Bank Swallows. Sometimes when they did not respond to the food call of the parent the old bird gave a very soft, high-pitched call that was exceedingly musical and sweet. As they become older they call more loudly and gradually their notes come to resemble the characteristic reedy twitter of the adults of the species.

Dayton Stoner (1936b) writes of the notes of the bank swallow as follows: "Nest-building as well as burrow-digging is accompanied by much harsh, unmusical twittering. Thereafter comparative quiet reigns during the period of egg-laying and incubation; vocalization is resumed as soon as the young are hatched.

"When alarmed, the bank swallow gives a shrill '*ke-a-g-h*' or '*te-a-r-r*,' which, though much feebler, reminds one of the note of the common tern. Apparently this is a warning note. When uttered by a swallow in flight as, for example, a frightened individual leaving the nest, or the bander's hand, it acts as a signal to other members of the colony."

In regard to the voice of the young, Stoner makes the following observations:

Almost immediately after hatching, young bank swallows may be heard to utter a fine, high-pitched "cheep"; but the nestlings are, as a rule, inclined to be silent, although hunger calls are uttered when the adults have not visited them for some time. When a family of four or five nestlings utter this call together, the sound reminds one of the buzzing of a swarm of bees. Occasionally, when well-fledged young are forcibly removed from the burrow, they utter a few loud, shrill cries of distress,—a signal for the adults in the immediate vicinity to congregate and add their own excited notes to the clamor. But unlike many other passerine birds, the distress calls of the young usually are not long continued, even when the bird is roughly handled. Only a sharp pain then seems to provoke an outcry. Ordinarily, the young individuals that occasionally drop into the water likewise remain silent.

For a few days immediately preceding abandonment of the nest, the young often appear at the mouth of the burrow, twittering and warbling much after the manner of the adults. At this age their voice is lower and harsher than that of the adult, but within a few days no pronounced difference in timbre was noted.

Enemies.—The most important enemies of the bank swallow are those predatory animals that are able to gain access to the nesting burrows. In the Oneida Lake region of New York State, Stoner (1936b) found the skunk (*Mephitis nigra*) to be one of the chief offenders. He cites a case where on June 23, 1931, a skunk had dug through the thick turf and into the soil to a depth of a foot to reach the young in a bank swallow's nest. He notes several additional cases in which the skunk was blamed for similar depredations on eggs and young. Stoner states that the predatory activities of the skunk can probably be attributed only to a comparatively small number of individuals living in the immediate vicinity of the bank swallow colonies. In general the skunk is not a serious enemy of the swallow.

William Brewster (1903) describes in detail a very interesting case where a mink destroyed a colony of bank swallows located at Lancaster, Mass. On May 24, 1902, there were 108 holes, but on June 19 all but one pair had been destroyed by a mink. In a colony of

22 nests located on the banks of the Androscogin River, Brunswick, Maine, nine were destroyed, presumably by a mink or a weasel.

At a large colony of bank swallows near Topsham, Maine, where many of the nesting holes were near the edge of the turf at the top of a gravel pit, I saw a cat several times during June 1924 attempting to capture the swallows as they darted near her or attempted to enter the nesting burrows. No captures were noted, but the cat's frequent presence there indicated that she had been successful. Certainly young leaving their nests prematurely would be subject to capture by cats visiting a colony at such times.

Laurence B. Potter (1924) on July 20, 1924, found many nests destroyed by a badger in a colony located on Frenchman River, East-end, Saskatchewan, Canada. The badger dug down from above until within reach of the nestling swallows.

Dayton Stoner (1938) writes that adult bank swallows were killed and partially devoured in their nesting burrows by house rats (*Rattus norvegicus*). Stoner (1936b) found a deer mouse (*Peromyscus leucopus noveboracensis*) in a bank swallow's nest that had a deep feather lining. There was no evidence that the mouse had destroyed eggs or young. He also records finding a garter snake, a northern flicker, a prairie harvest mouse, and various insects in the nesting burrows of bank swallows.

Predatory birds probably are not so serious enemies of the active fast-flying swallows as of more sluggish species, but some meet an untimely death in this manner. P. L. Errington (1932) reports a bank swallow being eaten by a barred owl (*Strix varia*), and Stoner (1938) states that the crow is responsible for the destruction of a few bank swallows.

In the past perhaps more than in recent years thoughtless boys have been guilty of excavating numbers of bank swallows' nests.

The automobile is one of the modern enemies of the bank swallow. In Maine where colonies are located in sand pits near the highways I have frequently seen bank swallows that had been killed by the fast-moving cars. O. A. Stevens (1932) writes that on July 23, 1931, he found 10 or 12 dead bank swallows on a graveled road about a small lake 25 miles east of Fargo, N. Dak. He also cites a report of W. E. Brentzel, who saw three places in a roadway near Pelican Rapids, Minn., that were black with dead swallows, estimated at a total of 1,000. Such losses when frequent are more disastrous than predatory animals and birds.

In places where the bank swallows dig their nesting holes in the sides of gravel pits entire colonies may be destroyed by slides caused when the gravel is removed for road building or other purposes. A colony of about 20 nests located on the banks of the Androscogin River near Brunswick, Maine, was destroyed by a slide following a flooded condition of the river. Similar instances have been cited by others.

Bank-swallow burrows are sometimes appropriated by other birds as nesting sites. Forbush (1929) states that the house sparrow and the starling sometimes drive out the bank swallows and utilize the burrows for their own nests. Stoner (1925) found a burrow that had been excavated by a bank swallow in a colony on the Little Sioux River containing four fledging house sparrows in good condition and nearly ready to leave the nest. A. Dawes DuBois (MS.) writing concerning a colony of bank swallows located at French Village, St. Clair County, Ill., states: "We estimated the number of holes in the bank to be about 400, in two groups or tiers; but only the holes of the upper group were being used by the swallows. The lower holes, now enlarged, were being used by a great many English sparrows." One of the bank swallow tunnels examined by DuBois on May 22 contained a large blacksnake.

During prolonged cold rainy spells of weather numbers of bank swallows may perish from exposure and lack of food. Macoun (1909) cites a case of this nature:

On June 5th, 1902, there was a severe and very cold storm, and at one colony the birds evidently crowded into the partially completed burrows for shelter, to such an extent that those at the end were crushed or smothered to death; almost every burrow had three or four dead birds, rammed hard against the end; one hole had six, jammed into a mass which held together, so strongly I was able to drag it out by pulling on one wing; some holes contained but one bird, and in these cases the little bodies were not so badly jammed; one of these solitary corpses proved to be that of a barn swallow; I presume these single birds died from the cold as doubtless had the several found on the ground at the foot of the bank; altogether some 30 or 40 swallows perished in this colony.

It is not unusual to find bank swallows infested with mites and lice. Harold Peters (1936) has found the mites *Liponyssus sylviarum* and *Atricholaelaps glasgowi* on bank swallows taken in Pennsylvania and the louse *Myrsidea dissimilis* on a South Carolina bank swallow. Stoner (1936b) reports mites (*Atricholaelaps*) as not uncommon on bank swallows examined by him in the Oneida Lake region, New York. Many of the young swallows in the nest late in June and July harbored these mites, the numbers seemed to increase somewhat as the season advanced. Lice (*Myrsidea dissimilis*) were according to Stoner (1936b) on most of the swallows examined by him. These parasites attained their maximum abundance on nestlings after mid-July.

The above-mentioned species of lice and mites have been present on many of the birds examined in the vicinity of Brunswick, Maine.

These parasites are not a serious menace to the birds but in cases of heavy infestations may cause a certain degree of discomfort especially to young confined in the narrow quarters of their nest.

The fleas *Ceratophyllus riparius* and *C. celsus* have been found in the burrows and nests as well as on the adult birds. Fleas may be found in burrows and nests not occupied and probably persist throughout the year in such situations. Leonard K. Beyer (1938) states that fleas (*C. riparius*) bred abundantly in the material of nests he had under daily observation at Sodus Bay on the southern shore of Lake Ontario. He states that the fleas were often seen crawling over the birds. Mites also infested them and the birds' frequent scratching and picking at themselves were doubtless caused by the attacks of these parasites.

The most injurious external parasite of the nestling bank swallow is the muscoid fly *Protocalliphora splendida*, which is common among birds examined in Maine colonies. Stoner (1936b) found that scores of individuals examined in the Oneida Lake region, New York, in 1931 carried the larvae in various stages of development. These larvae become attached to almost any part of the host's body by means of a suctorial disk. They commonly enter the nasal and ear openings where the skin is delicate and the blood supply generous. They are also attracted by the rich supply of blood about the bases of the rapidly growing tail and large wing feathers. They are also to be found in the region of the vent and oil glands. According to Stoner, the late or second broods are more heavily infested than earlier or first broods of the season.

In 10 out of 1,200 young and adult bank swallows examined by Stoner there were diseased external parts. Four juvenal swallows taken in June and July exhibited abnormal and seriously diseased feet or tarsi or both. It is possible that there is a relation existing between this infection and the occurrence of the *Protocalliphora* larvae.

No extensive study has been made of the internal parasites of the swallows, but Dr. Eloise B. Cram (1927) records the nematode parasite *Acuaria attenuata* as being found between the tunics of the gizzard in the bank swallow. The swallow served as the primary host of this parasite.

The bank swallow, which nests in such a well-hidden situation, would not be expected to be a victim of the cowbird; however, Friedmann (1929) reports a set of bank swallows' eggs taken at Lacon, Ill., by William E. Loucks, which contains six eggs of the swallow and one of the cowbird. No details of the location of the nest, whether exposed or in an exceptionally short burrow, are given. In all events it represents a very unusual and purely accidental case.

Fall.—The flocking of the bank swallow with other species of swallows is one of the most striking features of their behavior during autumn. Because such concentrations are so impressive they have been noted and reported by various observers in all parts of the country.

Abby F. C. Bates (1895) presents a very interesting and lucid account of a swallow roost located at the junction of the Mussalonskee stream and the Kennebec River near Waterville, Maine. Extracts of her report are as follows:

The willow trees along the banks of this stream, particularly a close row some five or six hundred feet in length, form the roosting place of vast numbers of Swallows. During the forenoon and early afternoon very few swallows are to be seen in the sky,—indeed they are conspicuous by their absence,—but a little before sunset the birds begin to arrive in the vicinity, flying, sailing, chasing each other around in the upper air, everywhere within eye's reach. * * * Shortly after sunset they gather more nearly in the region directly above the trees, incomers from every point of the horizon still joining them and toward the last exhibiting great hurry and intentness, as if fearful of being "late to meeting."

Then begin the movements that are the most interesting feature of this gathering. At intervals *clouds* of Swallows will evolve something like order out of their numbers and perform *en masse* some of the most fantastic curves, spirals, counter-marches, snake-like twists and turns, with the sky as a background, that ever a company of genus homo executed on a finely polished floor. For instance, one evening they separated into two parts, one going to the right, the other to the left, each division making a grand circle outward, then joining again for a forward movement. There were some stragglers, but the figure was distinct and was twice performed, with other evolutions interspersed. Then a long snake-like movement from the upper air down very slightly inclined from the vertical, with two twists in it, a loop around a tall tree farther down the stream, and back, brought them into the tree-tops for roosting. * * * Occasionally they drop down into the trees like pieces of paper, but oftener the final alighting is a combined movement, sometimes in the shape of an inverted cone,—usually in a grand sweep after their most elaborate evolution. Frequently they swoop out of the trees company after company, several times before the last settling, their wings not only making a tremendous whirring, but a perceptible movement of the air. Their chattering keeps up from half to three quarters of an hour after they settle in the trees and their dark little bodies against the sunset sky look as numerous as the leaves. Often they weigh down a branch and then a great chattering, scolding and re-adjustment ensues. * * * The noise which they make is suggestive of the whirring of looms in a cotton mill, heard through the open windows,—or of some kinds of waterfalls.

They leave the trees in the morning a little before sunrise. August 26 we watched them go out. At 4:15 there were sounds as if of awakening and gradually the noise increased. At 4:25 they began to arise in companies at intervals of two or three minutes. They did not remain long in the locality and by five o'clock not one was to be seen.

The above unusual colony consisted of bank and barn swallows with a large number of martins. The concentrations began about the middle of July and reached their maximum during the latter part of August. By September 9 the swallows had departed and only martins were observed at the roost. On September 26 practically all the birds were gone, there being only about 40 martins at that time. The roost at Waterville had been observed for a great many years before Mr. Bates made his observations and probably continued long after that date.

E. J. Sawyer (1918) records a large roost of swallows at Black River Bay on Lake Ontario, Jefferson County, N. Y. At this place the flocks consisted of bank, tree, and barn swallows, and incidentally a single rough-winged swallow was noted. The birds roosted in large marshy area where cattails, iris, arrowhead, and pickerel weed predominated. Sawyer also refers to the swarms of swallows near the Hackensack Meadows near New York City.

Louis W. Campbell (1932) writes concerning unusually large flocks of bank swallows observed near Toledo, Ohio, as follows:

Because of their proximity to the lake marshes, Toledo and its suburbs extending to the east play host to varying numbers of swallows each year. Up to 1931 the greatest flock I had ever witnessed was a flight of about 10,000 Bank Swallows (*Riparia riparia*) seen on July 15, 1928. This year, probably as the result of the drying up of all inland ponds, Bank Swallows were unusually common. The greatest number appeared on August 8, 1931, when the Little Cedar Point marsh, about ten miles east of Toledo, was visited by a huge flock of Bank Swallows, the number of which I estimated to be more than 250,000.

This congregation marked the climax of the flight, although large flocks were seen both before and after that date, roosting on telephone wires along the roadside. This habit of roosting makes estimating numbers a very easy task for the observer. The count for the entire season is: July 12—100; July 18—200; July 26—3,000; August 2—8,000; August 8—250,000; August 15—30,000; August 23—50,000; August 30—500; September 7—50; and September 13—8.

According to O. Widmann (1907) bank swallows begin to collect in immense flocks in the Mississippi River bottoms at St. Louis as early as July 1. Migration is well under way by the first of August and continues during the entire month. All the birds have gone by the middle of September.

John H. Sage (1895) gives an interesting account of a swallow roost made up of bank, tree, barn, and cliff swallows that he observed near Portland, Conn.:

On the opposite side of the Connecticut River from Portland are what are locally known as the "Little River" meadows. These meadows contain several hundred acres and through them flows Sebethe (Little) River which empties into the Connecticut. Along the banks of this "little river" and its tributaries, water oats (*Zizania aquatica*) grow in abundance, giving food and shelter to the Rail, Marsh Wrens, and many other birds. These oats are the roosting place of thousands of Swallows, the birds spending the night clinging to the upright reeds, one above another. * * *

The Swallows commence to congregate in the marshes early in August, and a small number of them may be found there the last week in October; the bulk, however, are seen from the middle of August until late in September. During the day they leave the meadows and only a few are seen in the vicinity, but at half past four in the afternoon they begin to appear from all directions, the flight ceasing about 6 P. M. My house is situated on high ground some two miles east of the marsh, and the flight of these birds over my premises, and toward the meadow, is so regular (from 4:30 to 6 o'clock) each afternoon, that a watch is hardly necessary to tell the time of day.

On July 29, 1921, at Hankinson, N. Dak., F. C. Lincoln (1925) saw 3,000 bank swallows, mostly young, dusting in a road. At times for a distance of 25 or 30 yards the dusty road was literally a mass of wriggling swallows. When flushed from the road they alighted so thickly on the telephone wire that it resembled a great string of beads, and for a distance of several yards no wire was visible. The bulk of the flock finally took wing and circled steadily upward in a confused and orderless mass until they had gained three or four hundred yards of altitude. The swallows were still circling when he left the vicinity.

A. D. Henderson writes that on August 21, 1923, he watched a number of bank swallows migrating along the Pembina River, Alberta, where there were many nesting holes of the swallows in one of the banks. Many of them entered the holes; a few came out and flew on southward; but, as he heard twittering in the holes and as no more had come out by 8 o'clock, he concluded that they were spending the night in the holes.

The great mass of the bank swallows leave the northern section of their breeding range during the month of August but a certain number linger over until September. It is very unusual to find a bank swallow in Canada or Northern United States after the first of October. A few bank swallows spend the winter along our southern border, but the great bulk of them pass farther south, migrating via the Bahamas and West Indies but chiefly via Mexico to Brazil, northern Argentina, and central Chile.

Bird banding.—Much valuable information on the migratory movements of bank swallows has been obtained by means of bird banding. Most of the returns from very extensive banding have been limited to birds recaptured in their nesting colonies, but Mr. Lincoln (1939) has reported that one banded at Clear Lake, Ind., on June 12, 1932, was found, probably dead, in a house near Yquitos, Peru, in June 1936; and another banded at Ephraim, Wis., on June 8, 1933, was found dead at Manistique, Mich., on June 30, 1934.

Dr. Stoner (1937) had, up to that time, banded 4,925 bank swallows, of which 3,044 were young and 1,881 were adults taken at various nesting colonies in Iowa and New York. Of the total of 3,044 *young* bank swallows banded, 31 were captured as returns, and of these 6 (19.3 percent) were recovered in the same colony as reared and banded. Of the 1,881 *adult* bank swallows banded, 68 were captured as returns and of these 51 (75 percent) were recovered in the same colony as originally banded. Stoner concludes from the results of his records that "bank swallows commonly breed when less than one year old, that is, the first season after hatching. * * * The first year bank swallows do not often return to breed in the gravel pit in which they were reared; a somewhat larger number return to the general region to nest. A very *much larger* proportion of adults, however, return to a gravel pit in which they have nested previously while a *still larger* proportion return to the general region where they previously have nested." The bank swallows usually, if not always, have a different mate each year. Certain individuals change mates before a first brood is reared; others have had different mates for the first and second matings of a season.

Bird banding has contributed definite knowledge concerning the life span of birds, information concerning birds living a normal life that cannot be derived by any other known method. Dayton Stoner (1938) has banded 5,904 bank swallows, of which 2,834 were adults and 3,070 were young. It is obvious that the age of adults is not certain, but they are at least one year old, possibly older when banded. Of 169 of Stoner's returns 36 (21.3 percent) were banded as young and 133 (78.6 percent) were banded as adults. Eighteen of the returns were one year old, 104 at least two years old, 33 at least 3 years, 10 at least 4 years, and 3 at least 5 years old. One individual banded as a nestling was 6 years old.

Dr. Stoner concludes that the average life span of the bank swallow is two or three years and for the territory under consideration the bank swallow does not ordinarily exceed three years. A number of the bank swallows attain the age of four years, and a few continue life for five years, but it is very unusual for a bank swallow to attain the age of six years or more.

DISTRIBUTION

Range.—Circumpolar; breeding over the entire Northern Hemisphere and wintering in South America, Africa, and India.

Breeding range.—In North America the breeding range of the bank swallow extends **north** to northern Alaska (Kowak Delta, Hogatza River, Fort Yukon, and Hulahula River); northern Yukon (Johnson Creek); Mackenzie (Fort McPherson, Fort Goodhope, Fort Anderson, and the Lower Hanbury River); northern Manitoba (Hayes River); northern Quebec (forks of the Koksoak); northern Labrador (Okak); and Newfoundland (St. Georges Bay). **East** to Newfoundland (St. Georges Bay); thence south along the Atlantic coast to southeastern Virginia (Petersburg). **South** to southern Virginia (Petersburg, Lynchburg, and Blacksburg); southeastern Tennessee (Athens); central Alabama (Washington Ferry); southern Texas (Santa Maria, Laredo, Del Rio, and Langtry); southern

Arizona (Fort Lowell and Tucson); and southern California (Ocean-side). The **western** limits extend north along the Pacific coast from southern California (Oceanside and Huntington Beach) to Puget Sound, Wash. (Smith Island); thence through the interior of British Columbia (Ashcroft, Pack River, Klappan River, and Atlin); south-western Yukon (Lake Marsh and Fifty-mile River); and Alaska (Lake Aleknagik, Bethel, Iditarod, and Kobuk Delta).

Winter range.—From available information it appears that in winter the bank swallows of the Western Hemisphere are concen-trated in central South America. Most of the winter stations are for Brazil (Joazeiro, Tapirapoan, and Caicara), but it also has been recorded at this season from Bolivia (Chilon), and northern Peru (Nauta). A specimen was taken at Cali, Colombia, on February 6, 1912; one was obtained at Encontrados, Venezuela, on February 13, 1908; and one was collected on March 2, 1925, at Concepcion, Tucu-man Province, Argentina. One found dead in a house at Yquitos, Peru, in June 1936 had been banded (C30270) on June 12, 1932, at Clear Lake, Ind.

Spring migration.—Early dates of spring arrival are: Florida—St. Petersburg, April 10. Georgia—Macon, April 13. Virginia—Variety Mills, April 11. District of Columbia—Washington, April 4. Pennsylvania—Philadelphia, April 8. New Jersey—Mahwah, April 3. New York—New York City, April 10. Connecticut—Hartford, April 17. Massachusetts—Boston, April 28. Vermont—St. Johnsbury, April 19. Maine—Lewiston, April 18. New Bruns-wick—Grand Manan, April 27. Quebec—Montreal, May 6. Louisi-ana—New Orleans, February 16. Arkansas—Helena, March 26. Tennessee—Nashville, March 31. Missouri—Fayette, April 8. Illi-nois—Joliet, April 15. Indiana—Bloomington, April 6. Ohio—Oberlin, April 6. Michigan—Detroit, April 11. Ontario—Ottawa, April 17. Iowa—Iowa City, April 5. Wisconsin—Madison, April 12. Minnesota—Minneapolis, April 9. Manitoba—Aweme, April 28. Kansas—Topeka, April 10. Nebraska—Lincoln, April 13. South Dakota—Yankton, April 21. North Dakota—Larimore, April 28. Saskatchewan—Indian Head, April 27. Arizona—Tucson, April 5. Colorado—Durango, April 8. Wyoming—Laramie, April 23. Montana—Fortine, April 18. Alberta—Alliance, May 1. Califor-nia—Santa Barbara, April 8. Oregon—Yamhill County, April 3. Washington—Clallam Bay, April 1. British Columbia—West Sum-merland, April 24. Alaska, Fairbanks, May 14.

Fall migration.—Late dates of fall departure are: Alaska—Kobuk River, August 19. British Columbia—Okanagan Landing, Septem-ber 27. Oregon—Harney County, September 12. Alberta—Belve-dere, August 19. Montana—Fortine, September 14. Wyoming—Laramie, September 4. Colorado—Denver, September 12. Saskatchewan—Eastend, August 26. North Dakota—Argusville, August 31. South Dakota—Forestburg, September 10. Nebraska—Lincoln, September 27. Kansas—Fort Hays, September 21. Oklahoma—Tulsa, October 12. Texas—Somerset, October 22. Manitoba—Aweme, August 31. Minnesota—Minneapolis, September 28. Wisconsin—Madison, September 20. Iowa—Sabula, September 26. Ontario—Ottawa, September 15. Michigan—Detroit, September 11. Ohio—Oberlin, October 7. Illinois—Chicago, September 16. Missouri—Concordia, September 29. Tennessee—Athens, September 29. Quebec—Montreal, September 9. Maine—Pittsfield, September 5. Vermont—Wells River, September 9. Massachusetts—Dennis, September 15. New York—Rhinebeck, September 13. District of Columbia—Washington, September 21. Georgia—Savannah, October 17.

Casual records.—On the Arctic coast of Alaska a flock of seven was seen at Wainwright on August 10, 1922, and a specimen in immature plumage was collected at Point Barrow on August 26, 1929. One was found dead in a rain-water barrel on St. George Island, Alaska, on June 20, 1923. A pair was reported as seen on June 12, 1820, at Liddon Gulf on the north side of Melville Island, Northwest Territories.

Egg dates.—Alaska: 7 records, June 10 to July 3.

California: 48 records, April 29 to June 12; 24 records, May 13 to June 3, indicating the height of the season.

Illinois: 14 records, May 18 to June 22; 8 records, May 24 to June 9.

Massachusetts: 21 records, May 21 to July 3; 11 records, May 27 to June 15.

STELGIDOPTERYX RUFICOLLIS SERRIPENNIS Audubon

ROUGH-WINGED SWALLOW

PLATES 61, 62

HABITS

CONTRIBUTED BY EDWARD VON SIEBOLD DINGLE

The rough-winged swallow was discovered by John James Audubon in Louisiana, but the description of the bird in his "Ornithological Biography" is based rather on specimens collected many years later at Charleston, S. C., which city is, therefore, the type locality. Audubon's (1838) description of his first meeting with this swallow is as follows:

On the afternoon of the 20th of October 1819, I was walking along the shores of a forest-margined lake, a few miles from Bayou Sara, in pursuit of some Ibises, when I observed a flock of small Swallows bearing so great a resemblance to our common Sand Martin, that I at first paid little attention to them. The Ibises proving too wild to be approached, I relinquished the pursuit, and being fatigued by a long day's exertion, I leaned against a tree, and gazed on the Swallows, wishing that I could travel with as much ease and rapidity as they, and thus return to my family as readily as they could to their winter quarters. How it happened I cannot now recollect, but I thought of shooting some of them, perhaps to see how expert I might prove on other occasions. Off went a shot, and down came one of the birds, which my dog brought to me between his lips. Another, a third, a fourth, and at last a fifth were procured. The ever-continuing desire of comparing one bird with another led me to take them up. I thought them rather large, and therefore placed them in my bag, and proceeded slowly toward the plantation of William Perry, Esq., with whom I had for a time taken up my residence.

The naturalist examined his specimens carefully and saw that they were different birds from the sand martin, or bank swallow, but he continues. "At this time my observations went no further."

Then, "about two years ago, my friend the Rev. John Bachman, sent me four Swallow's eggs accompanied with a letter, in which was the following notice—'Two pairs of Swallows resembling the Sand Martin, have built their nests for two years in succession in the walls of an unfinished brick house at Charleston, in the holes where the scaffolding had been placed. It is believed here that there are two species of these birds.' * * *

"I have now in my possession one pair of these Swallows procured by myself in South Carolina during my last visit to that State."

The roughwing enjoys a very extended range in the Western Hemisphere. Essentially a bird of the Austral Zone, it does not hesitate to establish itself in mountainous country thousands of feet above sea level. According to Miller (1930) the bird breeds in the heart of the Pocono Mountains of Pennsylvania, at about 2,000 feet; in western North Carolina Brewster (1886) found it up to 2,500 feet. James B. Dixon (MS.) says that in California it breeds from sea level up to 6,500 feet. Grinnell, Dixon, and Linsdale (1930) report specimens collected at Red Rock P. O., Calif., at an elevation of 5,300 feet; also, birds observed at Petes Valley, 4,500 feet; Secret Valley, 4,500 feet; and Jones, 5,400 feet.

This bird, also locally known as the sand, or gully, martin, is rather solitary in habits and usually does not congregate during the breeding season, as does its near relative the bank swallow. However, as Dawson (1923) says, "favorable conditions may attract several pairs to a given spot, as a gravel pit, but when together they are little given to community functions."

Courtship.—Grinnell and Storer (1924) write: "From time to time the males were seen in pursuit of the females and, while so engaged, to make rather striking use of their seemingly plain garb. They would spread the long white feathers (under tail coverts) at the lower base of the tail until they curled up along either side of the otherwise brownish tail. The effect produced was of white outer tail feathers, such as those of the junco or pipit. Males can by means of this trick be distinguished from the females at a distance of fully 50 yards. An examination of specimens in hand reveals the fact that the under tail coverts of the males are broader and longer than those of the females."

Nesting.—Burrows, excavated in precipitous banks of clay, sand, or gravel by the birds themselves, are the usual nesting sites of the roughwing. The length of the burrow depends, as H. H. Bailey (1913) says, "much on the character of the soil in which it is started. Weather conditions also make a moist or hard soil for them to work in." Minimum depth of burrow is about 9 inches; and, in these shallow excavations, the nest can be sometimes seen from the outside. The greater number of tunnels, however, are long enough to keep the nest from view and protect it from driving rains. Under ideal working conditions, tunnels 4 and 5 feet long are often excavated, sometimes reaching even a distance of 6 feet.

Bailey further says: "The height of the nesting cavity in the bank also varies greatly, the nature of the soil strata affecting the drilling of the hole, which is made by the birds using their feet to scratch with, and push the dirt backward out of the tunnel. Unlike the kingfisher, their beaks play a secondary part in the drilling of their home, so they usually select a place in the soft strata where the roof will be the under side of a hard strata of soil, and so eliminate the chances of a cave-in."

Dawson (1923) writes that "in open country, where the cover is scarce but the food supply attractive" he found them nesting "along irrigating ditches with banks not over two feet high." Weydemeyer (1933) found nests in Montana in banks 1 to 50 feet up.

This swallow is an excellent example of a species that can readily adapt itself to conditions and utilize any kind of cavity for the reception of its nest. It builds in holes in masonry, sides of wooden buildings, adobe walls, quarries and caves; crannies and ledges under bridges, culverts, and wharfs; and gutters, drainpipes, and sewer-pipes. Deserted burrows of the kingfisher (*Megaceryle alcyon*) also are frequently used, and in the West holes of ground squirrels and other small mammals. According to Tyler (1913), these holes are thoroughly renovated before occupancy "as is evidenced by the small mounds of dust, leaves and trash that are to be seen below the entrances to occupied cavities."

A nesting site near the village of Mount Pleasant, S. C., used occasionally by rough-winged swallows was in the end of a hole in a bank of burnt oystershell—location of an antebellum lime kiln facing Copahee Sound. A round piece of wood had been buried in the lime, and when it decayed it left a tunnel 3 inches wide and several feet deep. The late Arthur T. Wayne first showed it to me. He related that, on one occasion, upon his approach, the bird left the hole and was immediately pursued closely by a sharp-shinned hawk (*Accipiter velox*). It eluded its pursuer, however, and dived back into the hole, where it remained.

Howell (1924) writes: "A most remarkable site selected by one or more pairs of these birds for their nest was on a buttress beneath the deck of a transfer steamboat which made daily trips on the Tennessee River from Guntersville to Hobbs Island, a distance of 24 miles, leaving at 10 A. M. and returning at 6 P. M. The birds, of course, followed the boat all the way to feed their young. A nest examined on the boat, June 19, 1913, contained young."

Hollow trees, it seems, are rarely used, but Eifrig (1919) says: "June 10, 1915, I saw a pair * * * nesting in a dead cottonwood on the top of a dune at Millers. * * * The female looked out of the hole and the male perched as close by as he could." Observers agree that the entrance hole of the roughwing's tunnel differs from that of the bank swallow; S. F. Rathbun (MS.) says: "Quickly I detected the difference that existed in the shape of the entrance of the nesting tunnel used by the rough-winged swallow, by contrast with that of the bank swallow; for in the case of the former the shape of the entrance was elliptical, sometimes much so; it was larger and appeared carelessly made. But the bank swallow would make the entrance more circular, especially if the digging was easy; it was decidedly smaller, neater in its outline. And a person could readily see these differences even when some distance from the bank."

Most of the birds that nest in cavities, tunnels, or crevices build either no nest at all or one of indifferent construction; the rough-winged swallow is no exception. S. S. Dickey's (MS.) description of the nests as "loose, crude foundations" is a good one.

The bulk of the nest depends largely on the size of the cavity that holds it. Nests I have taken from sand banks along the South Carolina coast are 1½ to 1¾ inches thick and are composed of grasses and rootlets.

A nest collected by J. F. Freeman (MS.) from a timber under a wharf, where there was plenty of room, is a rather bulky affair, built on a foundation of large chips and pieces of bark deposited during construction of the wharf. The nest proper is made of grasses and a few leaves of live oak (*Quercus virginiana*) and is lined with fine grasses. The distance from the top of the nest to the beam is 5 inches.

According to locality various materials are used, as grasses, pine needles, straw, weeds, roots, and, as Dickey (MS.) says, "shells of chicken eggs and now and then bud scales, panicles, seed tops, petals of such flowers as dogwood (*Cornus florida*), Carices, and *Juncus*. Into their composition go pieces of deciduous leaves and petioles, notably those of the black willow (*Salix nigra*) and heartleaf willow (*Salix cordata*). A number of nests curiously contained moist horse dung; we wonder why. Perhaps the vile smell tends to ward off vermin."

R. F. Mason, Jr. (MS.), reports the wide use of holly leaves in Maryland. In coastal Virginia H. H. Bailey says that seaweed is largely used in nest construction.

In Florida, according to Howell (1932), nests are made of dried rootlets, grass, weed stems, and a few dried beans and are lined with dried or partly burnt grass.

Dickey (MS.) writes: "Curiously, the parents supply broods daily with beds of fresh green leaves of the common locust (*Robinia pseudoacacia*). Soiled leaves are removed, with the dung."

A departure from the usual type of nest construction is described by Goss (1886), who says: "Nest in holes in banks of streams, constructed of the same material as the Barn Swallow." He describes the nest of the latter bird as "constructed of layers of mud and grasses, and lined with fine grasses and downy feathers."

"In the vicinity of Fortine, Mont.," says Weydemeyer (1933), "I have been able to determine the stage of nesting, at some time during the season, shown by thirty-four nests of the Rough-winged Swallow. * * * I give below the range of dates, for different stages of nesting, which these records show. Nest under construction: May 8, 1931, to June 15, 1929. Eggs (seven nests), June 14, 1928, to July 6, 1923."

Eggs.—[AUTHOR'S NOTE: The rough-winged swallow lays anywhere from four to eight eggs to a set, but the set usually consists of six or seven eggs; thus the sets will average larger than those laid by the bank swallow. They are more elongated, as a rule, than the eggs of other swallows, usually elliptical-ovate. They are somewhat glossy, pure white, and unmarked. The measurements of 50 eggs average 18.3 by 13.2 millimeters; the eggs showing the four extremes measure **20.6** by 14.7, 17.3 by **15.0**, **16.5** by 12.7, and 17.8 by **12.2** millimeters.]

Young.—A. F. Skutch (MS.) says that the female incubates the eggs for 16 days, while Dickey (MS.) gives 12 days. Apparently the male occasionally helps his mate with incubation duties, and Blake (1907) mentions a nest under his observation in Vermont where the birds took turns at sitting on the eggs.

Dickey (MS.) says: "The young at first are mere weak infants, gray and yellow, with the blood vessels and organs showing somewhat through their skins. They are coated with streaks of gray down. They develop rapidly. Within one week they assume somewhat the aspect and plumage of the adults. When they are ready to leave the nest, at the lapse of 12 days, they are pale brown but cannot well be differentiated from the adults while on the wing. I went to the trouble to collect and examine young just out of the nests. Superficially their forms seemed more like bank swallows than like their adult parents."

Skutch (MS.) continues: "When 13 days old the nestlings were well feathered, but they remained in the burrow a full week longer, gaining strength to fly." Thus, he considers the nestling period to be 20 or 21 days. Weydemeyer (1933), out of a total of 34 nests under observation, gives the following dates of young in nest: June 8, 1921, to July 9, 1928. In nine other nests, the young left the nest by July 22, 1931, to July 29, 1930.

The roughwing raises one brood during the season.

Plumages.—At the time of leaving the nest the young birds are similar to their parents in size, feathering, and length of wing and tail, but the first primary lacks the roughness of the adult feather; indeed, it is probable that nearly a year passes before the young birds acquire this saw edge that gives them their name. Also, the plumage is tinged with rufous or cinnamon, especially on the throat and upper breast; the wing coverts and tertials are margined with the same ruddy tint.

Dwight (1900) says: "First winter plumage acquired by a complete postjuvenal moult after the birds have migrated southward in September, or very likely while they move leisurely along in flocks."

The first nuptial plumage is apparently acquired by wear. Adults have a complete postnuptial molt after they have migrated southward, mainly in September or later. The sexes are alike in all plumages.

Food.—Howell (1924) says: "The food of the rough-winged swallow consists principally of insects, with a few spiders. Flies composed nearly one-third (32.89 per cent) of the total. Ants and other Hymenoptera are extensively eaten, and bugs to a lesser extent. Beetles amounted to nearly 15 per cent of the food and included the cotton-boll weevil, alfalfa weevil, rice weevils and flea beetles. A few moths, caterpillars, dragonflies, Mayflies, and an occasional grasshopper make up the remainder of this bird's food."

Behavior.—In the field the roughwing appears as a sober-colored little bird, plain grayish brown above and lighter below. At a distance it can easily be confused with the bank swallow, but when the latter bird sweeps over the observer the breast band is readily detected. If the two birds are seen together, the larger size of *serripennis* and its more brownish appearance are at once apparent.

Lynds Jones (1912) says: "The more deliberate flight of the Rough-Wing as compared with the Bank was always noticeable. The flight also tended to be more straight-away, with fewer abrupt turnings. The Rough-Wing gives one the feeling of great reserves of energy."

Theed Pearse (MS.) mentions that the "flight differs from other species of swallow, stroke of wing being higher."

When its nest is approached the bird glides out and is soon joined by its mate; then the two usually wheel back and forth at a short distance away. If bare branches or telegraph wires happen to be near at hand the birds will perch upon them and wait for the intruder to go.

Dickey (MS.) writes that the "parent, not seemingly uneasy, tended to hover half-concealed behind a screen of black willows, 200 feet away. It would, however, glide out, to see what was taking place, then disappear."

In describing a pair of breeding swallows, Brewster (1907) writes: "Once they alighted on a large, flat-topped boulder at the water's edge where they moved about by a succession of short, quick runs, reminding me of Semipalmated Plover feeding on a sand beach. I have never before seen Swallows of any kind move so quickly by the aid of their feet alone."

Henshaw (1875) says that on the Provo River, Utah, "they roost in large numbers upon the dead bushes along the banks. So numerous are they and so closely do they sit huddled together that six individuals were secured at a single shot."

Voice.—The roughwing is, generally speaking, a silent bird; its notes, rather weak and inaudible at a distance, are described as "harsh" or "squeaky" by observers.

Dickey (MS.) writes: "They give vent to a kind of rasping squeak, difficult to describe in mere words. The exclamations are vented while the species glides upstream or when it is approached near the nest; *quiz-z-z-zeep; quiz-z-z-zeep* is what it sounds like." Cooper (1870) writes: "They have only a faint twittering note when flying."

Grinnell, Dixon, and Linsdale (1930) refer to the call of the rough-wing as "their sputtery notes, *pssrt, pssrt.*"

Enemies.—The rough-winged swallow does not appear to be greatly victimized by predatory birds and mammals; strong powers of flight, more or less inaccessible nesting sites, and generally solitary habits combine to keep it out of danger.

H. H. Bailey (1913) says: "The mortality in this section is great, their chief enemy being the black snake."

Probably the greatest cause of destruction to eggs and young is the flooding of the burrows by spring tides and river freshets. According to Wayne (1910) this condition is quite prevalent in the flat, sandy coastal country of the Southeast. It also often happens elsewhere, owing to the fact that, while this swallow usually burrows near the top of the bank, it often excavates nearer the base. In building under bridges and culverts the bird sometimes places its nest so near the water that even a slight rise would engulf it. Dickey (MS.) says: "From potholes in sandstone cliffs near Worely, Monongalia County, W. Va., I have known anglers to extract the young of rough-winged swallows. These, they contended, proved to be excellent bait in bass fishing in local creeks."

Peters (1936) lists specimens of this swallow from Maryland and Virginia as being found infected with the mites *Liponyssus sylviarum* and *Atricholaelaps* sp.

Without positive proof I believe that the common sand crab (*Ocypode albicans*) might, to a limited extent, prey on eggs and nestings of the roughwing. This crustacean abounds on the south Atlantic coast, excavating its burrows in sand hills and the bases of sand banks, as do the swallows. It causes much damage by burrowing into turtle nests on the Carolina coast and consuming the eggs. It is ever on the alert for anything edible that the waves might bring ashore. Terns and shearwaters washed up after hurricanes are quickly ruined as specimens, as I have several times sorrowfully experienced.

Winter.—While this swallow is highly migratory and the great majority of individuals winter south of the United States, records from five States designate it as a winter visitant within our borders. It is possible that some of the so-called spring arrivals are birds that have wintered in the neighborhood. Wayne (1910) says: "The birds of this species which winter along the coast, generally, if not invariably, confine themselves to large bodies of water adjacent to wooded lands."

Griscom (1932) says that "the Rough-winged Swallow is a common winter visit to the whole of Guatemala, except the Pacific coast." He quotes from Mr. Anthony's notes as follows: "Common during the winter months to about 8000 feet altitude. The first were noted at Progreso about September 8, with mixed flocks of Cliff and Barn Swallows. A considerable flight of these species appeared at this station on the above date and hundreds were seen along the telegraph wires for a day or two, when they became much less common but not rare until the following May. In the altitudes, *Stelgidopteryx* is apt to be seen with *Tachycineta* which is equally common."

DISTRIBUTION

Range.—North and South America.

Breeding range.—The rough-winged swallow breeds **north** to central British Columbia (Kispiox Valley); Alberta (probably Jasper Park, Camrose, and Lake Newell); northeastern Montana (Bowdoin Lake); North Dakota (Bismarck and Grafton); southeastern Manitoba (Indian Bay); Wisconsin (Danbury and Orienta); northern Michigan (Blaney and Mackinac); southern Ontario (Barrie, Kingston, and Ottawa); Vermont (Norwich); and New Hampshire (Ashland and Snowville). **East** to New Hampshire (Snowville and Boscawen); Rhode Island (Charlestown and Westerly); Massachusetts (Swansea); and south along the Atlantic coast to Florida (St. Augustine, New Smyrna, and Osteen); northeastern Colombia (Fonseca); and central Brazil (Rio Xingu, Rio Jamauchim, and Rio de Janeiro). **South** to southern Brazil (Rio de Janeiro, Descalvados, and Matto Grosso) and southern Ecuador (Juntas de Tamana, Zamora, and Casanga). **West** to western Ecuador (Casanga, Duran, and Esmeraldas); western Colombia (Barbacoas, Las Lomitas, and San Jose); and north along the Pacific coast of Central America, California, Oregon, and Washington, to British Columbia (Alberni, Comox, Hazelton, and Kispiox Valley).

The range as outlined is for the entire species, of which only one race (*Stelgidopteryx ruficollis serripennis*) is found north of Central America. The southern forms are apparently nonmigratory.

Winter range.—During the winter season, the North American form is found **north** to southern Sonora (Alamos); Durango (Chacala); Veracruz (Orizaba and Tlacotalpan); and Quintana Roo (Tulum). **East** to Quintana Roo (Tulum and Camp Mengel); and Costa Rica (San Jose). **South** to Costa Rica (San Jose); Guatemala (Progreso and San Lucas); and Guerrero (Ometepec and Acapulco). **West** to Guerrero (Acapulco); Colima (Manzanillo); Nayarit (Gavilan); Sinaloa (Mazatlan); and Sonora (Alamos).

The species has been recorded at San Diego, Calif., on January 27; "a number" were reported as seen at New Roads, La., on December 15, 1917, and one was collected at Baton Rouge, on January 26, 1938; while a single bird was recorded at St. Marks, Fla., on January 2, 1917.

Spring migration.—Early dates of spring arrival are: Florida—Pensacola, March 18. Georgia—Athens, March 20. South Carolina—Columbia, March 20. North Carolina—Raleigh, March 28. Pennsylvania—Philadelphia, March 30. New Jersey—Morristown, April 14. New York—New York City, April 12. Connecticut—Fairfield, April 17. Vermont—Bennington, April 25. Mississippi—Biloxi, March 30. Arkansas—Helena, March 13. Tennessee—Nashville, March 20. Missouri—Monteer, March 27. Indiana—Richmond, April 11. Ohio—Oberlin, April 16. Michigan—Ann Arbor, April 17. Ontario—London, April 17. Iowa—Keokuk, March 27. Wisconsin—Madison, April 8. Minnesota—Minneapolis, April 8. Texas—San Antonio, February 15. Oklahoma—Tulsa, April 6. Kansas—Manhattan, April 5. Nebraska—Red Cloud, April 16.

South Dakota—Vermillion, April 19. North Dakota—Grafton, April 27. Arizona—Tucson, February 19. Colorado—Denver, April 18. Wyoming—Laramie, May 1. Montana—Missoula, April 25. California—Santa Barbara, February 24. Oregon—Mercer, April 7. Washington—Tacoma, April 3. British Columbia—Chilliwack, April'5.

Fall migration.—Late dates of fall departure are: British Columbia—Okanagan Landing, September 28. Washington—Pullman, September 11. Oregon—Newport, September 6. California—San Diego, November 9. Montana—Fortine, September 9. Wyoming—Laramie, September 4. Arizona—Fort Verde, September 20. South Dakota—Yankton, August 24. Nebraska—Red Cloud, September 11. Kansas—Onaga, September 16. Oklahoma—Tulsa, October 5. Minnesota—Minneapolis, September 20. Wisconsin—New London, September 20. Iowa—Indianola, September 27. Ontario—Toronto, September 13. Ohio—Oberlin, September 23. Indiana—Richmond, October 1. Missouri—St. Louis, October 11. Tennessee—Athens, October 13. Arkansas—Hot Springs National Park, October 14. Mississippi—Edwards, October 5. Louisiana—New Orleans, November 5. Vermont—Wells River, August 4. Massachusetts—Ipswich, September 5. Connecticut—Fairfield, September 2. New York—Rhinebeck, September 9. New Jersey—Hackettstown, September 11. Pennsylvania—McKeesport, September 4. District of Columbia—Washington, September 11. Georgia—Savannah, October 15. Florida—Pensacola, October 11; Key West, October 24.

Casual records.—One was seen at Eastend, Saskatchewan, on July 20, 1930.

Egg dates.—California: 23 records, April 15 to July 9; 13 records, May 10 to June 3, indicating the height of the season.

Illinois: 9 records, May 17 to June 6.

Pennsylvania: 14 records, May 3 to June 15; 8 records, May 21 to June 9.

Washington: 6 records, June 11 to July 4.

CHELIDONARIA URBICA URBICA (Linnaeus)

EUROPEAN MARTIN

HABITS

Since the European martin has been taken as an accidental straggler on the east coast of Greenland, we add it to our list of North American birds. It, with its various subspecies, is found over much of the Old World as a breeding bird or a migrant.

Although its distribution in general is much the same as the European swallow, Capt. A. W. Boyd (1935) says that "the House

Martin is far more a bird of the village or town than the Swallow and is wont to concentrate in colonies in restricted areas rather than breed in isolated pairs in many separate buildings, though isolated pairs often occur. Thus in an area of 8,000 acres in Lancashire 19 pairs were concentrated in two groups; an area of about 4,000 acres in Norfolk contained 158 pairs of which 110 were in one small village." This does not mean that the martin never nests elsewhere, for it is often found in rural districts and builds its nest on farm buildings. Yarrell (1876–1882) says that "the Martin yet retains in this country some of its original seats, for it still chooses its breeding-place in cliffs, generally on the coast, but sometimes inland, and quite apart from any human habitation."

Nesting.—The best account of the nesting habits of the martin seems to have been given by Macgillivray (1840) and his contributors, from which I shall quote freely. He writes:

Most commonly the nest of this swallow is placed in the upper corner of a window, often also under the eaves of out-houses, and in similar situations, where it is sheltered from above, sometimes on the face of a rock, whether on the sea-shore or inland. * * * When in the corner of a window, it is of rounded form externally, flat on the adhering sides, rectangular above, and has a roundish or transversely oblong aperture at the top, almost always on the sheltered side, or that next the middle of the window. I have seen several instances in which the aperture was at the outer edge of the window, and sometimes it has a kind of neck, or the mouth projects an inch or more. The nest is usually large, having an external diameter of from six to eight inches. The outer part of one examined by me * * * consisted of pellets of friable sandy mud, not in the least glutinous, intermixed with small, generally angular pebbles or gravel. Into this outer crust were thrust numerous straws or fragments of stems of grasses, which became free internally, and were circularly disposed. Within was a layer of wool, partly interwoven with the straw, and lastly a thick bed of large feathers of the domestic fowl. Another nest from a village near Edinburgh, is six inches in diameter externally. The outer shell is a solid mass of fine loam, which has been built of pellets in the form of soft mud, so that the outer surface presents horizontally compressed mammillae. The average thickness of this crust is seven-twelfths of an inch. It is quite friable, and if any glutinous matter has ever been intermixed with the mud, it has entirely disappeared, but it is in some measure held together by a considerable intermixture of short straws. The next layer is of straws of various kinds, mostly decayed. This is followed by a thick layer of wool, which is succeeded by a great quantity of hogs' bristles, cows' hair, human hair, a piece of linen, a bit of tape, and a number of feathers, chiefly of the domestic fowl.

As for the method of gathering material for the nest, he says that "they alight by the edge of a pool, or brook, often on the street or road after rain, select a portion, seize it with their bill, fly off to their nest, and apply it in its wet state to the edge of the unfinished crust, which they thus gradually build from the bottom upwards, until it is completed. * * * Straws and feathers they pick up in the same manner, that is by alighting, and seldom while on the wing."

Mr. Hepburn wrote to him: "Besides the places above-mentioned, in which they fix their nests, I have seen them in arches, once against a rafter, immediately under the ridge hole of a shed, on another occasion, in the south-east angle of the wall of a house, at the distance of eighteen feet from the ground, and about six or eight feet from the eaves. Although quite exposed, it bore with immunity all the storms by which it was assailed, and its owners raised two broods in it. * * * I have never seen its nest placed so low that you could reach it when standing. * * * It is usually finished in twelve or fourteen days. When built in a sheltered situation, it will last for years, and is occupied every season probably by the same pair."

J. A. Harvie-Brown (1908) has published a photograph of a martin's nest built "against a flat pane of glass" in a window.

Oliver G. Pike (1912) says: "It seems a custom of the House-Martins when they begin nest-building, to build the nests as quickly as possible, and I have seen a whole colony of fourteen birds set to work on *one* nest, and when this was finished, to help with the others."

Eggs.—The number of eggs laid by the European martin varies from three to six, but the commonest is four or five. The eggs are pure white and are slightly glossy. The measurements of 54 British eggs average 19.4 by 13.2 millimeters (Witherby, 1920).

Young.—Incubation is shared by both parents and is said to last about 14 days. Two broods are ordinarily raised in a season and sometimes three. Macgillivray (1840) writes:

During incubation both the parents are frequently seen in the nest, and at all times they repose there at night. When the young are nearly fledged, they frequently appear at the edge of the aperture, and are occasionally clamorous for food, which is brought to them sometimes with astonishing rapidity. For some days after they have left the nest, they fly about in its vicinity, and are fed by their parents. It is amusing to see the expedients to which recourse is often had to induce them to leave the nest, the parent birds sometimes pushing or dragging them out, but more frequently enticing them by showing them how easy it is to fly away. Very frequently the young betake themselves to the upper corner of a window to rest, clinging for a short time by means of the feet and tail; and in this situation I have seen them fed. For some time after they have come abroad, they return to the nest at night, reposing there with their parents.

As to the periods of feeding, Th. Durham Weir, Esq., wrote to Macgillivray (1840):

In one of the bedrooms of my house, on Friday the 28th of July 1837, I made the following observations. At 25 minutes after 4 o'clock in the morning, the old martins began to feed their young ones which were four in number.

From that time until 5 o'clock, they fed them four times; from 5 to 6 o'clock eleven times; from 6 to 7 o'clock twenty-four times; from 7 to 8 o'clock fifteen times; from 8 to 9 o'clock twenty-three times; from 9 to 10 o'clock twenty-five times; from 10 to 11 o'clock twenty times; from 11 to 12 o'clock twenty-six times; from 12 to 1 o'clock twenty-six times; from 1 to 2 o'clock twenty-seven times; from 2 to 3 o'clock twenty-eight times; from 3 to 4 o'clock twenty times; from 4 to 5 o'clock twenty times; from 5 to 6 o'clock twenty-seven times; from 6 to 7 o'clock ten times; and from 7 to 8 o'clock only once, making in all 307 times. At 10 minutes after 8 o'clock, having ceased from their labours, they went into their dormitory. They brought to their nestlings at each time, sometimes two, at other times three, four, five, and even more flies of different sizes.

Referring to the departure from the nest, Macgillivray (1840) quotes Mr. Hepburn as follows:

The following morning, about eight o'clock, immediately after my return from the fields, I observed the old birds dashing up to the window, then describing short curves in the air, and repeating a note, the meaning of which could not be misunderstood. I knew from experience the young were about to take their flight. One of them balanced itself in the entrance, looked timidly into the void, considered the risk for some time, and then allowed its fellow to take its place. During all this time the parents kept diving about, within a few feet of the nest, and often fluttering within a few inches of the entrance, and endeavouring by many winning gestures to induce their charge to follow them. The remaining bird also, after sitting for some time, distrustful of its powers, retired, and the first one once more appeared. Opening and shutting his wings, and often half preparing to retire, he at length summoned up all his resolution, sprung from his perch, and with his self-taught pinions winnowed the air. He and the parents, who were in ecstasies, returned to the window, and, being joined by the other young bird, they all day long sported chiefly about the tree tops, till seven in the evening, when they re-entered the nest. The following day they were again sporting about, and the young were repeatedly fed by their parents. * * *

The first brood, which is generally abroad by the middle of June, live apart. The second brood is fledged by the end of August. They and their parents join the first brood and their companions, at Linton Distillery, when some hundreds of the species are to be seen. The remainder of their stay is spent in short aërial excursions, in sunning themselves on house-tops, in feasting and song; until, about the third week of September, when they bid farewell to the scenes of their youth, which many of them are never again to behold, and away they speed in a body far towards the noontide sun.

Several observers have noted that four, or more, birds often assist in the feeding of a single brood of young; in some cases some of these birds may have been the young of a previous brood.

Plumages.—The nestling martin is scantily clothed in white down, according to Witherby's Handbook (1920). A full description of each plumage may be found in this excellent work, from which it appears that the young martin in juvenal plumage is much like the adult, but browner and duller in color. The sexes are practically alike in all plumages. The juvenal body plumage begins to molt in August, but the wings and tail feathers are not molted until November. Of the adult winter plumage, the Handbook says: "This plumage is acquired by moult which commences in Europe in Aug. (exceptionally July) with body-feathers, sometimes also wing-coverts and some secondaries are moulted, but, usually wings and tail are not moulted until winter. Body-plumage is moulted again Jan.–April but material available is insufficient to decide if wings and tail are moulted completely in autumn as they are in spring."

Food.—According to Witherby's Handbook (1920) the food of the martin consists of "insects taken on the wing, chiefly Diptera, especially the smaller species (Chironomidae, etc.), but also Coleoptera (*Tachyporus*, *Aphodius*, *Coccinella*, etc.), occasionally Lepidoptera (*Pieris*, *Triphaena*, *Agrotis*, etc.) and Neuroptera (Oldham)."

Herbert Massey (1917) observed martins and swallows settling and feeding on flowers, of which he says: "On the top of the knoll there were tall plants of ragwort and thistle plainly to be seen standing out against the sky; the birds were flying low over the plants and sometimes hovering like a moth at the flowers, and now and again Swallows and Martins would settle for a few seconds on the flat heads of the ragwort and pick off the insects. On examining the plants I found a quantity of little black flies."

Behavior.—Macgillivray (1840) writes:

The ease and rapidity of its flight, however marvellous, excite no astonishment, as we are daily in the habit of witnessing them; but a true lover of nature can nevertheless contemplate its airy windings for hours with delight. The evolutions of this species resemble in all respects those of the Red-fronted Swallow; but its flight is perhaps somewhat less rapid, although it is certainly very difficult to decide with accuracy in a comparison of this kind. Its sweeps and curves however seem to me to be less bold, or rather less extended; but its dexterity is equally remarkable. It mingles in its sportive-like pursuits with both the other species, although each kind seems to give some preference to the society of its own members. The influence of the weather on the flight of insects causes it to observe the same selection of places as they; so that in calm and cloudless days it flies more in the open air, in windy weather more in the shelter of hedges and walls, and in damp evenings it skims over the grass and corn. * * *

Although the feet of this bird are very small, it can settle without difficulty on a wall top, a roof, the branch of a tree, or on the ground, and is capable of walking, although in an uneasy and rather ungraceful manner.

Voice.—The same author says of the martin's voice:

Its ordinary cry is a rather loud chirp, which it frequently emits, more especially when it flies in the vicinity of its nest. It has been called a twitter, but the syllables which it most resembles are *chir-rup*. When flying over a field, or under the shelter of trees, with its young, it has a softer and more pleasant chirp, which is responded to by them; and in calling to one to come up and receive an insect, it utters a repetition of its notes, so as to produce a low chitter. Its song is loud enough to be heard in calm weather at the distance of three hundred yards, and is cheerful, although not remarkable for melody. It is often emitted at intervals while the bird is on the wing in the neighbourhood of its nest, and is sometimes heard more continuously, in fact for ten minutes or more, when perched on the roof of a house or other elevated place.

Witherby's Handbook (1920) says: "Song, a contented twittering often warbled whilst clinging to, or actually inside, nest. Call-notes, a soft 'preet-a-preet' and a coarse 'screet,' often uttered on wing. Alarm-note, sharp, shrill 'prt.'"

Enemies.—C. J. Pring (1929) publishes the following account of young martins being killed and eaten by woodpeckers:

After the Martins had begun to hatch this year (1929), several nests were discovered with large holes in the bottom, and the contents missing. More nests were discovered thus destroyed, until, anxious to discover the reason, Col. Heneage ordered two of his gardeners to keep watch on the nests. This soon disclosed the fact that a pair of Great Spotted Woodpeckers (*Dryobates m. anglicus*) were destroying the nests and eating the young Martins—or at least portions of them. The procedure was to fly to a nest and, clinging on to it, hammer it with their bills until part of it broke away. In the case of those nests containing young, the young sometimes fell out in a heap on to the ground below, or else were dragged out of the nest by the Woodpeckers. They were then taken one at a time (either from the nest or the ground) on to the branch of a pear tree and hammered to pieces, the breast being the chief part which was then eaten. * * *

In the case of nests containing eggs, if the eggs were fresh, they were broken and left on the ground below the nests, but the contents of those eggs which contained a well-formed embryo were eaten. Allowing an average of four eggs for each nest, quite 100 eggs or young Martins must have been destroyed in about a week.

The female woodpecker was caught, and its gizzard was found to contain a number of pieces of small feathers, some with blood quills attached, which evidently came from the underparts of the young martins.

House sparrows (*Passer domesticus*) cause the martins considerable trouble by forcibly ejecting them and appropriating their nests. The martins are usually unable to evict the sparrows after they have taken possession, but Macgillivray (1840) mentions three, apparently authentic cases, where a number of martins joined forces and entombed the unwelcome sparrow alive by walling up the entrance to the nest with mud; this sounds like a fairy tale, however, for it seems hardly likely that the sparrow could not succeed in breaking out before the mud had time to harden.

Fall.—Macgillivray (1840) says: "Towards the end of September, the House Martins collect into large flocks, which for several days perform long excursions in the neighbourhood of their residence, and are seen settling on the house-tops. At length, in the beginning of October, they disappear, although here and there a few individuals may be seen flying about for some weeks later. Instances of their occurrence even in November are mentioned by several persons, but none are seen after that month."

Range.—The martin is generally distributed, as a summer resident, in the British Isles, but it is more local, especially in Ireland, than the swallow; it is a scarce breeder in the Orkneys; it only occasionally nests and is not common, as a migrant even, in the Shetlands, and it is a rare vagrant to the Outer Hebrides. In Europe it breeds from latitude 70° N. in Scandinavia southward to the Mediterranean and eastward in Asia to the Yenisei and Turkestan. It is replaced by other forms in other parts of Asia and in northwestern Africa. It winters in southeastern Africa and Mossamedes and in northwestern India. It is casual in Iceland, Greenland, and Madeira.

Spring migration.—The main arrival in England of summer residents begins ordinarily early in April, and lasts through May, the later birds merging with the passage migrants. There are, however, a nu..nber of earlier dates running all through the last week in March, the earliest being March 23, 1913.

Fall migration.—In England, summer residents begin to move south about the middle of July; the departure from the south coast begins the first week in August and lasts until the third week in October; stragglers are frequent up to the middle of November and occasionally up to the first week in December; there is a late record for January 10, 1912. The main movement of passage migrants occurs in September and October.

All the above data on distribution and migration are taken from Witherby's Handbook (1920) and do not claim to be complete.

HIRUNDO ERYTHROGASTER Boddaert

BARN SWALLOW

PLATES 63, 64

HABITS

The familiar barn swallow is widely distributed over most of North America, breeding from northwestern Alaska to Mexico and from the Atlantic to the Pacific coast, including much of Canada and the United States.

Everybody who notices birds at all knows, admires, and loves the graceful, friendly barn swallow. No bird in North America is better known as a welcome companion and a useful friend to the farmer, as it comes each spring to fly in and out of the wide-open barn door, delighting him with its cheerful twittering, or courses about the barnyard in pursuit of the troublesome insects that annoy both man and beast. The peaceful beauty of the rural scene would lose much of its charm without this delightful feature. But such a charming rural scene is not so common as it used to be. The old-fashioned barn, with its wide-open doors, never closed, its lofty hay-mow, and the open sheds where the farm wagons stood are being replaced by modern structures, neatly painted buildings, with tightly closed doors and no open windows through which the birds can enter. Horses are replaced to a large extent by automobiles and tractors; cattle are housed in modern dairy barns; and the open haymow is disappearing. There is no room for the swallow in modern farming. Must it return to its primitive style of nesting or will other means of encouraging it to nest in our farmyards be employed? The birds will stay with us if we supply them with supports for their nests; a two-by-four joist, rough and not planed, nailed to the outside of a building, flat wide side against the wall, and placed well up under the eaves with about 5 inches of clearance, will accomplish the desired results.

Spring.—From its winter range in South America the barn swallow evidently migrates to North America over widely separated routes, through the West Indies and the Bahamas to Florida and through Central America and Mexico to more western points. By the former route the vanguard reaches Florida early in April, but the migration continues well into May and a few are seen even in June. Alexander F. Skutch writes to me: "In Central America, the barn swallow occurs chiefly if not solely as a transient, passing southward in great numbers during September, and again migrating toward the north from mid-March until mid-May, with a maximum of abundance during April. In the more settled parts of the country they are frequently seen resting in long rows on the telegraph wires and power lines, often in company with resident or migratory rough-winged swallows, and sometimes with cliff swallows. They seem to migrate by day, for I have sometimes seen great numbers of them fly overhead in loose flocks in a direct, undeviating course, northward in the spring and south or southeast in the autumn."

The hardy tree swallow often arrives in New England fairly early in March; but the date depends largely on the weather, which is often cold enough to cause a retreat. It comes as a harbinger of spring, but we never forget the old saying that one swallow does not make a summer. But when the more delicate barn swallow appears, a full month or more later, and we see it gracefully skimming low over the fields and ponds or inspecting its former nesting sites, we begin to feel that welcome spring is here, or at least near at hand. Mr. Forbush (1929) writes:

During the latter part of April the pioneers of the Barn Swallow host usually appear in Massachusetts. Sometimes they come too early and are met by cold and storm and so, unable to obtain food, they seek shelter in some building or huddle together behind a closed blind or window sill on the south side of a house until the sky clears and the temperature moderates. By snuggling together in their nest, some of these birds have been able to survive two or three cold days, when morning outdoor temperatures were as low at 15 to 17 degrees above zero mark, but such temperatures may be fatal, even when the birds are well protected.

If, when the Swallows arrive, they find the building closed in which they are accustomed to breed, they sometimes approach the house and fly about it, or about any inmate who appears, twittering and calling until someone takes pity on them and opens a door or window, when they immediately enter, showing their gratification by happy excited twittering. Many farmers cut a small hole in a barn gable to accommodate the birds.

At the northern end of the migration route the arrival is, of course, much later. Dr. E. W. Nelson (1887) says that, in northern Alaska, "this swallow arrives as soon as mild spring weather sets in, generally from the 18th to the 23d of May. The sea is still covered with an unbroken surface of ice as far as the eye can reach, and winter appears to be hardly gone when the first arrivals reach Saint Michaels and come fluttering about their former nesting sites." Lucien M. Turner (1886) places its arrival at a somewhat later date:

The Barn Swallow arrives at Saint Michael's about the 7th of June. A few of the more intrepid ones may arrive some few days earlier. By the 15th of the month as many as forty pairs have been counted in the dusk of twilight, which is light enough to see to read by at midnight during this season of the year. * * *

In the spring of 1876 snow squalls and frosty weather held until late in June. The poor birds had had no opportunity to recover their exhausted condition, resulting from their long flight to the north. Many of them succumbed to the chilling weather, while others, benumbed by the cold, permitted themselves to be handled and seemed to enjoy the warmth given out by the hand, as they nestled closely between them, without evincing any fear.

Courtship.—The actions of mating barn swallows are most pleasing to watch, as much of their love-making is done on the wing. Long, graceful pursuit flights carry the birds in and out of the barns, back and forth through the barnyard, around the buildings where they are accustomed to nest, or out over the open fields. Their flight is swift and hurried, accompanied by constant twittering or louder excited outcries, as if each were trying to outdo the other in their expression of springtime ardor. Finally, the female, satisfied that her pursuer is worthy of her, comes to rest on some convenient perch, the roof of the barn or a telephone wire. By way of invitation she twitches her wings and tail and turns her head from side to side, as the male flutters up and alights close beside her. There they rub their heads and necks together, interlock their bills, or preen each other's plumage in happy love-making. Copulation may take place at once, or they may fly off together to choose which of their old nests or nesting sites they will use.

Nesting.—We naturally think of the barn swallow as nesting in barns, and such has seemed to be their preference, where suitable old-fashioned barns can still be found. But when the red men roamed this continent, before the white men invaded it, there were no barns, and the swallows had to build their nests in such natural conditions as would suit their needs, giving them security for their nests and protection from their enemies. They found security and protection in rocky caves, in crevices in rocky cliffs, on shelves of projecting rocks where some protection from above was afforded, and even in holes or natural cavities in cutbanks. Dawson and Bowles (1909) thus describe one such primitive nesting site at the head of Lake Chelan, Wash.:

The shores of the lake near its head are very precipitous, since Castle Mountain rises to a height of over 8,000 feet within a distance of two miles. Along the shore-line in the side of the cliffs, which continue several hundred feet below the water, the waves have hollowed out crannies and caves. In one of these latter, which penetrates the granite wall to a depth of some twenty feet, I found four or five Barn Swallows' nests. * * * Other nests were found in neighboring crannies outside the cave. * * *

Mr. F. S. Merrill, of Spokane, reports the Barn Swallow as nesting along the rocky walls of Hangman's Creek, in just such situations as Cliff Swallows would choose; and back in '89, I found a few associated with Violet-greens along the Natchez Cliffs, in Yakima County.

A colony of some twenty pairs may be found yearly nesting on Destruction Island, in the Pacific Ocean. A few of them still occupy wave-worn crannies in the sand-rock, overlooking the upper reaches of the tide, but most of the colony have taken refuge under the broad gables of the keepers' houses.

O. J. Murie has sent me the following quotation from Ernest Ingersoll's "Knocking Round the Rockies," describing the primitive nesting habits of the barn swallow, as he found them at Hot Sulphur Springs, Colo., in 1874:

The niches in the rocks were occupied by large colonies of barn-swallows. * * * Sometimes the niches in the lime-rock (the whole mass of which had been built up of deposits from the mineral waters) were so close together that there would be half a dozen in a square yard; yet every one had its burnt-breast tenants, and the twittering silenced the gurgle and sputter of the rapid stream at the ledge's base. The floor of each niche was hollowed out, so that it only required to be softly carpeted to constitute it a perfect nest. For this grass-stems and a few large feathers were used, precisely as in our Eastern barns. But here the birds had greatly economized labor by occupying the niches, for they needed not to build the firm underpinning and stout high walls which become necessary in the barn, or on an exposed rock shelf, to prevent the eggs and young from rolling out; all these happy birds had to do was to furnish a home already made.

Although the above observations were made many years ago, I have no doubt that in many remote or thinly settled localities barn swallows still continue to nest under similar natural conditions. James B. Dixon writes to me that in San Diego County, Calif., as late as 1925, he found them nesting in caves in the ocean wall.

Barn swallows were quick to take advantage of the superior nesting sites offered by various man-made structures; and probably they also felt a sense of added security in their close association with friendly human beings. As civilization advanced, the swallows gradually learned to adopt such suitable nesting sites as they could find in, on, or under various types of buildings, in or on the outside of barns, sheds, or other farm buildings, inside of vacant houses, under the eaves of dwelling houses, under wharves and boat houses, and under bridges and culverts. In the rural districts of old New England, the old-fashioned barn, with its wide-open doors and its lofty haymow, seems to be the commonest and most characteristic nesting place. Here their mud nests may be plastered on the vertical surface of a beam or rafter, if it is not too smooth, but preferably where it can get some support on the flat surface of a beam or shelf, in the angle of a cross brace, or in a corner where two sides will be supported; the iron rail of a hay track, a loop of hanging rope, an iron hook, or wire loop, or even a projecting knot on a rough timber may offer the necessary initial support. We seldom find more than six or eight nests in a barn, but I have known of as may as 27 in a single barn here in eastern Massachusetts; and Dr. Charles W. Townsend (1920b) "counted fifty-five nests of this bird in a large barn at Ipswich, nearly all of which were occupied."

A. Dawes DuBois has sent me some notes on nests under bridges in Illinois. Five of these nests were attached to the vertical face of a floor joist, over water. Another was placed on top of a cross beam at the end of the bridge, fitted into a corner formed by the vertical faces of a joist and another beam at right angles to it. The tops of the nests were about 2 inches below the floor boards of the bridge. A nest under an iron bridge was stuck to the vertical web of an I-beam, but was so placed that the bottom of the nest rested on the lower flange of the beam.

Frank L. Farley tells me that, near Camrose, Alberta, "the barn swallow has apparently departed entirely from its natural nesting situations. Here they have not taken to the barns, but now they are increasing in numbers and nest entirely under bridges and culverts." James B. Dixon says in his notes that in San Diego County, Calif., they nest on the highway bridges that cross the estuaries along the ocean front; and in the San Joaquin Valley he has noted large colonies under bridges and beet-loading platforms.

Dr. Samuel S. Dickey writes to me that while staying at a small frame summer hotel built for the use of fisherman, at Beach Haven, N. J., he noted a number of barn swallows flying about and disappearing under the hotel building. "Since the building was erected on piles, 6 feet over deep water, it left a space about 5 feet high underneath. Here the swallows found ample room for breeding purposes. They were enabled to withdraw on three open sides. When we clambered down among the piles and planks, we caused a shower of birds to evacuate. And there, plastered upon crossed sticks next to the floor, were 15 nests." He visited a colony at Nantasket Beach, near Boston, where they bred under wharves and boat houses overlooking the sea.

In northern Alaska, about St. Michael and Nome, barn swallows are fairly common and seem to appreciate the advantages of nesting on what few buildings they can find; if buildings were more numerous they would probably increase in numbers. As it is they build their nests under the eaves of buildings, on projecting beams, inside empty houses, and even in the deserted sod houses of the Eskimos. The nests are profusely lined with white ptarmigan feathers, which are plentifully scattered over the tundra after the spring molt. In the Aleutian Islands we did not see any barn swallows west of Unalaska.

A few unusual nesting sites are worthy of mention. It is hardly to be wondered at that the swallows would continue to incubate and rear their young in a building that was moved after they had started nesting, but that they should continue to build and occupy their nests repeatedly on moving trains or boats is a remarkable illustration of their persistency and confidence, or of the scarcity of more suitable and stationary situations. At least two instances of nesting on moving trains have been reported. Harry S. Swarth (1935) reports that a narrow-gauge train, which carries passengers and freight over a 2-mile portage from Lake Tagish to Lake Atlin, British Columbia, provides a home for a pair of barn swallows. He says:

About the buildings at the Tagish Lake end of the line innumerable Barn and Cliff swallows nest. Under the eaves around one of the larger sheds there is an uninterrupted frieze of the Cliff Swallows' mud nests. But the really interesting feature of this colony lies in the action of a pair, or more properly a succession of pairs, of the Barn Swallow (*Hirundo erythrogaster*), in nesting on a moving train. For many years past one pair of swallows have built their nests and raised their broods on some part of the train that crosses the portage. They were first commented upon by E. M. Anderson, who, in the annual report of the Provincial Museum of Natural History (Victoria) for 1914, describes the nest as he saw it in one of the coaches. I have seen it on each of the several years that I have visited the region, and in all probability the nesting is an annual occurrence. The train crew take a personal interest in their guests, and for some years the swallows occupied an open cigar box that was fastened for their use under the roof of the open-sided passenger coach. In 1934 the nest was supported near the center and immediately under the roof of the baggage van, the sides of which are protected only by canvas curtains. I had occasion to cross the portage on the evening of June 21. When we embarked at the eastern end the swallows were not at home, but as soon as we arrived at the Tagish terminus both birds swooped into the car. There they settled down for the night, despite the fact that baggage was being piled beneath them to within a few inches of the roof.

Nesting year after year on a moving boat is another unusual habit, yet Burton W. Gates (1903), who had previously noted it, writes:

I recently wrote to Captain Harris, formerly of the steamer Horicon, on Lake George, New York, inquiring if the Swallows which, in the summer of 1900, nested beneath the guard-rails of his steamer had, in the three succeeding years, nested in similar places. His prompt reply was to the effect that "the Swallows have built their nests under the guard-rails of the various steamers which I have been running [I judge upon Lake George] for the past fifty-five years." The Captain is now retired from duty, but inquired of his son, the pilot of the new steamer Sagamore, regarding the habits of the birds in the past two seasons. To this, the Captain further wrote: "My son says that the Swallows were still with him this summer." Thus it would seem that the Swallows of Caldwell, New York, have, for generations, had a nesting habit peculiar to that locality.

It is perhaps not strange that the barn swallow should occasionally appropriate an old nest of the phoebe, as these birds often nest under bridges, or that the swallow should sometimes nest in a well or an abandoned mining tunnel, which somewhat resemble caves. Illustrating their confidence in humanity, Mr. Forbush (1929) mentions the following cases: "A pair built their nest close by a blacksmith's forge and reared their young, regardless of wheezing bellows, clanging hammers and showering sparks. Another pair in Falmouth, Massachusetts, built their nest in a room on a large farm, where agricultural products were daily prepared for market. They threaded their way in and out among the busy workers, industrious and fearless in their care for their growing young. A pair in Westborough, Massachusetts, took for their nesting-place a narrow shelf, in a barn, five feet above the floor, almost over a cow, where the milker could look directly into the nest. They stayed there and raised their young."

In building their nests the barn swallows show themselves to be expert masons, but unlike the cliff swallows and like the ancient Egyptians they cannot make bricks without straw. The first requisite is some clayey or otherwise sticky mud, which the birds obtain from the shore of some body of water or from mud puddles in roads or fields. Professor Herrick (1935), who has made a careful study of the nesting habits of this swallow, feels confident that the bird's saliva is not a factor in making the mud more adhesive. Authorities differ as to how the mud is carried. Dr. Dickey (MS.) says that "they waddle about on their short, weak legs and roll pellets of mud, which they place on top of the upper mandible and so bear to the nesting site." Lucien M. Turner (1886) confirms this:

"I have watched them for hours at a time, and when my eyes were not to exceed four feet from the birds at work. * * * The neck is stretched out nearly its full length and the head kept with the bill at a right angle to the neck. A slight pressing of the beak into the earth and a tugging twist of the body gently pulls toward the bird a small pellet of mud. The bird then lowers its neck to the ground with the beak on the opposite side of the pellet (or on the side next the bird.) The beak is now thrust under the pellet until the mass of mud is pushed onto the top of the bill and rests against the forehead. This is the manner in which it obtains the mud and is in position to enable the bird to deposit it. The mud is also smeared with the top of the beak.

Professor Herrick (1935), on the other hand, says:

The mud, which, as we have seen, is always carried in the barn swallow's ample mouth and throat, is pressed out through the partially opened mandibles as a semiliquid mass, at first in small and later in larger globules, a little here and a little there. These globules are eventually laid down in tiers or rows, and their uniformity is due to a uniform method of production and disposition.

This semifluid building mud or clay is made to adhere to a previously moistened surface of the wood by a peculiar *tapping movement* of the bill. * * * To secure this all-important contact a small area is first thoroughly moistened by repeatedly bringing the bill to it simultaneously with a visible muscular contraction in the throat region which presses out of the mouth drops of a liquid which, from the freedom with which it runs along the grain of the wood, must be mainly, if not wholly, water.

The masonry is started with a mud disk, or plaque, attached to the vertical surface, around and on which the structure is built up; the layers of mud are alternated or mixed with pieces of grass or straw; these are short at first and worked in with the mud, but later, longer pieces are used and the ends left hanging to some extent. "The method of depositing and fixing these little spears was interesting. A bird would come with a delicate blade or stem held crosswise in its bill, and the moment it touched the beam, a small pellet of wet mud would issue from the mouth and, like a drop of sealing-wax, fix it promptly to the vertical surface."

Both sexes work together, industriously and harmoniously, in the construction of the nest. Professor Herrick's birds took eight full days of upwards of 14 hours each to complete their nest; they "began their working-day as early as five in the morning, and they did not wholly rest from their labors until eight o'clock at night." They brought a load of material about every two or three minutes.

Dr. Harold B. Wood, who has sent me some notes on this subject, evidently agrees substantially with Professor Herrick on all the above points. A nest that he watched near Harrisburg, Pa., was built in six days. Dr. Dickey tells me that, from his observations, a pair of barn swallows will average 12 days in building a nest and will sometimes delay laying in it for two weeks after it is finished. They seem to prefer to repair and use an old nest rather than build a new one, so that often the same nest is occupied for a number of years, or a new one is built on the remains of the old one, or at least in exactly the same spot.

The size and shape of the nest vary greatly with its location; on a vertical surface the nest is roughly in the shape of a reversed half cone, the top being somewhat more than half a circle, and the lower end pointed; if built on a flat surface, the nest is more circular, like a phoebe's nest, and much shallower than the cone-shaped nests, but in such cases the birds often build mud walls along the flat beam for a foot or more, apparently for perches or to prevent the young from falling while exercising their wings. Nests built in corners or in the angles of braces are smaller and made to fit the spaces. The outer diameter of the circular top of the nest will average about 5 inches and the inner diameter is, fairly constantly, about 3 inches.

The composition of the nest consists mainly of dried mud mixed with grass and straws, and it is profusely lined with poultry feathers, mainly white ones. A nest dissected by Dr. Wood (1937b) "was found to contain, besides nearly seven and one-half ounces of dried earth, 1635 rootlets over one-half inch in length, 139 white pine needles, 450 pieces of dried grass, 10 chicken feathers, 4 pieces of wood, 2 human hairs and a piece of leaf and cotton, and a tablespoonful of minute pieces of rootlets and grass."

Horse hairs enter into the composition of many nests. Dr. Dickey tells me that he has seen swallows hover around stalls and disentagle the hairs lodged in cracks; these hairs sometimes make trouble for the birds.

Some nests are made without the use of mud; nests in narrow crannies or holes, with supporting floor and sides, do not need the mud foundation and are made of grasses, straws, feathers, and other available materials. James B. Dixon says in his notes: "The nests in the caves of the ocean walls are very unusual in that the birds are hard pressed for suitable mud and use a great deal of sea weed, with the results that the nests look like some old man with a beard, as the sea weed stringers hang down from the nest." These nests are lined with seaweed instead of feathers.

Eggs.—The barn swallow's set usually consists of four or five eggs; six eggs are fairly common, and seven are rare; as many as nine have been found in a nest, but these were probably the product of two females. The eggs are practically indistinguishable from those of the cliff swallow. They vary from ovate, the commonest shape, to elliptical-ovate, or rarely to elongate-ovate. The ground color is white, and the markings are in shades of bright reddish brown or darker browns, with sometimes a few underlying spots of pale lilac or drab. Some eggs are evenly covered with fine dots, more often sparingly than thickly, and others show numerous larger spots or small blotches scattered over the egg, or concentrated in a ring about the larger end; rarely an egg is nearly immaculate. The measurements of 50 eggs average 18.8 by 13.5 millimeters; the eggs showing the four extremes measure 21.1 by 13.5, 20.8 by 15.2, 17.3 by 13.2, and 19.0 by 12.2 millimeters.

Young.—Much study has been given to the development, care, and behavior of young barn swallows, and a number of excellent articles have been published on the subject. Since space will not permit extensive quotations from these, I shall attempt merely a condensed summary of the facts, with a few direct quotations. Readers who would like to make a more detailed study of the subject, are referred to the following more important papers: E. M. Davis (1937), Dayton Stoner (1935 and 1936a), Wendell P. Smith (1933 and 1937), and Harold B. Wood (1937a and 1937b).

Most observers give the period of incubation as either 15 or 17 days, but some have placed it as low as 13 days; it probably varies some, with 15 a fair average. Both parents share this duty, changing places at frequent intervals, averaging about every 15 minutes. Dr. Wood found that after the third day of incubation they changed at intervals of between 4 and 15 minutes; on the eleventh day the intervals ranged from 6 to 36 minutes. William Brewster (1938) says: "The change which took place on an average of over every fifteen minutes was effected with singular adroitness. The incoming bird, twittering loudly, flew directly to the nest always aiming for the point where its partner's tail projected over the rim. * * * So quickly was it done that I doubt if a person looking down on the nest from above could have got more than the briefest possible glimpse of the eggs."

The female apparently incubates during the night, with the male perched near her. Two broods seem to be raised generally in a season in the more temperate portions of the bird's range. Dean Amadon observed about a dozen pairs nesting in a barn in New York State, and tells me that "almost exactly 50 percent raised a second brood. Usually pairs that used an old nest, constructed in previous years, would raise two broods; those that built a new one only one."

The young remain in the nest 18 days (Smith), 19 days (Wood), or 23 days (Herrick) and are fed and cared for constantly by both parents. They are not always fed in regular rotation; generally the one that seems most anxious for food is fed first. Probably the youngest birds are fed on regurgitated food, though this does not seem to have been definitely observed; they are, however, known to be fed on fresh insects at an early age, mostly very small insects.

The parents eat the fecal sacks for the first few days, but later they carry them away until the young are about 12 days old, when the young are able to turn around and void their droppings over the edge of the nest.

Among the developments observed by Mr. Smith (1933) "was the opening of the eyes, a gradual process, beginning on the 5th day and being completed on the 8th. The use of the wings progressed from their being fluttered on the 9th day to short flights when removed from the nest on the 15th, although even on the latter date dependence was still largely placed upon crawling and hiding as a means of escape. Fear first became manifest at nine days of age, when, at the alarm note of a parent the young would retreat from the rim of the nest and crouch down in the bottom."

As the plumage begins to grow, the young swallows become more active, wagging their heads about and hanging them out over the edge of the nest, as if in imminent danger of falling, though few such accidents occur. When about 15 days old they begin active preening, scraping off the scales of the feather sheaths. At length the time comes for them to leave the nest; the parents urge them to do so by refusing to feed them in the nest and by flying back and forth near the nest, enticing them to leave. After leaving the nest they remain in the vicinity for several days, returning to the nest each night.

Dr. Charles W. Townsend (1920c) writes of some young birds that he watched: "At times they rested in trees and were fed by the parents, sometimes they were fed in mid-air, but doubtless they did some insect catching on their own hook. At six o'clock they were all back in the nest and being fed by the parents. For four more days this was repeated. The young left in the morning but returned to the nest at night, generally going and coming together. On the fifth night only two returned and after that they occupied the nest no more. I imagined I saw the family party several times, however, as a group of six or seven barn swallows flew past, and occasionally they would fly around under the porch, the adults pouring forth their souls in song."

Plumages.—The young nestling is scantily covered with "smoke gray," or darker, natal down on the forehead, occiput, scapular region, and middle of the back; this down persists later on the tips of the juvenal plumage and has not wholly disappeared on the head when the birds leave the nest.

The juvenal plumage is much like that of the adult but is duller and paler. The sexes are alike. The upper parts are dull iridescent green, more brownish on the crown, with reddish brown edgings on the nape, rump, and wing coverts; the under parts are pale cinnamon, tinged with russet on the throat, with an incomplete dusky or black band across the jugulum; the tail feathers have large subterminal white spots, the outer pair broadly rounded at the tips and reaching less than an inch beyond the central pair. This plumage is worn until after the birds leave for the south. Apparently a complete molt takes place while the birds are in the south, probably beginning with the body plumage in the fall and completed with the molt of the flight feathers late in winter or early in spring; but we do not know much about how this takes place, for adequate material is lacking. We *do* know that the birds return in spring in a plumage that is practically adult, with the long, attenuated lateral tail feathers extending fully one and one-quarter inches beyond the middle pair, with the metallic purplish feathers of the jugular band, the chestnut throat, and the darker cinnamon abdomen. All these colors are more or less variable, even in adults, where the sexes are often very much alike, though the female is usually much paler than the male. Adults have one complete annual molt, the postnuptial, after they leave for the south; this is probably accomplished somewhat earlier than that of the young.

The barn swallow has been known to hybridize with the cliff swallow, as explained under that species.

Food.—Professor Beal (1918) analyzed 467 stomachs of the barn swallow and found that the food was made up of 99.82 percent animal matter and 0.18 percent vegetable. "Diptera are evidently the choice food of the barn swallow. They average 39.49 percent of the food, or more than twice that of any order of insect." They are mostly allied to the common house fly, but include long-legged crane flies (Tipulidae), horse flies (Tabanidae) and several robber flies (Asilidae), which are said to be very destructive to honeybees.

Beetles of various families amount to 15.63 percent. The useful beetles, those that prey upon other insects, predaceous ground beetles (Carabidae) and ladybirds (Coccinellidae), amount to only 3.4 percent. All the remainder of the 80 species of beetles are "most of them harmful and some exceedingly so." These include the May-beetle family (Scarabaeidae), mostly small dung beetles, various weevils, including the cotton boll weevil and the rice weevil, and the "destructive engraver beetles that do so much damage to timber."

Hymenoptera other than ants make up 12.82 percent, consisting of bees and wasps, but only one honeybee was found, a drone. "Ants are eaten by the barn swallow to the extent of 9.89 percent of the food, some stomachs being entirely filled with wingless species." Hemiptera formed 15.1 percent, including stink bugs, leaf bugs, plant lice, and chinch bugs. Other items include a few Lepidoptera, 2.39 percent, mostly adults, still fewer grasshoppers and crickets, 0.51 percent; dragon flies, 4 percent; and a few May flies, spiders, and snails.

"All the vegetable matter found was contained in six stomachs, but it was real food in only four. One of these revealed seeds of the elderberry (*Sambucus*) and of *Cornus sericea*. Vegetable food in this stomach made up 75 percent of the contents. The second stomach held a single kernel of buckwheat, the third a root or bulb, and the fourth two seeds of *Croton texensis*."

Swallows probably spend more time in the air than any other group of passerine birds, not excepting the flycatchers; consequently nearly all their food is captured while they are on the wing. They catch more flies than the flycatchers because, in their swift and tireless flight, they cover much more open country, over fields, meadows, marshes, and ponds, whereas the flycatchers are limited to such insects as happen to fly near their perches.

Barn swallows follow the farmer while he is plowing to catch such insects as he stirs up and even alight on the ground to pick up others. They may often be seen coursing back and forth among moving cattle or sheep, gleaning the insects that these animals disturb. Anyone who has walked through a field of waving grass has noted how the swallows follow him, catching insects in the air or picking them deftly from the grass tops. Mr. Forbush (1907) says: "It is particularly serviceable about grass fields. The moths of the smaller cutworms, those of Arctians and Crambids, are among the injurious insects that it gleans when flying low over the grass. * * * Codling moths, cankerworm moths, and Tortricid or leaf-rolling moths are gathered from the orchard." Thus in many ways these swallows are most useful to the farmers.

Dr. Dickey (MS.) says: "There are known instances of vegetation having been saved on farms where barn swallows had breeding colonies, while neighboring farms, without swallows, were about denuded of plant life," in West Virginia, during a recent drought.

Behavior.—What is more graceful or more pleasing to watch than the flight of the barn swallow? The flight of the albatross, as it mounts from the trough between seas over the crest of a wave, is effortless, powerful, majestic; the soaring flight of the eagle and the spirited dash of the falcon are inspiring; and the swallow-tailed kite charms the observer with its grace. Few of us are privileged to see these masters of the air except on rare occasions, but, if we could only learn to appreciate our familiar little friend of the barnyard, we would see that his powers of flight compare favorably with the best of them. His flight is fully as graceful as that of the kite, which after all is only a glorified swallow; it is as swift for his own purposes as that of the falcon; and, in dashing through the narrow openings to his nesting places, he shows as much control of his wings as does the albatross in skimming the waves. His poise and grace on the wing are unsurpassed.

Wilson (1831) made some interesting calculations as to the number of miles that the barn swallow flies in its lifetime; assuming that it flies at the rate of a mile a minute, for ten hours a day, and lives ten years, it would fly 2,190,000 miles, or over 87 times around the earth; this is doubtless too high an estimate, but it is impressive, even if greatly discounted.

The barn swallow is a gentle, harmless creature that attends strictly to its own affairs and seldom troubles its neighbors. Like all birds, it defends its nesting territory against intrusion by phoebes or other swallows that attempt to occupy its nesting site; and it doubtless attempts to drive away predatory birds or mammals that threaten its eggs or young. I once saw a pair of barn swallows chasing and attacking a sharp-shinned hawk that came too near their nesting place; the hawk retreated, but the swallows followed it high up into the air and it finally tired of their attacks and disappeared. Dr. Townsend (1905) "heard a great outcry among some swallows and found a company of Barn Swallows mobbing a jack rabbit as he bounded off on the upper part of the beach," at Ipswich, Mass.

Swallows seem to show a playful spirit in many of their activities; they seem to enjoy the sport of gathering feathers for their nests; many a country boy has tried the fun of throwing feathers into the air and watching the birds scramble for them; the swallow seldom misses its aim, but, if the feather is dropped, it is immediately seized by another bird; often the feather changes "hands" several times before it reaches the barn; sometimes it seems as if the feather were dropped again and again in a spirit of mere play.

Swallows are sociable birds and often gather in large mixed flocks—barn, bank, tree, and cliff swallows—and seem to enjoy the sport of flying about and showing their mastery of the air, apparently for the pure joy of flying. Dr. Townsend (1920c) writes: "One September day at sunset a flock many hundreds if not thousands of these birds were alighted on the bushes, fence rails and wires near the waters of Sagamore Pond. They arose with the roar of many wings, and turning first their dark then their white surfaces to the observer, swirled about in irregular groups. Then they all flew close to the water, and every now and then hurled themselves at it so that the quiet surface of the pond was pitted with splashes as from a bombardment. Their heads, backs and wings were soused in the water, which they shook off in showers as they arose. At times they would dip lightly several times in succession."

It is a familiar sight late in summer to observe long lines of swallows of several species sitting on the telephone wires along the roads; here they sit and rest or preen their plumage, stretching one long wing down until its tip reaches beyond the tail as the plumage is dressed beneath it; sometimes one will scratch its head with its foot; but for the most part they sit quietly, enjoying one another's company and seldom quarreling unless the wires become overcrowded.

After the breeding season is over swallows become highly gregarious, as indicated above, and gather in immense flocks to roost at night in marshes or thickets of trees or bushes. W. E. Saunders (1898) describes such a gathering in a thick stand of willows near London, Ontario, as follows:

Passing these on the evening of August 4th of this year, I was attracted by the large number of Barn Swallows circling near it, which, as the night drew on, became more and more numerous, until I judged there were about 5,000 birds,—almost all Barn Swallows—in the flock. They flew at random until about 8 o'clock, only a few alighting in the roost before that time, but at 8.04 my note-book records them "falling like leaves," and by 8.05 half were settled. Their manner of descent was both interesting and beautiful, especially of those from the upper strata, for they were flying at all elevations from those just skimming the ground, to those so far up that they could with difficulty be seen, and these latter, in descending at an angle of only 20 degrees from the perpendicular, performed the most beautiful aerial evolutions it has been my fortune to witness. Setting their wings for the drop, they would waver from side to side as they came, much as a leaf wavers, but of course with many times greater speed. * * *

Within five minutes of the time of the first general movement, barely a tenth remained in the air, and their voices, which are so liquid and soft when heard singly, became one of the harshest dins imaginable—English Sparrows could be no worse—and it certainly sounded as if they were all talking at once.

At 8.12 only a few are recorded as remaining, and at 8.19 the last one went in.

Swallows have been seen at sea, far out from land, and may occasionally be forced down; they cannot swim, of course, but seem to be able to rise from the surface. Dr. Wood tells me that his son, Merrill Wood, saw a young barn swallow alight on the water of Narragansett Bay and, after a short rest, rise from the surface and fly away.

Voice.—Dr. Winsor M. Tyler contributes the following: "The notes of the barn swallow always seem to express happiness, in keeping with its joyous flashing across the sky. *Kvik-kvik, wit-wit,* he says, in short, delicate, but energetic syllables, as he doubles and turns and twists through the air. When he shoots through the open doorway of the barn to his nest, he greets his mate and young with a friendly *kivik,* accented at the end—almost pronouncing the word, quick. Out of doors again, either coursing along on the wing, or perched on roof or wire, he entertains his family, or lets his irrepressible energy go, with a long pleasing song of many jumbled, bubbling, rapid notes, culminating with a queer, ecstatic trilling sound which Ralph Hoffmann (1904) aptly terms 'a very curious rubbery note.' "

To the above pleasing account might be added the tributes of many other admirers. Dr. Townsend (1920c) writes: "To my mind the

Barn Swallow is one of our most delightful singers. His song is always full of charm, soft and lovely, devoid of all roughness. Besides delivering an individual song, he delights in singing in chorus. It is a sweet and cheerful song full of little trills and joyful bubbles of music, at times clear and sparkling, at times oozing and rubbery. Like the music of a brook it flows on indefinitely. At times the old barn is permeated with its melody. * * * From the first day of their arrival in late April till the end of August and even into September this charming bird sings. Very few birds have such a long and continuous song season."

The barn swallow not only sings delightfully throughout a long season, but also through a long summer day. Horace W. Wright (1912) says that the twitter of this swallow is one of the earliest sounds in the morning awakening, and seems "to proceed at first from the birds on their night perches." His earliest record is 2.51 A. M., and the average of 11 records is 3.04 A. M.

Field marks.—The barn swallow is one of the best known and most easily recognized of our swallows. It is the only North American swallow that has a long, deeply forked tail, with white spots on the tail feathers, and rich brown or buffy under parts. The females are usually duller colored than the males, but not always. The young birds are much duller in color and the lateral tail feathers are shorter than in the adults, but the under parts are never white and the tail is always more deeply forked than in other species.

Enemies.—Probably more swallows perish from the effects of inclement weather than from any other cause; a prolonged, cold rainstorm drives insects to cover, the swallows are unable to obtain the necessary amount of their accustomed food, and old birds, as well as young in the nests, die from hunger and cold. The use of horsehair in the nests, in which the wings, feet, or necks of the birds become entangled, results in the death of some. House wrens have been known to puncture the eggs of barn swallows and appropriate their nests. Cowbirds may very rarely deposit their eggs in the swallows' nests. English sparrows and phoebes contend with them for nesting sites and even drive them from their nests. As the swallow skims low over the surface of a pond, it may be caught by some large fish or bullfrog; Forbush (1929) gives an authentic account of such a case, from Mrs. Chester Bancroft, of Tyngsboro, Mass.: "There is a brook flowing through the Bancroft yard, in which lived an enormous bullfrog, which Mrs. Bancroft's daughter had been watching with interest during the summer of 1927. One day she saw the tips of a bird's wings protruding from the corners of his mouth. The frog was finally caught and relieved of what he had swallowed. It was a full-grown Barn Swallow."

I once had a large, black cat that was a great bird catcher; he would lie hidden in the tall grass and watch for passing swallows; one day I saw him leap into the air and catch a swallow, as it swooped low over the grass tops. Forbush (1929) says: "The appearance of a strange cat, a weasel or a Sharp-shinned Hawk, when the Swallows have young, is the signal for a concerted assault. I have even seen a lone pair of breeding birds drive both cat and weasel from the neighborhood of their helpless young."

The young in the nest are often attacked by parasites, such as the larvae of *Protocalliphora;* the maggots of these flies are very destructive to young birds of many species, attaching themselves to the eyelids, or entering the throats or nostrils, and probably causing the death of many young. Mr. DuBois tells me of how a barn swallow met its death by a very strange accident; in spite of its dexterity on the wing, it darted into the path of a flying golf ball and was killed by it.

Fall.—In the northern portion of its breeding range, where only one brood is raised, the fall migration begins early, or soon after the young are strong on the wing, probably early in August. Wendell P. Smith (1937) says of the birds that he studied at Wells River, Vermont: "Flocking evidently began about July 15, as flocks larger than family units were first seen on that date. Migration seemed under way by August 2, and local breeding birds seemed largely gone by August 18." In southern New England barn swallows are common enough in flocks all through August and part of September, but many of these may be birds that have come from farther north. The migration seems to be largely coastwise; we always see plenty of them on outer Cape Cod, during late summer and early fall; they scatter out over the marshes, coursing about independently to feed on the millions of mosquitoes and flies; on windy days, I have often seen them resting in compact flocks on the "back of the beach" (ocean side), squatting low in the sand and all facing the wind.

They seem to prefer to migrate by day and, strangely enough, often against a strong south wind. Some of them fly high in loose formation, but oftener they move along within 100 feet of the ground or lower. They may often be seen moving along the coast line, or the bank of a river, only a few feet above the ground or water, in a steady open stream, all following the same general direction but without any definite flock formation. They have been seen at sea, and must regularly migrate across the Gulf of Mexico to reach their winter range in South America.

Dr. Winsor M. Tyler contributes the following observation: "In the inland country of New England we see the last of the barn swallows during the first week of September, often in small parties, gathered on telegraph wires, resting on their journey, or sometimes we see them in actual migration—a few birds, inconspicuous, wide apart and silent, flying past on their way to the south. I know a little valley, 10 miles from the sea, drained by a narrow brook, which must lie directly in the swallows' path of migration, for if I spend half an hour there early in September and watch the sky to the north, I am almost sure to see a barn swallow coming toward me. It cruises along at the height of a tall tree, veering sometimes far to one side or the other, but holding, in the main, a straight course, following the dip of the land. Sometimes it is a lone bird, but sometimes before it is out of sight, mounting up over the hill at the southern end of the valley, a second or a third bird will appear out of the north and follow, far behind, on the same invisible track. These are the last barn swallows of the year. We shall see no more until, in early May, they come jubilantly back to the farmlands."

DISTRIBUTION

Range.—North and South America.

Breeding range.—The breeding range of the barn swallow extends **north** to Alaska (Port Clarence, Cape Blossom, Kobuk River, and Circle); Mackenzie (Fort Goodhope, Fort Franklin, Lake Fabre, and Hill Island Lake); southern Saskatchewan (Wiseton and Quill Lake); southern Manitoba (Lake St. Martin); Ontario (Kenora District, Port Arthur, Chapleau, and Cobalt); southern Quebec (Quebec City and Godbout); and rarely southern Labrador (Northwest River). **East** to rarely Labrador (Northwest River); Nova Scotia (Baddeck, Halifax, and Seal Island); and south along the Atlantic coast to North Carolina (Bodie Island, Pea Island, and Wilmington Beach). **South** to North Carolina (Wilmington Beach and Valle Crucis); Tennessee (Athens and Bell Buckle); northwestern Alabama (Tuscambia); Arkansas (Clinton and Clear Creek); Oklahoma (Redland, Norman, and Wichita Forest); southern Texas (Del Rio); Veracruz (Las Yigas and Perote); Puebla (Huamantla and Atlixco); Michoacan (Patzcuaro); and southern Jalisco (Tuxpan). **West** to Jalisco (Tuxpan, Guadalajara, and Etzatlan); Nayarit (Santiago); Durango (El Salto); northwestern Baja California (San Quintin and Los Coronados Islands); and north along the Pacific coast to Alaska (St. Lazaria Island, English Bay, Bethel, St. Michael, and Port Clarence).

Winter range.—The winter range of the barn swallow is not thoroughly defined, but it appears certain that any records for this season north of South America must be considered as accidental. During January and February 1893, Dr. E. W. Nelson noted individuals of this species at Mexico City and at two or three points in the State of Morelos. No other definite records are known for this season in Central America. From present information the winter range may be outlined as **north** to probably Colombia (Juntas de Tamana and Santa Marta region) and British Guiana (Abary River and Blairmont). **East** to British Guiana (Blairmont); southwestern Brazil (Corumba); Paraguay (Lambare); and eastern Argentina (Corrientes, Ocampo, Barracas al Sud, and Lavalle). **South** to central Argentina (Lavalle and Lezama) and probably rarely central Chile (Santiago). **West** to probably rarely Chile (**Santiago and Arica**); Peru (Ica, Chorillos, and Callao); probably rarely Ecuador (Bucay); and probably western Colombia (Juntas de Tamana). A bird banded (C–3371) at Muscow, Saskatchewan, in July 1927 was recovered eight years later at El Carmen, in northwestern Bolivia.

Spring migration.—Early dates of spring arrival in the United States and Canada are: Florida—Pensacola, April 2. Georgia—Atlanta, April 10. South Carolina—Charleston, April 3. North Carolina—Raleigh, April 2. Maryland—Cambridge, April 3. Pennsylvania—Berwyn, April 7. New Jersey—Morristown, April 3. New York—New York City, April 5. Connecticut—Fairfield, April 10. Massachusetts—Taunton, April 6. Vermont—Rutland, April 10. Maine—Ellsworth, April 12. New Brunswick—Scotch Lake, April 18. Quebec—Sherbrooke, April 18. Louisiana—New Orleans, March 20. Arkansas—Roberts, April 2. Tennessee—Nashville, April 1. Missouri—St. Louis, April 7. Illinois—Chicago, April 4. Indiana—Lafayette, April 3. Ohio—Youngstown, April 2. Michigan—Vicksburg, April 2. Ontario—Ottawa, April 8. Iowa—Keokuk, April 3. Minnesota—Minneapolis, April 9. Texas—Eagle Pass, February 9. Kansas—Manhattan, March 29. Nebraska—Whitman, April 15. South Dakota—Yankton, April 15. North Dakota—Marstonmoor, April 25. Saskatchewan—McLean, April 30. Arizona—Tombstone, March 20. New Mexico—State College, April 7. Colorado—Denver, April 17. Wyoming—Cheyenne, April 21. Montana—Terry, May 5. Alberta—Flagstaff, April 29. California—Fresno, March 4. Oregon—Portland, April 13. Washington—Tacoma, April 24. British Columbia—Courtenay, April 25. Alaska—Ketchikan, May 15.

Fall migration.—Late dates of fall departure are: Alaska—Ketchikan, September 3. British Columbia—Courtenay, October 3. Washington—Seattle, October 24. California—Fresno County, October 7. Alberta—Camrose, September 16. Montana—Fortine, September 25. Colorado—Denver, October 3. New Mexico—head of Mimbres River, October 13. Saskatchewan—Eastend, September 22. North Dakota—Fargo, September 28. Kansas—Onaga, October 5. Oklahoma—Copan, October 21. Texas—Somerset, October

20. Manitoba—Aweme, September 30. Minnesota—Hutchinson, October 5. Iowa—Wall Lake, October 3. Ontario—Guelph, September 10. Michigan—Vicksburg, September 17. Ohio—Oberlin, October 15. Indiana—Richmond, October 13. Illinois—Port Byron, October 8. Missouri—Concordia, October 11. Kentucky—Eubank, October 6. Mississippi—Biloxi, October 19. Louisiana—New Orleans, November 3. Prince Edward Island—North River, September 15. New Brunswick—St. John, September 29. Maine—Phillips, September 21. Massachusetts—Harvard, September 22. Connecticut—Hartford, October 9. New York—New York City, October 14. Pennsylvania—Berwyn, October 5. District of Columbia—Washington, September 23. Georgia—Savannah, October 27. Florida—Pensacola, November 23. One banded (F–35418) at North Eastham, Cape Cod, Mass., on June 28, 1931, was retaken at Panama City, Fla., on August 26, 1931.

Casual records.—Barn swallows in migration have been recorded on several occasions from islands of the Galapagos Archipelago and also from Bermuda. One was taken at Lake Blanca, in Patagonia, on November 19, 1901.

In North America, one was taken at Hopedale, Labrador, on May 15, 1934, and evidence obtained to indicate that the species may occasionally breed in that area. One was taken on Mansel Island, Northwest Territories, on June 14, 1930, and one was seen 25 miles south of Cape Eskimo, Keewatin, on August 13, 1901. In Greenland, a specimen was obtained about 1830 at Fiskenaesset, and another about 1856 at Nanortalik. One was seen on St. George Island, Alaska, on May 28, 1890, and on St. Paul Island, on June 4, 1890, while three specimens were taken at Point Barrow, July 1, 1929, June 4, 1930, and July 1931.

Egg dates.—Alaska: 5 records, June 7 to July 5.

California: 71 records, April 9 to July 24; 35 records, May 16 to June 16, indicating the height of the season.

Illinois: 17 records, May 12 to July 17; 9 records, June 11 to July 2.

Massachusetts: 24 records, May 17 to July 6; 12 records, May 27 to June 11.

New Jersey: 25 records, May 16 to July 27; 13 records, May 22 to June 7.

Virginia: 17 records, May 21 to July 15; 9 records, May 21 to June 23.

HIRUNDO RUSTICA RUSTICA Linnaeus

EUROPEAN SWALLOW

HABITS

The European swallow owes its place on our list to its casual occurrence on both coasts of Greenland; it has, apparently, never been taken in continental North America. It is now regarded as only subspecifically distinct from our common barn swallow. Our bird is smaller than the European bird and more intensely colored on the under parts; and the European bird has a broad band of glossy blue-black across the chest, while in our bird this band is completely interrupted or only faintly connected.

It seems to fill the same place in European bird life as our beloved barn swallow does in America, for Yarrell (1876–82) writes: " 'The swallow,' says Davy in his 'Salmonia,' 'is one of my favourite birds, and a rival of the nightingale; for he cheers my sense of seeing as much as the other does my sense of hearing. He is the glad prophet of the year—the harbinger of the best season; he lives a life of enjoyment amongst the loveliest forms of nature; winter is unknown to him; and he leaves the green meadows of England in autumn, for the myrtle and orange groves of Italy, and for the palms of Africa.' "

Like our barn swallow, which it closely resembles in all its habits, the European bird favors the open rural districts, where it is associated with scattered farm buildings and domestic animals, rather than the more thickly settled towns and villages. About the farms it finds suitable nesting sites in the buildings occupied by the animals and an abundant food supply in the numerous insects attracted to such places. The companionship and protection offered by sympathetic human friends may also be a factor. Capt. A. W. Boyd (1935) says that "downland, heathland and moorland support very few pairs, just as they include few buildings; the highest altitude at which breeding pairs in the areas under observation have been found is 900 feet. In industrial and urban districts their density is low."

Spring.—Although there are a number of earlier dates recorded, the average time for the arrival of the main body of the swallows in England seems to be around the first week in April; they continue to arrive, however, all through that month merging with the coming of the passage migrants at the end of April, which migration lasts all through May and even into June, according to Witherby (1920).

Nesting.—Capt. Boyd (1935) says: "It has been found that the nesting sites of the great majority of Swallows are associated with domestic animals, of which cows are first favourites; pigs are very attractive and horses also, though recent reductions in the number of stables occupied by horses have made them of less importance to the bird; hen-houses are often occupied, but the number of pairs nesting in dwelling-houses is comparatively small." Dr. A. G. Butler (1896) writes:

It places its nest in various situations—on joists of barns, out-houses, boat-houses, in which case the form of the nest varies from an oval to a half- or quarter-cup; against perpendicular walls under eaves of barns; inside chimneys, wells, and mines; in corners of pillared porticoes to large houses; under rustic bridges; cases have even been recorded of nests built in a hole in a tree about thirty feet from the ground, and in the forking branch of a sycamore. In its wild state in mountainous or rocky localities this bird naturally builds against the sides of cliffs under overhanging ledges of rock, or in caves; but in Great Britain it usually seeks the habitations of men for nesting-sites.

The nest is always open above; the walls thick, and formed of mud-pellets mixed with straw, hay, or hair; the lining consists of fine grass-stems, usually almost concealed by a quantity of feathers, though in some instances these are absent.

In addition to the above-mentioned nesting sites, most of which may be considered as fairly normal, many unusual sites have been recorded.

One of the most remarkable is Yarrell's (1876–82) time-honored account of a nest "built on the wings and body of a dead Owl, hanging from a rafter in a barn"; he also mentions "a bracket, a picture-frame and a bell-crank. * * * the half-open drawer of a table, and the loop of a chain in a boathouse." Stephen J. White (1908) tells of a nest built on a glass shade hanging over a gas burner in a stable; and there are other records of nests on lamp shades, hanging lamps and electric light bulbs. F. K. Staunton (1929) shows a photograph of three nests on a hat, two on the brim and one inside, hanging on the wall of a disused stable. Evidently, when normal nesting sites are scarce, these swallows will use almost any suitable receptacles for their nests that will offer reasonable support.

Eggs.—Dr. Butler (1896) says of the eggs that "the ground color is pure white, appearing rosy when not incubated; speckled, spotted, and occasionally heavily blotched with deep pitchy brown, often intermixed with sienna reddish spots, and with lavender grey shell-spots; in some eggs the spots are small and tolerably evenly distributed over the entire surface, sometimes they are larger, and principally collected at the larger end; sometimes there is an imperfect zone of spots, and even large blotches near the larger end; some eggs are elongated ovals, others short and well formed."

According to Witherby's Handbook (1920), the usual clutch consists of 4 or 5 eggs, sometimes 3 or 6, and as many as 8, or even 9, have been recorded. The measurements of 50 British eggs average 20.2 by 13.9 millimeters.

Young.—The incubation period varies from 14 to 15 days, more or less. Both sexes share the duties of incubation and the care of the young, which remain in the nest 20 to 22 days, if undisturbed. Generally two and sometimes three broods are reared in a season. Capt. Boyd's (1935) inquiry for 1934 reported that "in all, 2,665 young were ringed or counted, giving an average of 4.01 for all broods recorded."

Dom Ethelbert Horne (1924) gives an interesting account of how young swallows are fed; he had removed the nest, with the small

young, to a box that he could move, and eventually the old birds
became so tame that he could watch them at close range; he writes:

For several days, I had the old Swallows feeding the young with the box
resting on my lap. The bird would come to the nest with two, three and
sometimes four, small "blue-bottles" in its beak, arranged transversely, one in
front of the other. It put one fly into the open mouth of each young one,
with extraordinary rapidity, and was off again directly unless it had any
excreta to carry away, which delayed it a moment or two. How the bird
managed to catch and arrange the flies in its beak in the way it did was a
puzzle, but a much greater one was how it got rid of the fly's wings at the last
moment before putting it into the young one's mouth. The two wings would
fall from the beak of the old bird as if they had been nipped off with something, but as the beak was holding other flies and could not close, it was probably the tongue that did this wonderful trick. Time after time as we bent right
over the bird, we saw the wings fall from the fly, but we never saw how they
were removed. * * * The bright blue or green, rather small "blue-bottle"
was the only insect caught. The old birds used to skim about over a grass
field at the back of the summer-house, and catch them there.

After the young leave the nest, they are fed by their parents for
a while, sometimes on the wing, but they soon disperse more or less
widely over the surrounding country, and seldom return to breed in
the vicinity of their birth, though commonly within a few miles.

Plumages.—The plumages and molts of the European swallow
follow the same sequence as in our barn swallow, though the postjuvenal and postnuptial molts begin before the birds leave for the
south. Witherby's Handbook (1920) states that in both adults and
young the winter "plumage is acquired by a very gradual moult,
commencing in Europe with body feathers and sometimes median
and lesser wing-coverts in Aug. (exceptionally July) and continuing
in Africa, whence specimens in every month from Sept. to March
(exceptionally April) are in various stages of complete moult."

Food.—Dr. Butler (1896) says: "The food of the Swallow consists
largely of gnats, small flies, and ephemerae; but it frequently settles
on the roads, or on manure heaps, to search for small dung-beetles;
owing to its short legs, its progression on the earth is somewhat
awkward, and when hurried it uses its wings to help it along; it
usually drinks on the wing, skimming the surface of the water as it
glides over."

Witherby's Handbook (1920) lists: "Insects taken on the wing,
chiefly Diptera (Chironomidae, *Tipula*, *Empis*, *Borborus*, *Calliphora*,
etc.), but also Coleoptera (*Curculio*, *Helophorus*, *Tachinus*, *Aphodius*,
etc.), Hymenoptera (winged ants), Neuroptera (*Calopteryx*), and
exceptionally Lepidoptera."

Behavior.—I cannot find in the literature any indication that the
behavior of the European swallow differs in any respect from that
of our American representative of the species. It has the same

graceful, pleasing flight and the same gentle, friendly, and confiding manners. In fact, the habits of the two birds are almost exactly alike in every way; what has been written about one would apply almost equally as well to the other.

Voice.—It has similar twittering calls to those of the barn swallow and a pleasing song. Dr. Butler (1896) says: "The Swallow is an admirable singer, and I shall not easily forget the pleasure with which I first heard it, as it poured forth its sweet melody from the girders of a large railway station in Switzerland, in 1869; I have heard it several times since, both in Kent and Norfolk, singing from a telegraph wire; the song is very varied and, to my mind, far more melodious than that of a Linnet."

Enemies.—Rats and mice do some damage to the eggs and young of swallows; cats and weasels, and perhaps owls, kill some birds. There is a long list of parasites that infest the nests and attack the young. House sparrows and wrens often usurp the nests of swallows, but these are usually old nests and cases of eviction of the swallows are not common. Capt. Boyd (1935) mentions, in addition to the above statements, that robins and spotted flycatchers have been known to use old swallows' nests. As swallows often like to use their nests for a second or third brood, the usurpation of even the old nests often interferes with this habit and may provoke controversies.

DISTRIBUTION

Range.—According to Witherby's Handbook (1920), the European swallow is generally distributed as a summer resident throughout the British Isles, occasionally staying throughout the winter, but it breeds rarely in the extreme west of Ireland, the northwest of Scotland and the Orkneys, and very rarely in the Shetlands and the Outer Hebrides. Its range includes Europe, northwestern Africa, and the western parts of Asia. It is casual on Spitsbergen, Jan Mayen, Greenland, and Novaya Zemlya and occasional on the Faroes and Iceland. It winters in tropical and southern Africa and in India and its islands. It is "replaced by allied forms in Syria (? Asia Minor), Egypt, north Asia to Japan, and North America, all being migrants, wintering far south."

Spring migration.—In the British Isles "early arrivals of summer residents begin end of third week March (early dates Feb. 29, 1912, Cornwall, Mar. 2, 1912, Cardigan, Mar. 5, 1918, Lancs., Mar. 6 and 13, 1913, Scilly); main arrival variable, from April 1 to end second week, merging end April into arrival of passage-migrants and continuing to fourth week May" (Witherby, 1920).

Fall migration.—The same authorities state that the summer residents begin to move south during the last week in July, and emigration from the south coasts begins the second week in August; passage migrants from the north begin to arrive early in September. Stragglers are frequent in November, up to the end of the third week. "(Late dates Dec. 1, 1908, Norfolk, Dec. 20, 1911, Berks., Jan. 3, 1913, Dorset, Jan. 23, 1913, Kerry)."

Egg dates.—Eggs have been recorded in Great Britain through practically all the five months from May to September.

PETROCHELIDON ALBIFRONS ALBIFRONS (Rafinesque)

NORTHERN CLIFF SWALLOW

PLATES 65–68

HABITS

CONTRIBUTED BY ALFRED OTTO GROSS

The generic name of the cliff swallow, *Petrochelidon*, is derived from the Greek *petra*, a rock, and *chelidon*, a swallow, and its specific name is from the Latin *albus*, white, and *frons*, forehead; hence a rock swallow with a white forehead.

There are two common names of this swallow alluding to its nesting site that vie with each other for popularity; they are cliff swallow and eaves swallow. The former, although less appropriate in many sections of its nesting range today, is the one adopted by American ornithologists. Less commonly used names originating from the character of the nests or building material are jug swallow, mud swallow, pipe swallow, and mud dauber. It is not surprising that the plumage of this well-marked swallow has given rise to such appellations as crescent swallow, white-fronted swallow, and square-tailed swallow. Another common name, which was widely used by the earlier ornithologists but seldom applied today, is republican swallow. This unusual name originated with Audubon (1831) when he first discovered the species at Henderson on the banks of the Ohio River in the spring of 1815. He writes: "I drew up a description at the time, naming the species *Hirundo republicana*, the republican swallow, in allusion to the mode in which individuals associate, for the purpose of forming their nests and rearing their young."

The cliff swallow ranks well among our most abundant birds and has a breeding range extending from Mexico to Alaska and across the continent from the Pacific to the Atlantic coast lines. In spite of its present-day abundance and wide distribution, it was unknown to Wilson and other early ornithologists. Its history although obscure is of unusual interest.

This swallow was first brought to our attention by John Reinhold Forester who refers to it as *Hirundo* 35 in an account of the birds sent from Hudson's Bay published in the Philosophical Transactions in 1772. Forester gave it no name, and it was left for Thomas Say to name and describe the species from a type specimen taken in 1820 on Long's Expedition to the Rocky Mountains, an account of which, compiled by Edwin James, was published in 1823.

In the same year (1820) it was also discovered in the Rocky Mountains by Sir John Franklin's party between Cumberland House and Fort Enterprise and on the banks of Point Lake in latitude 65°. In June 1825 a number of these birds made their appearance at Fort Chippewyan and built their nests under the eaves of the house. This fort had then existed many years, and trading posts had been in existence a century and a half, and yet this was the first instance, according to Baird, of the bird placing itself under the protection of man throughout that wide extent of territory. Audubon (1831) first saw this bird at Henderson in 1815, and two years later he found a colony breeding at Newport, Ky., which dated back to the same year. In 1837 Brewer (Baird, Brewer, and Ridgway, 1905) received the eggs of this swallow from Coventry, Vt., where they were known as "eave" swallow, perhaps the first instance of the application of this common name. The date at which the Vermont colony appeared could not be determined. Brewer saw them for the first time in 1839 at Jaffrey, N. H., where they appeared the year before. The same year they were noted in Burlington, Vt., where they had been known for only three years. A large colony was observed in Attleborough in 1842. In the same year they also appeared apparently for the first time in Boston, Hingham, and other places in the neighborhood.

In 1824 DeWitt Clinton stated he had met them at Whitehall, N. Y., at the southern end of Lake Champlain in 1817 about the time of their first appearance on the Ohio noted by Audubon. At about this time they were seen for the first time at Randolph, Vt. These swallows were seen for the first time in Winthrop, Maine, in 1830 and at Carlisle, Pa., in 1841.

In 1861 Prof. A. E. Verrill discovered a large colony of these birds building on the high limestone cliffs of Anticosti Island, in the Gulf of St. Lawrence. Since this appeared to be a long-established colony and far removed from civilization, Verrill was led to believe that the extension of the range of the cliff swallow was not of a recent date. An inquiry that he conducted revealed that they were known in Maine long before they were discovered in the West, and he concluded that they were not indigenous in the West. The fact that they were not seen by the early ornithologists proves nothing; there is little doubt that they were nesting in remote situations of eastern North America when civilized man appeared.

One thing is apparent: as the land of New England was cleared for fields and pastures, and as barns with wide eaves were erected, the cliff swallows, finding an abundance of food and sheltered places for their nests, left their primitive environment of isolated cliffs to come in close association with man. Under these new conditions the birds multiplied and spread from place to place where they had not been seen before. It is also possible that there was an eastward movement later from the cliffs and bluffs of the West, but for this there is no concrete evidence.

In more recent years the prosperity of the cliff swallows in their newly acquired environment has suffered. Such factors as the invasion of English sparrows, the improvement and painting of barns, and the desire of owners to rid their buildings of the mud nests have affected the local fluctuations in the number of these valuable and attractive birds.

So serious was the decrease of these birds in New Jersey where formerly they were abundant that the New Jersey Audubon Society under the leadership of W. DeWitt Miller proposed a campaign by the society for increasing the summer resident cliff swallows in New Jersey (Bowdish, 1930). Miller considered modern barns offering poor support for nests and molestation by thoughtless boys and men as factors contributing to the decrease. The plans of the society were announced in its bulletin and made public through news items sent to the press. A survey of nesting colonies was made, and suggestions, such as nailing plates to the ends of the rafters of barns to afford more dependable supports for the nests and providing a readily available mud supply, were made. The value of the swallows was also impressed on owners of buildings where colonies were located. Furthermore, prizes were offered for the three largest colonies. I have seen no reports of the effectiveness of the campaign, but it is at least an interesting project in connection with the history of this species.

Spring.—The first cliff swallows reach northern Mexico and our Southwestern States during the last week of March. From this region the wave of migration is exraordinary from the standpoint of its speed up the Pacific coast and its lag in the southeastern part of the United States. It has long been known that the earliest records always come from California where it usually becomes common long before it has been observed in Texas, the Gulf States, and Florida. By March 20, when the vanguard has not quite reached the lower Rio Grande in Texas, the species is already north of San Francisco, Calif. Lincoln (1939) presents a plausible explanation of this ornithological puzzle by pointing out that the cliff swallow goes around the Gulf of Mexico rather than across it. He states further: "The reason for this circuitous route around the Gulf of Mexico lies in the fact that the cliff swallow is a day migrant and catches its daily ration of winged insects as it travels slowly along in the proper direction, which obviously it could not do if its route were across the Gulf of Mexico." By the first week of May the cliff swallows have reached New England, and at about the same time they appear in Alaska again, forcibly emphasizing the rapidity of the migration wave along the Pacific coast and the comparatively slow progress along the Atlantic seaboard. Lincoln makes a very illuminating comparison of the migration of the cliff swallow and the blackpoll warbler as follows:

During the winter months these species are together in South America. Upon the approach of spring, bringing with it the impulse to start northward toward their respective breeding grounds, the warblers strike straight across the Caribbean Sea to Florida, while the swallows begin their journey by a flight of several hundred miles westward to Panama. Thence they move leisurely along the western shore of the Caribbean Sea to Mexico, and as if to avoid a long trip over water, go completely around the western side of the Gulf of Mexico. By making this long circuitous flight, swallows that nest in Nova Scotia add more than 2,000 miles to the length of their migratory journey. The question may be asked—Why should the strong-winged cliff swallow use a route that is so much longer and more roundabout than that followed by the Black-Poll Warbler? The explanation is simple as the swallow is a day migrant while the warbler travels at night. The migration of the warbler is made up of a series of long nocturnal flights, alternated with days of rest and feeding. The swallow, on the other hand, starts its migration several weeks earlier and catches each day's ration of flying insects during a few hours of aerial evolutions, which at the same time carry it slowly in the proper direction. Flying along the insect-teeming shores of the Gulf of Mexico, the 2,000 extra miles that are added to its migration route are but a fraction of the distance that these birds actually cover in pursuit of their food.

While a considerable number of cliff swallows have been banded there are not sufficient returns to assist in plotting the actual migration routes. One record, however, is of interest: A cliff swallow banded at Dell Rapids, S. Dak., on June 14, 1937, was captured and released with its band at Ghent, W. Va., on July 16, 1937. This bird in the course of a month flew approximately 1,200 miles in a southeasterly direction. If we take into account its erratic and circuitous flights in quest of insects during its journey, it probably averaged more than 100 miles a day. Its flight to the eastward is interesting, since it migrates to its winter home in South America via Mexico and Central America. One record, however, is not sufficient to indicate that many individuals travel so far out of bounds of a direct route.

The cliff swallow migrates in flocks, and practically all the reports of the large numbers seen throughout the migration route mention the association of the cliff swallow with barn and tree swallows as well as other members of the family. At Rowland's Marsh in the Sierras, Ray (1918) saw on May 27, 1912, at least a thousand cliff swallows in migration. The birds rested for a time in the same grove of dead pines where large numbers of tree swallows had been observed. Knight (1905) describes an interesting observation he made of fatigued migrating cliff swallows at Bangor, Maine, on May 17. "They were perched by hundreds in low shrubs and bushes between the Bicycle Path and a flooded meadow just beyond. Some small willow bushes were each occupied by fifty or more Swallows, whose drooping, half-spread wings and open beaks, through which they drew spasmodic gasps of air, were a most eloquent testimonial of the fact they had just arrived from a long journey and were exceedingly fatigued."

The route of migration takes the cliff swallow over thousands of arduous miles of travel, but nevertheless their arrival is remarkably punctual, sometimes varying but a day or two from the expected time, a fact readily verified by any one who observes the appearance of the birds at the nesting colonies. The arrival of the cliff swallows at the colony located on the walls and arches of the old San Juan Capistrano Mission in southern California has been given great publicity and in recent years has even been the subject of radio broadcasts. It has been claimed that the birds never vary from March 19; even the hour of their arrival has been said to be constant. Ornithologists who have checked these assumptions have found that the swallows are not infallible to such extravagant claims.

Courtship.—The courtship of cliff swallows is carried on simultaneously with their early nesting activities and especially in places where the birds are busily engaged in gathering the mud used in the construction of their homes. The two following accounts of observations made in widely separated sections of the country will serve to illustrate their interesting behavior at such times. Brewster (1938) in his journal of the birds of the Umbagog region vividly describes the activities of cliff swallows he observed in that retreat on June 8, 1909, as follows:

This morning as I was watching the birds I saw two come together in the air and whirl around and around straight down to the ground, where they remained for more than a minute, in what I took to be sexual union, waving and fluttering their wings like butterflies. The other members of the colony seemed actively interested in the affair, and, indeed not a little excited about it, for they collected over the prostrate birds and dashed down almost to them, with loud cries. When the pair finally separated, one bird flew off in one direction and the other in another. I do not think it could have been a fight, for Eave Swallows are among the most peaceable and social of all birds and I have never known them to show the slightest tendency to quarrel.

Wetmore (1920) describes the activities of cliff swallows he observed in New Mexico as follows:

On June 11 they were building nests on the sandstone cliff above the Laguna de la Puerta. The birds came down to the lake shore in little bands of ten or a dozen and alighted close together with trembling wings extended at an angle from their backs, standing high on their legs to avoid soiling their feathers. After alighting they leaned over, filled the mouth with mud with one or two sharp digs and then rose to fly back up the steep slopes to the colony. Males frequently alighted on the backs of the females as they gathered mud and copulation took place while the birds were on the ground.

Nesting.—The nests of the cliff swallow are cleverly constructed of pellets of mud or clay, are roofed over, and generally assume a flask, retort, or bottle shape with a narrow entrance leading into an enlarged chamber. The protruding neck of the nest varies greatly, from some 5 or 6 inches long, or even longer, to others that have no neck, or even to an open structure approaching the barn swallow type of nest. The cliff swallow may strengthen the walls of its plastic home by the use of straws and horsehair, but such materials are not so freely used as they are in the nests of the barn swallow. As a result the former are more friable and are subject to a greater amount of erosion and breaking, especially if they become dampened during times of heavy rainstorms. The chamber of the nest is scantily lined with a few dried grass stems to which a few feathers and other materials are sometimes added.

The cliff swallows are gregarious in their nesting habits, and it is exceptional to find isolated nests far distant from others of the species. The primitive nesting sites of this swallow are in various situations afforded by bluffs, cliffs, and the perpendicular walls of deep gorges in remote mountainous sections of its nesting range, and many of them continue to nest in regions where such situations prevail. Many of the largest colonies, some of them comprising thousands of nests, are located on cliffs. The cliff swallow, however, is a plastic, adaptable species that has been quick to accept new and very different situations provided by civilized man and thus has materially increased its range as well as its numbers. They have accepted the great concrete dams, erected by man in the construction of power projects, which are the nearest approach to the primitive sites used by these cliff dwellers. Today, especially in the eastern parts of its nesting range, this swallow is associated with the eaves of houses and barns, where its friable nests are better protected from rainstorms, but this has also made them subject to greater molestation by such enemies as the English sparrow. In New England it is unusual to find them nesting on cliffs. Although I have visited more than 50 populous colonies in New England and the Maritime Provinces I have seen only one located on a cliff. So general has this change in habitat been established that the name eaves swallow is considered a much more appropriate designation of the species. In cases where the birds have been given human encouragement the colonies located on barns often vie in numbers with those located on cliffs. Christensen (1927) has published a photograph of a colony located on his barn in Newark, Nebr., that contained 800 nests. Other colonies of similar size have been noted. The nests located on buildings are usually on old unpainted structures where it would be expected that the pellets of mud would adhere much better to the rough surface. Nevertheless, painted buildings are utilized in numerous instances.

Not only has the cliff swallow sought out the sites afforded by buildings of isolated farms and camps, but it has even invaded the cities and there built its nests on private homes and public buildings located on busy thoroughfares. Gould (1906) reports that cliff swallows built their nests on the store of Woodman, True & Co. of Portland and also under the eaves of the Portland Savings Bank; and I counted 42 nests located under the gables and eaves of a church at Popham Beach, Maine. Old missions in California, such as the famous San Juan Capistrano, constantly visited by swarms of tourists, are occupied by these confiding swallows. Nor are they adverse to the environment afforded by a large university campus. Grinnell (1937) and others have had much to report concerning the existence and life of a large colony that has taken up its residence on the concrete walls of the Life Sciences Building of the University of California at Berkeley. Burleigh (1930) reports their nesting on one of the buildings of the campus of the University of Washington.

The further adaptability of this bird is shown in its ability to meet a situation when the usual nesting sites are not available. Dawson (1923) writes: "Both Grinnell and Willett have recorded how the Cliff Swallows of Bear Valley, hard put to it for nesting sites in an otherwise delectable country, attached their retorts to the sides and under surfaces of great pine trees." A photograph published by Dawson (1923) shows a colony of nests on a yellow pine taken at Bear Lake in the San Bernardino Mountains of California. The nests are restricted to clusters concentrated on the trunk beneath the larger branches of the tree where they are better protected from rains. Nuttall (1832) states that cliff swallows build in trees along the Columbia River.

When the cliff swallows resort to buildings they usually plaster their nests under the protecting eaves on the outside of the structures but not infrequently they may enter the building. Evermann (1886) writes: "A colony of more than a hundred pairs nested in a shed in Santa Paula (California). The nests were fastened to the rafters, much after the manner of the Barn Swallow. Many horse-hairs were plastered into the nests and these often caused the death of the builders. I took from this shed some six or eight birds which I found hanging about the nests, they having gotten entangled in the hairs." Johnson (1900) found the retort-shaped nests against or under the beams inside a barn. Some of the nests were built against the hay which served as a back for the nest. Johnson states: "The season was a very wet, rainy one which accounts for the change from eaves outside, where they usually built." McCann (1936) found cliff and barn swallows nesting together, one nest of the cliff swallow was within 12 inches of a barn swallow's nest. Goodsell (1919) also noted cliff and barn swallows nesting together using old foundations of one another's nests. Whittle (1922) describes an interesting case in which a cliff swallow utilized a robin's nest built on two telephone wires entering a building. The nest of the swallow was not in contact with the building but merely began at the rim and domed over the old robin's nest. At Frenchman's River, Canada, Potter (1932) discovered that cliff swallows had taken possession of the interior of an old freight car.

Not only do these birds select sites on the outside and inside of buildings and various structures but they may also nest underneath them. An enormous colony of cliff swallows nested on the piling under an old building at the edge of the water at Clear Lake, Calif. Kobbé (1900) found many cliff swallows nesting in caves on the ocean side of Cape Disappointment, Wash. Dice (1918) found the nests in a road tunnel under the railroad tracks near Lamor, Wash. Townsend (1917) states that cliff swallows have been known to breed in the abandoned burrows of bank swallows, and Carpenter (1918) found several cases of typical bottleneck mud nests built over entrances to old rooms of bank swallows at Oceanside, Calif. In one instance the eggs of the cliff swallow were found at the end of a 2-foot tunnel lying in a typical seaweed nest of the bank swallow.

The cliff swallow has also proved its great versatility in coping with unusual situations. Hales (1904) tells of an attempt of cliff swallows to build under the eaves of a barn where there was no projecting object to serve as a foundation for a nest, which resulted in repeated failures. One resourceful pair took advantage of a weatherstrip over a door and built an enormous cylinder tube, which they filled with straw to a height of 18 inches, allowing just enough room for a beautifully shaped nest at the top. The birds were awarded success in bringing forth their brood of young. The following two instances further illustrate how the cliff swallow is able to meet with unusual conditions. Reed (1927) relates a case of a cliff swallow's nest built over a back door of a home where there was a constant traffic of persons and a slamming of the screen door. After the young had hatched the nest crumbled and fell with the young to the floor. The young were placed in a strawberry box hung up where the nest had been plastered. The adult birds at once proceeded to cover the box with mud to protect and conceal the young, which were successfully reared. The following year the pair returned and repaired the strawberry basket, but after an egg was mysteriously destroyed they made over a phoebe's nest and successfully reared their brood there. Wright (1924) cites a similar case in which the young of a fallen nest were placed in a tin can nailed to the barn. The birds returned at once and built a neck of a nest over the opening of the can.

Cliff swallows are not averse to nesting in the proximity to the nests of other larger birds. Forbush (1929) records a case where the nests were built among the outer sticks of a great blue heron's nest. Coues (1878), Taverner (1919), and Forbush (1929) cite cases where nests of the cliff swallow were built on the cliffs about prairie falcons' nests. Taverner also found them about a duck hawk's nest. Apparently they lived in harmony and were not disturbed by these larger predacious species.

It has been observed by Taverner (1928) and others that the nests of large colonies of cliff swallows located on cliffs are situated in sites least affected by heavy rains. Mr. Taverner's account of such a colony located on the cliffs of the Red Deer River, Alberta, Canada, is as follows:

This river bed is sunk from two to three hundred feet below the main prairie level and in many cases the valley walls approach the river bank in irregular, and more or less sheer, cliffs. Many wall spaces thus formed are sites of large colonies of nesting cliff swallows that plaster their gourd-like nests closely together making continuous mud-incrustations over considerable surfaces. The boundaries of these aggregations of nests are often very definite but to the casual observer often erratically arbitrary in outline and extent. Many groups of nests are obviously under sheltering overhangs, but others seem well out into the open and subject to the inclemencies of every weather. Often there is no apparent reason why nests should be huddled closely together as if space were very precious and then cease at an imaginary line beyond which conditions seem equally, or even more, desirable.

When the rains come, however, darkening the exposed faces of the cliffs with their wetness, much of the mystery is explained. In practically every case it is then seen that the nest colonies occupy only the dry spots of the irregular, and generally wet, surfaces and that the soluble, fragile nest-structures often cease almost on the line of moisture. Most of these colonies seem to be occupied only for a single season or a short series of years and new sites are selected at frequent intervals; consequently there are everywhere old and deserted nest groups in various stages of delapidation whose obvious age bespeaks their permanency and the good judgment with which they were founded as regards prevailing weather, wind and rain. One mild shower and wetting would be sufficient to dissolve their clayey structure into its constituent gumbo to drop with unctuous splash to the talus below or to flow away in stalactites of sluggish mud.

It does not seem that this safety of situation is generally achieved by a system of trial and error for few ruins of recent and obvious errors are noted as would be were that the case. By some means Cliff Swallows after nesting under such condition for countless generations have evolved methods of nest-site selection that nearly unerringly pick out amid the multitudinous wall exposures and tricky wind currents of the canyons the safe situations. There is no necessity here at least to defer nest making until a mud-making rain supplies structural material and coincidentally marks out the safety zones on the cliffs for the river supplies them constantly, irrespective of weather, a source of the choicest mud, and building can be undertaken at any time. Swallows also generally like to carry on their work in bright sun and a drying atmosphere that fixes the growing sub-structure before further accretions are added. It may be however that the site is decided upon during or immediately after a shower and perhaps the foundation is outlined before the drying cliffs lose their tell-tale moisture and is then left to be completed in better building weather. However it is, whether occult instinct or empirical methods guide the birds, there is here shown a most interesting adaptation to environment.

The activities about a colony of cliff swallows during the time of nest building are fascinating. Such a scene has been admirably described by Coues (1878) in his account of the cliff swallows observed in the Colorado Valley:

Suddenly they appear—quite animated and enthusiastic, but undecided as yet; an impromptu debating society on the fly, with a good deal of sawing the air to accomplish, before final resolutions are passed. The plot thickens; some Swallows are seen clinging to the slightest inequalities beneath the eaves, others are couriers to and from the nearest mud-puddle; others again alight like feathers by the water's side, and all are in a twitter of excitement. Watching closely these curious sons and daughters of Israel at their ingenious trade of making bricks, we may chance to see a circle of them gathered around the margin of the pool, insecurely balanced on their tiny feet tilting their tails and ducking their heads to pick up little "gobs" of mud. These are rolled round in their mouths till tempered, and made like a quid into globular form, with a curious working of their jaws; then off go the birds, and stick the pellet against the wall, as carefully as ever a sailor, about to spin a yarn, deposited his chew on the mantel-piece. The birds work indefatigably; they are busy as bees, and a steady stream flows back and forth for several hours a day, with intervals for rest and refreshment, when the Swallows swarm about promiscuously a-flycatching. In an incredibly short time, the basement of the nest is laid, and the whole form becomes clearly outlined; the mud drys quickly, and there is a standing place. This is soon occupied by one of the pair, probably the female, who now stays at home to welcome her mate with redoubled cries of joy and ecstatic quivering of the wings, as he brings fresh pellets, which the pair in closest consultation dispose to their entire satisfaction. In three or four days, perhaps, the deed is done; the house is built, and nothing remains but to furnish it. The poultry-yard is visited, and laid under contribution of feathers; hay, leaves, rags, paper, string—swallows are not very particular— may be added; and then the female does the rest of the "furnishing" by her own particular self. * * *

Seeing how these birds work the mud in their mouths, some have supposed that the nests are agglutinated, to some extent at least, by the saliva of the birds. It is far from an unreasonable idea—the chimney swift sticks her bits of twigs together, and glues the frail cup to the wall with viscid saliva; and some of the Old World Swifts build nests of gummy spittle, which cakes on drying, not unlike gelatine. Undoubtedly some saliva is mingled with the natural moisture of the mud; but the readiness with which these Swallows' nests crumble on drying shows that saliva enters slightly into their composition— practically not at all—and that this fluid possesses no special viscosity. Much more probably, the moisture of the birds' mouths helps to soften and temper the pellets, rather than to agglutinate the dried edifice itself.

The details of mud gathering and nest building are described by Audubon (1831):

About day-break they flew down to the shore of the river, one hundred yards distant, for the muddy sand of which their nests were constructed, and worked with great assiduity, until near the middle of the day, as if aware that the heat of the sun was necessary to dry and harden their moist tenements. They then ceased from labour for a few hours, amused themselves by performing aerial evolutions, courted and caressed their mates with much affection, and snapped at flies and other insects on the wing. They often examined their nests to see if they were sufficiently dry, and as soon as they appeared to have acquired the requisite firmness, they renewed their labours. Until the females began to sit, they all roosted in the hollow limbs of the Sycamores (*Platanus occidentalis*) growing on the bank of the Licking River, but when incubation commenced, the males alone resorted to the trees. A second party arrived, and were so hard pressed for time, that they betook themselves to the holes in the wall, where bricks had been left out for the scaffolding. These they fitted with projecting necks, similar to those of the complete nests of the others.

Brown (1910) describes briefly but well the mud-gathering habits of cliff swallows he observed at Grand Pré, Nova Scotia:

In one place was a trench dug some five feet deep, and with a most inviting bed of soft sticky clay at the bottom. The Swallows were making the most of the opening of such a mine, and, through the entire forenoon that I observed them, they flocked in numbers and worked most conscientiously. * * * They came in eager succession, fluttering down, feet dropped, ready to settle lightly on the soft mud. The moment the feet touched ground, the body and tail were well up, so as not to soil those sleek feathers, and the wings extended straight over the back, continually fluttering to keep the feet from sinking or sticking. Mouthfuls of clay were quickly gathered, the wings continually shaking, and soon the Swallow was off. Every one was busy, mostly mindful only of his own affairs but now and then a tiff occurred, where two wanted the same spot. Every newcomer called softly, and those flying above and across were musically happy.

Knowlton (1881) made some interesting observations on cliff swallows at Brandon, Vt., in which it is clear that in the life of the swallows incidents occur that parallel those in a human society. He says:

One day, while watching them, I noticed one bird remained in her half-finished nest, and did not appear to be much engaged. Soon a neighbor, owning a nest a few feet away, arrived with a fresh pellet of clay and, adjusting it in a satisfactory manner, flew away for more. No sooner was she out of sight than the quiet bird repaired to the neighbor's nest, appropriated the fresh clay and moulded it into her own nest! When the plundered bird returned, no notice was taken of the theft, which was repeated as soon as she was again out of sight. I saw these movements repeated numerous times, but was called away, and when I again returned both nests were completed.

In the same place a nest remained undisturbed, and was occupied by probably the same pair of birds for several seasons. This spring they returned to the old nest, and all appeared prosperous, until one day I noticed a number of Swallows engaged in walling up the entrance of this old nest. This, and the outline of a new nest over the old, was soon completed. I then broke open the closed nest and found within the dead body of a Swallow. This bird had probably died a natural death, and the friends being unable to remove the body, and knowing it would soon become offensive, adopted this method of sealing it up.

Occasionally three swallows may be seen engaged in building one nest. E. O. Grant (Forbush, 1929) watched a nest at Long Lake, Allegash, Maine, that was built by three birds, and he says that they all took turns in incubating the eggs. He believed them to be two males and a female. Brewster (1906) in his study of cliff swallows in the Cambridge region of Massachusetts found as a rule two birds in each nest, but several of the nests he examined sheltered three birds each. Brewster concludes that there is good reason to suspect that cliff swallows sometimes practice polygamy or polyandry.

The time required to build a nest varies a great deal with different pairs of birds, at least as reported by various observers. Knight (1901) states that 10 to 14 days are required to finish the nest. In my own experience in watching the building of a nest in a colony at Kent Island, Bay of Fundy, 5 days elapsed from the time the first pellets of mud were applied to the time the nest was ready for its lining. Three days later the first egg was laid. Ordinarily it requires a week for nest construction but the weather conditions, mud supply, and amount of disturbance are factors that will materially affect the length of time.

Most all observers agree that cliff swallows in general raise two broods of young during any one breeding season. Hatch states that even three broods are sometimes reared, but I am inclined to believe that is very exceptional.

The incubation period varies from 12 to 14 days according to various authorities. The time that elapsed from the date of the laying of the last egg to the hatching of the first young in a nest which I observed at Kent Island was 13 days.

Eggs.—[AUTHOR'S NOTE: The number of eggs in a complete set varies from three to six, but four or five eggs constitute the usual complement. They vary in shape from ovate to elliptical-ovate, or rarely to elongate-ovate. The ground color is white, creamy white, or pinkish white. Some eggs are evenly, either sparingly or thickly, marked with fine dots or small spots, or a mixture of both; and some have a few small blotches. In some the markings are concentrated at the larger end or coalesced elsewhere into a dense wreath. The markings are in various shades of light and dark browns, or "brownish drab," with underlying spots or small blotches in the paler shades of "Quaker drab." Very rarely the eggs are entirely unmarked. The measurements of 50 eggs average 20.3 by 13.9 millimeters; the eggs showing the four extremes measure 22.9 by 14.2, 22.4 by 15.2, 17.3 by 13.2, and 18.8 by 12.7 millimeters.]

Plumages.—Brewster (1878) has described the juvenal or first plumage from a male specimen taken at Upton, Maine, July 27, 1874: "Top of head, back and scapulars dark brown; collar around nape, dull ashy, tinged anteriorly with rusty. Rump as in adult, but paler; forehead sprinkled with white, and with a few chestnut feathers. Secondaries broadly tipped with ferruginous. Throat white, a few feathers spotted centrally with dusky. Breast and sides ashy, with a rusty suffusion, most pronounced on the latter parts. A very small area of pale chestnut on the cheeks." This plumage is retained by the birds until they have migrated to their winter quarters.

According to Dwight (1900) the first winter plumage is acquired by a complete postjuvenal molt in the winter habitat. When the birds return the following spring wear is evident in the wings and tail although the resistant metallic feathers show little of it. In this plumage the birds acquire a glossy blue head and back and the rich chestnut of the chin and auriculars and a black throat spot. The breast and throat feathers are streaked, and they have a conspicuous crescent on the forehead.

The first nuptial plumage is acquired by wear. The adult winter plumage is acquired by a complete postnuptial molt in the winter quarters, as was the first winter plumage. This plumage is also similar to the first winter plumage, and the sexes are similar.

Since the cliff swallow is closely associated with other swallows it is not surprising that occasionally hybrids appear, although it is truly remarkable that hybridism occurs here in birds not only of different species but also of different genera. Mearns (1902) describes a hybrid between a barn swallow and a cliff swallow taken June 14, 1893, at Fort Hancock, Tex. He found a pair of swallows that were mated and had almost completed a nest attached to a rafter of a barn; the nest was similar to that of the barn swallow, having the entrance at the top. Both birds were shot; the male was a typical barn swallow, but the female, which was about to lay eggs, was a hybrid between the barn and cliff swallows. The characters were intermediate between those of the two species. Arthur H. Norton informs me that he observed a hybrid between the barn and cliff swallow in a colony near Portland, Maine, in 1883.

Chapman (1902) describes a very interesting male hybrid between the cliff and tree swallows taken by Leon C. Holcomb at Springfield, Mass., on August 20, 1902. It was a bird of the year, and in addition to presenting hybridism it also exhibited albinistic characters, though Chapman states this may have been the result of hybridity. In general this hybrid resembled the tree swallow below and the cliff swallow above, the rusty and buff markings of the cliff swallow, however, being in this supposed hybrid white. Chapman gives a complete comparative description of normal adults and hybrid and then comments as follows:

It is of course well known that in the Tree Swallow both birds of the year and adults moult before leaving us for the South while the Cliff Swallow migrates before moulting. It is consequently of interest to observe that in this hybrid moult has begun normally with the innermost primaries.

This fact is also of importance in determining the bird's age and, in connection with the unworn condition of the wing-feathers, it leaves no doubt that the specimen is in post-natal plumage.

The radical differences in the character of the nests of the supposed parents of this bird lead one to speculate on the type of nest-structure in which it was reared, but, unfortunately, our curiosity in this direction cannot be gratified.

Food.—Beal (1918) has reported on the examination of the stomach contents of 375 swallows taken in every month from March to September. The food consisted almost entirely of animal matter; the small amount of vegetable matter (0.66 percent) was taken accidentally with food. However, two stomachs of birds taken in May in Texas were entirely filled with the fruit of *Juniperus monosperma*, which was undoubtedly taken intentionally as food.

The animal food consists of insects with a few spiders. The insects consisted of beetles (26.88 percent), and of these members of the May-beetle family, chiefly the small dung beetles, which are easily taken on the wing as they fly in swarms near the ground, are important. Snout beetles or weevils, including the destructive cotton-boll weevil, were taken every month in large quantities, but in September weevils constituted 50 percent of the food. Thirty-five cliff swallows collected in the vicinity of cotton fields in Texas contained 687 boll weevils, an average of 19 to each stomach. Besides those mentioned above Beal found many other beetles, representing 113 species, in the stomach contents of the 375 cliff swallows. McAtee (1926) found the chestnut, chinquapin, and acorn weevils of the genus *Balanimus* and also *Ops pinus*, a special enemy of the white pine, during his special studies of the relation of birds to woodlots in New York State.

Beal found that ants, mostly winged forms, were eaten every month except March. Other Hymenoptera eaten were wild bees and wasps and some parasitic species. The remains of 35 honeybees (*Apis mellifera*) were identified in 13 stomachs.

Hemiptera comprised 26.23 percent of the food, the most important representative being the well-known and harmful chinch bug (*Blissus leucopterus*).

Diptera comprised 13.95 percent. Green flesh flies were present in large numbers, but crane flies, which are eaten by other swallows, were not found in the stomachs of the cliff swallows.

Lepidoptera were not important; very few birds are fond of adult butterflies or moths though they may relish their larvae. Orthoptera, according to Beal, are lightly regarded by the cliff swallow, but Bryant (1914) includes the cliff swallows as destroyers of grasshoppers in California. He reports that an individual cliff swallow eats on the average of three grasshoppers a day. Beal found a few other insects, such as dragonflies, Mayflies, and lace-winged flies, represented in the food. McAtee reports that 18 specimens of the southern corn rootworm (*Diabrotica duodecimpunctata*) in the stomach of a single cliff swallow. Murie (1935) examined a cliff swallow killed by an automobile; it had its crop filled with brine flies (*Ephydra millbrae*). The cliff swallow is one of the birds responsible for the destruction of mosquitoes.

It is evident from the above determinations that the cliff swallow is distinctly beneficial to the best interests of man as far as its food habits are concerned. It is also obvious from the nature of the insects eaten that the cliff swallows capture the greater part of their food during flight.

Beal (1907) in a special investigation conducted in California identified the food of 22 nestlings varying in age from two days to those just leaving the nest. They were taken from May 30 to July 2, inclusive. Beal found animal matter in the food of the young to be similar to that eaten by the adults but differing in its relative proportions. Soft-bodied insects, such as Hymenoptera and Diptera, ranked highest, amounting to three-fourths of the food eaten by the young. Hemiptera and Coleoptera, hard-bodied insects, which are freely eaten by the adults, represented a relatively small percentage of the food of the young. An unaccountable difference existed in the fact that a considerable amount of gravel was taken from the stomachs of the young, while none was found in the stomachs of the adult cliff swallows examined.

Brown (1910) made observations of the feeding activities at colonies of cliff swallows in Washington County, Maine. He counted the number of visits of the adults and, finding that they carried on an average of three insects each trip, estimated that 900 insects were destroyed on an average for each day the young were in the nests. Such figures should convey a distinct meaning to those who may be ignorant of or doubt the value of the cliff swallow in its relations to the farmer on whose buildings it seeks a nesting site.

Voice.—The swallows are not distinguished as far as their songs are concerned, and in this the cliff swallow is no exception. Roberts (1932) states: "The notes consist of a rapid twittering or chattering, sometimes rather harsh and creaky, but during the nesting-season these acquire a more musical quality and fill the place of a more pretentious nuptial song." In Baird, Brewer, and Ridgway (1905) the song is described as "an unmusical creak, rather than a twitter, frequent rather than loud, and occasionally harsh, yet so earnest and genial in its expression that in effect is far from being unpleasant." In an unpublished manuscript S. S. Dickey states: "These sounds are less pronounced and are evener in their outflow than are the sharper ones of the barn swallow. There is not so much metallic 'squeak' in their quality as that of the tree swallow."

Enemies.—The most serious enemy of the cliff swallow, especially of those nesting on barns and other buildings, is the English sparrow. The ousting of cliff swallows from their nests by English sparrows and their importance as a factor in the local fluctuations of the number of these swallows have been a subject of comment by scores of observers of which the following are typical:

Brewster (1906) states "the Eave Swallow suffers directly and very seriously from the encroachments of the House Sparrows who destroy its eggs and young and take possession of its nests whenever opportunity offers." Barrows writes as follows: "About the larger cities and towns in Michigan the English Sparrow has been a potent factor in reducing the numbers of Cliff Swallows. The mud nests of swallows form convenient receptacles for the eggs of Sparrows and they often take possession of the nests and drive the swallows away entirely." Forbush (1929) writes: "Thus the 'Cliff' Swallows under man's protection became 'Eaves' Swallows and waxed fat and numerous until the decade beginning in 1870, when the House Sparrow began to increase and spread over New England. Then the nests built by the industrious Cliff Swallows were appropriated, after a struggle, by the swarming and ubiquitous Sparrows, whose clumsy and bustling occupancy soon resulted in the destruction of their stolen domiciles. As the Sparrows increased in southern New England, they spread northward and eastward until the greater part of the Cliff Swallows had been driven into Maine." In Maine, where I have had an opportunity to observe many nesting colonies, there has been but little evidence of the molestation by English sparrows. It is becoming more and more apparent that, as the English sparrow population decreases, the cliff swallows are again becoming established where formerly they were ousted by the marauders. In the far West the English sparrow still remains a menace of increasing importance.

Grinnell (1937) gives a very interesting account of the invasion of the English sparrows into a large colony of cliff swallows that built their nests on the concrete walls of the Life Sciences Building at Berkeley, Calif. This building, completed in 1929 was first occupied by cliff swallows in 1935. This colony was closely observed by Grinnell and others, who reported on its progress from time to time. In 1936 the sparrows arrived and Grinnell's comments are as follows:

On May 27, I first became aware of the presence of English Sparrows (*Passer domesticus*). At the southeast corner of the building I saw males of this aggressive species behaving as if evicting cliff swallows from the latter's nests. Male sparrows, calling loudly, were either perched on the coping under the nests or were actually ensconced within the entrances of the swallows' nests. Furthermore, on the southeast steps at 1 p. m. of May 27, I found remains of eggs. On May 30, at 1 p. m., in the same location, I saw on the steps two dead young swallows of about hatching age, a splashed, incubated egg, and many fragments of eggshells—the latter definitely of the swallows because of the umber spotting on the shell fragments. Directly above, along the row of 48 swallows' nests, were several male English Sparrows. * * *

When I made a census of swallows' nests on June 4, I counted 11 English Sparrows, 7 of which were males perched on the coping near swallows' nests from which protruded wisps of nesting material; the birds were singing, if English Sparrows can be said to sing, as though sitting females were inside certain of the nests. Again and again, a sparrow was seen to enter a swallow's nest; but I did not see a swallow and a sparrow enter the same nest. On the southeast steps, I saw more, dead, naked, seemingly just hatched swallows, and also two more splashed eggs. The same sort of observations were made on June 5 and 6. On the latter date, sparrows were seen carrying nesting material into swallow's nests at the southwest corner of LSB, so that all of the three corners of the building patronized by the swallows had also attracted sparrows. * * *

I was amazed at the aggregate large numbers of English Sparrows around LSB, where before 1936 I could not recall ever having seen any, since its construction, in the summer time. Again we have illustrated how the fortunes of one kind of animal may be influenced favorably by some circumstance in the economy of another. The swallows, by furnishing appropriate nest sites, brought the sparrows into a territory new for them. Partial supplantation, or succession, was in evidence.

Dayton Stoner (1939b) cites an interesting case, which he terms parasitism, in which English sparrows confiscated the nest of a cliff swallow and later were driven away or abandoned it after the female sparrow had deposited an egg. Three eggs of the cliff swallow were added. The sparrow egg hatched first, followed several days later by the hatching of the swallow eggs. The young swallows perished, but the young sparrow thrived through the diligent care of the foster parents, the cliff swallows.

Friedmann (1931) writes as follows concerning the parasitism by the cowbird: "Mr. J. Hooper Bowles informs me that he has a parasitized set of four eggs of the Swallow and one of the Cowbird from La Anna, Pennsylvania, June 30, 1914, and that three nests of this Swallow [Cliff Swallow] at that place contained Cowbird's eggs."

Cliff swallows, which nest in situations well above level ground, would not be expected to be molested by snakes, but nevertheless cases on record indicate that snakes may sometimes prove to be a menace to nesting birds. Sawyer (1907) reports that an adder was discovered under the eaves of a barn where it was observed to be busily engaged in devouring cliff swallow eggs, and Cameron (1908) writes that a rattlesnake climbed the veranda poles of the Cross S ranch on Mispah Creek and devoured all the nestlings in reach.

Ants have proved very destructive to cliff swallows and other forms of life at San Diego, Calif. According to Anthony (1923) a colony of cliff swallows that had nested in one of the towers at the San Diego Museum was destroyed by the so-called Argentine ant introduced presumably from South America. Twenty-five dead nestlings were found in the deserted nests examined. The young were covered with the ants, while a steady column of the insects marched from the top of the tower to the ground.

The cliff swallows are more or less infested with Mallophaga and other external parasites, but so far as I know no comprehensive study has been made of the parasites and diseases of the cliff swallow. Knight (1908) states the nests often contain bird lice, ticks, and bed bugs. Several observers have reported finding bed bugs in the nests of cliff swallows. Cameron (1908) states that in Montana the ranch owners destroy cliff swallows because their presence brings bed bugs into the houses. Thus indirectly through man bed bugs become destructive to the best interests of the cliff swallow. According to some observers the prejudice against the cliff swallow in certain sections of the Middle West is so great that these birds may have to return to their primitive nesting sites on cliffs if they are to be perpetuated. Such an attitude toward a very beneficial species of bird is unfortunate and unwarranted. (See Widmann, 1907, p. 201.)

Predaceous birds cannot be considered as serious enemies of the cliff swallow. Indeed there are instances where the cliff swallows have nested in apparent harmony with prairie falcons and duck hawks. The appearance of a hawk in the vicinity of a colony of cliff swallows never creates any evidence of excitement, whereas other passerine birds may exhibit great concern.

Unseasonable weather, especially sudden drops in temperature after the swallows have arrived in spring, have sometimes proved destructive. Audubon (1831), writing concerning cliff swallows observed at Henderson on the Ohio River in the spring of 1815 states: "It was an excessively cold morning, and nearly all were killed by the severity of the weather." Kimball (1889) made an investigation of the mortality of cliff swallows in June 1889 at Rockford, Ill., after it was discovered that very few of the birds were in evidence although they had come in their usual numbers earlier in the season. He says:

An examination of the newly completed nests revealed dead birds in nearly every nest. Large numbers were also found dead on the ground in the vicinity of the buildings frequented by them. Twenty-two nests were examined on one barn, about six miles northeast of this city, and thirty-seven dead birds were found in the nests. About two miles from this barn one hundred nests were examined on a large barn and dead birds were found, one or more in each nest. Five or six miles northwest of the city a like condition of affairs was reported. An investigation in a section of country six or eight miles north and fifteen to twenty miles south of the city revealed a similar destruction of Eave Swallows; the ground about the barn on one farm was reported "covered with dead birds."

Mr. Kimball attributes this great mortality to the unusual weather conditions prevailing in that section of the country. There was a week of warm weather at the time the swallows arrived, followed by a prolonged cold period. The warm weather probably brought the birds north in advance of their usual time. Kimball estimates that in the northern section of Illinois over 90 percent of the swallows died.

From the foregoing and other similar reports it is evident that unseasonable cold weather may prove most disastrous to the cliff-swallow population. In addition to the cold to which the swallows are subjected and to which they are not adapted they also suffer privation from the lack at such times of their all essential food, flying insects.

Fall.—After the nesting season the adults and offspring of a colony cling together for some time as a group. It is then that their presence becomes conspicuous and attracts attention even from the casual observer, who is impressed by their great numbers as they congregate on telephone wires, buildings, or treetops. It is a common experience when traveling along the New England highways late in July or early in August to encounter many groups ranging in size from a few dozen to 75 or 100 birds. Later the colony groups join with others of the same and other species of swallows to roost in favorable traditional places creating large concentrations numbering many hundreds and even thousands of individuals.

Jackson (1923) saw 300 cliff swallows at Bent's Camp, Mamie Lake, Wis., from August 23 to 28, 1917. On August 29 there were 40, and by August 31 there were only six left. Frances Schneider (1922) saw 1,000 cliff swallows congregated near a ranch reservoir near Artesia, Calif., on July 26, 1922. She noted another assemblage at the same place numbering several hundreds on July 29. According to A. S. Allen (1921) enormous numbers of cliff and other swallows assemble each year in the Suisun Marshes, Calif., after the middle of July. Cooke

(1888) reported "thousands and thousands" of cliff swallows resting in the latter part of July on a marsh near Red Rock, Indian Territory (Oklahoma). Widmann (1907) writes: "The only time the species is present in great numbers is from the middle of August to the middle of September. At this period of southward migration thousands and thousands gather at night at the common roosts in the Spartina marshes of north Missouri. All are gone before the end of September."

Such flocks as mentioned above do not migrate en mass but travel in relatively small groups. Cooke (1914) made extensive observations of the migration of cliff swallows at Caddo, Okla., late in August and early in September when swallows were in sight almost continuously moving southward. Late in August the flight of swallows began about at 5:30 A. M. and lasted an hour. On September 1 the first group passed at 5:10 and the last at 6:10. At 8 A. M. on September 25 about 30 cliff swallows passed in one flock going rapidly south; and for the next two weeks cliff swallows were seen about one-half of the evenings and one-third of the mornings in numbers from five birds to 200, about nine-tenths of them heading straight south and the rest flying about in search of food. The last cliff swallow was seen on October 9.

DISTRIBUTION

Range.—North and South America.

Breeding range.—The breeding range of the cliff swallow extends **north** to Alaska (Holy Cross, Rampart, and Bettles); northern Mackenzie (Rat River, Fort Goodhope, Lockhart River, Kendall River, and Artillery Lake); northern Manitoba (Cochrane River and possibly rarely Churchill); northern Ontario (Martin Falls and Moose Factory); Quebec (Godbout, Mont Louis River, Percé, and the Magdalen Islands); and Nova Scotia (Baddeck). The **eastern** limits extend south along the Atlantic coast from Nova Scotia (Baddeck, Halifax, and Barrington) to Virginia (Aylett). **South** to central Virginia (Aylett, Lexington, and Blacksburg); northern Alabama (Fort Deposit); southern Texas (Corpus Christi and Fort Brown); Puebla (Atlixco); Jalisco (La Barca); Durango (Rio Sestin); and northern Baja California (San Quintin). The **western** limits extend north along the Pacific coast from northern Baja California (San Quintin and San Isidro), to Alaska (Mount McKinley, Flat, and Holy Cross).

Winter range.—The winter range is imperfectly known but available information indicates that it extends from southern Brazil (Itarare) and northern Argentina (Tucuman) south to central Argentina (Cape San Antonio, Flores, and Zelaya).

The range as outlined is for the species of which three subspecies are currently recognized. The typical form, the northern cliff swallow (*Petrochelidon albifrons albifrons*), is the race breeding in Canada, on the west coast of Mexico, and in the United States south to Arizona, New Mexico, and Texas; the lesser cliff swallow (*P. a. tachina*) breeds in southern Texas and eastern Mexico; the Mexican cliff swallow (*P. a. melanogaster*) occupies the Mexican tableland north to southern Arizona and New Mexico.

Spring migration.—Early dates of spring arrival in the United States and Canada are: Florida—Coconut Grove, March 4. Georgia—Macon, April 11. District of Columbia—Washington, April 10. Pennsylvania—Coatesville, April 17. New York—Geneva, April 15. Connecticut—Hartford, April 17. Massachusetts—Boston, April 19. Vermont—Bennington, April 13. Maine—Ellsworth, April 21. New Brunswick—Grand Manan, April 14. Quebec—Montreal, April 19. Kentucky—Bowling Green, April 2. Missouri—St. Louis, April 10. Illinois—Chicago, April 11. Indiana—Bloomington, April 12. Ohio—Columbus, April 14. Michigan—Ann Arbor, April 11. Ontario—London, April 19. Iowa—Davenport, April 7. Wisconsin—Milwaukee, April 7. Minnesota—Minneapolis, April 13. Manitoba—Aweme, April 28. Texas—Kerrville, April 4. Kansas—Manhattan, April 13. Nebraska—Red Cloud, April 23. South Dakota—Yankton, April 21. North Dakota—Larimore, April 30. Saskatchewan—Indian Head, May 6. New Mexico—State College, April 4. Arizona—Tucson, April 5. Colorado—Loveland, April 24. Wyoming—Wheatland, May 1. Montana—Big Sandy, May 2. Alberta—Banff, May 8. Mackenzie—Peace River, May 20. California—Los Angeles, February 26. Oregon—Narrows, March 11. Washington—Tacoma, April 4. British Columbia—Okanagan Landing, April 19. Alaska—Nulato, May 10.

Fall migration.—Late dates of fall departure are: Alaska—Ramparts of Porcupine River, August 28. British Columbia—Okanagan Landing, September 15. Washington—Seattle, October 9. Oregon—Klamath County, September 18. California—Kernville, October 28. Mackenzie—Pelican Rapid, August 24. Alberta—Banff, September 3. Montana—Fortine, September 8. Wyoming—Yellowstone Park, September 6. Colorado—Denver, October 3. Arizona—Salt River Bird Reservation, October 14. New Mexico—junction of Beaver Creek and the Gila River, October 16. Saskatchewan—Eastend, September 6. South Dakota—Sioux Falls, September 10. Nebraska—Valentine, October 4. Kansas—Onaga, October 8. Texas—Austin, October 1. Manitoba—Aweme, September 24. Minnesota—Jackson County, October 8. Wisconsin—Madison, October 4. Iowa—Giard, September 30. Ontario—Ottawa, September 30. Michigan—Sault

Ste. Marie, September 27. Ohio—Youngstown, September 27. Illinois—Rantoul, September 19. Missouri—Concordia, October 6. Tennessee—Athens, September 26. Mississippi—Biloxi, October 23. Prince Edward Island—North River, September 4. New Brunswick—St. John, September 9. Maine—Phillips, September 9. Vermont—Wells River, September 15. Massachusetts—Woods Hole, September 28. Pennsylvania—State College, October 3. District of Columbia—Washington, September 7. North Carolina—Lake Eden, September 12. Alabama—Booth, September 20. Florida— St. Augustine, November 11; Key West, November 27. One banded (34–95949) at Dell Rapids, S. Dak., on June 14, 1937, was recaptured at Ghent, W. Va., on July 16, 1937.

Casual records.—A specimen was taken by a native on St. Paul Island, Alaska, about June 10, 1918.

Egg dates.—Arizona: 8 records, June 3 to August 10.

California: 109 records, April 27 to July 5; 55 records, May 8 to June 20, indicating the height of the season.

Illinois: 16 records, May 20 to July 5; 8 records, May 22 to June 20.

Massachusetts: 13 records, May 30 to June 20; 7 records, June 2 to 11.

Pennsylvania: 10 records, May 31 to June 17.

Texas: 60 records, April 15 to July 19; 30 records, May 6 to June 12.

PETROCHELIDON ALBIFRONS TACHINA Oberholser

LESSER CLIFF SWALLOW

HABITS

Dr. H. C. Oberholser (1903), in naming this swallow, describes it as similar to the northern cliff swallow but "decidedly smaller, the forehead ochraceous instead of cream color." He says further that it "is intermediate between *lunifrons* [=*albifrons*] and *melanogastra*, approaching in size very close to the latter. In respect to the color of the forehead, as well, its aberration from *lunifrons* is in the direction of *melanogastra*, with which also it may be found to intergrade. After due allowance has been made for individual variation which, however, does not exist to an unusual degree, the characters exhibited by this new race seem to be very constant, at least in the considerable series available for examination."

The "considerable series" he refers to seems to have consisted of 7 specimens of *tachina* and 8 of *melanogastra*, not a very impressive number. Mr. Ridgway (1904) recognized both forms on what was practically the same series, with one more *tachina* and two more *melanogaster* (=*melanogastra*). These two forms are practically alike in size, the only conspicuous difference being in the color of the forehead. Mr. Ridgway calls this "fawn color, dull cinnamon, or wood brown" in *tachina*, and "chestnut or cinnamon-rufous" in *melanogaster*.

The breeding range given for *tachina*, in the 1931 Check-list, is "western Texas, the Rio Grande Valley, and through eastern Mexico to Vera Cruz."

The general habits of the lesser cliff swallow apparently do not differ materially from those of the northern cliff swallow, except that it seems more inclined to nest under primitive conditions, on cliffs and canyon walls, and less inclined to nest on buildings than the northern bird.

The eggs of the lesser cliff swallow are similar to the eggs of the other cliff swallows. The measurements of 58 eggs, reported by Col. John E. Thayer (1915), average 20.2 by 13.7 millimeters; the eggs showing the four extremes measure 23.1 by 14.5, 20.7 by 14.8, and 17.5 by 13.0 millimeters.

PETROCHELIDON ALBIFRONS MELANOGASTER (Swainson)

MEXICAN CLIFF SWALLOW

HABITS

The Mexican cliff swallow is readily separable from the northern cliff swallow by its smaller size and by its "chestnut" or "cinnamon-rufous" forehead and rump; but it is very similar to the lesser cliff swallow in size, and the latter is intermediate between the other two races in the color of the forehead. The breeding range of the Mexican cliff swallow extends from southern Arizona and southwestern New Mexico southward over the Mexican tableland to Guatemala.

Nesting.—While I was collecting with Frank C. Willard along the San Pedro River in the vicinity of Fairbanks, Ariz., we visited the home ranch of the Boquillas Cattle Co. to hunt for swallows' nests in their large 2-story barn. We found a small colony of barn swallows nesting on the joist braces over the carriageway in the lower story. In the empty hayloft above a few Mexican cliff swallows were flying in and out of the building, just beginning to construct their mud nests on the rafters in the peak of the roof. At that date, May 17, 1922, there were no completed nests of the cliff swallows, and most of the barn swallows' nests held incomplete sets. Mr. Willard investigated this colony later, of which he (1923a) says: "On June 9, in company with Mr. Ed. C. Jacot, I again visited the colony and found that it consisted of eight pairs. The birds looked out at us from each of the six completed nests. Two nests were placed at the peak of the roof by each of three adjacent pairs of rafters. Two incomplete nests were farther down the line. Four of the nests held complete sets, two of four and two of five eggs. Incubation was barely noticeable."

In the same paper Mr. Willard tells of finding, in August 1915, a somewhat numerous colony of these swallows at Fort Huachuca, Ariz., "nesting in scattered locations all over the fort." Eight pairs had nests "under the eaves of the railroad station and of the section foreman's house." One nest was "under the roof of the open coal storage shed." Several nests were found in second-story rooms of unfinished buildings. He estimated that there were about 25 pairs in the entire colony.

With the exception of the pairs nesting in the new quarters, all the nests were attached to well painted woodwork. In the quarters, they were attached to the plastered walls close up against the ceilings. They were the usual gourd-shaped nests of mud pellets, with a few bits of grass for lining and a very few feathers. In the dry atmosphere of the mountains, the pellets of mud dried very quickly and it was surprising to see how fast a pair of birds could build up the walls of their abode. Both birds took part in the building. On arriving at the nest with a pellet of wet mud, the bird would press it into its appointed place and hold it there for several seconds until it was "set." I never saw a pellet thus held in place drop off when the bird loosened its hold.

Mrs. Bailey (1928) mentions several large nesting colonies in New Mexico on information furnished by J. S. Ligon; these colonies were located in the more primitive fashion, on cliffs along the banks of rivers; the smallest colony contained 60 occupied nests, another was occupied by 100 pairs, another "colony was composed of about four hundred and fifty occupied nests extending for a distance of a hundred feet," and in the largest group there were about 500.

Eggs.—The four or five eggs that usually comprise the full set for this swallow show all the variations seen in eggs of the species elsewhere. These are fully described under the northern cliff swallow. The measurements of 30 eggs average 19.8 by 14.1 millimeters; the eggs showing the four extremes measure 21.8 by 14.0, 21.5 by 15.0, and 18.1 by 13.4 millimeters.

In all other respects the habits of the Mexican cliff swallow do not seem to differ materially from those of the northern subspecies.

PETROCHELIDON FULVA CAVICOLA Barbour and Brooks

CUBAN CLIFF SWALLOW

HABITS

The Cuban cliff swallow owes its place on our list to the capture of two specimens by W. E. D. Scott (1890) on Garden Key, Dry Tortugas, Fla., on March 22 and 25, 1890. It was evidently only a straggler there, as it apparently has not been seen anywhere in

Florida since, except for one seen by Mr. Scott at the same place a few days later in a flock of tree swallows. It is apparently confined to Cuba and the Isle of Pines.

In naming this race, Barbour and Brooks (1917) describe it as "similar to *Petrochelidon fulva fulva* from Santo Domingo, but a little larger and differently colored. The Cuban birds show a much greater extension of the fulvous area below and a consequent restriction of the white area on the belly. In the Cuban birds the throat and chest are usually more richly colored than in the individuals of true *fulva*. They also have the rufescent or fulvous area changing gradually into the white or whitish of the mid-ventral region, whereas in the Haitian birds the white is clearer and purer and the boundary of the fulvous zone is quite sharply defined."

Dr. Barbour (1923) writes of its habits:

In Cuba they arrive in late February and gather in large flocks about the caves in which they nest. Occasionally abandoned buildings are occupied, or even the recesses of a deep veranda, but caves, sometimes open but equally often deep and dark, are the usual breeding-places chosen. A favorite spot is where the river disappears into a limestone cavern right in the town of San Antonio de los Banos. This was an impossible place to shoot, but Brooks and I found that, when we crept into the cave at night and then flashed an electric torch, the birds came in swarms clinging to our hats and clothes, as phototropic as moths. We soon had plenty, chosen by hand. A nesting-place near Bolondron is in a deep, steep, almost perpendicular, tubular cave mouth, which at first looked like a haunt for bats but nothing else. The old wooden hotel at Herradura had a few nesting under the eaves, and swarms inhabit the great caverns under Moro Castle, perched at the mouth of the bottle harbor of Santiago de Cuba. The nest is of mud, mixed with grasses and feathers, and is not so enclosed as with our Cliff Swallows.

W. E. Clyde Todd (1916) regards it as "a summer resident only" on the Isle of Pines, "of which the winter habitat is still unknown." He says that in the Caballos Mountains, as early as April 6, "the birds were observed going in and out of holes in the cliffs near the tops of the mountains, where they evidently had eggs or young. These nesting-places were quite inaccessible by ordinary means, but a little later, in the Casas Mountains, some pairs were found with nests only about twenty feet up the face of an exposed cliff."

I have seen only three eggs of this subspecies; these are quite evenly sprinkled with very fine dots, a type often seen in other eggs of this species and in eggs of our common cliff swallow. The measurements of these three eggs are 20.9 by 14.6, 20.9 by 14.6, and 20.8 by 15.0 millimeters.

DISTRIBUTION

Range.—This species is not regularly migratory and appears to be confined to the Caribbean region and eastern Mexico, casually to southern Texas and accidental in southern Florida.

The range extends **north** to southern Texas (Kerr County); Cuba (Trinidad and Guantanamo); northern Haiti (Cap-Haïtien and Tortue Island); northern Dominican Republic (Monte Cristi and San Lorenzo); and northern Puerto Rico (Aguadilla, Manati, and Bayamon). **East** to Puerto Rico (Bayamon and Guayama). **South** to southern Puerto Rico (Guayama and Cabo Rojo); southern Dominican Republic (Ciudad Trujillo); Jamaica (Port Henderson); Yucatan (Izamal and Calcehtok); and Chiapas (Ocozocuantla). **West** to Chiapas (Ocozocuantla); Tamaulipas (Miquihuana); Coahuila (Saltillo, Sabinas, and Monclova); and Texas (Kerr County).

Several subspecies are recognized. The typical race known as the Hispaniolan cliff swallow (*Petrochelidon fulva fulva*) is confined to the island of Hispaniola. The Cuban cliff swallow (*P. f. cavicola*) is found in Cuba and on the Isle of Pines. Two specimens of this race were taken at Dry Tortugas, Fla., on March 22 and 25, 1890. The Coahuila cliff swallow (*P. f. pallida*) is the form found in northeastern Mexico. A pair were taken in Kerr County, Tex., on April 23 and 24, 1910, and this race was later found to be nesting in that area.

Egg dates.—Cuba: 1 record, June 10.

Texas: 8 records, June 7 and 8.

PETROCHELIDON FULVA PALLIDA Nelson

COAHUILA CLIFF SWALLOW

HABITS

This is a pale race of a West Indian species that is found in northeastern Mexico, Tamaulipas and Coahuila, and north to Kerr County, Tex. Additional races of the species *fulva* occur in Cuba, the Isle of Pines, Jamaica, Haiti, Puerto Rico, the Greater Antilles, and perhaps Yucatan and parts of southern Mexico.

Nesting.—Col. John E. Thayer (1915) gives us the only account we have of the nesting habits of this subspecies, as follows:

The Mexican form of Cliff Swallow (*Petrochelidon fulva pallida*), described by Nelson, was found nesting by my collector near Japonica in Kerr County, Texas, during the month of June, 1914. He collected a series of birds and eleven sets of eggs. There was rather a large colony nesting in a cave. The entrance of this cave was like a mine shaft. The ceiling was covered with holes where the water had once eroded into the limestone rock. The Swallows nest in these holes, plastering a little mud like a balcony to hold the eggs in. A forty foot ladder was used to get up to them. The cave was poorly lighted and very damp. It was 50 feet from the floor of the cave to the ground, where the entrance was. The opening was about 8 ft. in diameter. About 10 feet down, the cave widened out into a spacious chamber. The only light was from the shaft-like entrance. To enter the birds pitched head first and diverged into the semi-dark chamber and began a detour of circles to check the impetus of their plunge."

Eggs.—The series of eggs taken by Colonel Thayer's collector contained two sets of three, eight sets of four, and one set of five eggs. I have examined these eggs and cannot see that they differ materially from the eggs of our common cliff swallow, except for a slight average difference in size. The measurements of the 43 eggs average 19.5 by 14.0 millimeters; the eggs showing the four extremes measure 21.6 by 14.3, 20.5 by 14.8, 17.3 by 13.7, and 20.2 by 13.5 millimeters.

PROGNE SUBIS SUBIS (Linnaeus)

PURPLE MARTIN

PLATES 69, 70

HABITS

CONTRIBUTED BY ALEXANDER SPRUNT, JR.

It has always seemed to me that literature has been somewhat chary of the purple martin. Song and story have long stressed the advent of robin, bluebird, and goose as heralds of spring, and so they are, but is the martin any less so? True, it comes somewhat later than these others, but who can fail to thrill when, on waking early one morning, one hears the rich, gurgling calls of the first martin! It is a signal that spring is really at hand, indeed, at one's very door. When the martins come, can summer be far behind?

This largest of the swallows, in its handsomely glossy livery, whether slurred by literature or not, has been a favorite with humanity for many generations. Even before the white man came to America's shores it was a dooryard bird in Indian villages, and its status as such is unchanged today. It is, beyond all doubt, *the* "bird-box" species of this country. Its range is extensive, almost universal indeed, and it occurs from coast to coast and border to border. Young and old admire it, encourage it, and protect it, and those who have a word of criticism for it are few and far between. Alexander Wilson said that, in his day, he never found but one man who disliked the martin, and many a modern ornithologist will have had the same experience, if indeed it can be matched!

Some birds occupy high pedestals in human regard, typified by the robin in the North and the mockingbird in the South, but in North and South the purple martin comes and goes as a welcome arrival and regretful departure; an always invited avian neighbor. Few are those anywhere who would fail to subscribe heartily to the wish—may its tribe increase.

Spring.—The martin makes its appearance in the United States from late in January on through April. The vanguard of the migratory hosts from South America cross the Gulf of Mexico and make landfall in the south from January 20 (Florida, Howell, 1932) to 29 (Louisiana, Oberholser, 1938). It is, therefore, one of the very earliest of migrants.

There is not, however, the marked regularity of appearance that is characteristic of many other northward-bound species. Early February would fit the Florida arrivals, on the average, better than late January. For instance, at Melrose, Fla., the arrival date for 12 years averaged February 9 (Howell).

In the Charleston, S. C., area, Washington's Birthday is about the time to expect this welcome summer resident, though the earliest record is February 6, and Arthur T. Wayne (1910) has also noted them on the 7th and 16th of that month. There are at least three dates for the 16th. Strangely enough, they appear not to reach North Carolina until "the latter part of March, while in most of the State, the earliest dates are about the middle of April, with only an occasional March record" (Pearson, Brimley, and Brimley, 1919). In the vicinity of Raleigh, N. C., it has been seen in March but once in 25 years! At Cape May, N. J., the arrival varies from March 27 to May 2, with the great majority of first dates occurring in April, throughout the month (Stone, 1937). In New York and the Hudson Valley, late April sees them arrive (Chapman, 1912), while the species appears in Massachusetts and other parts of New England from April 14 on (Forbush, 1929). Taverner (1934) gives no specific dates for arrival in Canada but mentions in another connection that they arrive "early in spring" and are often caught by unseasonable frosts and cold rains. The spring migration appears to be about parallel in dates across the country; in Minnesota, for instance, the birds sometimes arrive late in March but mostly early in April (Roberts, 1932).

It will be seen, therefore, that the martin moves northward rather leisurely for a range from late in January to early in May would cover the extremes from Florida to the northern portions of the range. April appears to be *the* time throughout much of the northern half of the whole range across the country. As a general rule, the males arrive in advance of the females and spend the interim in establishing themselves at old nesting boxes, feeding and preening, all these activities being accompanied by a vociferous indulgence in vocal effort.

Nesting.—Before the advent of the white man the martin used natural cavities of trees and cliffs for nesting sites. But even in those distant days there was some bird-house nesting, for the Indians were fond of these birds and, as Wilson (1831) says, "even the solitary Indian seems to have a particular respect for this bird." He gives an account of the methods used by the "Choctaws and Chickasaws" who "cut off all the top branches from a sapling near their cabins, leaving the prongs a foot or two in length, on each of which they hang a gourd, or calabash, properly hollowed out for their convenience." Forbush (1929) adds that "when saplings were not conveniently situated the Indians set up poles, fastened cross-bars to them and hung the gourds to these cross-bars."

This custom was taken up by the southern Negroes on the plantations and is continued to this day. Ever since childhood I have seen, on the old rice plantations, and about Negro cabins in the Carolina Low Country, martin gourds hung on cross-arms from a tall pole. The rural Negroes think highly of the martins and are sure that they chase away hawks from poultry.

Instances of strictly primitive nesting are still to be seen in remote parts of the country. Roberts (1932) gives an account of martins breeding among large boulders on Spirit Island, Lake Milles Lacs, Minn. Howell (1932) mentions two or three examples in Florida, one near La Belle and another at Naples. A unique situation came under his observation on Anna Maria Key in May 1918, when he found a pair using a hole in a palmetto piling over water, the cavity being about 3 feet from the surface.

I have seen one instance of primitive nesting in Florida, that of a small colony of about five pairs utilizing a tall, dead pine perforated with woodpecker holes. This tree stands near the banks of the Kissimmee River, near the hamlet of Cornwell, in Highlands County, Fla., and martins were using it late in March 1940. Shown to several participants in the Wildlife Tours undertaken in that region during the early part of 1940 by the Audubon Association, it never failed to elicit the greatest interest. Flickers and bluebirds, as well as a red-bellied woodpecker, were also using this avian apartment house. I have had it reported that martins use the hollows in very old cypresses in some of the large river swamps of South Carolina, along with chimney swifts, which is certainly very likely, though I have not seen this association personally.

Today, the purple martin is unknown to the great majority of people in this country except as a dooryard bird. Its popularity is tremendous, and nesting houses throughout the various States must run into the thousands. It is not a "choosy" species as regards the type of box, for anything from a boy-manufactured cigar-box home to the most elaborate miniature mansion is utilized, and, in the South at least, the martin is as partial to gourds as anything else. The number of rooms in a martin house is simply up to the owner who maintains it, these vary from 1 to 20 or 30 as a rule, but some houses have as many as 200.

After the males have arrived and located themselves, they await the arrival of their consorts, and when they appear mating takes place soon. This may vary by locality, those birds breeding in the southern part of the range not mating so soon as those occupying the northern sectors. The immature birds mate later than the adults anywhere. Both sexes build the nest, and it is a most animated sight when a colony is engaged in such construction. A great deal of vocal effort accompanies it, and the gurgling chatter goes on throughout most of the daylight hours.

Nest material differs rather widely. In most parts of the South, it is often confined simply to grass and leaves. In other localities, twigs, feathers, mud, rags, paper, string, straw, and shreds of bark have been noted. S. F. Rathbun (MS.) mentions a nest on the watertower of a building in Seattle, Wash., which was "composed entirely of bits of rubber insulation from electric wires; this was lined with pieces of wood." Occasionally, a rim of dirt is placed in front of the nest to keep the eggs from rolling out of the entrance. F. W. Rapp (MS.) states that he has "seen some of these mounds built up to 2½ inches, while others make hardly any attempt to do so." Both sexes incubate the eggs, the female, however, assuming the greater part of this duty. The incubation period varies with locality apparently, being variously noted as from 12 to 20 days.

Forbush states that only one brood is raised in New England, and this is the case throughout the greater part of the range. It holds for the Carolinas certainly. Audubon (1840) insisted that two broods were raised in his day and that in Louisiana three were brought forth. In coastal South Carolina, though martins are abundant by the end of the first week in March, nest-building does not commence until the last of April, and it is usual that eggs are not laid until the middle of May. Since the birds are not through until early July, there would be time for only the one brood.

The usual nesting box is placed upon a pole, the height of which is commonly 15 or 20 feet. Many of the poles are hinged to facilitate cleaning, painting, etc. The measurements recommended are: Individual compartments, about 8 by 8 inches; entrance, about 2 inches in diameter and 1½ inches above the floor. The house is placed in fairly open situations. On erection in spring it is well to put up the house as the martins arrive; otherwise it gives English sparrows and starlings opportunity to become established therein. Forbush (1929) gives an instance of a friend of his erecting the house as the martins arrived, and "they were so glad to see it that they could not wait until it was up. While it was going up they flew around it, singing and fluttering about it, and when it was half-way up, they all alighted upon it and rode up with it."

The homing instinct in the martin must be very strong, as the foregoing incident well illustrates. One of the most striking examples of a returning martin colony I ever heard was related to me by Alston Clapp, of Houston, Tex. While in his yard on one occasion he showed me his colony and said that the year previous he had taken down the house to paint it. Something delayed him, and it was not up when the martins arrived. Attracted by a great chattering one morning, he went out into the garden and saw the birds fluttering and circling about in the air *where the house should have been*, at the exact elevation occupied by it when placed!

Elevations vary, of course, some houses being placed much lower (or higher) than others. The lowest I ever saw was a large, rather elaborate house holding a thriving colony in Warrenton, Va. It was only 9 feet from the ground. The primitive nesting site noted by Howell (1932) on Anna Maria Key, Fla. (3 feet), may be regarded as the extreme in low elevation. The highest that has come to my attention is recorded by S. F. Rathbun (MS.) in Seattle, Wash. Speaking of it he says that "a rather high brick building in the lower business district was surmounted by a tall flagpole capped with a ball. This ball was at a height of about 130 feet above the street. One day when I was watching martins glide above the building one of them flew directly to the ball and disappeared. By the use of a pair of field glasses a check or crack could be seen in the side of the ball, which accounted for the bird's disappearance; it was using the ball as a nesting place. Use of it continued for a number of years, until the ball was replaced by another."

Walter Faxon (1897) records martins nesting on top of street lights in Cambridge, Mass.

Another unusually low nesting site was seen by S. S. Dickey (MS.) at Seth, a mining village on the Coal River, Boone County, W. Va. He found that some boys had erected crudely made, one-room boxes, 8 to 9 feet above ground, and had painted them brightly, one being red, white, and blue! This same correspondent has also witnessed, as have others, the use of buildings by martins as nesting sites. "While I was in Madison, Wis.," he writes, "I was entertained by several colonies of martins. They were building in cracks in the wooden eaves of business buildings in the city square, facing the capital building."

It is entirely possible that in some residential districts martins are disturbed by the actions of small boys, who are thoughtless enough to cause the birds trouble. For instance, in the case of the low-nesting house in Warrenton, Va., mentioned above, the owner had noticed that the birds used the rooms on the *far* side from his residence, the yard usually being "infested" with boys and dogs! Wishing to see more of the birds' activities, he turned the box so these occupied rooms could be seen from his own house, but the martins promptly deserted their already occupied nests and rebuilt in rooms on the far side!

There are, throughout the country, several particularly noteworthy colonies of martins, but one of the best known, as well as a very populous one, is established in Greencastle, Pa. It is hardly correct to say "one," however, for the martin population is spread here and there over town, in many boxes, but, to the citizens, who take great interest and pride in the birds, it is the Greencastle Martin Colony! Those who have never seen it have something in store for them. The outstanding attraction of this colony is the tameness of the birds and the intense interest of the citizens in their welfare. The houses are by no means elaborate, simply plain wooden boxes as a rule, with many compartments. They are at low elevations, about 8 or 9 feet, attached to telephone poles around the square in the heart of the business district, as well as on hotels and stores. The birds seem to prefer the rush and bustle of the retail district rather than the quieter residential sections.

These Greencastle martins go back into history. There are records to show that they have been there at least since 1840. One curious lapse occurred in this long tenure, 10 or 15 years after the Civil War. The martins did not return to Greencastle for nearly 15 years! It was a mystery that has never been completely explained, but G. F. Ziegler (1923), whose article on these birds should be read by everyone interested, considers that the hiatus occurred at the time the English sparrow was most rapidly multiplying and that these two events are connected. Now, however, the martins are again, and have long been, the town's great attraction. The first arrivals appear about the middle of March, and by the second week in April most of the boxes are occupied.

Eggs.—[AUTHOR'S NOTE: The purple martin lays from 3 to 8 eggs, usually 4 or 5; the larger numbers are rare. The eggs vary from ovate to elliptical-ovate. They are pure, dead white and practically without gloss. The measurements of 50 eggs average 24.5 by 17.5 millimeters; the eggs showing the four extremes measure **26.4** by 16.3, 25.4 by **20.3**, 22.7 by 16.0, and 24.9 by **15.8.**]

Young.—The period of incubation has been recorded by various observers as 12, 13, 15, or even 20 days; probably the normal period is 12 or 13 days. Incubation apparently is performed by the female only, but both parents assist in the feeding and care of the young. Under favorable circumstances two broods are raised in a season; Audubon (1840) says that sometimes three broods are raised in a season in Louisiana.

The young usually remain in the nest 24 to 28 days, but Forbush (1929) says that "the young sometimes remain in the nest for about six weeks. * * * Many of them return to the nest night after night for a week or ten days, especially if the weather be windy and stormy." Of the feeding, he says:

Among the insects brought were some large dragon-flies; some were brought by the wings, and the young bird leaning forward snatched the insect and swallowed it, often with difficulty, leaving the wings in the beak of the parent. Some were held by the body in the beak of the adult bird and were swallowed wings and all by the young bird, though the ends of the wings stuck out of its mouth for some time afterward. In some cases small snails and egg-shells are fed to the young along with their insect food.

Excessive heat and swarming parasites in summer often cause the death of young Martins in the nest, or they are killed by falling to the ground, in their attempts to escape from suffocation or the tormenting parasites in the nest. When a young bird falls to the ground it is soon deserted by its parents, who give up the attempt to preserve its life, and if not killed by the fall it is soon picked up by some cat or other prowler.

Charles Macnamara (1917) writes:

By the first of July most of the doors are crowded with little heads, and the whole front of the house blossoms suddenly with enormous yellow mouths whenever an old bird sweeps in with a beak full of insects. Numerous counts made at different times of the day during the first two weeks of July, 1917, showed that, with remarkable regularity, a parent arrived with food every thirty seconds. This year nine pairs occupied the house, and assuming that each pair had four young, and that they were fed in turn, then each nestling was fed every eighteen minutes. A similar count for a whole day, from 4 a. m. to 8 p. m., cited in Chapman's Handbook of Birds of Eastern North America, when reduced to the same basis as my results, gives a feeding every twenty minutes. * * * As the young grow up, however, they are not fed so often. After the middle of July the pace slackens considerably, and the old birds have more time to sit around on the verandahs and nearby trees, and gossip and scold.

Plumages.—[AUTHOR'S NOTE: The sexes are much alike in the juvenal plumage, except that the young female has the whole top of the head gray, whereas in the young male the forehead only is gray. Dwight (1900) describes the juvenal male as "above, including wings and tail, sooty or clove-brown, the forehead and a nuchal band grayish, the feathers of the head and back indistinctly dull steel-blue. Feathers of the wings with very narrow whitish edgings. Below, white, mouse-gray on chin, throat, breast, sides and tibiae, the feathers of chin, lower breast and abdomen with narrow dusky shaft streaks." This is worn until after the birds leave for the south, where a molt, probably complete, produces a first winter plumage.

In the first winter plumage the sexes are more readily distinguished. The male has acquired considerable steel-blue plumage on the upper parts, is generally darker, and is much like the adult female in coloration, but the chin, throat, breast, and sides are pale gray and the abdomen whitish. The young female is duller, the upper parts less glossed with steel-blue, and the under parts are more extensively white. This plumage is worn, apparently without much change, for practically a year, or until the next postnuptial molt; this complete molt takes place after the birds have gone south and produces the well-known adult plumage of each sex.]

Food.—The whole diet of the purple martin can be fully covered by one word—insects! When that is said, all is said, for that is what the bird subsists upon and nothing else. However, since the same can be said for other birds, some elaboration is necessary in regard to specific kinds of insects. Prof. F. E. L. Beal made an exhaustive study of the martin's food (1918) and found only a few spiders besides true insects. These creatures are so close to insects, however, that, in many minds, they are identical. The Hymenoptera composed the greatest item, amounting to 23 percent, ants and wasps figuring mostly, with a few bees. To accusations that martins destroy honeybees, he had a definite answer that in only 5 out of 200 stomachs did honeybees appear, and every one of them was a drone.

Flies amount to 16 percent of the total food and include some of the house-fly family as well as numerous long-legged tipulids. The Hemiptera, or bugs, amounted to 15 percent and included stink bugs, treehoppers, and negro bugs. Beetles composed 12 percent and are represented by May, ground, dung, cotton-boll, and clover weevil beetles. Moths and butterflies were found to some extent. Dragonflies seem general favorites and were found in 65 stomachs, some of which contained nothing else.

In connection with this habit of eating dragonflies, Forbush (1929) states that "adult dragonflies are considered to be useful, as they destroy harmful smaller insects, including mosquitoes, but the young of dragonflies are destructive to small fishes, and this habit may neutralize the beneficial habits of these insects. As Martins are said to feed heavily at times on mosquitoes, their destruction of dragonflies may be immaterial."

He says further that "in some instances a great decrease of mosquitoes is said to have followed the establishment of Martin colonies, but I have had no opportunity to investigate these reports." Certainly, it would be logical to suppose that the area about a thriving martin colony would be freer of mosquitoes than one without these birds. T. S. Roberts (1932), after listing such insect prey as ants, wasps, daddy-long-legs, horse flies and robber flies (which prey on honeybees), bugs, beetles, moths, dragonflies, and spiders, ends with the somewhat remarkable statement that the martin is "rather neutral from an economical standpoint but worthy of protection." He appears to be in an isolated position among most writers, who are entirely commendatory of the martin's economic value. Junius Henderson (1927) quotes some interesting data from Attwater in saying that a quart of wing covers of cucumber beetles were found in one martin nesting-box. Henderson says rather vividly, in comparison with Roberts' opinion above, that since "Martins are very active, requiring a large amount of food, and a considerable part of each insect is indigestible, the number of insects they destroy in order to get sufficient nourishment is 'not only beyond calculation, but almost beyond comprehension.' The food is often compressed into a hard mass, so it is wonderful how much a stomach may contain. The mass of insects contained in a Swallow or Martin, would before compression, equal or exceed the bulk of the bird's body."

Audubon (1840) says little of the martin's food, mentioning only that "large beetles" figure in it, and that the birds "seldom seize the honey-bee." Alexander Wilson (1831) devotes more space to this phase and states that he "never met with more than one man who disliked the martins, and would not permit them to settle near his house. This was a penurious close-fisted German, who hated them because, as he said, 'they eat his *peas*.' I told him he must certainly be mistaken, as I never knew an instance of martins eating *peas;* but he replied with coolness, that he had many times seen them himself 'blaying near the hife; and going *schnip, schnap*' by which I understood that it was his *bees* that had been the sufferers; and the charge could not be denied"!

Relative to the enormous numbers of insects destroyed by this species, as well as the assiduous care of the young in providing them with food, is the now classic example given by Widmann (1884). He watched a colony of 16 pairs of these birds from 4 a. m. to 8 p. m., and during that time the parents came to the young 3,277 times, or an average of 205 times for each pair. The females made 1,823 visits, the males 1,454.

John A. Farley (1901b) records that about the cranberry bogs of Plymouth and Barnstable, Mass., the martin devours numbers of the imagoes of the fireworm (*Rhopobota vacciniana*), which is a highly beneficial act, since cranberry growers estimate that over a term of years, they lose 50 percent of their crops by insects, chiefly the fireworm.

F. L. Farley (MS.) writes from Camrose, Alberta, that martins are very fond of bits of egg-shells, so much so that "they are as crazy for these shells as are cedar waxwings for ripe fruit." He continues: "Mrs. Farley saves most of her eggshells for one of our friends who has about 30 pairs of martins nesting. He just breaks them up and throws them down on the ground under his boxes and before he reaches the house there are numbers of martins on the ground, feeding on them and even taking bits up to the young. The first time it was noted that martins liked shells was when a man saw them holding on to a stucco house and pulling away at oyster-shells that were protruding from the cement. The party told me he tried to feed them eggshells at once, and from that time on all the martin men in town have been doing this." No doubt it was the lime that attracted the birds. Farley adds the interesting item that "our purple martins have increased now (1939) to more than 200 pairs in our little town, from a single pair that nested here in 1918."

The food is, of course, procured mostly on the wing and in the usual swallow fashion of darting, swooping, and wheeling in erratic flight, but graceful in the extreme. Sometimes, late in the after-noon, or early in the morning, martins skim the surfaces of ponds and rivers, dipping down expertly for drinks. Occasionally they pick up food from the ground by walking about. In any summation of the martin's food habits and economic value Taverner's (1934) statement is eminently fitting. Under the heading "Economic Status" he says: "The Martin like the other swallows is a bird with no bad habits, and with so many good ones that every effort should be made to aid its increase." Here is no betwixt and between state-ment, but a straight declaration of a fact that should be apparent to every student of this valuable species.

Behavior.—The purple martin essentially typifies the grace that makes the swallows famous. Beautifully proportioned, trim, and streamlined, it looks like a miniature plane as it sails overhead on outstretched, sable wings. Master of the air, as are all the family, it is not so spectacular in aerial evolutions as are some much larger birds, but this is by no reason of inability. Because of its small size much of its performance aloft is not easy to see and watch, as with a bird of larger wing expanse. Martins do not usually fly at great speed but are perfectly capable of such at need. Any bird that catches such swift insects as dragonflies must, of necessity, be a fin-ished flier. Feeding is accomplished largely on the wing, but martins can and do resort to the ground at times, where they feed on ants and other terrestrial insects.

Francis H. Allen (MS.) says that the "flight consists of a rapid flapping of wings, alternating with periods of sailing, either in a straight line or in a long, sweeping curve or arc. The bird often flies high, and his mellow, staccato song can be heard distinctly when the singer is hard to find in the expanse of sky."

Witmer Stone (1937) gives a graphic picture of the martin's flight as follows:

The martin on the wing is deserving of careful study: a glorified swal-low. * * * His flight is at all times a wonderfully graceful performance.

I stood on the edge of the meadows, one day in June, while twelve males were skimming the tops of the salt grass, tilting now to this side now to that to maintain balance in the air. Now one of them turns in his course and passes close to my head, swift as an arrow and uncanny in his blackness, which has no relieving spot of white, not even on the belly * * * His call uttered either in greeting or protest as he passes is a harsh *zhupe, zhupe*. Again he will mount upward with rapid strokes of his narrow pointed wings only to return again to the lower level on a long sloping sail. Sometimes at the very summit of the ascent he will come about into the wind and remain stationary, on rapidly beating wings, before sliding away on the long downward sail. When flying high over the town, late in the summer, the Martins' mastery of the air is particularly noticeable. They come in against the wind on set wings like small three-cornered kites, steading themselves now and then with two or three short wingbeats, and then, apparently tiring of this sport, they will drop through considerable distances and, flapping rapidly, regain their former altitude. While Martins flying low over the meadows are undoubtedly engaged in seeking food many of their aërial evolutions, like those of other expert fliers, seem to be for the shear joy of flight. * * * I watched a single Martin associated with a band of Swifts maintaining a position directly over Congress Hall hotel, in the face of a strong south wind, for at least half an hour. The birds would often remain absolutely stationary in the air for several minutes at a time, evidently supported by the upward currents of air deflected by the walls of the building. Here was no search for food but some sort of enjoyment or play.

While at rest the martin sits rather erectly and with an alert carriage, often uttering the characteristic note of contentment. Preening occupies considerable time, and several may be engaged in this at once, offering a never-ending variety of poses. They are fond of bathing and often do so on the wing. Audubon (1840) describes this process unusually well: "They are very expert at bathing and drinking while on the wing, when over a large lake or river, giving a sudden motion to the hind part of the body, as it comes into contact with the water, thus dipping themselves in it, and then rising and shaking their body, like a water spaniel, to throw off the water."

The behavior of martins in storms has been commented upon by early and recent writers. Audubon says:

The power of flight possessed by these birds can be best ascertained * * * when they encounter a violent storm of wind. They meet the gust, and appear to slide along the edge of it * * * The foremost front the storm with pertinacity, ascending or plunging along the skirts of the opposing currents, and entering their undulating recesses, as if determined to force their way through * * * all huddled together in such compact masses as to appear like a black spot. Not a twitter is then to be heard from them by the spectator below; but the instant the farther edge of the current is doubled, they relax their efforts, to refresh themselves, and twitter in united accord, as if congratulating each other on the successful issue of the contest.

S. F. Rathbun (MS.) writes that on the night of August 1, 1931, at Seattle, Wash., "not long after midnight there was a heavy thunderstorm with a rather strong wind. I arose to close a window that faced the storm, and the outdoors showed a pitchy darkness. To my surprise I heard the calls of martins, and for a short time whenever there were flashes of lightning I could see a number of these birds playing in the air in advance of the dark clouds. The martins drifted by with the storm, which was only of short duration."

The purple martin is a fearless bird in defending its territory. This characteristic has been noted for as long as man has had to do with the species and accounts for his original desire to have the birds as neighbors. The Indians and later the Negroes both induced martins to nest about their wigwams and cabins because of their readiness to drive away any winged intruder that might attack poultry. Crows, hawks, eagles, and vultures are quickly set upon and driven away, the whole colony combining in a mass attack that rapidly puts the invader to rout. Wilson (1831) says that the martin "also bestows an occasional bastinading on the king bird when he finds him too near his premises; though he will, at any time, instantly co-operate with him in attacking the common enemy."

Another trait of the martin that has long attracted attention and produced much writing is its communal roosting habit late in summer, when the species gathers in great flocks preparatory to and during migration. Concentrations up to 100,000 birds have been noted, and the attendant noise sometimes results in such a nuisance to people that direct efforts are made against the birds and many killed through various methods. To some degree these roosts are a parallel to those of the vanished passenger pigeon in that branches of trees are broken by the weight of the birds and, as Arthur T. Wayne (1910) puts it, "the noise produced by such a multitude resembled the sound of escaping steam." In 1905 a huge roost at Wrightsville Beach (near Wilmington), N. C., was attacked by irate citizens and 8,000 to 15,000 birds were killed. The North Carolina Audubon Society succeeded in convicting 12 of the offenders, who were fined.

G. Clyde Fisher (1907) describes a roost near Quincy, Fla., which he estimated to contain 5,000 birds and, like Wayne, was impressed with the noise, which he also described as being "much like escaping steam."

A typical roost, and a very well known one, was that at Cape May, N. J., written of in detail by Witmer Stone (1937). Students should peruse his account with great interest. It is too long to quote here, and since 1936 the roost has been deserted not only by martins but by robins, starlings, and grackles. However, it may sometime again be instituted, and extracts of Stone's account are given herewith:

For many years it [the roost] was located on the Physick property on the principal street of the town. Here there is a grove of silver maples about thirty feet in height and covering an area of some two acres, growing so close together that their tops join one another, making a dense canopy with constant shade. * * * Were it not for this roost, the only one in South Jersey so far as I know, Martin history at Cape May would come to a close early in August when the last of the fledglings become self dependent and sail away with their parents. But as it is, though there may be many days in August when practically no martins are to be found for miles around Cape May from sunrise to sunset, they will gather in ever increasing numbers to pass the night in this small grove which, so far as our eyes can detect, offers no advantages over hundreds of similar groves past which the birds must have flown. It would seem that most of these Martins must have come from areas far to the north of New Jersey, as the local breeding Martins could not have yielded such a crop of young. I estimate that there are not more than fifty pairs of the birds in Cape May and perhaps twice that number elsewhere in the peninsula and these hundred and fifty pairs could not produce more than six hundred offspring, making some nine hundred Martins in all, and yet at least 15,000 of the birds come to Cape May every night to roost. In the New York area, including northern New Jersey, Ludlow Griscom states that the Martin colonies are very locally distributed and that the birds are rare as transients, which further complicates the question of where our Martins come from! Another fact of interest is that on July 23, 1926, before any of the young had left the Cape May nesting boxes one thousand Martins had already assembled at the roost.

Later in Stone's account of this same roost he says that while the birds gather together and settle on the twigs of the trees for the night, "their calls produced a constant twittering like *escaping steam*, now swelling loudly and then dying away again." The italics are mine, this being the third author to refer to escaping steam!

Voice.—Though unable to lay any claim to being a musician, and therefore incompetent to judge music, I cannot but feel some slight resentment toward writers who characterize the voice of the martin as "unmusical"! True, some of the notes are such, but there are many others that are decidedly pleasing. All writers seem to agree that on the whole the vocal efforts of the species are "pleasing," which, in itself, denotes that they are musical rather than otherwise.

Certainly, they *are* pleasing. This fact has undoubtedly a great deal to do with the attraction of a martin colony about one's home. People generally enjoy hearing them. The bird has no specific "song," such as many species possess, but the varied medley of notes has a definite quality of imparting satisfaction and enjoyment to human ears.

The outstanding impression always left upon me is that of *contentment*. In few birds is this so typified. The gurgling chatter of a group about its nesting box gives one the complete assurance that, so far as the martins are concerned, "all's well with the world"! There is a restfulness about the notes that is distinctly relaxing; they can be listened to at a time when a person is reclining on a porch in an easy chair in springtime and fit perfectly with the droning of bees about a flowering vine, the sighing of a breeze through moss-hung oaks, and the distant calling of willets over the salt marshes.

The notes of the martins simply "fit" a spring and summer picture, and while such a description may be exactly what a nonmusical ear would produce in words, i. e., very inadequate, it is certainly something of a tribute to the birds!

As has been so often and rightly said, it is difficult, not to say impossible, to render any idea of a bird's notes by written words. Yet that is about the only way to be specific, and while one's own interpretation may differ from another's, it continues to be done, and has to be. In the case of the martin, Audubon (1840) did not attempt a word translation. He simply says that "the note of the martin is not melodious, but is nevertheless very pleasing." Wilson (1831) went farther and produced the following: "Loud musical 'peuo, peuo, peuo'." Forbush (1929) quotes W. M. Tyler as describing a "loud, rich chirrupping." Many authors describe some of the lower notes as "guttural," which is accurate enough. The alarm note is stated by Tyler as "kerp," and he adds a "low-toned 'kroop' song and several throaty notes followed by a spluttering trill."

Francis H. Allen (MS.) says that the "call notes uttered in flight are a low, mellow but somewhat husky *chip;* a *ye(r)p,* and a *kew.*" The similarity between Tyler's *kerp* and Allen's *ye(r)p* is at once apparent, and this note can hardly be described in a better way.

S. S. Dickey (MS.) has this to say: "Purple martins, which scarcely can be called songsters, usually utter loud, penetrative twitterings. They give vent to exclamations of singular delight." This last is a rather happy expression and fits my idea exactly! He adds that "as the nesting season advances they, by spells, break almost into song; *spick-spack-spitter-spee-spack* are the syllables. When troubled by English sparrows, bluebirds, and starlings, they swoop in downward curves in pursuit of the nuisances, and utter buzzing sounds as *spiz-spiz-spiz.*" Dickey concludes with the statement that these "outcries are not to be confused with those of any other swallow. The utterances have a character of their own, and it is good to hear them if you chance to be one fond of past association and summer excursions." One can easily subscribe to the last thought; it at once brings to mind a dreamy summer afternoon in the side yard, but some of Dickey's "sp" interpretations are rather difficult to follow. Evidently this sound has impressed him considerably, for he uses it consistently.

During the great gatherings at roosts late in summer the birds are extremely vociferous, and their notes at such times have been likened to "escaping steam" by some authors. This imparts a rather hissing impression that is certainly characteristic of these communal gatherings, but the simile leaves one a little cold. At such times the vocal efforts of the birds are neither musical nor pleasing and often result in becoming a nuisance because of monotony and volume.

While the martin is not, in any sense, nocturnal, its notes are sometimes heard after dark. One such instance is recorded by Abby F.
C. Bates, of Waterville, Maine (1901). She heard martins plainly
about 10 P. M. on the moonlight night of August 8, 1900. Mr. Bates
heard martins on June 15, 1930, between 2 and 3 A. M. as he was
returning home from a late train. These auditors seem to consider
this a highly unusual thing, but in the South I have frequently heard
martins at night. I once lived across the street in Charleston from
a friend who maintained a large martin colony, while just over the
back fence was another. I recall hearing the birds now and then,
but I never made any series of observations or records of it. However, on looking up my notes under this species, I do find the following: "Heard martins 'singing' in E. A. Williams' bird-house at 11
P. M. tonight." This entry was under date of May 1, 1933. These
night notes are of a lazy, sleepy character, which one might expect
from birds aroused by some slight disturbance. In no case have I
ever heard them in an alarmed or excited nature, simply a low
chuckle, or gurgle, indicative of restlessness or temporary wake-
fulness.

Enemies.—The enemies of the purple martin appear to be confined to a few other birds and the weather. Certainly man is extraor-
dinarily absent as such unless the occasional outbursts of impatience
at the roosts can be so construed. Though in so many cases man
is Enemy No. 1 to bird life, this is a happy exception indeed.

The natural enemies of this species are those importations among
avian circles, the English sparrow and the starling. Bluebirds have
been listed as enemies in that they sometimes compete for nesting
boxes, but it is difficult to conceive of a bluebird being an "enemy"
to anything, and such interference is inconsequential. However, that
the sparrow and starling are, is beyond all doubt. Both of these
interlopers cause endless trouble to martins and human friends of
the latter who dislike to see their favorites usurped. I often have
questions put to me as to how to get rid of them!

The quarrelsome dispositions of these trouble-makers are too well
known for elaboration. There are instances by the legion where they
have appropriated nesting boxes. Specifically, J. K. Jensen, of
Wahpeton, N. Dak. (1918), mentions that in that area English sparrows entered the compartments of a martin colony freely and destroyed so many eggs that few of the swallows could be raised. Many
others in various parts of the country have seen similar occurrences.

The rapid spread of the starling southward and westward has resulted in its becoming pretty much Enemy No. 2. Its larger size
and even more efficient methods make it as much if not a greater

menace. Various means are resorted to in attempting to drive out these invaders such as shooting, noises, water-cures, etc., but it is often hard to make an impression, and more than one martin colony has been lost.

Adverse weather bulks rather largely in a martin's life at times. Martins are very susceptible to cold, and unseasonable spells of it play havoc with them. After a cold spell has depleted a colony, it is usually a long while before they return to that locality. It is, of course, a lack of food supply as well as the weather itself that reacts detrimentally on the birds in these cases. An insectivorous bird's digestion is very rapid and demands that it be more or less constantly eating, and two or three days of severe cold so eliminates insects that starvation not infrequently occurs. These spells sometimes take place as late as mid-April.

In the Charleston area, where I have lived all my life, sudden cold, which is very rare in spring, sometimes affects the martins. My old friend Arthur T. Wayne (1910) states that he has known the species to be affected seriously only once. He says that "on Tuesday morning, February 14, 1899, the temperature registered 6° above zero at Charleston * * * followed again by a very severe cold wave accompanied by snow. * * * On April 14 and 15, 1907, however, large numbers died from cold and starvation during the prevalence of gales and cold weather." One of these spells provided Wayne with a very beautiful albinistic specimen of this species, which he found dead under his "swallowhouse" as he invariably called it.

Rains, cold ones or even protracted ones in warm weather, occasionally wreak havoc. F. B. Horton, of Brattleboro, Vt. (1903), writes that in June 1903 long rains resulted in the death of 30 young and 2 adults in a colony there. The remaining martins deserted the place leaving 12 unhatched eggs.

Forbush (1929) quotes Dr. Brewer as describing a cold rain spell that eliminated martins in "eastern Massachusetts" and as a result none have returned there "to this day." The rain of June 1903, mentioned in the Horton note above, extended into Massachusetts and is mentioned also by Forbush, who says that it destroyed "most of the Martins in Massachusetts and contiguous parts of New England." He brings out the fact that when martins do come back after such a spell to the locality they find their houses occupied by English sparrows. He also adds that excessive heat and vermin constitute enemies of this species. The latter at times kill the young birds outright. Heat, in the restricted space of a martin house, utterly exposed to the sun as it is, must be a factor certainly. One would think it unbearable at times. Forbush (1929) states that the "parents have a habit of collecting many green leaves and placing them in the nest, a practice which may tend by evaporation to reduce the heat." This is a very interesting observation and one that has been made also by P. A. Taverner (1933).

Apparently there is rather little specific data on migration casualties, but there is little doubt that such occur. Wintering in South America, the northward migration of the birds is very deliberate, as has been pointed out, covering the time from late January until into June. Thus, there are perhaps less mass movements over the Gulf of Mexico than are true of other birds, and it follows that casualties would be fewer, but bad weather over that body of water must result in the loss of some. Lighthouse victims do not figure largely.

S. S. Dickey (MS.) writes that in his observation "screech owls are the worst foe" of martins. He states that he "made it a practice to frequent the vicinity of occupied boxes well into the night, and every once in a while a screech owl would come shadowlike, alight for a moment on the runway or porch of a bird-house, then begin a scrambling kind of a noise. By the use of a strong flashlight I was enabled to get some conception of how *Asio* performs his pillaging. He peers intently into an occupied room, then leans sideways, lifts his face skyward, and reaches as far as he possibly can with one leg into the orifice. Shortly he creates consternation. The squeals of the inmates resound. He brings forth squirming birds. With the prey in his talons, he flits to the nearest tree or shed roof, there to devour his bill-of-fare. I have known him to continue such ravages for an hour at least, taking as many as a half dozen martins to appease his ravenous appetite."

Fall.—The martin is an early migrant. Sometimes the roosting habit of late summer is noted on the part of certain birds before young have left the nest boxes. Stone (1937) mentions this, and again speaks of barren or nonbreeding birds "roosting" as early as June 25. After the young have flown the southward movement really takes place. This seems to be very general over much of the martin's range. Forbush (1929) gives September 30 as the latest date for New England. F. M. Chapman (1912) gives departure dates from Cambridge, Mass., as August 25; Ohio, September 5; Illinois, September 10; Minnesota (SE.), September 9. In Michigan, Barrows (1912) states that "it is one of the first of our swallows to move southward in autumn, usually disappearing about the middle of August and rarely seen as late as September 1." The migration in New Mexico is even earlier, Mrs. Bailey (1928) stating that "only stragglers being left after the first week in August." At Cape May, N. J., that great funnel point of migration, Stone (1937) has noted the last individuals from September 3 to 15 over a long period. About Charleston, S. C., the last martins are seen approximately at the same time, though a marked reduction in the numbers of the species occurs from mid-July on, when many begin to migrate.

Proceeding southward, the last ones leaving Florida are recorded by Howell (1932) as from late September to October 2 about Pensacola and Tallahassee, while down in the Keys, at the jumping-off place, they were noted at Sombrero Key on October 6. A rather late flight was recorded at St. Marks (south of Tallahassee on the Gulf) on November 10, 1912. Again, in another November flight, Howell states that Mrs. Hiram Byrd saw "great flocks of swallows over the glades" on the 11th, of which "some were Martins." A flight seen by this observer on November 4 was described by Howell as being "between Royal Palm Hammock and Homestead." Here again Howell confuses Royal Palm Hammock with Royal Palm *Park*, the latter being Paradise Key in Dade County, while the former lies across the Tamiami Trail in Collier County, a hundred miles or so to the westward. Though in this case he says "hammock," he means "park" for he quotes Mrs. Byrd as saying that the martins were "circling over a glade at the edge of the Park." She estimated this flock to cover 9 miles of territory and to contain "anywhere from a hundred thousand to a million or more."

H. W. Ballantine noted three birds at Orlando on December 18, 1915, which Howell says "may be considered wintering birds." This is a parallel case to his statement that gray kingbirds must sometime winter in Florida because of a single specimen seen in Royal Palm Park in December. Rather would it appear that, like the kingbird, these three martins at Orlando were belated migrants, for in six years of constant field work in southern Florida I have never seen a single specimen, not even in the keys. It is my confirmed belief that the birds leave the country entirely, and there is nothing to offset this except the single exception of the December Orlando birds above, if that can be considered an exception. Oberholser (1938) gives the departure date (latest) from Louisiana as October 22. Definite knowledge of its winter home in Brazil has resulted from banding, and Lincoln (1939) gives an example of a specimen "found" in December 1936 .(the same month in which the Orlando "wintering" birds were seen) near Ttaituba, Para, Brazil, which had been banded at Winona, Minn., on May 30, 1934.

Thus, it will be seen that the autumnal migration of the martin is almost as leisurely a matter as is the spring movement. It covers a rather long period.

DISTRIBUTION

Range.—North and South America.

Breeding range.—The purple martin nests **north** to southern British Columbia (Nanaimo and Vancouver); central Alberta (Edmonton and Camrose); central Saskatchewan (Prince Albert and Quill Lake); southern Manitoba (Lake St. Martin and Shoal Lake); southern Ontario (Kenora, Sault Ste. Marie, and Ottawa); southern Quebec (Montreal, Quebec, and Kamouraska); New Brunswick (Chatham); and Prince Edward Island (Brackley Point and Pictou). The **eastern** limits of the range are the Atlantic coast from Prince Edward Island (Pictou) south to southern Florida (Fort Lauderdale). On the **south** the range extends westward from southern Florida (Fort Lauderdale and Fort Myers) along the Gulf coast to southern Texas (Columbia and Rio Grande City); Guanajuato (Guanajuato and Celaya); Jalisco (Lagos, Guadalajara, and Etzatlan); Nayarit (Santiago); and southern Baja California (San Jose del Cabo). The **western** limits are along the Pacific coast from Baja California (San Jose del Cabo, La Paz, and El Rayo) north to southwestern British Columbia (Victoria and Nanaimo).

Winter range.—During the winter season these birds are apparently concentrated chiefly in the Amazon Valley of Brazil (Manaqueri, Barra do Rio Negro, and Itaituba). They have been found also at this season in the coastal regions of Brazil (San Luiz, Rosario, Rio de Janeiro, and Iguape). One banded (B–219327) at Winona, Minn., in May 1934 was recovered in December 1936 near Para, Brazil.

The range as outlined is for the species, which has been separated into eastern and western forms. The common purple martin (*Progne subis subis*) is the race found in the United States and Canada; the western martin (*P. s. hesperia*) breeds only in Baja California. Its winter range has not yet been discovered.

Spring migration.—Early dates of spring arrival in the United States and Canada, are: Florida—Orlando, January 27; Pensacola, February 5. Georgia—Savannah, February 21. South Carolina—Charleston, February 16. North Carolina—Raleigh, March 16. District of Columbia—Washington, March 9. Pennsylvania—Philadelphia, March 11. New Jersey—Vineland, March 29. New York—Shelter Island, April 12. Connecticut—Jewett City, April 2. Massachusetts—Boston, April 6. Vermont—Rutland, April 10. Maine—Lewiston, April 12. New Brunswick—Scotch Lake, April 21. Quebec—Quebec City, April 30. Louisiana—New Orleans, January 31. Arkansas—Helena, February 18. Tennessee—Knoxville, March 15. Missouri—St. Louis, March 15. Illinois—Olney, March 3. Indiana—Fort Wayne, March 19. Ohio—Youngstown, March 16. Michigan—Ann Arbor, March 16. Ontario—Toronto, April 6. Iowa—

Iowa City, March 7.   Wisconsin—Madison, March 29.   Minnesota—Minneapolis, March 31.   Manitoba—Aweme, April 25.   Texas—Grapevine, January 20.   Oklahoma—Norman, March 7.   Kansas—Topeka, March 18.   Nebraska—Beatrice, March 24.   South Dakota—Vermillion, April 2.   North Dakota—Fargo, April 19.   Saskatchewan—Indian Head, May 4.   Arizona—Huachuca Mountains, April 22.   Colorado—Loveland, April 21.   Montana—Great Falls, May 10.   Alberta—Red Deer, May 7.   California—Stockton, March 1.   Washington—Tacoma, April 9.   British Columbia—Burrard Inlet, April 20.

Fall migration.—Late dates of fall departure are: Washington—Seattle, September 16.   Alberta—Glenevis, August 30.   Colorado—Lost Park, September 1.   Arizona—Pima County, September 20.   North Dakota—Fargo, September 4.   South Dakota—Aberdeen, September 14.   Kansas—Onago, September 28.   Oklahoma—Oklahoma City, September 16.   Texas—Grapevine, October 10.   Manitoba—Aweme, September 21.   Minnesota—Minneapolis, October 21.   Wisconsin—Sheboygan, September 27.   Iowa—New Sharon, September 24.   Ontario—Toronto, September 20.   Michigan—Newberry, September 27.   Ohio—Columbus, October 7.   Indiana—Hobart, September 30.   Illinois—Chicago, September 27.   Missouri—Concordia, September 13.   Tennessee—Knoxville, October 5.   Mississippi—Biloxi, October 9.   Louisiana—New Orleans, October 22.   Quebec—Montreal, September 25.   New Brunswick—Scotch Lake, September 12.   Maine—Pittsfield, September 12.   Vermont—Rutland, September 15.   Massachusetts—Harvard, September 12.   New York—Geneva, September 12.   Pennsylvania—Doylestown, September 23.   District of Columbia—Washington, September 23.   North Carolina—Chapel Hill, September 10.   South Carolina—Charleston, October 6.   Georgia—Atlanta, September 23.   Florida—St. Augustine, October 19.

Casual records.—North of the normal range in Canada the purple martin has been recorded on a few occasions. Among these are: Quebec, recorded at Godbout on May 20, 1896; and northern Alberta, several reported seen on the White Mud River, in June 1903, and ten or a dozen at Fort McMurray on June 7, 1928.   One was taken at Cape Prince of Wales, Alaska, on June 3, 1929.   A specimen was captured alive during the winter of 1899–1900 at Nassau, New Providence, Bahama Islands, and one was collected at Palmira, Argentina, on December 15, 1901.

Egg dates.—California: 33 records, May 5 to July 14; 17 records, June 2 to 6, indicating the height of the season.

Florida: 7 records, March 26 to May 19.

Baja California: 1 record, May 3.

New Jersey: 9 records, May 31 to June 1.

Ontario: 3 records, May 30 to June 18.
Texas: 6 records, April 30 to June 2.
Wisconsin: 9 records, May 18 to July 19.

PROGNE SUBIS HESPERIA Brewster

WESTERN MARTIN

HABITS

William Brewster (1889) described and named this race of the purple martin from a large series collected in southern Lower California, where the race shows its most pronounced characters. The males of this race are indistinguishable, he said, from the males of the eastern race, only the females showing the characters on which the subspecies is based. He described the female as "differing from female *subis* in having the abdomen, anal region, crissum, and under tail-coverts pure white, nearly or quite immaculate, the throat, breast, flanks, forehead, fore part of crown and nuchal collar grayish white, the feathers of the back and rump conspicuously edged with grayish or pale brown, the bend of the wing and the under wing-coverts mottled profusely with whitish." He gave as its habitat: "California (Ojai Valley) and Lower California (Sierra de la Laguna)." He included the California locality in the range because, he said, "Mr. Batchelder has two females from the Ojai Valley, California, which are practically identical with my Lower California specimens."

Later Mr. Ridgway (1904), in describing this race, called attention to its slightly smaller size but did not emphasize the pure white of the posterior under parts. He extended its range still farther northward to British Columbia, apparently including the whole intervening Pacific slope. Based on the findings of these two eminent ornithologists, the name *hesperia* has been applied for many years to the purple martins breeding west of the Rocky Mountains. This is not surprising when we consider the somewhat difficult problem in systematics involved, which Dr. Joseph Grinnell (1928a) attempted to solve and whose solution has been accepted in the 1931 Check-list. Grinnell made a careful study of a series of 99 specimens of purple martins from California, Lower California, and eastern localities and discovered that there was a gradual diminution in size between the most northern-breeding birds and those from southern Lower California; and he noted that there was also a progressive increase in the amount of white on the under parts between the northern and the southern birds. The problem was whether to include the Pacific coast birds under the eastern race, *subis*, to leave them as they were, under *hesperia*, or to describe a new subspecies to include the birds from California northward. To adopt the last course would be naming a third subspecies, which is obviously intermediate; so he wisely accepted the first alternative, which restricts the name *hesperia* to the birds that breed "in the peninsula of Lower California from the Cape district north at least to lattitude 31." As for comparison with eastern birds, he was "unable to find any satisfactory mensural differences between upper Californian and Eastern birds, not even as to forking of tail; * * * there is greater aggregate difference between birds of upper California and those of southern Lower California than there is between the former and those of the eastern United States; the 'slightly paler' coloration of California females is a difficult diagnostic character to use, because of the great range of variation in intensity and extent of the whiteness— many Eastern and Californian birds being indistinguishable; in the material examined, all Cape district females are distinguishable from Eastern females on basis of color, while nearly all males, as well as females, are distinguishable on basis of size."

Mr. Brewster (1902) had this to say about the habits of the western martin:

Regularly each afternoon, during May and the first week of June, a few congregated over an open space in front of a hunter's cabin. They usually flew at a considerable height, but the males every now and then pitched downward nearly to the earth, descending with great velocity and making a booming noise very like that of the eastern Nighthawk. This remarkable habit, unknown in the common Martin, was constantly practised here, but, curiously enough, it was not once observed at Triunfo, where Mr. Frazar found the Western Martins abundant during the last three weeks of June. Belonging to the mine at this latter place, was an immense wood-pile covering over three acres and harboring great numbers of long-horned beetles upon which the Martins and Texan Nighthawks feed greedily. The Martins appeared every afternoon, a little before sunset, to the number of two or three hundred, and skimmed back and forth over the wood-pile until twilight fell. * * * They disappeared suddenly and totally, immediately after a succession of heavy showers early in July, and were not afterwards met with excepting at San José del Cabo, where a few, evidently migrating, were seen passing southward in late August and early September.

Mr. Bryant records the Western Martin from several places in the northern portions of Lower California, and says that it has been found nesting by Mr. Belding in dead pines at Hansen's. Mr. Anthony states that in the neighborhood of San Fernando, it is "not uncommon at the mission and an occasional pair was seen in other localities, nesting in Woodpecker holes in the giant cactus."

The measurements of the only two eggs, from Lower California, that I have been able to locate are 25.4 by 17.3 and 24.9 by 17.3 millimeters, hardly differing any in size or in other respects from eggs of the species elsewhere.

PROGNE CRYPTOLEUCA Baird

CUBAN MARTIN

HABITS

The name *cryptoleuca*, meaning hidden white, was aptly applied to this species on account of the concealed bands of white on the feathers of the lower abdomen in the adult male. The male is otherwise like our common purple martin. The female has much white on the posterior under parts, but it is not concealed, the immaculate white of the breast, abdomen, anal region, and under tail coverts being sharply contrasted with the grayish brown of the chest and sides.

This martin, a straggler from Cuba, owes its place on our list to two accidental occurrences in Florida. There are two specimens in the United States National Museum: one was taken at Cape Florida on May 18, 1858, and the other at Clearwater on an unknown date, but in summer plumage. It was formerly supposed to breed in southern Florida, but all breeding specimens have been proved to be our common purple martin. The Cuban martin is now supposed to breed only in Cuba and on the neighboring Isle of Pines. Dr. Thomas Barbour (1923) says of its haunts in Cuba:

The Cuban Martins arrive in the cities in large numbers, and from late February to late August they swarm about their chosen belfries. Santo Domingo church tower in Havana is a great favorite. Saledad in Camaguey, the old church in the plaza at Santa Clara, the Ayuntamiento at Matanzas, the *parroquia* at Guane, the eaves of an old apothecary shop at Sumidero, are all favorite haunts. It readily will be seen that these are not advantageous collecting-grounds. We got four one morning at Sumidero; they fell on the sidewalk and much uncomplimentary comment resulted. One I shot flying over woods at Palo Alto, from a few pairs that seemed to be preparing to nest in a great dead ceiba. The people are fond of the birds, and have transferred to this species the familiar legend of the Crossbill. The crosses which surmount all Latin churches are constantly preferred perches.

W. E. Clyde Todd (1916) says of its status on the Isle of Pines:

Mr. [A. C.] Read appears to be the only observer to have met with it in any numbers. He states that it is a summer resident only, appearing as early sometimes as February 8 (1914), March 12 (1912), and March 28 (1910), and remaining until about the first of November. This agrees with what is known concerning its seasonal status in Cuba, where Gundlach says that it disappears towards the end of August and does not return until February. What becomes of it in the intervening months remains an unexplained mystery, since it is a species scarcely known outside of its recognized breeding-range. Mr. Read has also had the good fortune to find it breeding. The nest appears to be built in an old woodpecker's-hole in a bottle-palm or pine-tree, and the four or five white eggs are laid in May.

The molts and plumages apparently follow the same sequence as in the purple martin. Nothing seems to have been published on the food of this species. The adult male cannot be distinguished in the field from the purple martin, but the abrupt contrast of the white belly with the gray of the sides and chest is a good field mark for the adult female.

I have been unable to locate any eggs of this species but have no reason to think that they differ in any respect from those of the purple martin.

DISTRIBUTION

Range.—During the breeding season the Cuban martin is confined to Cuba and the Isles of Pines. The winter range is unknown, but in spring migration it has been recorded from Jamaica and also from Guatemala (Quirigua and Gualan).

Casual records.—A specimen was taken at Cape Florida on May 18, 1858, and another (date?) at Clearwater, Fla.

PROGNE CHALYBEA CHALYBEA (Gmelin)

GRAY-BREASTED MARTIN

HABITS

The gray-breasted martin, a well-marked species, enjoys a wide distribution in Central America and northern South America, breeding from the lower Rio Grande Valley in Texas southward through Mexico and Central America to Peru and northern Brazil. An allied race occurs in Bolivia and southern Brazil. Throughout all this wide range it seems to show a decided preference for the cities and towns or the vicinity of human habitations, nesting in or on various buildings, or in bird-boxes. Referring to El Salvador, Dickey and van Rossem (1938) write:

Gray-breasted martins have taken as kindly to civilization as have their northern relatives, and during the breeding season are to be found chiefly in the vicinity of towns and villages. There were, in 1912, 1925, 1926, and 1927, several colonies scattered about the city of San Salvador, where birds could be seen entering openings under eaves of some of the taller buildings. In that city also there was for several years, and probably still is, a populous martin roost in the trees over the band-stand in Parque Barrios, where the birds were in no wise disturbed by the nightly concerts. In rural districts they congregate about the village churches, since such usually offer the most secure nesting sites. They were also observed, at Santo Tomás, to enter crevices under the roofing tiles of low, one-story buildings. After the breeding season there seems to be a general dispersal over the entire country within the limits of the Arid Lower Tropical Zone.

And G. Inness Hartley (Beebe, Hartley, and Howes, 1917) says: "In all civilized districts from Rio northward this is the first bird to greet the traveler. As the steamer warps up to the pier there are always a few perched on the ridge of a nearby roof or garrulously hovering over the deck. Proceeding inland by water or on foot one will see them always present, where human habitations exist."

Mr. Hartley has written a full chapter on the habits of the gray-breasted martin, as observed by him at Kalacoon House, Hills Estate, Mazaruni River, British Guiana. Most of what follows is quoted or condensed from what he wrote.

Courtship.—A female, he says, "would be sitting alone, awaiting her mate by the prospective nest. Suddenly, after many beautiful evolutions in the air, he would join her, and their admiration for one another was shown by wide open bills and a perfect babble of warbles. They would sit thus for a few moments each with its mouth open, or they snapped their bills at imaginary insects, as if one were urging the other to feed it. Then each would seek to relieve its feelings in flight, only to return later and repeat the whole performance."

Nesting.—The nestings operations are described as follows:

A small box with four compartments had been erected a short time before, on a pole, with the hope that some of the birds would take advantage of it. Immediately a pair of palm tanagers took possession. This was too much for our pair of martins, which at once—incited by jealousy and need for a new home—drove away the tanagers and appropriated the partially completed nest as their own. The occupation was not accomplished, however, without many a scuffle with the original tenants and other pairs of martins who had nesting ideas of their own. * * *

In the meantime other pairs had commenced to build, selecting various portions of the huge beams that acted as plates for Kalacoon house. The nests were composed of sticks, straws, dried grass, string, cloth and anything that would act as building material. They were placed back from the edge of the beam usually in a corner next to a floor joist. * * * Unlike the purple martin, the too near company of others was not desired and it went hard with the individual who inadvertently overstepped his neighbor's territory. * * *

The Kalacoon martins commenced to lay about the first of April. Every bird had been busy for the two preceding weeks collecting material, courting, and fighting. Sometimes a dozen or more would gather on the ground in front of the house and sort over the little twigs and dried grass blades lying there. This always was attended with perfect harmony until two birds would decide that they both liked the same stick. They resorted then to force in the dispute that followed, and the fight would go on up in the air or down on the ground, until both were exhausted. In the meantime the object of their differences was usually spirited away by a third party. At any rate they always forgot what they were fighting about and never returned to the spot to look for it.

He tried the experiment of marking, with blotches of black ink, a set of eggs to see if the parents would recognize them; they seemed quite a little disturbed, but one of them finally settled on the eggs.

The following day he removed "both nest and eggs, putting them in a prominent spot, only a few feet away from their original nesting place." The birds were much excited, flying about and returning again and again to the original spot; but, though the eggs were in plain sight, they never found them; evidently, it was the location and not the nest that they recognized. These martins apparently raise two broods in a season.

Eggs.—The gray-breasted martin lays three to five eggs. These are like purple martin's eggs, pure unmarked white, ovate, and with very little if any gloss. The measurements of 4 eggs are 22.3 by 15.2, 22.2 by 15.0, 22.0 by 15.5, and 22.0 by 15.0 millimeters.

Young.—Mr. Hartley says: "During the period of incubation, which lasted from fifteen to sixteen days, the male showed much solicitude for his mate. He sat for hours by her side near the nest and chirped and twittered in low sweet tones as if striving to enliven the monotony of her somewhat irksome position. Several times each day, though only for a few minutes, she took journeys in search of food."

Both parents were busy in feeding the young, which "went on all day long, from early morning till late at night. * * * After every third or fourth trip, one of the parents cleaned nest with its bill, carrying away the excrement incased in its thin shiny sack, to drop it at a safe distance from the house."

The food of the young "consisted entirely of insects—flying ants, termites, ant-lions and dragon-flies." Mr. Hartley continues:

Sometimes a dragon-fly was brought of too large dimensions to be easily swallowed whole. Then the wings were severed, one by one, from the body, which was well crushed by the bill of the parent. The youngster would seize it fiercely and swallow it with incredible rapidity, undergoing terrible contortions, gasping and choking for several minutes after it had gone down.

The young birds were lined up at the edge of the beam, twenty-two days after hatching, ready to begin their trials of flight. They returned to their nests for a few nights and then, having partly learned to care for themselves, departed elsewhere to roost. * * *

The art of catching their meal did not come quite so easily as the first flights. They had to be fed for a week or more after they were dodging and darting about in the air, and some even clamored for food after their parents were nesting again. * * * It gradually dawned upon them, as time went on, that they might secure their food themselves, as well as from their parents. But this came only after the elders had dropped one or two insects which made the youngsters scramble to secure them before they escaped.

Plumages.—The sequence of molts and plumages seems to be the same as in the purple martin. According to Ridgway (1904) the sexes are alike in the first year plumage, being "similar to the adult female, but much duller in color, the upper parts dark grayish sooty or dull grayish black, slightly if at all glossed, the gray of anterior under parts fading gradually into the white of posterior portions, and inner web of exterior pair of rectrices with an indistinct dull whitish or pale grayish subterminal spot on edge; sides and flanks sometimes faintly tinged with pale brown."

He says that the second year young male is "exactly like the adult female in coloration," which, in turn, is not strikingly different from the adult male, being duller and paler with the steel-blue gloss of the upper parts less continuous and the forehead and crown more sooty.

Food.—Nothing seems to have been published on the food of this martin, except what is mentioned above in the food of the young. But it is fair to assume that, like other martins, it will capture whatever flying insects it can find available within its habitat.

Behavior.—Mr. Hartley (Beebe, Hartley, and Howes, 1917) says of the gray-breasted martins, as he observed them in British Guiana:

They are extremely tame and unafraid and because of this courage and pugnacity they are one of the most useful birds that gather about the homestead. No low flying hawk will for long withstand the vicious onslaughts of the many martins that gather about him. Thus the life of many a seed-eating finch and caterpillar-destroying wren has been preserved.

The windows of Kalacoon house always remained open and soon after our arrival several martins took advantage of this to roost on the rafters over our heads, entering through a window close beneath the peak of the roof. On the rare occasions when it had to be closed on account of the rain which poured through in gusts, the birds gathered outside in numbers, some on the sill and others on the eaves above, and tried to express their troubles in a loud bubbling and chatter. Though there were other open windows nearby, they never used them, but always, if their own private entrance were closed, sought other roosting places for the night. They roosted in pairs and never allowed a third to encroach upon what they considered their own territory.

DISTRIBUTION

Range.—South and Central America; casual in the Rio Grande Valley, Texas.

This is a more or less nonmigratory species (of several subspecies) that ranges **north** to Nayarit (San Blas); Coahuila (Sabinas); Tamaulipas (Ciudad Victorio); Yucatan (Shkolak, Izamal, and Chichen-Itza); British Guiana (Island of Trinidad and Georgetown); Dutch Guiana (Paramaribo and Maroni River); French Guiana (Cayenne); and northwestern Brazil (Island of Mixiana and Para). **East** to Brazil (Para, Goyaz, and Rio Araguaya); Uruguay (Rocha and Montevideo); and eastern Argentina (La Plata and Cape San Antonio). **South** to central Argentina (Cape San Antonio, Cordoba, and Tucuman). **West** to northwestern Argentina

(Tucuman); Bolivia (Santa Cruz and Ignacito); Peru; Ecuador (Bucay and Esmeraldas); Colombia (Caqueta and Cali); Panama (Colon, Barro Colorado Island, and Volcan de Chiriqui); Costa Rica (Boruca, San Jose, and Bagaces); Guatemala (Cahabon and San Geronimo); Chiapas (Huehuetan); Oaxaca (Tehuantepec and Santo Domingo); Guerrero (Egid Nuevo); and Nayarit (San Blas).

Casual records.—A specimen was taken at Rio Grande City, Tex., on April 25, 1880, and another was collected at Hildago, Tex., on May 18, 1889.

Egg date.—Mexico: 1 record, April 23.

LITERATURE CITED

ABBOTT, CLINTON GILBERT.
   1922. The friendly phoebe.   Bird-Lore, vol. 24, pp. 75–79.
ALEXANDER, HORACE GUNDRY.
   1935. A chart of bird song.   British Birds, vol. 29, pp. 190–198.
ALLEN, AMELIA SANBORN.
   1921. The season.—San Francisco region.   Bird-Lore, vol. 23, p. 255.
ALLEN, FRANCIS HENRY.
   1913. More notes on the morning awakening.   Auk, vol. 30, pp. 229–235.
   1922. Some little known songs of common birds.   Nat. Hist., vol. 22, pp. 235–242.
ALLEN, FRANCIS HENRY, and GRISCOM, LUDLOW.
   1932. The gray kingbird in Massachusetts.   Auk, vol. 49, pp. 87–88.
ALLEN, GLOVER MORRILL.
   1903. A list of the birds of New Hampshire.   Proc. Manchester Inst. Arts and Sci., vol. 4, pp. 23–222.
AMERICAN ORNITHOLOGISTS' UNION.
   1931. Check-list of North American birds.   Ed. 4.
ANTHONY, ALFRED WEBSTER.
   1923. Ants destructive to bird life.   Condor, vol. 25, pp. 132–133.
AUDUBON, JOHN JAMES.
   1831. Ornithological biography, vol. 1.
   1834. Ornithological biography, vol. 2.
   1838. Ornithological biography, vol. 4.
   1840. The birds of America, vol. 1.
   1841. The birds of America, vol. 3.
AUSTIN, OLIVER LUTHER, Jr., and LOW, SETH HASKELL.
   1932. Notes on the breeding of the tree swallow.   Bird-Banding, vol. 3, pp. 39–44.
BAERG, WILLIAM J.
   1930. The song period of birds of northwest Arkansas.   Auk, vol. 47, pp. 32–40.
BAILEY, ALFRED MARSHALL.
   1928. Notes on the winter birds of Chenier au Tigre, Louisiana.   Auk, vol. 45, pp. 271–282.
BAILEY, ALFRED MARSHALL, and WRIGHT, EARL GROVER.
   1931. Birds of southern Louisiana.   Wilson Bull., vol. 43, pp. 190–219.
BAILEY, FLORENCE MERRIAM.
   1896. Notes on some birds of southern California.   Auk, vol. 13, pp. 115–124.
   1902a. The scissor-tailed flycatcher in Texas.   Condor, vol. 4, pp. 30–31.
   1902b. Handbook of birds of the Western United States.
   1918. A return to the Dakota lake region.   Condor, vol. 20, pp. 110–114.
   1928. Birds of New Mexico.
BAILEY, HAROLD HARRIS.
   1913. The birds of Virginia.
   1925. The birds of Florida.

BAILEY, HARRY BALCH.
    1883. Memoranda of a collection of eggs from Georgia. Bull. Nuttall Orn.
        Club, vol. 8, pp. 37–43.
BAILEY, WILLIAM LLOYD.
    1915. Young kingbirds on a cherry and dragon-fly diet. Auk, vol. 32, pp.
        368–369.
    1916. Yellow-bellied flycatcher (*Empidonax flaviventris*) breeding on the
        Pocono Mountain, Pa. Auk, vol. 33, p. 200.
BAIRD, SPENCER FULLERTON; BREWER, THOMAS MAYO; and RIDGWAY, ROBERT.
    1905. A history of North American birds. Land birds, vol. 2.
BAIRD, SPENCER FULLERTON; CASSIN, JOHN; and LAWRENCE, GEORGE NEWBOLD.
    1860. The birds of North America.
BAIRD, WILLIAM McFUNN, and BAIRD, SPENCER FULLERTON.
    1843. Descriptions of two species, supposed to be new, of the genus *Tyran-
        nula* Swainson, found in Cumberland County, Pennsylvania. Proc.
        Acad. Nat. Sci. Philadelphia, vol. 1, pp. 283–285.
BALL, WILLIAM HOWARD, and WALLACE, ROBERT BROWNE.
    1936. Further remarks on birds of Bolling Field, D. C. Auk, vol. 53, pp.
        345–346.
BANCROFT, GRIFFING.
    1930. The breeding birds of central Lower California. Condor, vol. 32, pp.
        20–49.
BANGS, OUTRAM.
    1914. The Bahama swallow in Cuba. Auk, vol. 31, p. 401.
BARBOUR, THOMAS.
    1923. The birds of Cuba. Mem. Nuttall Orn. Club, No. 6.
BARBOUR, THOMAS, and BROOKS, WINTHROP SPRAGUE.
    1917. Two new West Indian birds. Proc. New England Zool. Club, vol. 6,
        pp. 51–52.
BARLOW, CHESTER.
    1901. A list of the land birds of the Placerville-Lake Tahoe stage road.
        Condor, vol. 3, pp. 151–184.
BARROWS, WALTER BRADFORD.
    1912. Michigan bird life.
BATES, ABBY FRANCES CALDWELL.
    1895. A swallow roost at Waterville, Maine. Auk, vol. 12, pp. 48–51.
    1901. Maine bird notes. Auk, vol. 18, pp. 400-401.
BEAL, FOSTER ELLENBOROUGH LASCELLES.
    1897. Some common birds in their relation to agriculture. U. S. Dept. Agr.
        Farmers' Bull. 54.
    1907. Birds of California in relation to the fruit industry, pt. 1. U. S. Dept.
        Agr. Biol. Surv. Bull. 30.
    1910. Birds of California in relation to the fruit industry, pt. 2. U. S. Dept.
        Agr. Biol. Surv. Bull. 34.
    1912. Food of our more important flycatchers. U. S. Dept. Agr. Biol. Surv.
        Bull. 44.
    1918. Food habits of the swallows, a family of valuable native birds. U. S.
        Dept. Agr. Bull. 619.
BEEBE, CHARLES WILLIAM.
    1905. Two bird-lovers in Mexico.
BEEBE, WILLIAM; HARTLEY, GEORGE INNESS; and HOWES, PAUL GRISWOLD.
    1917. Tropical wild life in British Guiana. Zool. Contr. Tropical Research
        Stat. New York Zool. Soc.

BELKNAP, B. H.
1938. Phoebes versus tent caterpillars. Univ. New York, Bull. to Schools, vol. 24, pp. 121–124.

BENDIRE, CHARLES EMIL.
1895. Life histories of North American birds. U. S. Nat. Mus. Spec. Bull. 3.

BENT, ARTHUR CLEVELAND, and COPELAND, MANTON.
1927. Notes on Florida birds. Auk, vol. 44, pp. 371–386.

BEYER, LEONARD K.
1938. Nest life of the bank swallow. Wilson Bull., vol. 60, pp. 122–137.

BICKNELL, EUGENE PINTARD.
1879. Capture of two rare birds at Riverdale, N. Y. Bull. Nuttall Orn. Club, vol. 4, pp. 60–61.
1885. A study of the singing of our birds. Auk, vol. 2, pp. 249–262.

BISHOP, LOUIS BENNETT.
1896. Descriptions of a new horned lark and a new song sparrow, with remarks on Sennett's nighthawk. Auk, vol. 13, pp. 129–135.
1900a. Descriptions of three new birds from Alaska. Auk, vol. 17, pp. 113–120.
1900b. Birds of the Yukon region, with notes on other species. North Amer. Fauna 19.

BLAKE, FRANCIS GILMAN.
1907. The nesting of *Stelgidopteryx serripennis* in Norwich, Vt. Auk, vol. 24, pp. 103–104.

BOLLES, FRANK.
1890. Snake skins in the nests of *Myiarchus crinitus*. Auk, vol. 7, p. 288.

BOND, JAMES.
1936. Birds of the West Indies.

BOWDISH, BEECHER SCOVILLE.
1903. Birds of Porto Rico. Auk, vol. 20, pp. 10–23.
1930. An attempt to restore the cliff swallow to New Jersey. Auk, vol. 47, pp. 189–193.

BOWLES, JOHN HOOPER.
1900. Nesting of the streaked horned lark. Condor, vol. 2, pp. 30–31.

BOWLES, JOHN HOOPER, and DECKER, FRANK RUSSELL.
1927. A comparative field study of Wright's and Hammond's flycatchers. Auk, vol. 44, pp. 524–528.

BOYD, ARNOLD WHITWORTH.
1935. Report on the swallow enquiry, 1934. British Birds, vol. 29, pp. 3–21.

BRAND, ALBERT RICH.
1938. Vibration frequencies of passerine bird song. Auk, vol. 55, pp. 263–268.

BRANDT, HERBERT.
1940. Texas bird adventures, in the Chisos Mountains and on the northern plains.

BREWSTER, WILLIAM.
1878. Descriptions of the first plumage in various species of North American birds. Bull. Nuttall Orn. Club, vol. 3, pp. 56–64.
1882a. Impressions of some southern birds. Bull. Nuttall Orn. Club, vol. 7, pp. 94–104.
1882b. On a collection of birds lately made by Mr. F. Stephens in Arizona. Bull. Nuttall Orn. Club, vol. 7, pp. 193–212.
1886. An ornithological reconnaissance in western North Carolina. Auk,

BREWSTER, WILLIAM—Continued.

1888. Descriptions of supposed new birds from Lower California, Sonora, and Chihuahua, Mexico, and the Bahamas. Auk, vol. 5, pp. 82–95.

1889. Descriptions of supposed new birds from western North America and Mexico. Auk, vol. 6, pp. 85–98.

1891. Descriptions of seven supposed new North American birds. Auk, vol. 8, pp. 139–149.

1895. Notes on certain flycatchers of the genus *Empidonax*. Auk, vol. 12, pp. 157–163.

1897. The Bahaman swallow in Florida. Auk, vol. 14, pp. 221–222.

1898. Revival of the sexual passion in birds in autumn. Auk, vol. 15, pp. 194–195.

1902. Birds of the Cape region of Lower California. Bull. Mus. Comp. Zool., vol. 41, No. 1, pp. 1–241.

1903. A tragedy in nature. Bird-Lore, vol. 5, pp. 151–152.

1906. The birds of the Cambridge region of Massachusetts. Mem. Nuttall Orn. Club, No. 4.

1907. Breeding of the rough-winged swallow in Berkshire County, Massachusetts. Auk, vol. 24, pp. 221–222.

1936. October Farm.

1937. The birds of the Lake Umbagog region of Maine, pt. 3. Bull. Mus. Comp. Zool., vol. 66, pp. 408–521.

1938. The birds of the Lake Umbagog region of Maine, pt. 4. Bull. Mus. Comp. Zool., vol. 66, pp. 525–620.

BRIMLEY, CLEMENT SAMUEL.

1889. Nesting of the Acadian flycatcher at Raleigh, N. C. Ornithologist and Oologist, vol. 14, pp. 136–137.

BRIMLEY, HERBERT HUTCHINSON.

1934. Unusual actions of a phoebe (*Sayornis phoebe*). Auk, vol. 51, pp. 237–238.

BROOKS, ALLAN.

1909. Some notes on the birds of Okanagan, British Columbia. Auk, vol. 26, pp. 60–63.

BROWN, FRANK ARTHUR.

1910. Cliff swallows. Bird-Lore, vol. 12, pp. 137–138.

BROWNELL, WILL C.

1887. Nesting of the Trail's [sic] and Acadian flycatchers. Oologist, vol. 4, pp. 96, 97.

BRYANT, HAROLD CHILD.

1914. Birds as destroyers of grasshoppers in California. Auk, vol. 31, pp. 168–177.

BRYANT, HENRY.

1859. A list of birds seen at the Bahamas, from Jan. 20th to May 14th, 1859, with descriptions of new or little known species. Proc. Boston Soc. Nat. Hist., vol. 7, pp. 102–134.

BRYENS, OSCAR MCKINLEY.

1932. Two pairs of tree swallows mated during two seasons. Bird-Banding, vol. 3, pp. 177–178.

BURLEIGH, THOMAS DEARBORN.

1930. Notes on the bird life of northwestern Washington. Auk, vol. 47, pp. 48–63.

BURNS, FRANKLIN LORENZO.
1915. Comparative periods of deposition and incubation of some North American birds. Wilson Bull., vol. 27, pp. 275–286.
1921. Comparative periods of nestling life of some North American Nidicolae. Wilson Bull., vol. 33, pp. 90–99.

BUTLER, AMOS WILLIAM.
1897. The birds of Indiana. 22d Ann. Rep. Indiana Dept. Geol. and Nat. Resources, pp. 515–1197.

BUTLER, ARTHUR GARDINER.
1896. British birds with their nests and eggs.

CAMERON, EWEN SOMERLED.
1907. The birds of Custer and Dawson Counties, Montana. Auk, vol. 24, pp. 389–406.
1908. The birds of Custer and Dawson Counties, Montana. Auk, vol. 25, pp. 39–56.

CAMPBELL, LOUIS WALTER.
1932. A large flock of bank swallows near Toledo, Ohio. Wilson Bull., vol. 44, pp. 118–119.
1934. A family of Arkansas kingbirds near Toledo, Ohio. Auk, vol. 51, p. 85.
1936. An unusual colony of alder flycatchers. Wilson Bull., vol. 48, pp. 164–168.

CARPENTER, NELSON KITWOOD.
1918. Observations in a swallow colony. Condor, vol. 20, pp. 90–91.

CHANEY, RALPH WORKS.
1910. Summer and fall birds of the Hamlin Lake region, Mason County, Mich. Auk, vol. 27, pp. 271–279.

CHAPMAN, FRANK MICHLER.
1898. Notes on birds observed at Jalapa and Las Vigas, Vera Cruz, Mexico. Bull. Amer. Mus. Nat. Hist., vol. 10, pp. 15–43.
1900. Bird studies with a camera.
1902. A hybrid between the cliff and tree swallows. Auk, vol. 19, pp. 392–394.
1912. Handbook of birds of eastern North America.
1917. Notes on the plumage of North American birds. Bird-Lore, vol. 19, pp. 330–331.

CHERRIE, GEORGE KRUCK.
1892. A preliminary list of the birds of San José, Costa Rica. Auk, vol. 9, pp. 322–329.

CHILDS, JOHN LEWIS.
1900. Tree swallows by the million. Auk, vol. 17, pp. 67–68.

CHRISTENSEN, RASMUS.
1927. [Photograph of a barn with 800 cliff swallow nests.] Bird-Lore, vol. 29, p. 246.

CHRISTY, BAYARD HENDERSON.
1932. The kingbird's crown-patch, *Tyrannus tyrannus* (Linnaeus). Cardinal, vol. 3, p. 69.
1940. Mortality among tree swallows. Auk, vol. 57, pp. 404–405.

COLLINGE, WALTER EDWARD.
1924–1927. The food of some British wild birds.

COOK, ALBERT JOHN.
1896. Food of woodpeckers and flycatchers. Auk, vol. 13, pp. 85–86.

COOKE, WELLS WOODBRIDGE.

    1888. Report on bird migration in the Mississippi Valley in the years 1884–1885. U. S. Dept. Agr., Div. Economic Orn., Bull. 2.

    1914. Some winter birds of Oklahoma. Auk, vol. 31, pp. 473–493.

COOPER, JAMES GRAHAM.

    1870. Geological survey of California. Ornithology, vol. 1. Land birds.

COTTAM, CLARENCE.

    1938. Nesting of an eastern kingbird in a deserted oriole nest. Condor, vol. 40, p. 259.

COTTAM, CLARENCE, and HANSON, HAROLD CARSTEN.

    1938. Food habits of some Arctic birds and mammals. Field Mus. Nat. Hist., zool. ser., vol. 20, pp. 405–426.

COTTAM, CLARENCE, and KNAPPEN, PHOEBE.

    1939. Food of some uncommon North American birds. Auk, vol. 56, pp. 138–169.

COUES, ELLIOTT.

    1878. Birds of the Colorado Valley. U. S. Geol. Surv. Terr. Misc. Publ. 11.

    1880. On the nesting in Missouri of *Empidonax acadicus* and *Empidonax trailli*. Bull. Nuttall Orn. Club, vol. 5, pp. 20–25.

CRAIG, WALLACE.

    1926. The twilight song of the wood pewee: A preliminary statement. Auk, vol. 43, pp. 150–152.

    1933. The music of the wood pewee's song and one of its laws. Auk, vol. 50, pp. 174–178.

CRAM, ELOISE BLAINE.

    1927. Bird parasites of the nematode suborders Strongylata, Ascaridata, and Spirurata. U. S. Nat. Mus. Bull. 140.

CRIDDLE, STUART, and CRIDDLE, NORMAN.

    1917. Horned larks at Aweme, Manitoba. Ottawa Nat., vol. 30, pp. 144–148.

DANIEL, JOHN WARWICK, Jr.

    1902. Summer birds of the Great Dismal Swamp. Auk, vol. 19, pp. 15–18.

DAVIS, EDWARD M.

    1937. Observations on nesting barn swallows. Bird-Banding, vol. 8, pp. 66–72.

DAWSON, WILLIAM LEON.

    1923. The birds of California. Vols. 1 and 2.

DAWSON, WILLIAM LEON, and BOWLES, JOHN HOOPER.

    1909. The birds of Washington.

DEARBORN, NED.

    1903. The birds of Durham and vicinity, New Hampshire.

DECK, RAYMOND S.

    1934. Feathered philosophers. Bird-Lore, vol. 36, pp. 226–231.

DICE, LEE RAYMOND.

    1918. The birds of Walla Walla and Columbia Counties, southeastern Washington. Auk, vol. 35, pp. 40–51 and 148–161.

DICKEY, DONALD RYDER, and VAN ROSSEM, ADRIAAN JOSEPH.

    1938. The birds of El Salvador. Field Mus. Nat. Hist., zool. ser., vol. 23.

DIXON, JOSEPH SCATTERGOOD.

    1920. Nesting of the olive-sided flycatcher in Berkeley, California. Condor, vol. 22, pp. 200–202.

DOAN, WILLIAM D.

    1888. Birds of West Virginia. West Virginia Agr. Exp. Stat. Bull. 3.

DuBois, Alexander Dawes.
1935. Nests of horned larks and longspurs on a Montana prairie. Condor, vol. 37, pp. 56–72.
1936. Habits and nest life of the desert horned lark. Condor, vol. 38, pp. 49–56.
Dwight, Jonathan, Jr.
1890. The horned larks of North America. Auk, vol. 7, pp. 138–158.
1900. The sequence of plumages and molts of the passerine birds of New York. Ann. New York Acad. Sci., vol. 13, pp. 73–360.
Eifrig, Charles William Gustave.
1905. Ornithological results of the Canadian *Neptune* expedition to Hudson Bay and northward, 1903–1904. Auk, vol. 22, pp. 233–241.
1919. Notes on birds of the Chicago area and its immediate vicinity. Auk, vol. 36, pp. 513–524.
Elliott, John J.
1939. Wintering tree swallows at Jones Beach fall and winter of 1937 and 1938. Bird-Lore, vol. 41, pp. 11–16.
Ellison, Lincoln.
1936. Unusual nesting site of the eastern kingbird. Condor, vol. 38, p. 216.
Errington, Paul Lester.
1932. Food habits of southern Wisconsin raptors. Condor, vol. 34, pp. 176–186.
Evans, William.
1891. On the periods occupied by birds in the incubation of their eggs. Ibis, 1891, pp. 52–93.
Evermann, Barton Warren.
1886. A list of the birds obtained in Ventura County, California. Auk, vol. 3, pp. 179–186.
1889. Birds of Carroll County, Indiana. Auk, vol. 6, pp. 22–30.
Eyles, Don Edgar.
1938. Great kingbird nesting in Georgia. Oriole, vol. 3, No. 3, pp. 24–25.
Fargo, William Gilbert.
1934. Walter John Hoxie. Wilson Bull., vol. 36, pp. 169–196.
Farley, John Austin.
1901a. The alder flycatcher (*Empidonax traillii alnorum*) as a summer resident of eastern Massachusetts. Auk, vol. 18, pp. 347–355.
1901b. Massachusetts bird notes. Auk, vol. 18, pp. 398–400.
Faxon, Walter.
1897. Purple martins (*Progne subis*) breeding in electric arc-light caps. Auk, vol. 14, pp. 407–408.
Faxon, Walter, and Hoffmann, Ralph.
1900. The birds of Berkshire County, Massachusetts. Coll. Berkshire Hist. and Sci. Soc., vol. 3, No. 2.
Fisher, Albert Kenrick.
1893. The Death Valley expedition. 1, Report on birds. North Amer. Fauna 7.
Fisher, George Clyde.
1907. A purple martin roost. Wilson Bull., vol. 19, p. 119.
Fletcher, Laurence Brown.
1926. An example of the tree swallow's marital relations. Bull. Northeastern Bird-Banding Assoc., vol. 2, p. 56.

FORBES, STEPHEN ALFRED, and GROSS, ALFRED OTTO.
  1922. Numbers and local distribution in summer of Illinois land birds of the open country. Bull. Illinois Nat. Hist. Surv. Div., Dept. Registr. and Educ., No. 14, pp. 187–218.
FORBUSH, EDWARD HOWE.
  1907. Useful birds and their protection.
  1927. Birds of Massachusetts and other New England States. Vol. 2.
  1929. Birds of Massachusetts and other New England States. Vol. 3.
FREER, RUSKIN SKIDMORE.
  1933. Nesting of the prairie horned lark in central Virginia. Wilson Bull., vol. 45, pp. 198–199.
FRIEDMANN, HERBERT.
  1925. Notes on the birds observed in the lower Rio Grande Valley of Texas during May, 1924. Auk, vol. 42, pp. 537–554.
  1929. The cowbirds: A study in the biology of social parasitism.
  1931. Additions to the list of birds known to be parasitized by the cowbirds. Auk, vol. 48, pp. 52–65.
GABRIELSON, IRA NOEL.
  1915. The home of the great crest. Wilson Bull., vol. 27, pp. 421–434.
  1922. Short notes on the life histories of various species of birds. Wilson Bull., vol. 34, pp. 193–210.
GALLUP, FRED.
  1917. A substitute for a hollow limb. Bird-Lore, vol. 19, pp. 139–141.
GARDNER, ASTON COLBROOK.
  1921. A kingbird's unusual nesting site. Auk, vol. 38, pp. 457–458.
GATES, BURTON W.
  1903. Swallow's nest on board boat. Bird-Lore, vol. 5, pp. 198–199.
GILMAN, MARSHALL FRENCH.
  1915. A forty acre bird census at Sacaton, Arizona. Condor, vol. 17, pp. 86–90.
GOODSELL, WILLIAM.
  1919. Helping barn and cliff swallows to nest. Bird-Lore, vol. 21, pp. 175–176.
GOSS, NATHANIEL STICKNEY.
  1886. A revised catalogue of the birds of Kansas.
GOULD, JOHN M.
  1906. Swallows in Portland. Journ. Maine Orn. Soc., vol. 8, p. 78.
GRINNELL, JOSEPH.
  1897. Report on the birds recorded during a visit to the islands of Santa Barbara, San Nicolas and San Clemente, in the spring of 1897. Pasadena Acad. Sci., Publ. 1.
  1908. The biota of the San Bernardino Mountains. Univ. California Publ. Zool., vol. 5, pp. 1–170.
  1914a. A second list of the birds of the Berkeley campus. Condor, vol. 16, pp. 28–40.
  1914b. An account of the mammals and birds of the lower Colorado Valley. Univ. California Publ. Zool., vol. 12, pp. 51–294.
  1915. A distributional list of the birds of California. Pacific Coast Avifauna, No. 11.
  1926. A new race of Say phoebe, from northern Lower California. Condor, vol. 28, pp. 180–181.
  1927. Six new subspecies of birds from Lower California. Auk, vol. 44, pp. 67–72.

GRINNELL, JOSEPH—Continued.

1928a. Notes on the systematics of west American birds. I. Condor, vol. 30, pp. 121–124.

1928b. Notes on the systematics of west American birds. III. Condor, vol. 30, pp. 185–189.

1937. The swallows at the Life Sciences Building. Condor, vol. 39, pp. 206–210.

GRINNELL, JOSEPH; DIXON, JOSEPH; and LINSDALE, JEAN MYRON.

1930. Vertebrate natural history of a section of northern California through the Lassen Peak region. Univ. California Publ. Zool., vol. 35.

GRINNELL, JOSEPH, and LINSDALE, JEAN MYRON.

1936. Vertebrate animals of Point Lobos Reserve, 1934–35.

GRINNELL, JOSEPH, and STORER, TRACY IRWIN.

1924. Animal life in the Yosemite.

GRISCOM, LUDLOW.

1923. Birds of the New York City region. Amer. Mus. Nat. Hist. Handbook Ser., No. 9.

1932. The distribution of bird-life in Guatemala. Bull. Amer. Mus. Nat. Hist., vol. 64.

1934. The ornithology of Guerrero, Mexico. Bull. Mus. Comp. Zool., vol. 75, pp. 367–422.

HALES, HENRY.

1904. An unusual nest of the cliff swallow. Bird-Lore, vol. 6, p. 67.

HANNA, WILSON CREAL.

1931. Odd nesting site of ash-throated flycatcher. Condor, vol. 33, pp. 216–217.

1933. House finch parasitized by dwarf cowbird and black phoebe nests occupied by house finch. Condor, vol. 35, p. 205.

HARDY, MANLY.

1885. Late occurrence of the phoebe (Sayornis fuscus) at Brewer, Maine. Auk, vol. 2, p. 108.

HARLOW, RICHARD CRESSON.

1912. The breeding birds of southern Center County, Pennsylvania. Auk, vol. 29, pp. 465–478.

1918. Notes on the breeding birds of Pennsylvania and New Jersey. Auk, vol. 35, pp. 18–29.

HARTERT, ERNST, and GOODSON, ARTHUR.

1917. Notes and descriptions of South American birds. Nov. Zool., vol. 24, pp. 410–419.

HARVIE-BROWN, JOHN ALEXANDER.

1908. Martin's nest built on a window-pane. British Birds, vol. 1, pp. 355–356.

HAUSMAN, LEON AUGUSTUS.

1925. On the utterances of the kingbird, Tyrannus tyrannus Linn., with especial reference to a recently recorded song. Auk, vol. 42, pp. 320–326.

HAVEN, HERBERT MAURICE WEST.

1926. Another Arkansas kingbird (Tyrannus verticalis), in Maine. Auk, vol. 43, p. 371.

HEAD, ANNA.

1903. Nesting habits of two flycatchers at Lake Tahoe. Bird-Lore, vol. 5, pp. 153–155.

HENDEE, RUSSELL W.
1929. Note on birds observed in Moffat County, Colorado. Condor, vol. 31, pp. 24–32.
HENDERSON, JUNIUS.
1927. The practical value of birds.
HENDRICKS, GEORGE BARTLETT.
1933. Squirrel and flycatcher. Bird-Lore, vol. 35, p. 209.
HENNINGER, WALTHER FRIEDRICH.
1916. Notes on some Ohio birds. Wilson Bull., vol. 28, pp. 86–88.
HENSHAW, HENRY WETHERBEE.
1875. Report upon the ornithological collections made in portions of Nevada, Utah, California, Colorado, New Mexico, and Arizona during the years 1871, 1872, 1873, and 1874. Wheeler's Rept. Expl. Surv. West 100th Merid., vol. 5, pp. 131–507.
1876. On two Empidonaces, *traillii* and *acadicus*. Bull. Nuttall Orn. Club, vol. 1, pp. 14–17.
1884. The shore larks of the United States and adjacent territory. Auk, vol. 1, pp. 254–268.
HERRICK, FRANCIS HOBART.
1905. The home life of wild birds.
1935. Wild birds at home.
HERSEY, FRANK SEYMOUR.
1933. Notes on tree swallows and bluebirds. Auk, vol. 50, pp. 109–110.
HESS, ISAAC ELNORE.
1910. One hundred breeding birds of an Illinois ten-mile radius. Auk, vol. 27, pp. 19–32.
HOFFMAN, EDWARD CARLTON.
1930. A phoebe nest in an abandoned house. Wilson Bull., vol. 42, p. 134.
HOFFMANN, RALPH.
1901. A chebec's second brood. Bird-Lore, vol. 3, pp. 160–162.
1904. A guide to the birds of New England and eastern New York.
1927. Birds of the Pacific States.
HOLLAND, HAROLD MAY.
1923. Black phoebes and house finches in joint use of a nest. Condor, vol. 25, p. 131–132.
HONYWILL, ALBERT WILLIAM, JR.
1911. Notes on summer and fall birds of the Crooked Lake region, Cass and Crow Wing Counties, Minn. Auk, vol. 28, pp. 229–237.
HORNE, DOM ETHELBERT.
1924. A swallow's method of feeding young with flies. British Birds, vol. 17, pp. 306–307.
HORTON, FRANCES B.
1903. Mortality of purple martins (*Progne purpurea*) at Brattleboro, Vt. Auk, vol. 20, pp. 435–436.
HOWARD, OZRA WILLIAM.
1899. Some of the summer flycatchers of Arizona. Bull. Cooper Orn. Club, vol. 1, pp. 103–107.
1904. The Coues flycatcher as a guardian of the peace. Condor, vol. 6, pp. 79–80.
HOWE, REGINALD HEBER, JR.
1902. Notes on various Florida birds. Contributions to North American Ornithology, vol. 1, pp. 25–32.

HOWELL, ARTHUR HOLMES.
    1924. Birds of Alabama.
    1932. Florida bird life.
HUDSON, WILLIAM HENRY.
    1920. Birds of La Plata. Vol. 1.
HUEY, LAURENCE MARKHAM.
    1927. Where do birds spend the night? Wilson Bull., vol. 39, pp 215–217.
INGERSOLL, ERNEST.
    1883. Knocking round the Rockies.
JACKSON, HARTLEY HARRAD THOMPSON.
    1923. Notes on summer birds of the Mamie Lake region, Wisconsin. Auk,
        vol. 40, pp. 478–489.
JACOBS, JOSEPH WARREN.
    1888. Acadian flycatcher. Oologist, vol. 5, p. 13.
    1924. The other egg in the nest. Oologist, vol. 41, pp. 52–54.
JENSEN, JENS KNUDSEN.
    1918. Notes on the nesting birds of Wahpeton, North Dakota. Auk, vol. 35,
        pp. 344–349.
JEWETT, FRANK BALDWIN.
    1899. Nesting observations on the black phoebe. Condor, vol. 1, p. 13.
JOHNSON, EVERETT E.
    1900. Notes on the ducks, swallows and thrushes. Journ. Maine Orn. Soc.,
        vol. 2, pp. 8–11.
JONES, CAROL.
    1922. A winter phoebe at Bennington, Vt. Bird-Lore, vol. 24, p. 94.
JONES, LYNDS.
    1910. The birds of Cedar Point and vicinity. Wilson Bull., vol. 22, pp.
        97–115.
    1912. A study of the avi-fauna of the Lake Erie islands. Wilson Bull.,
        vol. 24, pp. 171–186.
KELSO, LEON HUGH.
    1931. Some notes on young desert horned larks. Condor, vol. 33, pp. 60–65.
    1939. The violet-green swallow at the nest. Oologist, vol. 56, pp. 90–92.
KENNARD, FREDERIC HEDGE.
    1898. Unusual nesting site of kingbird. Auk, vol. 15, p. 268.
    1913. Arkansas kingbird in Massachusetts. Auk, vol. 30, pp. 112–113.
KENNEDY, CLARENCE HAMILTON.
    1915. Adaptability in the choice of nesting sites of some widely spread
        birds. Condor, vol. 17, pp. 65–70.
KERMODE, FRANCIS.
    1928. The Lichtenstein kingbird on Vancouver Island. Condor, vol. 30,
        p. 251.
KIMBALL, FRANK HENRY.
    1889. Mortality among eave swallows. Auk, vol. 6, pp. 338–339.
KINSEY, ERIC CAMPBELL.
    1935. Parental instincts in black phoebes. Condor, vol. 37, pp. 277–278.
KNIGHT, ORA WILLIS.
    1901. Some prehistoric cave and cliff dwellers and their descendents.
        Journ. Maine Orn. Soc., vol. 3, pp. 37–40.
    1905. Notes on the bobolink and cliff swallow. Journ. Maine Orn. Soc.,
        vol. 7, pp. 47–48.
    1908. The birds of Maine.

KNOWLTON, FRANK HALL.
  1881. Notes on the habits of the cliff swallow (*Petrochelidon lunifrons*).
        Bull. Nuttall Orn. Club, vol. 6, p. 55.
KOBBÉ, WILLIAM HOFFMAN.
  1900. The birds of Cape Disappointment, Washington. Auk, vol. 17, pp.
        349–358.
KOPMAN, HENRY HAZLITT.
  1915. List of the birds of Louisiana. Part 6. Auk, vol. 32, pp. 15–29.
KUMLIEN, LUDWIG, and HOLLISTER, NED.
  1903. The birds of Wisconsin.
LADD, SAMUEL B.
  1891. Description of nests and eggs of *Dendroica graciae* and *Contopus
        pertinax*. Auk, vol. 8, pp. 314–315.
LANGILLE, JAMES HIBBERT.
  1884. Our birds in their haunts.
LATHAM, ROY.
  1924. Kingbirds on Long Island. Bird-Lore, vol. 26, p. 252.
LAWRENCE, GEORGE NEWBOLD.
  1874. Birds of western and northwestern Mexico. Mem. Boston Soc. Nat.
        Hist., vol. 2, pp. 265–319.
LINCOLN, FREDERICK CHARLES.
  1925. Notes on the bird life of North Dakota with particular reference
        to the summer waterfowl. Auk, vol. 42, pp. 50–64.
  1926. Death of young phoebes due to over-feeding? Auk, vol. 43, pp.
        546–547.
  1939. The migration of American birds.
LOW, SETH HASKELL.
  1933. Further notes on the nesting of the tree swallows. Bird-Banding,
        vol. 4, pp. 76–87.
  1934. Nest distribution and survival ratio of tree swallows. Bird-Banding,
        vol. 5, pp. 24–30.
LUSK, RICHARD DEWITT.
  1901. In the summer home of the buff-breasted flycatcher. Condor, vol.
        3, pp. 38–41.
MACGILLIVRAY, WILLIAM.
  1840. A history of British birds. Vol. 3.
MACNAMARA, CHARLES.
  1917. The purple martin. Ottawa Nat., vol. 31, pp. 49–54.
MACOUN, JOHN.
  1909. Catalogue of Canadian birds. Ed. 2.
MAILLIARD, JOHN WARD.
  1921. Notes on the nesting of the Yosemite fox sparrow, calliope humming-
        bird and western wood pewee at Lake Tahoe, California. Condor,
        vol. 23, pp. 73–77.
MAILLIARD, JOSEPH.
  1881. Remarkable persistency in nesting of the western yellow-bellied fly-
        catcher. Bull. Nuttall Orn. Club, vol. 6, p. 119.
  1932. Birds and mammals from the Kootenay Valley, southeastern British
        Columbia. Proc. California Acad. Sci., ser. 4, vol. 20, pp. 269–290.
MASSEY, HERBERT.
  1917. Swallows and martins settling and feeding on flowers. British Birds,
        vol. 11, p. 92.

MAYNARD, CHARLES JOHNSON.
    1896. The birds of eastern North America.
MCATEE, WALDO LEE.
    1905. The horned larks and their relation to agriculture. U. S. Dept. Agr.
        Biol. Surv. Bull. 23.
    1926. The relation of birds to woodlots in New York State. Roosevelt
        Wildlife Bull., vol. 4, pp. 1–148.
MCBEE, C. E.
    1931. The dusky horned lark in eastern Washington. Murrelet, vol. 12, pp.
        43–45.
MCCANN, HORACE DOLBEY.
    1936. A colony of cliff swallows in Chester County, Pa. Auk, vol. 53, pp.
        84–85.
MCCAULEY, CHARLES ADAMS HOKE.
    1877. Notes on the ornithology of the region about the source of the Red
        River of Texas. U. S. Geol. and Geogr. Surv. Terr., vol. 3, No. 3.
MEARNS, EDGAR ALEXANDER.
    1890. Observations on the avifauna of portions of Arizona. Auk, vol. 7, pp.
        251–264.
    1902. Description of a hybrid between the barn and cliff swallows. Auk,
        vol. 19, pp. 73–74.
MERRILL, HARRY.
    1881. Notes on a few Maine birds. Bull. Nuttall Orn. Club, vol. 6, pp. 249–250.
MERRILL, JAMES CUSHING.
    1878. Notes on the ornithology of southern Texas, being a list of birds ob-
        served in the vicinity of Fort Brown, Texas, from February, 1876,
        to June, 1878. Proc. U. S. Nat. Mus., vol. 1, pp. 118–173.
MIDDLETON, RAYMOND JONES.
    1936. A pair of northern crested flycatchers (*Myiarchus crinitus boreus*)
        mated for three consecutive years. Bird-Banding, vol. 7, pp. 171–172.
MILLER, LOYE HOLMES.
    1939. Song of the western wood pewee. Auk, vol. 56, pp. 188–189.
MILLER, MARY MANN.
    1899. Birds feeding on hairy caterpillars. Auk, vol. 16, p. 362.
MILLER, OLIVE THORNE.
    1892. Little brothers of the air.
MILLER, RICHARD FIELDS.
    1915. Notes on the Acadian flycatcher in the vicinity of Philadelphia, Pa.
        Oologist, vol. 32, p. 199.
    1930. Nesting of the rough-winged swallow in the Pocono Mountains, Penn-
        sylvania. Auk, vol. 47, p. 260.
MINOT, HENRY DAVIS.
    1877. Land and game birds of New England.
MORRIS, ROBERT TUTTLE.
    1912. Kingbirds eating sassafras berries. Bird-Lore, vol. 14, p. 354.
MORRISON, CHARLES F.
    1888. A list of some birds of La Plata County, Col., with annotations. Orni-
        thologist and Oologist, vol. 13, pp. 70–75.
MORSE, CHARLES M.
    1931. Kingbirds and robins. Bird-Lore, vol. 33, p. 260.

MOUSLEY, WILLIAM HENRY.

1916. Five years personal notes and observations on the birds of Hatley, Stanstead County, Quebec—1911–1915. Auk, vol. 33, pp. 57–73, 168–186.

1934. A study of the home life of the northern crested flycatcher (*Myiarchus crinitus boreus*). Auk, vol. 51, pp. 207–216.

MUNRO, JAMES ALEXANDER.

1919. Notes on some birds of the Okanagan Valley, British Columbia. Auk, vol. 36, pp. 64–74.

1929. Male tree swallow feeding nestling robins. Condor, vol. 31, pp. 75–76.

MURIE, ADOLPH.

1935. Some feeding habits of the western sandpiper. Condor, vol. 37, pp. 258–259.

MURRAY, JOSEPH JAMES.

1934. Breeding of the prairie horned lark at Lexington, Virginia. Auk, vol. 51, p. 380.

NAUMAN, EMIL DANTON.

1924. Peculiar nesting sites. Bird-Lore, vol. 26, pp. 179–180.

NEHRLING, HENRY.

1882. List of birds observed at Houston, Harris Co., Texas, and in the Counties Montgomery, Galveston and Ford Bend. Bull. Nuttall Orn. Club, vol. 7, pp. 166–175.

NELSON, EDWARD WILLIAM.

1887. Report upon natural history collections made in Alaska, between the years 1877 and 1881.

1904. A revision of the North American mainland species of *Myiarchus*. Proc. Biol. Soc. Washington, vol. 17, pp 21–50.

NICE, MARGARET MORSE.

1928. The morning twilight song of the crested flycatcher. Wilson Bull., vol. 40, p. 255.

1931a. The birds of Oklahoma. Revised ed. Publ. Univ. Oklahoma, vol. 3, No. 1.

1931b. Notes on the twilight songs of the scissor-tailed and crested flycatchers. Auk, vol. 48, pp. 123–125.

NICHOLS, JOHN TREADWELL.

1920. The tree swallow on Long Island. Bird-Lore, vol. 22, pp. 279–281.

NORTON, ARTHUR HERBERT.

1916. Notes on some Maine birds. Auk, vol. 33, pp. 376–383.

NUTTALL, THOMAS.

1832. A manual of the ornithology of the United States and Canada.

OBERHOLSER, HARRY CHURCH.

1902. A review of the larks of the genus *Otocoris*. Proc. U. S. Nat. Mus., vol. 24, pp. 801–883.

1903. A new cliff swallow from Texas. Proc. Biol. Soc. Washington, vol. 16, pp. 15–16.

1907 Description of a new *Otocoris* from Lower California. Proc. Biol. Soc. Washington, vol. 20, pp. 41–42.

1918. New light on the status of *Empidonax trailli* (Audubon). Ohio Journ. Sci., vol. 18, pp. 85–98.

1938. The bird life of Louisiana. Louisiana State Dept. Conserv. Bull. 28.

OBERLANDER, GEORGE.

1939. The history of a family of black phoebes. Condor, vol. 41, pp. 133–151.

OLDYS, HENRY.
    1904. The rhythmical song of the wood pewee. Auk, vol. 21, pp. 270–274.
PEARSON, THOMAS GILBERT; BRIMLEY, CLEMENT SAMUEL; and BRIMLEY, HERBERT
        HUTCHINSON.
    1919. Birds of North Carolina. North Carolina Geol. and Econ. Surv.,
        vol. 4.
PERRY, TROUP D.
    1911. An oological paradise. Oologist, vol. 28, pp. 176–177.
PETERS, HAROLD SEYMOUR.
    1933. External parasites collected from banded birds. Bird-Banding, vol.
        4, pp. 68–75.
    1936. A list of external parasites from birds of the eastern part of the
        United States. Bird-Banding, vol. 7, pp. 9–27.
PETERS, JAMES LEE.
    1936. Records of two species new to Arizona. Condor, vol. 38, p. 218.
PHILLIPS, ALLAN ROBERT.
    1937. A nest of the olive-sided flycatcher. Condor, vol. 39, p. 92.
    1940. Two new breeding birds for the United States. Auk, vol. 57, pp.
        117–118.
PICKWELL, GAYLE BENJAMIN.
    1931. The prairie horned lark. Trans. Acad. Sci. St. Louis, vol. 27.
PIKE, OLIVER GREGORY.
    1912. A colony of house-martins building one nest. British Birds, vol. 5,
        p. 225.
    1932. The nightingale: Its song and story, and other familiar song-birds
        of Britain.
PINCKNEY, A. J.
    1938. A hand-reared Arkansas kingbird. Wilson Bull., vol. 50, pp. 290–291.
PORTER, LOUIS HOPKINS.
    1907. The breeding habits of *Empidonax virescens* in Connecticut. Auk.
        vol. 24, p. 99.
POTTER, JULIAN KENT.
    1923. Gray kingbird (*Tyrannus dominicensis*) at Cape May, N. J. Auk,
        vol. 40, p. 536.
POTTER, LAURENCE BEDFORD.
    1924. Badger digs for bank swallows. Condor, vol. 26, p. 191.
    1932. Unusual nesting sites. Canadian Field-Nat., vol. 46, p. 49.
PRICE, WILLIAM WIGHTMAN.
    1888. Xantus's becard (*Platypsaris albiventris*) in the Huachuca Mountains,
        southern Arizona. Auk, vol. 5, p. 425.
PRING, CHRISTOPHER JOHN.
    1929. Great spotted woodpecker destroying nests and eating young of
        house-martins. British Birds, vol. 23, pp. 129–131.
PURDIE, HENRY AUGUSTUS.
    1878. The nest and eggs of the yellow-bellied flycatcher (*Empidonax
        flaviventris*). Bull. Nuttall Orn. Club, vol. 3, pp. 166–168.
RAY, MILTON SMITH.
    1903. A list of the land birds of Lake Valley, central Sierra Nevada Moun-
        tains, California. Auk, vol. 20, pp. 180–193.
    1906. A-birding in an auto. Auk, vol. 23, pp. 400–418.
    1918. Six weeks in the high Sierras in nesting time. Condor, vol. 20, pp.
        70–78.

RAY, ROSE CAROLYN.
    1932. Nesting of the Hammond flycatcher in Eldorado County, California.
        Condor, vol. 34, pp. 71–74.
REED, CHESTER ALBERT.
    1904. North American birds' eggs.
REED, CLARA EVERETT.
    1927. More notes on cliff swallows. Auk, vol. 44, p. 110.
RHOADS, SAMUEL NICHOLSON.
    1892. The birds of southeastern Texas and southern Arizona observed
        during May, June and July, 1891. Proc. Acad. Nat. Sci. Phila-
        delphia, pp. 98–126.
RIDGWAY, ROBERT.
    1869. A true story of a pet bird. Amer. Nat., pp. 309–312.
    1877. United States geological exploration of the fortieth parallel. Part
        2: Ornithology.
    1904. The birds of North and Middle America. Part 3.
    1907. The birds of North and Middle America. Part 4.
    1912. Color standards and color nomenclature.
RILEY, JOSEPH HARVEY.
    1905. List of birds collected or observed during the Bahama expedition of
        the Geographic Society of Baltimore. Auk, vol. 22, pp. 349–360.
ROADS, KATIE MYRA.
    1931. Use of former nest sites. Auk, vol. 48, p. 622.
ROBERTS, THOMAS SADLER.
    1932. The birds of Minnesota. Vol. 2.
ROBERTSON, JOHN McBRAIR.
    1933. Black phoebe nesting in a tree. Condor, vol. 35, p. 166.
ROBINSON, HELEN J.
    1927. Tree swallow habits and behavior at Brewer, Maine. Bull. North-
        eastern Bird-Banding Assoc., vol. 3, pp. 89–93.
ROCKWELL, ROBERT BLANCHARD.
    1908. An annotated list of the birds of Mesa County, Colorado. Condor,
        vol. 10, pp. 152–180.
ROHWER, ROLF D.
    1933. An unusual nesting-site of a pair of kingbirds. Bird-Lore, vol.
        35, p. 267.
ROSS, ROLAND CASE.
    1933. Do black phoebes eat honey-bees? Condor, vol. 35, p. 232.
ROSSELL, BEATRICE SAWYER.
    1921. A friendly wood pewee. Bird-Lore, vol. 23, pp. 303–304.
RUSSELL, HENRY NORRIS, Jr., and WOODBURY, ANGUS M.
    1941. Nesting of the gray flycatcher. Auk, vol. 58, pp. 28–37.
SAGE, JOHN HALL.
    1895. A swallow roost near Portland, Conn. Auk, vol. 12, p. 83.
SAMUELS, EDWARD AUGUSTUS.
    1883. Our northern and eastern birds.
SASS, HERBERT RAVENEL.
    1921. News & Courier, Charleston, S. C., Mar. 17.
SAUNDERS, ARETAS ANDREWS.
    1924. Recognizing individual birds by song. Auk, vol. 41, pp. 242–259.
    1935. A guide to bird songs.
    1938. Studies of breeding birds in the Allegany State Park. New York
        State Mus. Bull. No. 318.

SAUNDERS, WILLIAM EDWIN.

1898. A swallow roost. Ottawa Nat., vol. 12, pp. 145–146.

1909. The Acadian flycatcher in Ontario. Auk, vol. 26, p. 430.

1910. Acadian flycatcher in Ontario. Auk, vol. 47, pp. 209–210.

SAWYER, CLARENCE E.

1907. Brunswick notes. Journ. Maine Orn. Soc., vol. 9, pp. 83–84.

SAWYER, EDMUND JOSEPH.

1918. Swallows flocking. Bird-Lore, vol. 20, pp. 296–297.

SCHNEIDER, FRANCES B.

1922. *In* The season—Los Angeles region. Bird-Lore, vol. 24, pp. 289–290.

SCHOEGER, ARLIE WILLIAM.

1927. Notes on the distribution of some Wisconsin birds. I. Auk, vol. 44, pp. 235–240.

SCOTT, FREDERICK CLEMENT.

1934. A winter phoebe. Bird-Lore, vol. 36, p. 108.

SCOTT, WILLIAM EARL DODGE.

1879. Notes on birds observed at Twin Lakes, Lake County, Colorado. Bull. Nuttall Orn. Club, vol. 4, pp. 90–96.

1887. On the avi-fauna of Pinal County, with remarks on some birds of Pima and Gila Counties, Arizona. Auk, vol. 4, pp. 16–24.

1890. Two species of swallow new to North America. Auk, vol. 7, pp. 264–265.

SENNETT, GEORGE BURRITT.

1878. Notes on the ornithology of the lower Rio Grande of Texas. Bull. U. S. Geol. and Geogr. Surv. Terr., vol. 4, pp. 1–66.

1879. Further notes on the ornithology of the lower Rio Grande of Texas, from the observations made during the spring of 1878. Bull. U. S. Geol. and Geogr. Surv. Terr., vol. 5, pp. 371–440.

1884. Nest and eggs of Couch's tyrant flycatcher (*T. melancholicus couchi*). Auk, vol. 1, p. 93.

SHELLEY, LEWIS OBMAN.

1934a. Two pairs of tree swallows mated during two successive seasons. Bird-Banding, vol. 5, p. 91.

1934b. Tree swallow tragedies. Bird-Banding, vol. 5, p. 134.

1935. Notes on the 1934 tree swallow breeding-season. Bird-Banding, vol. 6, pp. 33–35.

1936a. Known history of eastern phoebe B127877. Bird-Banding, vol. 7, pp. 47–48.

1936b. A tree swallow rears a cliff swallow. Bird-Banding, vol. 7, p. 49.

SHIRLING, A. E.

1935. Observations on the violet-green swallow. Wilson Bull., vol. 47, pp. 192–194.

SILLOWAY, PERLEY MILTON.

1923. Relation of summer birds to the western Adirondack forest. Roosevelt Wild Life Bull., vol. 1, pp. 397–486.

SIMMONS, GEORGE FINLAY.

1925. Birds of the Austin region.

SKINNER, MILTON PHILO.

1928. A guide to the winter birds of the North Carolina sand-hills.

SMITH, AUSTIN PAUL.

1909. Observations on some birds found in southern Mexico. Condor, vol. 11, pp. 57–64.

1916. Winter notes from southern Texas. Condor, vol. 18, p. 129.

SMITH, CHARLES PIPER.
1927. The olive-sided flycatcher and coniferous trees. Condor, vol. 29, pp. 120–121.
SMITH, WENDELL PHILLIPS.
1933. Some observations of the nesting habits of the barn swallow. Auk, vol. 50, pp. 414–419.
1937. Further notes on the nesting of the barn swallow. Auk, vol. 54, pp. 65–69.
SMITH, WILBUR F.
1905. An interesting phoebe's nest. Bird-Lore, vol. 7, p. 144.
SOPER, JOSEPH DEWEY.
1928. A faunal investigation of southern Baffin Island. Nat. Mus. Canada Bull. 53.
SOPER, WILBUR F.
1905. An interesting phoebe's nest. Bird-Lore, vol. 7, p. 144.
SPIKER, CHARLES JOLLEY.
1937. The alder flycatcher in upland situations. Wilson Bull., vol. 49, p. 48.
SPURRELL, JOHN A.
1919. An annotated list of the land birds of Sac County, Iowa. Wilson Bull., vol. 31, pp. 117–126.
STAUNTON, FRANCES KATHERINE.
1929. Three swallows' nests on a hat. British Birds, vol. 23, p. 190.
STEVENS, ORIN ALVA.
1932. Bank swallows killed by automobiles. Wilson Bull., vol. 44, p. 39.
STICKNEY, GARDNER PERRY.
1923. [Letter to F. E. L. Beal on yellow-bellied flycatcher.] Wilson Bull., vol. 35, pp. 125–126.
STOCKARD, CHARLES RUPERT.
1905. Nesting habits of birds in Mississippi. Auk, vol. 22, pp. 146–158, 273–285.
STONE, WITMER.
1937. Bird studies at Old Cape May. Vol. 2.
STONER, DAYTON.
1925. Observations and banding notes of the bank swallow. Auk, vol. 42, pp. 86–94.
1935. Temperature and growth studies on the barn swallow. Auk, vol. 52, pp. 400–407.
1936a. How fast does a barn swallow grow? Univ. State of New York Bull. to Schools, vol. 22, pp. 122–124.
1936b. Studies on the bank swallow, *Riparia riparia riparia* (Linnaeus) in the Oneida Lake region. Roosevelt Wild Life Ann., vol. 9, pp. 122–233.
1937. Ten years' returns from banded bank swallows. New York State Mus. Circ. 18.
1938. Longevity in the bank swallow. Bird-Banding, vol. 9, pp. 173–177.
1939a. Temperature, growth and other studies on the eastern phoebe. New York State Mus. Circ. 22.
1939b. Parasitism of the English sparrow on the northern cliff swallow. Wilson Bull., vol. 51, pp. 221–222.
STONER, EMERSON AUSTIN.
1922. Odd nesting site of phoebe. Wilson Bull., vol. 34, pp. 115–116.
1938. A black phoebe's nest with eggs of three species. Condor, vol. 40, p. 42.

SUTTON, GEORGE MIKSCH.
1927. Flocking, mating, and nest-building habits of the prairie horned lark. Wilson Bull., vol. 39, pp. 131–141.
1928. The birds of Pymatuning Swamp and Conneaut Lake, Crawford County, Pennsylvania. Ann. Carnegie Mus., vol. 18, pp. 19–239.
1932. The exploration of Southampton Island, Hudson Bay. Mem. Carnegie Mus., vol. 12, pt. 2, Zoology, sect. 2. The birds of Southampton Island.
1936. Notes from Ellis and Cimarron Counties, Oklahoma. Auk, vol. 53, pp. 432–435.

SWAINSON, WILLIAM, and RICHARDSON, JOHN.
1831. Fauna Boreali-Americana, vol. 2. Birds.

SWARTH, HARRY SCHELWALD.
1904. Birds of the Huachuca Mountains, Arizona. Pacific Coast Avifauna, No. 4.
1922. Birds and mammals of the Stikine River region of northern British Columbia and southeastern Alaska. Univ. California Publ. Zool., vol. 24, pp. 125–314.
1929. The faunal areas of southern Arizona: A study in animal distribution. Proc. California Acad. Sci., ser. 4, vol. 18, pp. 267–383.
1935. A barn swallow's nest on a moving train. Condor, vol. 37, pp. 84–85.

SWENK, MYRON HARMON, and DAWSON, RALPH W.
1921. Notes on the distribution and migration of Nebraska birds. Wilson Bull., vol. 33, pp. 132–141.

TATE, RALPH C.
1925. Some materials used in nest construction by certain birds of the Oklahoma Panhandle. Univ. Oklahoma Bull., vol. 5, pp. 103–104.

TAVERNER, PERCY ALGERNON.
1919. The birds of the Red Deer River, Alberta. Auk, vol. 36, pp. 248–265.
1927. Some recent Canadian records. Auk, vol. 44, pp. 217–228.
1928. Cliff swallow nests and rain. Canadian Field-Naturalist, vol. 42, pp. 148–149.
1933. Purple martins gathering leaves. Auk, vol. 50, pp. 110–111.
1934. Birds of Canada.

TAVERNER, PERCY ALGERNON, and SUTTON, GEORGE MIKSCH.
1934. The birds of Churchill, Manitoba. Ann. Carnegie Mus., vol. 33, pp. 1–83.

TAVERNER, PERCY ALGERNON, and SWALES, BRADSHAW HALL.
1907. The birds of Point Pelee. Wilson Bull., vol. 19, pp. 133–153.

TAYLOR, WALTER PENN.
1912. Field notes on amphibians, reptiles and birds of northern Humboldt County, Nevada. Univ. California Publ. Zool., vol. 7, pp. 319–436.
1925. The breeding and wintering of the pallid horned lark in Washington State. Auk, vol. 42, pp. 349–353.

TERRILL, LEWIS MCIVER.
1915. Notes from the Laurentian Hills. Wilson Bull., vol. 27, pp. 302–309.

THAYER, JOHN ELIOT.
1906. Nests and eggs of the beardless flycatcher (Ornithion imberbe). Auk, vol. 23, pp. 460–461.
1915. Two species of cliff swallows nesting in Kerr County, Texas. Auk, vol. 32, pp. 102–103.

TODD, WALTER EDMOND CLYDE.
1916. The birds of the Isle of Pines. Ann. Carnegie Mus., vol. 10, pp. 146–296.
1940. Birds of western Pennsylvania.

TOMKINS, IVAN REXFORD.
　　1934. Notes from Chatham County, Georgia. Auk, vol. 51, pp. 252–253.
TOMKINS, JOHN H.
　　1928. A winter phoebe. Bird-Lore, vol. 30, p. 187.
TORREY, BRADFORD.
　　1893. The foot-path way.
　　1896. Spring notes from Tennessee.
　　1901. Footing it in Franconia.
　　1903. Sand swallows (*Riparia riparia*) nesting in sawdust. Auk, vol. 20, pp. 436–437.
　　1904. Nature's invitation.
TOWNSEND, CHARLES WENDELL.
　　1905. The birds of Essex County, Massachusetts.
　　1920a. Courtship in birds. Auk, vol. 37, pp. 380–393.
　　1920b. Supplement to the birds of Essex County, Massachusetts.
　　1920c. On the nesting, song and play of the tree swallow and barn swallow. Bull. Essex County Orn. Club, vol. 2, pp. 31–36.
　　1923. A breeding station of the horned lark and pipit on the Gaspé Peninsula. Auk, vol. 40, pp. 85–87.
TOWNSEND, CHARLES WENDELL, and ALLEN, GLOVER MORRILL.
　　1907. Birds of Labrador. Proc. Boston Soc. Nat. Hist., vol. 33, pp. 277–428.
TOWNSEND, MANLEY BACON.
　　1917. Nesting habits of the cliff swallow. Bird-Lore, vol. 19, pp. 252–257.
　　1926. A phoebe tragedy explained. Bird-Lore, vol. 28, p. 123.
TRAFTON, GILBERT HAVEN.
　　1908. The nest in the gutter. Bird-Lore, vol. 10, pp. 72–76.
TUCKER, BERNARD WILLIAM.
　　1938. *In* The handbook of British Birds, vol. 1, by Witherby, Jourdain, Ticehurst, and Tucker.
TURNER, LUCIEN MCSHAN.
　　1886. Contributions to the natural history of Alaska.
TYLER, JOHN GRIPPER.
　　1913. Some birds of the Fresno district, California. Pacific Coast Avifauna, No. 9.
VAN ROSSEM, ADRIAAN JOSEPH.
　　1914. Notes on the Derby flycatcher. Condor, vol. 16, pp. 11–13.
　　1927. The Arizona race of the sulphur-bellied flycatcher. Condor, vol. 29, p. 126.
　　1929. A record of *Tyrannus melancholicus occidentalis* for the State of Washington. Condor, vol. 31, p. 182.
　　1930. The Sonora races of *Camptostoma* and *Platypsaris*. Proc. Biol. Soc. Washington, vol. 43, pp. 129–131.
　　1936. Notes on birds in relation to the faunal areas of south-central Arizona. Trans. San Diego Soc. Nat. Hist., vol. 8, pp. 121–148.
VAN TYNE, JOSSELYN.
　　1933. The Arkansas kingbird in Michigan. Auk, vol. 50, pp. 107–108.
WALKER, ALEXANDER.
　　1914. Nesting of the gray flycatcher in Oregon. Condor, vol. 16, p. 94.
WALKER, CHARLES FREDERIC, and TRAUTMAN, MILTON BERNHARD.
　　1936. Notes on the horned larks of the central Ohio region. Wilson Bull., vol. 48, pp. 151–155.
WARREN, BENJAMIN HARRY.
　　1890. Report on the birds of Pennsylvania. Ed. 2.

WAYNE, ARTHUR TREZEVANT.

1894. Notes on the capture of the gray kingbird (*Tyrannus dominicensis*) near Charleston, South Carolina. Auk, vol. 11, pp. 178–179.

1910. Birds of South Carolina.

1927. The gray kingbird (*Tyrannus dominicensis*) again on the coast of South Carolina. Auk, vol. 44, pp. 565–566.

WETHERBEE, Mrs. KENNETH BRACKETT.

1932. Two pairs of tree swallows mated during two successive seasons. Bird-Banding, vol. 3, pp. 72–73.

WETMORE, ALEXANDER.

1916. Birds of Porto Rico. U. S. Dept. Agr. Bull. 326.

1920. Observations on the habits of birds at Lake Burford, New Mexico. Auk, vol. 37, pp. 393–412.

1939. Notes on the birds of Tennessee. Proc. U. S. Nat. Mus., vol. 86, pp. 175–243.

WETMORE, ALEXANDER, and SWALES, BRADSHAW HALL.

1931. The birds of Haiti and the Dominican Republic. U. S. Nat. Mus. Bull. 155.

WEYDEMEYER, WINTON.

1933. Nesting of the rough-winged swallow in Montana. Auk, vol. 50, pp. 362–363.

1934a. Singing of the tree swallow. Auk, vol. 51, pp. 86–87.

1934b. Tree swallows at home in Montana. Bird-Lore, vol. 36, pp. 100–105.

1935. Efficiency of nesting of the tree swallow. Condor, vol. 37, pp. 216–217.

WHEATON, JOHN MAYNARD.

1882. Report on the birds of Ohio. Rep. Geol. Surv. Ohio, Part 1: Zoology, pp. 187–628.

WHEDON, A. D.

1906. A kingbird family. Bird-Lore, vol. 8, pp. 117–122.

WHEELOCK, IRENE GROSVENOR.

1904. Birds of California.

WHITE, FRANCIS BEACH.

1937. Local notes on the birds at Concord, New Hampshire.

WHITE, STEPHEN JOSEPH.

1908. Swallow's nest built on a glass gas shade. British Birds, vol. 1, pp. 354–355.

WHITTLE, CHARLES LIVY.

1922. Resourceful cliff swallows. Bird-Lore, vol. 24, pp. 214–215.

1926. Notes on the nesting habits of the tree swallow. Auk, vol. 43, pp. 247–248.

1927. The role of the snake skin. Auk, vol. 44, pp. 262–263.

WIDMANN, OTTO.

1884. How young birds are fed. Forest and Stream, vol. 22, p. 484.

1907. A preliminary catalog of the birds of Missouri. Trans. Acad. Sci. St. Louis, vol. 17.

WILLARD, FRANCIS COTTLE.

1923a. The Mexican cliff swallow in Cochise County, Arizona. Condor, vol. 25, pp. 138–139.

1923b. The buff-breasted flycatcher in the Huachucas. Condor, vol. 25, pp. 189–194.

WILLIAMS, JOHN RAYNESFORD.

1935. Kingbird (*Tyrannus tyrannus*) attacks airplane. Auk, vol. 52, p. 89.

WILLIAMS, ROBERT WHITE.
    1928. Further notes on the birds of Leon County, Florida.—Fourth Supplement. Auk, vol. 45, pp. 164–169.
WILSON, ALEXANDER.
    1810. American ornithology, vol. 2.
    1831. American ornithology, vol. 2.
WILSON, CHARLES E.
    1923. Insect pests of cotton in St. Croix and means of combating them. Virgin Islands Agr. Exper. Stat. Bull 3.
WINGE, HERLUF.
    1898. Grønlands Fugle.
WITHERBY, HARRY FORBES, et al.
    1920. A practical handbook of British birds. Vol. 1. (New ed., 1938.)
WOOD, HAROLD BACON.
    1937a. The growth of young barn swallows. Bird-Banding, vol. 8, pp. 31–34.
    1937b. Observations at a barn swallow's nest. Wilson Bull., vol. 49, pp. 96–100.
WRIGHT, ALBERT HAZEN, and HARPER, FRANCIS.
    1913. A biological reconnaissance of Okefinokee Swamp: The birds. Auk, vol. 30, pp. 477–505.
WRIGHT, HORACE WINSLOW.
    1912. Morning awakening and even-song. Auk, vol. 29, pp. 307–327.
WRIGHT, WILLIAM SHERMAN.
    1924. A unique swallow's nest. Condor, vol. 26, pp. 153–154.
YARRELL, WILLIAM.
    1876–1882. A history of British Birds, ed. 4. Vol. 2.
YOUNGWORTH, WILLIAM.
    1937. Terrestrial feeding kingbirds. Wilson Bull., vol. 49, p. 117.
ZIEGLER, GEORGE FREDERICK, Jr.
    1923. Notes on a purple martin colony. Auk, vol. 40, pp. 431–436

INDEX

Abbott, Clinton G., on eastern phoebe, 143, 148.
Acadian flycatcher, 183.
acadicus, Empidonax, 185, 188, 192, 194.
actia, Otocoris alpestris, 335, 361, 364, 366, 367, 368, 370, 371.
adusta, Otocoris alpestris, 335, 368, 370, 371.
aglaiae aglaiae, Platypsaris, 1, 2.
aglaiae albiventris, Platypsaris, 1.
aglaiae latirostris, Platypsaris, 3, 9, 10.
aglaiae richmondi, Platypsaris, 7, 9, 10.
aglaiae sumichrasti, Platypsaris, 3, 9.
Aiken, C. E., on Cassin's kingbird, 74.
    on western wood pewee, 281, 282.
Alauda arvensis arvensis, 314.
Alaudidae, 314.
albifrons albifrons, Petrochelidon, 463, 484.
albifrons melanogaster, Petrochelidon, 483, 484, 485.
albifrons tachina, Petrochelidon, 483, 484.
albiventris, Platypsaris aglaiae, 1.
Alder flycatcher, 204.
Alexander, H. G., on skylark, 318.
Alfaro, Don Anastasio, 78.
Allen, A. A., 56.
Allen, A. S., on northern cliff swallow, 481.
Allen, C. A., on western flycatcher, 248, 251.
Allen, Francis H., on alder flycatcher, 210.
    on Arkansas kingbird, 64.
    on eastern kingbird, 22.
    on eastern phoebe, 150.
    on northern crested flycatcher, 116, 118, 119.
    on olive-sided flycatcher, 298.
    on purple martin, 498, 502.
    on tree swallow, 386, 393, 395.
    on yellow-bellied flycatcher, 181.
Allen, Francis H., and Griscom, Ludlow, on gray kingbird, 47.
Allen, Glover M., on yellow-bellied flycatcher, 181.
Allen, Glover M., and Townsend, C. W., on northern horned lark, 328, 329, 332.
alnorum, Empidonax trailli, 218.
alpestris, Otocoris alpestris, 322, 323, 325, 343, 356, 357, 367.
alpestris actia, Otocoris, 335, 361, 364, 366, 367, 368, 370, 371.
alpestris adusta, Otocoris, 335, 368, 370, 371.

alpestris alpestris, Otocoris, 322, 323, 325, 343, 356, 357, 367.
alpestris ammophila, Otocoris, 335, 366, 370, 371.
alpestris arcticola, Otoeoris, 320, 322, 335, 337, 361.
alpestris arenicola, Otocoris, 368, 370.
alpestris chrysolaema, Otocoris, 364, 368, 370.
alpestris enertera, Otocoris, 335, 366.
alpestris enthymia, Otocoris, 338.
alpestris flava, Otocoris, 325, 343.
alpestris giraudi, Otocoris, 335, 337, 356.
alpestris hoyti, Otocoris, 321, 322, 327, 335, 356, 357.
alpestris insularis, Otocoris, 335, 361, 367.
alpestris leucansiptila, Otocoris, 335, 371.
alpestris leucolaema, Otocoris, 322, 335, 337, 366, 368, 370, 371.
alpestris merrilli, Otocoris, 321, 335, 337, 357, 358, 362, 367.
alpestris oaxacae, Otocoris, 367, 368.
alpestris occidentalis, Otocoris, 335, 368, 370, 371.
alpestris pallida, Otocoris, 371.
alpestris peregrina, Otocoris, 335.
alpestris praticola, Otocoris, 335, 337, 342, 356.
alpestris rubea, Otocoris, 335, 364, 367.
alpestris strigata, Otocoris, 335, 348, 357, 359, 362, 367.
Amadon, Dean, x.
    on barn swallow, 448.
    on beardless flycatcher, 311.
American Museum of Natural History, x.
ammophila, Otocoris alpestris, 335, 366, 370, 371.
Anderson, A. H., 56.
Anderson, E. M., 444.
Anthony, A. W., 90, 129, 254, 510.
    on northern cliff swallow, 480.
    on olive-sided flycatcher, 291.
    on rough-winged swallow, 431.
    on San Jose phoebe, 174.
    on San Quintin phoebe, 165.
arcticola, Otocoris alpestris, 320, 322, 335, 337, 361.
arenicola, Otocoris alpestris, 368, 370.
Arizona crested flycatcher, 123.
Arizona sulphur-bellied flycatcher, 98.
Arkansas kingbird, 57.
Armstrong, Frank B., 51, 128.
Arnow, Isaac, 31, 36.

Riley, J. H., x.
  on gray kingbird, 33, 42, 43.
Riparia riparia maximiliani, 401.
  riparia riparia, 400.
riparia, Riparia riparia, 400.
riparia maximiliani, Clivicola, 401.
riparia maximiliani, Riparia, 401.
riparia riparia, Riparia, 400.
Roads, Katie M., on eastern wood pewee, 269.
Roberts, T. S., on Arkansas kingbird, 58, 67.
  on eastern wood pewee, 268, 271.
  on least flycatcher, 220, 221.
  on northern cliff swallow, 477.
  on northern crested flycatcher, 114.
  on purple martin, 490, 491, 496.
Robertson, John McB., on black phoebe, 156.
Robinson, Helen J., on tree swallow, 394.
Rockwell, Robert B., on Cassin's kingbird, 71.
Rogers, C. H., 195.
Rohwer, Rolf D., on eastern kingbird, 15.
Rolfe, E. S., 404.
Ross, R. C., on black phoebe, 163.
Rossell, Beatrice S., on eastern wood pewee, 274.
Rossignol, G. H., 31, 36.
Rough-winged swallow, 424.
Rowley, J. S., x, 367.
  on San Lucas flycatcher, 254.
  on San Lucas phoebe, 166.
  on San Lucas swallow, 384.
  on scorched horned lark, 369.
rubea, Otocoris alpestris, 335, 364, 367.
rubinus mexicanus, Pyrocephalus, 302.
Ruddy horned lark, 367.
ruficollis serripennis, Stelgidopteryx, 424.
Russell, H. N., Jr., and Woodbury, A. M., on gray flycatcher, 244.
Rust, Henry J., on Wright's flycatcher, 234, 235.
rustica, Hirundo rustica, 458.
rustica rustica, Hirundo, 458.
Sage, John H., on swallow roost, 420.
salictaria, Sayornis nigricans, 164, 165.
Samuels, E. A., on least flycatcher, 221.
San Jose phoebe, 173.
San Lucas flycatcher, 253.
San Lucas phoebe, 165.
San Lucas swallow, 383.
San Quintin phoebe, 165.
Sass, Herbert Ravenel, on gray kingbird, 34.
satrapa, Tyrannus melancholicus, 50.
Saunders, A. A., on Acadian flycatcher, 194.
  on eastern wood pewee, 274.
  on least flycatcher, 218.
Saunders, H., 343.
Saunders, W. E., on Acadian flycatcher, 184, 188.
  on barn swallow, 453.

Say, Thomas, on northern cliff swallow, 464.
saya, Sayornis saya, 166.
saya quiescens, Sayornis, 172, 173.
saya saya, Sayornis, 166.
saya yukonensis, Sayornis, 174.
Saylor, L. W., 312.
Sayornis nigricans brunnescens, 164, 165.
  nigricans nigricans, 154.
  nigricans salictaria, 164, 165.
  phoebe, 140.
  saya quiescens, 172, 173.
  saya saya, 166.
  saya yukonensis, 174.
Say's phoebe, 166.
Sawyer, C. E., on northern cliff swallow, 480.
Sawyer, E. J., on bank swallow, 420.
Scissor-tailed flycatcher, 82.
Schneider, Frances B., on northern cliff swallow, 481.
Schorger, A. W., on Acadian flycatcher, 184.
Scorched horned lark, 368.
Scott, F. Clement, on eastern phoebe, 152.
Scott, W. E. D., 30.
  on Arizona crested flycatcher, 125.
  on Bahama swallow, 372.
  on Cassin's kingbird, 70.
  on Cuban cliff swallow, 486.
  on Hammond's flycatcher, 232.
  on western wood pewee, 280.
Seagle, George A., on eastern wood pewee, 273.
Sennett, George B., on Couch's kingbird, 51, 52.
  on Derby flycatcher, 92.
  on Mexican crested flycatcher, 127.
serripennis, Stelgidopteryx ruficollis, 424.
Seton, E. T., on Arkansas kingbird, 57.
Shelley, Lewis O., on eastern phoebe, 151.
  on tree swallow, 386, 394.
Sherman, Althea, on eastern phoebe, 140, 141, 143.
Shirling, A. E., on violet-green swallow, 374, 378.
Silloway, P. M., on alder flycatcher, 207.
Simmons, G. F., on Acadian flycatcher, 185, 195.
  on scissor-tailed flycatcher, 83, 84, 89.
Skinner, Milton P., on eastern phoebe, 152.
Skutch, A. F., on barn swallow, 440.
  on Coues's flycatcher, 265.
  on eastern kingbird, 11, 26.
  on fork-tailed flycatcher, 77, 79, 80, 81.
  on olive-sided flycatcher, 300.
  on rough-winged swallow, 428, 429.
  on scissor-tailed flycatcher, 90.
  on sulphur-bellied flycatcher, 99, 101, 102, 103, 104, 105.
  on Sumichrast's becard, 3, 6, 8, 9.

Oakland County, Mich.   F. N. Wilson

EASTERN KINGBIRD.

PLATE 5

Savannah, Ga.   I. R. Tomkins.

St. Johns County, Fla., June 1931.   S. A. Grimes.

ADULT AND NEST OF GRAY KINGBIRD.

California.   J. A. Calder.

Washington.   R. T. Congdon.

NESTS OF ARKANSAS KINGBIRD.

PLATE 7

Clear Lake, Calif., May 22, 1923.   J. E. Patterson.

Aravaipa Creek, Ariz., June 2, 1936.   J. S. Rowley.

NESTS OF CASSIN'S KINGBIRD.

PLATE 8

A. A. Allen.

Austwell, Tex., May 14, 1939. SCISSOR-TAILED FLYCATCHER AT NEST.

PLATE 9

Arizona.   F. C. Willard.

NESTING CAVITY AND NEST OF ARIZONA SULPHUR-BELLIED FLYCATCHER.

PLATE 1

Logan County, Ill., June 30, 1913.　　　　　　A. D. DuBois.

Adult at nest entrance.

Ithaca, N. Y.　　　　　　A. A. Allen.

Adult and young.

NORTHERN CRESTED FLYCATCHER.

PLATE 11

Duval County, Fla., June 1937.
S. A. Grimes.

Male approaching nest.

SOUTHERN CRESTED FLYCATCHER.

Ithaca, N. Y.
A. A. Allen.

Nest in a bird house.

NORTHERN CRESTED FLYCATCHER.

Arizona.   F. C. Willard

NESTING SITES OF ASH-THROATED FLYCATCHER.

PLATE 13

Claremont, Calif.   W. M. Pierce.

Adult.

Arizona.   F. C. Willard.

Nest in a saguaro, cut open.

ASH-THROATED FLYCATCHER.

PLATE 14

Logan County, Ill., June 21, 1913.　　　　　　　　　　　A. D. DuBois.

Nest under a cottage porch.

Hennepin County, Minn., June 13, 1935.　　　　　　　　A. D. DuBois.

Pair of adults.

EASTERN PHOEBE.

PLATE 15

Near Princeton, N. J., June 7, 1931.   Laidlaw Williams.

Nest on a hook in a well.

Toronto, Ontario, June 30, 1939.   W. V. Crich.

Nest under a bridge.

EASTERN PHOEBE.

PLATE 16

W. G. F. Harris.

Rehoboth, Mass.

Old and new nests on cliff.

Marshall County, W. Va., April 30, 1909.

S. S. Dickey.

Well-spotted eggs.

NESTS OF EASTERN PHOEBE.

PLATE 17

Fairbank, Ariz., April 16, 1922.　　　　　　　　　　　　　　A. C. Bent.

San Bernardino County, Calif., April 12, 1916.　　　　　　　W. M. Pierce.

NESTS OF BLACK PHOEBE IN DESERTED HOUSES

PLATE 18

Azusa, Calif., September 15, 1930.　　　　　　　　R. S. Woods.

Adult.

:sno, Calif.　　　　　　　　　　　　　　　　W. M. Pierce.

Young.

BLACK PHOEBE.

PLATE 19

Weld County, Colo., June 5, 1911.   E. R. Warren.

Natural nesting site.

Colorado Springs, Colo., June 21, 1905.   E. R. Warren.

Immature bird.

SAY'S PHOEBE.

PLATE 20

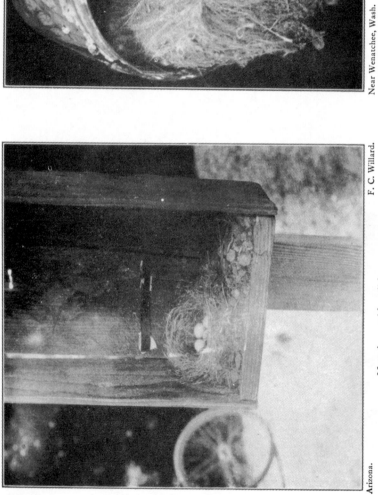

Arizona.   F. C. Willard.

Nest in an old mail box.

Near Wenatchee, Wash.   R. T. Congdon.

Nest in a lard pail in a stovepipe hole in a chimney.

NESTING SITES OF SAY'S PHOEBE.

PLATE 21

Montane County, Quebec, June 28, 1927.   W. J. Brown.

YELLOW-BELLIED FLYCATCHER.

PLATE 22

Montane County, Quebec, June 28, 1927.　　　　W. J. Brown.

Pocono Mountains, Pa.　　　　W. L. Bailey.

NESTS OF YELLOW-BELLIED FLYCATCHER.

PLATE 23

Sangamon County, Ill., June 16, 1908.   A. D. DuBois.

Preston County, W. Va, June 2, 1936.   S. S. Dickey.

NESTS OF ACADIAN FLYCATCHER.

PLATE 24

S. A. Grimes.

ACADIAN FLYCATCHERS AND NEST

Duval County, Fla., June 1936.

PLATE 25

Pocono Mountains, Pa.   W. L. Bailey.

Near Buffalo, N. Y., July 20, 1927.   S. A. Grimes.

NESTS OF ALDER FLYCATCHER

PLATE 26

ALDER FLYCATCHER.

Eliot Porter.

PLATE 27

Warren County, Pa., June 15, 1926.

S. S. Dickey.

Pennington County, Minn., June 1933.

S. A. Grimes.

NESTS OF LEAST FLYCATCHER

PLATE 2

LEAST FLYCATCHER.

PLATE 29

Near Wenatchee, Wash.

R. T. Congdon.

HAMMOND'S FLYCATCHER.

PLATE 3

San Bernardino Mountains, Calif., June 28, 1916.

W. M. Pierce.

WRIGHT'S FLYCATCHER.

PLATE 31

H. J. Rust.

Near Coeur d'Alene, Idaho.

NESTING OF WRIGHT'S FLYCATCHER.

PLATE 32

Inyo County, Calif.

W. L. Dawson
Courtesy National Audubon Society

NESTS OF WRIGHT'S FLYCATCHER

PLATE 33

Mono Flats, Calif. J. B. Dixon.

Mono County, Calif.
W. L. Dawson.
Courtesy National Audubon Society.

HABITAT AND NEST OF GRAY FLYCATCHER.

Mono County, Calif.   E. N. Harrison.

NESTS OF GRAY FLYCATCHER

PLATE 35

Laidlaw Williams.

Monterey County, Calif., May 29, 1940.

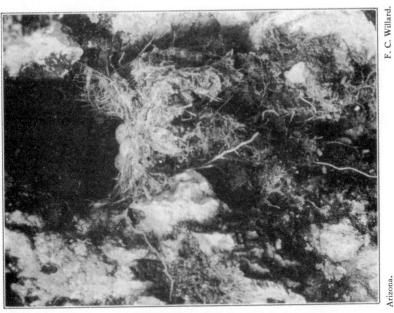

F. C. Willard.

Arizona.

NESTS OF WESTERN FLYCATCHER.

PLATE 3

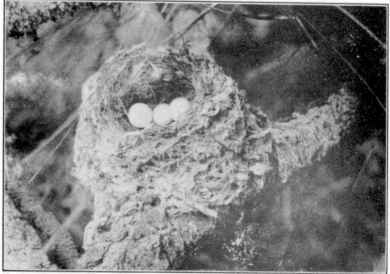

Huachuca Mountains, Ariz.   F. C. Willard.

NESTING TREE AND NEST OF BUFF-BREASTED FLYCATCHER.

PLATE 37

Huachuca Mountains, Ariz., May 26, 1922.   A. C. Bent.

Huachuca Mountains, Ariz.   F. C. Willard.

NESTING TREE AND NEST OF COUES'S FLYCATCHER.

PLATE 38

Hennepin County, Minn., July 22, 1935.

A. D. DuBois.

NEST OF EASTERN WOOD PEWEE.

PLATE 39

Hennepin County, Minn., August 1, 1935.

A. D. DuBois.

EASTERN WOOD PEWEE.

PLATE 40

Ithaca, N. Y.   A. A. Allen.

Adult and young.

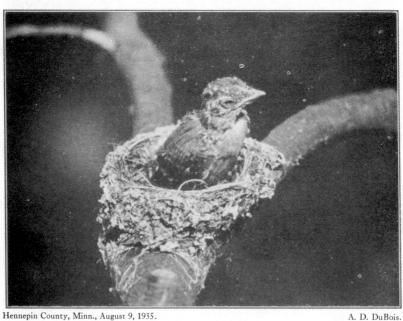

Hennepin County, Minn., August 9, 1935.   A. D. DuBois.

Young 12 days old.

EASTERN WOOD PEWEE,

PLATE 41

Steamboat Springs, Colo., July 28, 1911.   E. R. Warren.

Mono County, Calif., July 7, 1930.   J. S. Rowley.

JUVENILE AND NEST OF WESTERN WOOD PEWEE.

PLATE 42

Plymouth, Mass., June 8, 1901.   O. Durfee.

Plymouth, Mass., July 6, 1901.   A. C. Bent.

NESTING OF OLIVE-SIDED FLYCATCHER,

PLATE 43

Northumberland County, New Brunswick, June 22, 1925.   S. S. Dickey.

Old Forge, N. Y., June 28, 1920.   C. F. Stone.
                                                        Courtesy Verdi Burtch.

NESTS OF OLIVE-SIDED FLYCATCHER.

PLATE 44

J. S. Dixon.

Courtesy Museum of Vertebrate Zoology.

NESTING SITE AND NEST OF OLIVE-SIDED FLYCATCHER.

Berkeley, Calif., June 12, 1920.

PLATE 45

Arizona.

NEST OF VERMILION FLYCATCHER.

PLATE 46

A. M. Bailey.

Courtesy Colorado Museum of Natural History.

Nest of Northern Horned Lark.

Harrington Harbor, Quebec.

PLATE 47

Near Wolf Bay, Labrador, June 19.   A. A. Allen.

NEST OF NORTHERN HORNED LARK.

Teton County, Mont., June 21, 1915.   A. D. DuBois.

Young 7 days old.

DESERT HORNED LARK.

PLATE 48

Teton County, Mont., July 23, 1916.

A. D. DuBois.

MALE DESERT HORNED LARK.

PLATE 49

Gayle Pickwell.

Near Evanston, Ill.

NEST OF PRAIRIE HORNED LARK SHOWING ELABORATE "PAVEMENT."

PLATE 50

Near Milwaukee, Wis., April 16, 1937.

Peter Steig and Carl Kinzel.
Courtesy Murl Deusing.

Female incubating.

Near La Grange, Ill.

A. M. Bailey.
Courtesy Colorado Museum of Natural History.

Male at nest.

PRAIRIE HORNED LARK.

PLATE 51

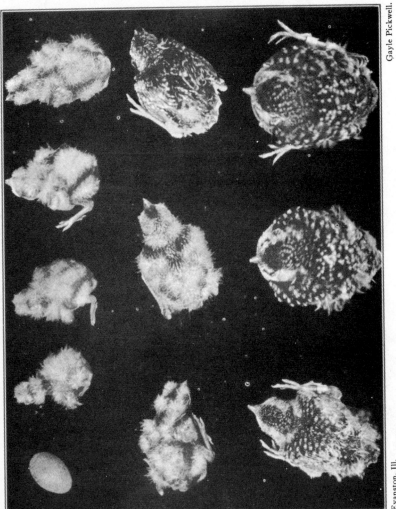

Evanston, Ill.   Gayle Pickwell.

GROWTH OF YOUNG PRAIRIE HORNED LARKS FROM EGG TO LEAVING NEST.

PLATE 52

Knox County, Ill., June 18, 1938.   H. M. Holland.

Nest in a cornfield.

Near St. Paul, Minn., April 1, 1929.   S. A. Grimes.

Well-made nest.

PRAIRIE HORNED LARK.

PLATE 53

Cochise County, Ariz., May 29, 1936.   J. S. Rowley.

NEST OF SCORCHED HORNED LARK.

Kern County, Calif.   W. M. Pierce.

NEST OF CALIFORNIA HORNED LARK.

PLATE 54

Crested Butte, Colo., June 11, 1900.   E. R. Warren.

Nesting site.

L. L. Haskin.

Adult male.

VIOLET-GREEN SWALLOW.

PLATE 55

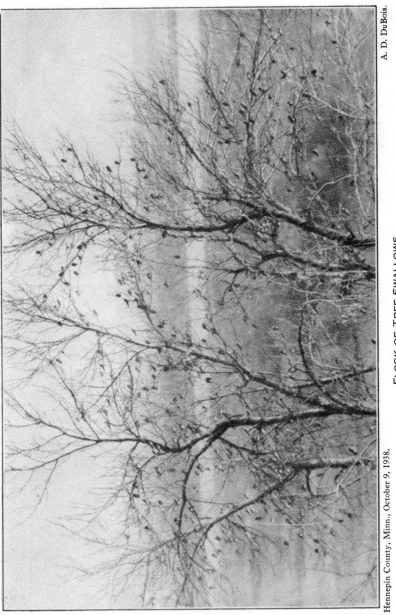

A. D. DuBois.

FLOCK OF TREE SWALLOWS.

Hennepin County, Minn., October 9, 1938.

PLATE 56

Hennepin County, Minn., June 30, 1936.

A. D. DuBois.

Riverside County, Calif., April 20, 1916.

W. M. Pierce.

ADULT TREE SWALLOW AND NEST.

PLATE 57

Branchport, N. Y., May 28, 1916.   Verdi Burtch.

Hennepin County, Minn., May 30, 1938.   A. D. DuBois.

BOX NESTING OF TREE SWALLOW.

PLATE 58

Topsham, Maine, June 2, 1932.

A. O. Gross.

Nesting site.

Bear Creek, Oreg., June 20, 1925.

J. E. Patterson.

Excavated nest.

BANK SWALLOW.

PLATE 59

Near their burrows.

Ithaca, N. Y.

A. A. Allen.

Adult.

BANK SWALLOWS.

PLATE 60

FALL GATHERING OF SWALLOWS.

PLATE 61

Littleton, W. Va., May 24, 1936.   S. S. Dickey.

Nest taken from a drain pipe in a retaining wall.

St. Marys County, Md., May 15, 1938.   R. F. Mason, Jr.

Nesting site in old kingfisher's hole.

ROUGH-WINGED SWALLOW.

PLATE 62

Adult and young.

Ithaca, N. Y.

A. A. Allen.

Adult.

ROUGH-WINGED SWALLOWS.

PLATE 63

A. D. DuBois.

Hennepin County, Minn., June 22, 1933.

Monongalia County, W. Va., June 10, 1936.

S. S. Dickey.

NEST OF, AND ADULT BARN SWALLOW.

PLATE 64

Branchport, N. Y. Adult feeding young. Young in nest. Verdi Burtch.

BARN SWALLOWS

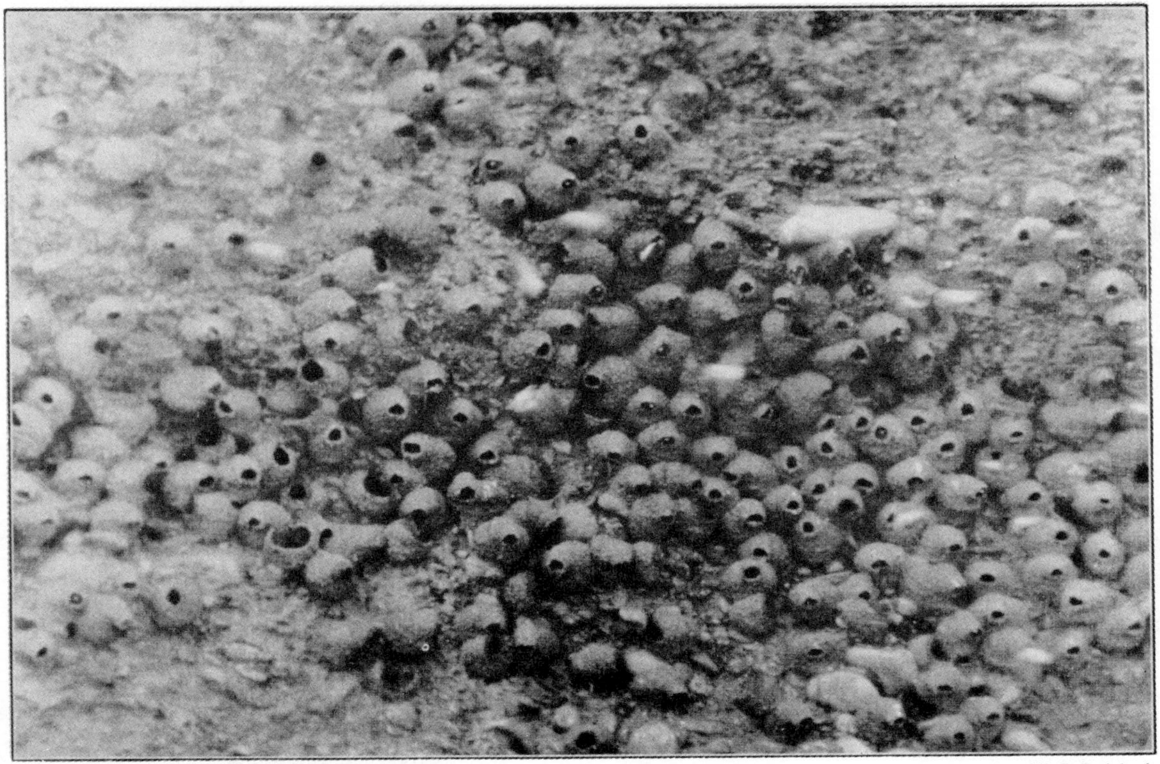

PLATE 65

CLIFF NESTS OF NORTHERN CLIFF SWALLOW.

Calaveras County, Calif., April 9, 1937.

PLATE 66

Kent Island, New Brunswick, July 15, 1932.   A. O. Gross.

Adult building nest.

Blue Island, Ill.
A. M. Bailey.
Courtesy Colorado Museum of Natural History.

Nests on a building.

NORTHERN CLIFF SWALLOW.

PLATE 67

San Bernardino Mountains, Calif., June 25, 1916.

W. M. Pierce.

CLIFF-SWALLOW NESTS ON A PINE TREE.

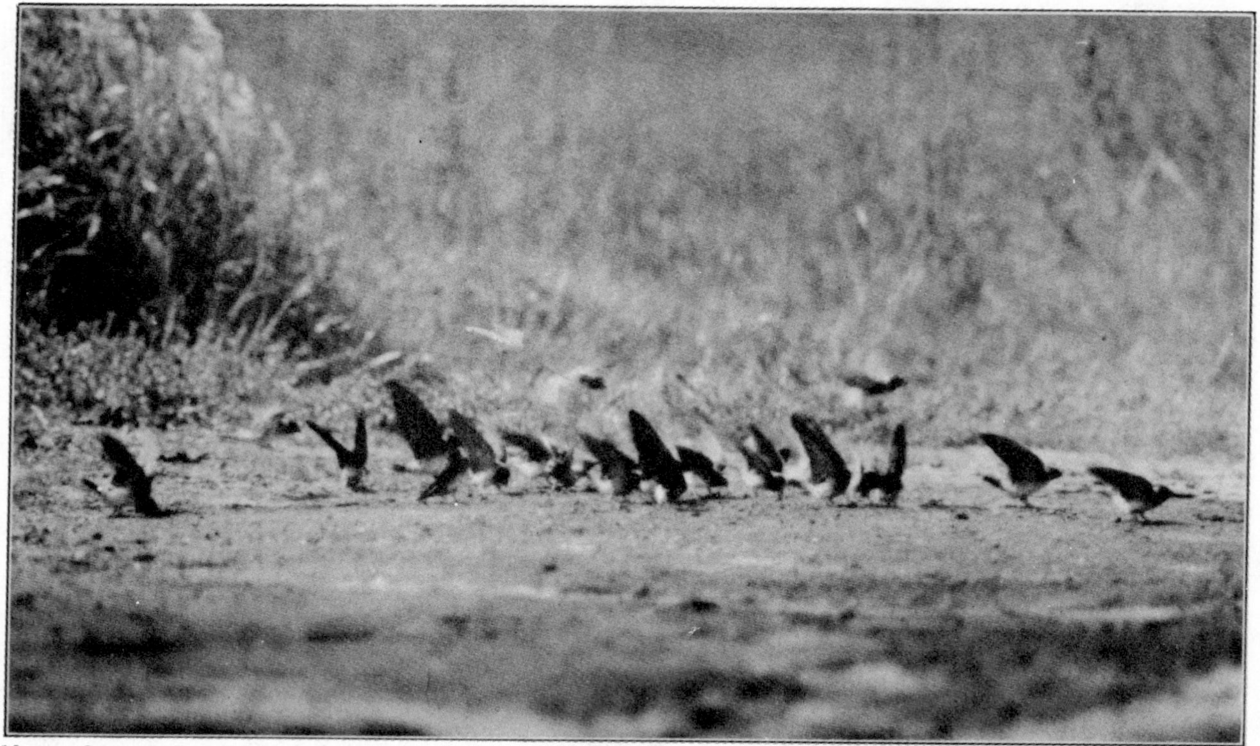

Monterey County, Calif., April 23, 1938. Laidlaw Williams.

CLIFF SWALLOWS GATHERING MUD.

PLATE 69

Duval County, Fla., June 1935.

S. A. Grimes.

GOURD NESTS OF PURPLE MARTINS.

PLATE 70

Jacksonville, Fla., May 30, 1930

S. A. Grimes.

Jacksonville, Fla., May 2, 1930.

S. A. Grimes.

BOX NESTS OF PURPLE MARTINS.

A CATALOGUE OF SELECTED DOVER BOOKS
IN ALL FIELDS OF INTEREST

A CATALOGUE OF SELECTED DOVER BOOKS IN ALL FIELDS OF INTEREST

AMERICA'S OLD MASTERS, James T. Flexner. Four men emerged unexpectedly from provincial 18th century America to leadership in European art: Benjamin West, J. S. Copley, C. R. Peale, Gilbert Stuart. Brilliant coverage of lives and contributions. Revised, 1967 edition. 69 plates. 365pp. of text.

21806-6 Paperbound $3.00

FIRST FLOWERS OF OUR WILDERNESS: AMERICAN PAINTING, THE COLONIAL PERIOD, James T. Flexner. Painters, and regional painting traditions from earliest Colonial times up to the emergence of Copley, West and Peale Sr., Foster, Gustavus Hesselius, Feke, John Smibert and many anonymous painters in the primitive manner. Engaging presentation, with 162 illustrations. xxii + 368pp.

22180-6 Paperbound $3.50

THE LIGHT OF DISTANT SKIES: AMERICAN PAINTING, 1760-1835, James T. Flexner. The great generation of early American painters goes to Europe to learn and to teach: West, Copley, Gilbert Stuart and others. Allston, Trumbull, Morse; also contemporary American painters—primitives, derivatives, academics—who remained in America. 102 illustrations. xiii + 306pp. 22179-2 Paperbound $3.50

A HISTORY OF THE RISE AND PROGRESS OF THE ARTS OF DESIGN IN THE UNITED STATES, William Dunlap. Much the richest mine of information on early American painters, sculptors, architects, engravers, miniaturists, etc. The only source of information for scores of artists, the major primary source for many others. Unabridged reprint of rare original 1834 edition, with new introduction by James T. Flexner, and 394 new illustrations. Edited by Rita Weiss. 6⅝ x 9⅝.

21695-0, 21696-9, 21697-7 Three volumes, Paperbound $13.50

EPOCHS OF CHINESE AND JAPANESE ART, Ernest F. Fenollosa. From primitive Chinese art to the 20th century, thorough history, explanation of every important art period and form, including Japanese woodcuts; main stress on China and Japan, but Tibet, Korea also included. Still unexcelled for its detailed, rich coverage of cultural background, aesthetic elements, diffusion studies, particularly of the historical period. 2nd, 1913 edition. 242 illustrations. lii + 439pp. of text.

20364-6, 20365-4 Two volumes, Paperbound $6.00

THE GENTLE ART OF MAKING ENEMIES, James A. M. Whistler. Greatest wit of his day deflates Oscar Wilde, Ruskin, Swinburne; strikes back at inane critics, exhibitions, art journalism; aesthetics of impressionist revolution in most striking form. Highly readable classic by great painter. Reproduction of edition designed by Whistler. Introduction by Alfred Werner. xxxvi + 334pp.

21875-9 Paperbound $2.50

VISUAL ILLUSIONS: THEIR CAUSES, CHARACTERISTICS, AND APPLICATIONS, Matthew Luckiesh. Thorough description and discussion of optical illusion, geometric and perspective, particularly; size and shape distortions, illusions of color, of motion; natural illusions; use of illusion in art and magic, industry, etc. Most useful today with op art, also for classical art. Scores of effects illustrated. Introduction by William H. Ittleson. 100 illustrations. xxi + 252pp.
21530-X Paperbound $2.00

A HANDBOOK OF ANATOMY FOR ART STUDENTS, Arthur Thomson. Thorough, virtually exhaustive coverage of skeletal structure, musculature, etc. Full text, supplemented by anatomical diagrams and drawings and by photographs of undraped figures. Unique in its comparison of male and female forms, pointing out differences of contour, texture, form. 211 figures, 40 drawings, 86 photographs. xx + 459pp. 5⅜ x 8⅜. 21163-0 Paperbound $3.50

150 MASTERPIECES OF DRAWING, Selected by Anthony Toney. Full page reproductions of drawings from the early 16th to the end of the 18th century, all beautifully reproduced: Rembrandt, Michelangelo, Dürer, Fragonard, Urs, Graf, Wouwerman, many others. First-rate browsing book, model book for artists. xviii + 150pp. 8⅜ x 11¼. 21032-4 Paperbound $2.50

THE LATER WORK OF AUBREY BEARDSLEY, Aubrey Beardsley. Exotic, erotic, ironic masterpieces in full maturity: Comedy Ballet, Venus and Tannhauser, Pierrot, Lysistrata, Rape of the Lock, Savoy material, Ali Baba, Volpone, etc. This material revolutionized the art world, and is still powerful, fresh, brilliant. With *The Early Work*, all Beardsley's finest work. 174 plates, 2 in color. xiv + 176pp. 8⅛ x 11. 21817-1 Paperbound $3.00

DRAWINGS OF REMBRANDT, Rembrandt van Rijn. Complete reproduction of fabulously rare edition by Lippmann and Hofstede de Groot, completely reedited, updated, improved by Prof. Seymour Slive, Fogg Museum. Portraits, Biblical sketches, landscapes, Oriental types, nudes, episodes from classical mythology—All Rembrandt's fertile genius. Also selection of drawings by his pupils and followers. "Stunning volumes," *Saturday Review*. 550 illustrations. lxxviii + 552pp. 9⅛ x 12¼. 21485-0, 21486-9 Two volumes, Paperbound $10.00

THE DISASTERS OF WAR, Francisco Goya. One of the masterpieces of Western civilization—83 etchings that record Goya's shattering, bitter reaction to the Napoleonic war that swept through Spain after the insurrection of 1808 and to war in general. Reprint of the first edition, with three additional plates from Boston's Museum of Fine Arts. All plates facsimile size. Introduction by Philip Hofer, Fogg Museum. v + 97pp. 9⅜ x 8¼. 21872-4 Paperbound $2.00

GRAPHIC WORKS OF ODILON REDON. Largest collection of Redon's graphic works ever assembled: 172 lithographs, 28 etchings and engravings, 9 drawings. These include some of his most famous works. All the plates from *Odilon Redon: oeuvre graphique complet,* plus additional plates. New introduction and caption translations by Alfred Werner. 209 illustrations. xxvii + 209pp. 9⅛ x 12¼.
21966-8 Paperbound $4.00

DESIGN BY ACCIDENT; A BOOK OF "ACCIDENTAL EFFECTS" FOR ARTISTS AND DESIGNERS, James F. O'Brien. Create your own unique, striking, imaginative effects by "controlled accident" interaction of materials: paints and lacquers, oil and water based paints, splatter, crackling materials, shatter, similar items. Everything you do will be different; first book on this limitless art, so useful to both fine artist and commercial artist. Full instructions. 192 plates showing "accidents," 8 in color. viii + 215pp. 8⅜ x 11¼. 21942-9 Paperbound $3.50

THE BOOK OF SIGNS, Rudolf Koch. Famed German type designer draws 493 beautiful symbols: religious, mystical, alchemical, imperial, property marks, runes, etc. Remarkable fusion of traditional and modern. Good for suggestions of timelessness, smartness, modernity. Text. vi + 104pp. 6⅛ x 9¼. 20162-7 Paperbound $1.25

HISTORY OF INDIAN AND INDONESIAN ART, Ananda K. Coomaraswamy. An unabridged republication of one of the finest books by a great scholar in Eastern art. Rich in descriptive material, history, social backgrounds; Sunga reliefs, Rajput paintings, Gupta temples, Burmese frescoes, textiles, jewelry, sculpture, etc. 400 photos. viii + 423pp. 6⅜ x 9¾. 21436-2 Paperbound $5.00

PRIMITIVE ART, Franz Boas. America's foremost anthropologist surveys textiles, ceramics, woodcarving, basketry, metalwork, etc.; patterns, technology, creation of symbols, style origins. All areas of world, but very full on Northwest Coast Indians. More than 350 illustrations of baskets, boxes, totem poles, weapons, etc. 378 pp. 20025-6 Paperbound $3.00

THE GENTLEMAN AND CABINET MAKER'S DIRECTOR, Thomas Chippendale. Full reprint (third edition, 1762) of most influential furniture book of all time, by master cabinetmaker. 200 plates, illustrating chairs, sofas, mirrors, tables, cabinets, plus 24 photographs of surviving pieces. Biographical introduction by N. Bienenstock. vi + 249pp. 9⅞ x 12¾. 21601-2 Paperbound $4.00

AMERICAN ANTIQUE FURNITURE, Edgar G. Miller, Jr. The basic coverage of all American furniture before 1840. Individual chapters cover type of furniture— clocks, tables, sideboards, etc.—chronologically, with inexhaustible wealth of data. More than 2100 photographs, all identified, commented on. Essential to all early American collectors. Introduction by H. E. Keyes. vi + 1106pp. 7⅞ x 10¾. 21599-7, 21600-4 Two volumes, Paperbound $11.00

PENNSYLVANIA DUTCH AMERICAN FOLK ART, Henry J. Kauffman. 279 photos, 28 drawings of tulipware, Fraktur script, painted tinware, toys, flowered furniture, quilts, samplers, hex signs, house interiors, etc. Full descriptive text. Excellent for tourist, rewarding for designer, collector. Map. 146pp. 7⅞ x 10¾. 21205-X Paperbound $2.50

EARLY NEW ENGLAND GRAVESTONE RUBBINGS, Edmund V. Gillon, Jr. 43 photographs, 226 carefully reproduced rubbings show heavily symbolic, sometimes macabre early gravestones, up to early 19th century. Remarkable early American primitive art, occasionally strikingly beautiful; always powerful. Text. xxvi + 207pp. 8⅜ x 11¼. 21380-3 Paperbound $3.50

ALPHABETS AND ORNAMENTS, Ernst Lehner. Well-known pictorial source for decorative alphabets, script examples, cartouches, frames, decorative title pages, calligraphic initials, borders, similar material. 14th to 19th century, mostly European. Useful in almost any graphic arts designing, varied styles. 750 illustrations. 256pp. 7 x 10.   21905-4 Paperbound $4.00

PAINTING: A CREATIVE APPROACH, Norman Colquhoun. For the beginner simple guide provides an instructive approach to painting: major stumbling blocks for beginner; overcoming them, technical points; paints and pigments; oil painting; watercolor and other media and color. New section on "plastic" paints. Glossary. Formerly *Paint Your Own Pictures*. 221pp.   22000-1 Paperbound $1.75

THE ENJOYMENT AND USE OF COLOR, Walter Sargent. Explanation of the relations between colors themselves and between colors in nature and art, including hundreds of little-known facts about color values, intensities, effects of high and low illumination, complementary colors. Many practical hints for painters, references to great masters. 7 color plates, 29 illustrations. x + 274pp.
20944-X Paperbound $2.75

THE NOTEBOOKS OF LEONARDO DA VINCI, compiled and edited by Jean Paul Richter. 1566 extracts from original manuscripts reveal the full range of Leonardo's versatile genius: all his writings on painting, sculpture, architecture, anatomy, astronomy, geography, topography, physiology, mining, music, etc., in both Italian and English, with 186 plates of manuscript pages and more than 500 additional drawings. Includes studies for the Last Supper, the lost Sforza monument, and other works. Total of xlvii + 866pp. 7⅞ x 10¾.
22572-0, 22573-9 Two volumes, Paperbound $10.00

MONTGOMERY WARD CATALOGUE OF 1895. Tea gowns, yards of flannel and pillow-case lace, stereoscopes, books of gospel hymns, the New Improved Singer Sewing Machine, side saddles, milk skimmers, straight-edged razors, high-button shoes, spittoons, and on and on . . . listing some 25,000 items, practically all illustrated. Essential to the shoppers of the 1890's, it is our truest record of the spirit of the period. Unaltered reprint of Issue No. 57, Spring and Summer 1895. Introduction by Boris Emmet. Innumerable illustrations. xiii + 624pp. 8½ x 11⅝.
22377-9 Paperbound $6.95

THE CRYSTAL PALACE EXHIBITION ILLUSTRATED CATALOGUE (LONDON, 1851). One of the wonders of the modern world—the Crystal Palace Exhibition in which all the nations of the civilized world exhibited their achievements in the arts and sciences—presented in an equally important illustrated catalogue. More than 1700 items pictured with accompanying text—ceramics, textiles, cast-iron work, carpets, pianos, sleds, razors, wall-papers, billiard tables, beehives, silverware and hundreds of other artifacts—represent the focal point of Victorian culture in the Western World. Probably the largest collection of Victorian decorative art ever assembled—indispensable for antiquarians and designers. Unabridged republication of the Art-Journal Catalogue of the Great Exhibition of 1851, with all terminal essays. New introduction by John Gloag, F.S.A. xxxiv + 426pp. 9 x 12.
22503-8 Paperbound $4.50

A HISTORY OF COSTUME, Carl Köhler. Definitive history, based on surviving pieces of clothing primarily, and paintings, statues, etc. secondarily. Highly readable text, supplemented by 594 illustrations of costumes of the ancient Mediterranean peoples, Greece and Rome, the Teutonic prehistoric period; costumes of the Middle Ages, Renaissance, Baroque, 18th and 19th centuries. Clear, measured patterns are provided for many clothing articles. Approach is practical throughout. Enlarged by Emma von Sichart. 464pp. 21030-8 Paperbound $3.50

ORIENTAL RUGS, ANTIQUE AND MODERN, Walter A. Hawley. A complete and authoritative treatise on the Oriental rug—where they are made, by whom and how, designs and symbols, characteristics in detail of the six major groups, how to distinguish them and how to buy them. Detailed technical data is provided on periods, weaves, warps, wefts, textures, sides, ends and knots, although no technical background is required for an understanding. 11 color plates, 80 halftones, 4 maps. vi + 320pp. 6⅛ x 9⅛. 22366-3 Paperbound $5.00

TEN BOOKS ON ARCHITECTURE, Vitruvius. By any standards the most important book on architecture ever written. Early Roman discussion of aesthetics of building, construction methods, orders, sites, and every other aspect of architecture has inspired, instructed architecture for about 2,000 years. Stands behind Palladio, Michelangelo, Bramante, Wren, countless others. Definitive Morris H. Morgan translation. 68 illustrations. xii + 331pp. 20645-9 Paperbound $3.00

THE FOUR BOOKS OF ARCHITECTURE, Andrea Palladio. Translated into every major Western European language in the two centuries following its publication in 1570, this has been one of the most influential books in the history of architecture. Complete reprint of the 1738 Isaac Ware edition. New introduction by Adolf Placzek, Columbia Univ. 216 plates. xxii + 110pp. of text. 9½ x 12¾.
21308-0 Clothbound $10.00

STICKS AND STONES: A STUDY OF AMERICAN ARCHITECTURE AND CIVILIZATION, Lewis Mumford.One of the great classics of American cultural history. American architecture from the medieval-inspired earliest forms to the early 20th century; evolution of structure and style, and reciprocal influences on environment. 21 photographic illustrations. 238pp. 20202-X Paperbound $2.00

THE AMERICAN BUILDER'S COMPANION, Asher Benjamin. The most widely used early 19th century architectural style and source book, for colonial up into Greek Revival periods. Extensive development of geometry of carpentering, construction of sashes, frames, doors, stairs; plans and elevations of domestic and other buildings. Hundreds of thousands of houses were built according to this book, now invaluable to historians, architects, restorers, etc. 1827 edition. 59 plates. 114pp. 7⅞ x 10¾.
22236-5 Paperbound $3.50

DUTCH HOUSES IN THE HUDSON VALLEY BEFORE 1776, Helen Wilkinson Reynolds. The standard survey of the Dutch colonial house and outbuildings, with constructional features, decoration, and local history associated with individual homesteads. Introduction by Franklin D. Roosevelt. Map. 150 illustrations. 469pp. 6⅝ x 9¼. 21469-9 Paperbound $4.00

THE ARCHITECTURE OF COUNTRY HOUSES, Andrew J. Downing. Together with Vaux's *Villas and Cottages* this is the basic book for Hudson River Gothic architecture of the middle Victorian period. Full, sound discussions of general aspects of housing, architecture, style, decoration, furnishing, together with scores of detailed house plans, illustrations of specific buildings, accompanied by full text. Perhaps the most influential single American architectural book. 1850 edition. Introduction by J. Stewart Johnson. 321 figures, 34 architectural designs. xvi + 560pp.

22003-6 Paperbound $4.00

LOST EXAMPLES OF COLONIAL ARCHITECTURE, John Mead Howells. Full-page photographs of buildings that have disappeared or been so altered as to be denatured, including many designed by major early American architects. 245 plates. xvii + 248pp. 7⅞ x 10¾.

21143-6 Paperbound $3.50

DOMESTIC ARCHITECTURE OF THE AMERICAN COLONIES AND OF THE EARLY REPUBLIC, Fiske Kimball. Foremost architect and restorer of Williamsburg and Monticello covers nearly 200 homes between 1620-1825. Architectural details, construction, style features, special fixtures, floor plans, etc. Generally considered finest work in its area. 219 illustrations of houses, doorways, windows, capital mantels. xx + 314pp. 7⅞ x 10¾.

21743-4 Paperbound $4.00

EARLY AMERICAN ROOMS: 1650-1858, edited by Russell Hawes Kettell. Tour of 12 rooms, each representative of a different era in American history and each furnished, decorated, designed and occupied in the style of the era. 72 plans and elevations, 8-page color section, etc., show fabrics, wall papers, arrangements, etc. Full descriptive text. xvii + 200pp. of text. 8⅜ x 11¼.

21633-0 Paperbound $5.00

THE FITZWILLIAM VIRGINAL BOOK, edited by J. Fuller Maitland and W. B. Squire. Full modern printing of famous early 17th-century ms. volume of 300 works by Morley, Byrd, Bull, Gibbons, etc. For piano or other modern keyboard instrument; easy to read format. xxxvi + 938pp. 8⅜ x 11.

21068-5, 21069-3 Two volumes, Paperbound $10.00

KEYBOARD MUSIC, Johann Sebastian Bach. Bach Gesellschaft edition. A rich selection of Bach's masterpieces for the harpsichord: the six English Suites, six French Suites, the six Partitas (Clavierübung part I), the Goldberg Variations (Clavierübung part IV), the fifteen Two-Part Inventions and the fifteen Three-Part Sinfonias. Clearly reproduced on large sheets with ample margins; eminently playable. vi + 312pp. 8⅛ x 11.

22360-4 Paperbound $5.00

THE MUSIC OF BACH: AN INTRODUCTION, Charles Sanford Terry. A fine, nontechnical introduction to Bach's music, both instrumental and vocal. Covers organ music, chamber music, passion music, other types. Analyzes themes, developments, innovations. x + 114pp.

21075-8 Paperbound $1.25

BEETHOVEN AND HIS NINE SYMPHONIES, Sir George Grove. Noted British musicologist provides best history, analysis, commentary on symphonies. Very thorough, rigorously accurate; necessary to both advanced student and amateur music lover. 436 musical passages. vii + 407 pp.

20334-4 Paperbound $2.75

JOHANN SEBASTIAN BACH, Philipp Spitta. One of the great classics of musicology, this definitive analysis of Bach's music (and life) has never been surpassed. Lucid, nontechnical analyses of hundreds of pieces (30 pages devoted to St. Matthew Passion, 26 to B Minor Mass). Also includes major analysis of 18th-century music. 450 musical examples. 40-page musical supplement. Total of xx + 1799pp.
(EUK) 22278-0, 22279-9 Two volumes, Clothbound $17.50

MOZART AND HIS PIANO CONCERTOS, Cuthbert Girdlestone. The only full-length study of an important area of Mozart's creativity. Provides detailed analyses of all 23 concertos, traces inspirational sources. 417 musical examples. Second edition. 509pp.
21271-8 Paperbound $3.50

THE PERFECT WAGNERITE: A COMMENTARY ON THE NIBLUNG'S RING, George Bernard Shaw. Brilliant and still relevant criticism in remarkable essays on Wagner's Ring cycle, Shaw's ideas on political and social ideology behind the plots, role of Leitmotifs, vocal requisites, etc. Prefaces. xxi + 136pp.
(USO) 21707-8 Paperbound $1.50

DON GIOVANNI, W. A. Mozart. Complete libretto, modern English translation; biographies of composer and librettist; accounts of early performances and critical reaction. Lavishly illustrated. All the material you need to understand and appreciate this great work. Dover Opera Guide and Libretto Series; translated and introduced by Ellen Bleiler. 92 illustrations. 209pp.
21134-7 Paperbound $2.00

HIGH FIDELITY SYSTEMS: A LAYMAN'S GUIDE, Roy F. Allison. All the basic information you need for setting up your own audio system: high fidelity and stereo record players, tape records, F.M. Connections, adjusting tone arm, cartridge, checking needle alignment, positioning speakers, phasing speakers, adjusting hums, trouble-shooting, maintenance, and similar topics. Enlarged 1965 edition. More than 50 charts, diagrams, photos. iv + 91pp.
21514-8 Paperbound $1.25

REPRODUCTION OF SOUND, Edgar Villchur. Thorough coverage for laymen of high fidelity systems, reproducing systems in general, needles, amplifiers, preamps, loudspeakers, feedback, explaining physical background. "A rare talent for making technicalities vividly comprehensible," R. Darrell, High Fidelity. 69 figures. iv + 92pp.
21515-6 Paperbound $1.25

HEAR ME TALKIN' TO YA: THE STORY OF JAZZ AS TOLD BY THE MEN WHO MADE IT, Nat Shapiro and Nat Hentoff. Louis Armstrong, Fats Waller, Jo Jones, Clarence Williams, Billy Holiday, Duke Ellington, Jelly Roll Morton and dozens of other jazz greats tell how it was in Chicago's South Side, New Orleans, depression Harlem and the modern West Coast as jazz was born and grew. xvi + 429pp.
21726-4 Paperbound $2.50

FABLES OF AESOP, translated by Sir Roger L'Estrange. A reproduction of the very rare 1931 Paris edition; a selection of the most interesting fables, together with 50 imaginative drawings by Alexander Calder. v + 128pp. 6½x9¼.
21780-9 Paperbound $1.50

AGAINST THE GRAIN (A REBOURS), Joris K. Huysmans. Filled with weird images, evidences of a bizarre imagination, exotic experiments with hallucinatory drugs, rich tastes and smells and the diversions of its sybarite hero Duc Jean des Esseintes, this classic novel pushed 19th-century literary decadence to its limits. Full unabridged edition. Do not confuse this with abridged editions generally sold. Introduction by Havelock Ellis. xlix + 206pp. 22190-3 Paperbound $2.00

VARIORUM SHAKESPEARE: HAMLET. Edited by Horace H. Furness; a landmark of American scholarship. Exhaustive footnotes and appendices treat all doubtful words and phrases, as well as suggested critical emendations throughout the play's history. First volume contains editor's own text, collated with all Quartos and Folios. Second volume contains full first Quarto, translations of Shakespeare's sources (Belleforest, and Saxo Grammaticus), Der Bestrafte Brudermord, and many essays on critical and historical points of interest by major authorities of past and present. Includes details of staging and costuming over the years. By far the best edition available for serious students of Shakespeare. Total of xx + 905pp. 21004-9, 21005-7, 2 volumes, Paperbound $7.00

A LIFE OF WILLIAM SHAKESPEARE, Sir Sidney Lee. This is the standard life of Shakespeare, summarizing everything known about Shakespeare and his plays. Incredibly rich in material, broad in coverage, clear and judicious, it has served thousands as the best introduction to Shakespeare. 1931 edition. 9 plates. xxix + 792pp. (USO) 21967-4 Paperbound $3.75

MASTERS OF THE DRAMA, John Gassner. Most comprehensive history of the drama in print, covering every tradition from Greeks to modern Europe and America, including India, Far East, etc. Covers more than 800 dramatists, 2000 plays, with biographical material, plot summaries, theatre history, criticism, etc. "Best of its kind in English," *New Republic.* 77 illustrations. xxii + 890pp. 20100-7 Clothbound $8.50

THE EVOLUTION OF THE ENGLISH LANGUAGE, George McKnight. The growth of English, from the 14th century to the present. Unusual, non-technical account presents basic information in very interesting form: sound shifts, change in grammar and syntax, vocabulary growth, similar topics. Abundantly illustrated with quotations. Formerly *Modern English in the Making.* xii + 590pp. 21932-1 Paperbound $3.50

AN ETYMOLOGICAL DICTIONARY OF MODERN ENGLISH, Ernest Weekley. Fullest, richest work of its sort, by foremost British lexicographer. Detailed word histories, including many colloquial and archaic words; extensive quotations. Do not confuse this with the Concise Etymological Dictionary, which is much abridged. Total of xxvii + 830pp. $6\frac{1}{2}$ x $9\frac{1}{4}$. 21873-2, 21874-0 Two volumes, Paperbound $6.00

FLATLAND: A ROMANCE OF MANY DIMENSIONS, E. A. Abbott. Classic of science-fiction explores ramifications of life in a two-dimensional world, and what happens when a three-dimensional being intrudes. Amusing reading, but also useful as introduction to thought about hyperspace. Introduction by Banesh Hoffmann. 16 illustrations. xx + 103pp. 20001-9 Paperbound $1.00

POEMS OF ANNE BRADSTREET, edited with an introduction by Robert Hutchinson. A new selection of poems by America's first poet and perhaps the first significant woman poet in the English language. 48 poems display her development in works of considerable variety—love poems, domestic poems, religious meditations, formal elegies, "quaternions," etc. Notes, bibliography. viii + 222pp.

22160-1 Paperbound $2.50

THREE GOTHIC NOVELS: THE CASTLE OF OTRANTO BY HORACE WALPOLE; VATHEK BY WILLIAM BECKFORD; THE VAMPYRE BY JOHN POLIDORI, WITH FRAGMENT OF A NOVEL BY LORD BYRON, edited by E. F. Bleiler. The first Gothic novel, by Walpole; the finest Oriental tale in English, by Beckford; powerful Romantic supernatural story in versions by Polidori and Byron. All extremely important in history of literature; all still exciting, packed with supernatural thrills, ghosts, haunted castles, magic, etc. xl + 291pp.

21232-7 Paperbound $2.50

THE BEST TALES OF HOFFMANN, E. T. A. Hoffmann. 10 of Hoffmann's most important stories, in modern re-editings of standard translations: Nutcracker and the King of Mice, Signor Formica, Automata, The Sandman, Rath Krespel, The Golden Flowerpot, Master Martin the Cooper, The Mines of Falun, The King's Betrothed, A New Year's Eve Adventure. 7 illustrations by Hoffmann. Edited by E. F. Bleiler. xxxix + 419pp. 21793-0 Paperbound $3.00

GHOST AND HORROR STORIES OF AMBROSE BIERCE, Ambrose Bierce. 23 strikingly modern stories of the horrors latent in the human mind: The Eyes of the Panther, The Damned Thing, An Occurrence at Owl Creek Bridge, An Inhabitant of Carcosa, etc., plus the dream-essay, Visions of the Night. Edited by E. F. Bleiler. xxii + 199pp. 20767-6 Paperbound $1.50

BEST GHOST STORIES OF J. S. LEFANU, J. Sheridan LeFanu. Finest stories by Victorian master often considered greatest supernatural writer of all. Carmilla, Green Tea, The Haunted Baronet, The Familiar, and 12 others. Most never before available in the U. S. A. Edited by E. F. Bleiler. 8 illustrations from Victorian publications. xvii + 467pp. 20415-4 Paperbound $3.00

MATHEMATICAL FOUNDATIONS OF INFORMATION THEORY, A. I. Khinchin. Comprehensive introduction to work of Shannon, McMillan, Feinstein and Khinchin, placing these investigations on a rigorous mathematical basis. Covers entropy concept in probability theory, uniqueness theorem, Shannon's inequality, ergodic sources, the E property, martingale concept, noise, Feinstein's fundamental lemma, Shanon's first and second theorems. Translated by R. A. Silverman and M. D. Friedman. iii + 120pp. 60434-9 Paperbound $1.75

SEVEN SCIENCE FICTION NOVELS, H. G. Wells. The standard collection of the great novels. Complete, unabridged. *First Men in the Moon, Island of Dr. Moreau, War of the Worlds, Food of the Gods, Invisible Man, Time Machine, In the Days of the Comet.* Not only science fiction fans, but every educated person owes it to himself to read these novels. 1015pp. (USO) 20264-X Clothbound $5.00

LAST AND FIRST MEN AND STAR MAKER, TWO SCIENCE FICTION NOVELS, Olaf Stapledon. Greatest future histories in science fiction. In the first, human intelligence is the "hero," through strange paths of evolution, interplanetary invasions, incredible technologies, near extinctions and reemergences. Star Maker describes the quest of a band of star rovers for intelligence itself, through time and space: weird inhuman civilizations, crustacean minds, symbiotic worlds, etc. Complete, unabridged. v + 438pp. (USO) 21962-3 Paperbound $2.50

THREE PROPHETIC NOVELS, H. G. WELLS. Stages of a consistently planned future for mankind. *When the Sleeper Wakes,* and *A Story of the Days to Come,* anticipate *Brave New World* and *1984,* in the 21st Century; *The Time Machine,* only complete version in print, shows farther future and the end of mankind. All show Wells's greatest gifts as storyteller and novelist. Edited by E. F. Bleiler. x + 335pp. (USO) 20605-X Paperbound $2.50

THE DEVIL'S DICTIONARY, Ambrose Bierce. America's own Oscar Wilde—Ambrose Bierce—offers his barbed iconoclastic wisdom in over 1,000 definitions hailed by H. L. Mencken as "some of the most gorgeous witticisms in the English language." 145pp. 20487-1 Paperbound $1.25

MAX AND MORITZ, Wilhelm Busch. Great children's classic, father of comic strip, of two bad boys, Max and Moritz. Also Ker and Plunk (Plisch und Plumm), Cat and Mouse, Deceitful Henry, Ice-Peter, The Boy and the Pipe, and five other pieces. Original German, with English translation. Edited by H. Arthur Klein; translations by various hands and H. Arthur Klein. vi + 216pp.
20181-3 Paperbound $2.00

PIGS IS PIGS AND OTHER FAVORITES, Ellis Parker Butler. The title story is one of the best humor short stories, as Mike Flannery obfuscates biology and English. Also included, That Pup of Murchison's, The Great American Pie Company, and Perkins of Portland. 14 illustrations. v + 109pp. 21532-6 Paperbound $1.25

THE PETERKIN PAPERS, Lucretia P. Hale. It takes genius to be as stupidly mad as the Peterkins, as they decide to become wise, celebrate the "Fourth," keep a cow, and otherwise strain the resources of the Lady from Philadelphia. Basic book of American humor. 153 illustrations. 219pp. 20794-3 Paperbound $1.50

PERRAULT'S FAIRY TALES, translated by A. E. Johnson and S. R. Littlewood, with 34 full-page illustrations by Gustave Doré. All the original Perrault stories—Cinderella, Sleeping Beauty, Bluebeard, Little Red Riding Hood, Puss in Boots, Tom Thumb, etc.—with their witty verse morals and the magnificent illustrations of Doré. One of the five or six great books of European fairy tales. viii + 117pp. 8⅛ x 11. 22311-6 Paperbound $2.00

OLD HUNGARIAN FAIRY TALES, Baroness Orczy. Favorites translated and adapted by author of the *Scarlet Pimpernel.* Eight fairy tales include "The Suitors of Princess Fire-Fly," "The Twin Hunchbacks," "Mr. Cuttlefish's Love Story," and "The Enchanted Cat." This little volume of magic and adventure will captivate children as it has for generations. 90 drawings by Montagu Barstow. 96pp.
22293-4 Paperbound $1.95

CATALOGUE OF DOVER BOOKS

THE RED FAIRY BOOK, Andrew Lang. Lang's color fairy books have long been children's favorites. This volume includes Rapunzel, Jack and the Bean-stalk and 35 other stories, familiar and unfamiliar. 4 plates, 93 illustrations x + 367pp.
21673-X Paperbound $2.50

THE BLUE FAIRY BOOK, Andrew Lang. Lang's tales come from all countries and all times. Here are 37 tales from Grimm, the Arabian Nights, Greek Mythology, and other fascinating sources. 8 plates, 130 illustrations. xi + 390pp.
21437-0 Paperbound $2.50

HOUSEHOLD STORIES BY THE BROTHERS GRIMM. Classic English-language edition of the well-known tales — Rumpelstiltskin, Snow White, Hansel and Gretel, The Twelve Brothers, Faithful John, Rapunzel, Tom Thumb (52 stories in all). Translated into simple, straightforward English by Lucy Crane. Ornamented with head-pieces, vignettes, elaborate decorative initials and a dozen full-page illustrations by Walter Crane. x + 269pp.
21080-4 Paperbound $2.00

THE MERRY ADVENTURES OF ROBIN HOOD, Howard Pyle. The finest modern versions of the traditional ballads and tales about the great English outlaw. Howard Pyle's complete prose version, with every word, every illustration of the first edition. Do not confuse this facsimile of the original (1883) with modern editions that change text or illustrations. 23 plates plus many page decorations. xxii + 296pp.
22043-5 Paperbound $2.50

THE STORY OF KING ARTHUR AND HIS KNIGHTS, Howard Pyle. The finest children's version of the life of King Arthur; brilliantly retold by Pyle, with 48 of his most imaginative illustrations. xviii + 313pp. 6⅛ x 9¼.
21445-1 Paperbound $2.50

THE WONDERFUL WIZARD OF OZ, L. Frank Baum. America's finest children's book in facsimile of first edition with all Denslow illustrations in full color. The edition a child should have. Introduction by Martin Gardner. 23 color plates, scores of drawings. iv + 267pp.
20691-2 Paperbound $2.50

THE MARVELOUS LAND OF OZ, L. Frank Baum. The second Oz book, every bit as imaginative as the Wizard. The hero is a boy named Tip, but the Scarecrow and the Tin Woodman are back, as is the Oz magic. 16 color plates, 120 drawings by John R. Neill. 287pp.
20692-0 Paperbound $2.50

THE MAGICAL MONARCH OF MO, L. Frank Baum. Remarkable adventures in a land even stranger than Oz. The best of Baum's books not in the Oz series. 15 color plates and dozens of drawings by Frank Verbeck. xviii + 237pp.
21892-9 Paperbound $2.25

THE BAD CHILD'S BOOK OF BEASTS, MORE BEASTS FOR WORSE CHILDREN, A MORAL ALPHABET, Hilaire Belloc. Three complete humor classics in one volume. Be kind to the frog, and do not call him names . . . and 28 other whimsical animals. Familiar favorites and some not so well known. Illustrated by Basil Blackwell. 156pp.
(USO) 20749-8 Paperbound $1.50

EAST O' THE SUN AND WEST O' THE MOON, George W. Dasent. Considered the best of all translations of these Norwegian folk tales, this collection has been enjoyed by generations of children (and folklorists too). Includes True and Untrue, Why the Sea is Salt, East O' the Sun and West O' the Moon, Why the Bear is Stumpy-Tailed, Boots and the Troll, The Cock and the Hen, Rich Peter the Pedlar, and 52 more. The only edition with all 59 tales. 77 illustrations by Erik Werenskiold and Theodor Kittelsen. xv + 418pp. 22521-6 Paperbound $3.50

GOOPS AND HOW TO BE THEM, Gelett Burgess. Classic of tongue-in-cheek humor, masquerading as etiquette book. 87 verses, twice as many cartoons, show mischievous Goops as they demonstrate to children virtues of table manners, neatness, courtesy, etc. Favorite for generations. viii + 88pp. 6½ x 9¼.
22233-0 Paperbound $1.25

ALICE'S ADVENTURES UNDER GROUND, Lewis Carroll. The first version, quite different from the final *Alice in Wonderland,* printed out by Carroll himself with his own illustrations. Complete facsimile of the "million dollar" manuscript Carroll gave to Alice Liddell in 1864. Introduction by Martin Gardner. viii + 96pp. Title and dedication pages in color. 21482-6 Paperbound $1.25

THE BROWNIES, THEIR BOOK, Palmer Cox. Small as mice, cunning as foxes, exuberant and full of mischief, the Brownies go to the zoo, toy shop, seashore, circus, etc., in 24 verse adventures and 266 illustrations. Long a favorite, since their first appearance in St. Nicholas Magazine. xi + 144pp. 6⅝ x 9¼.
21265-3 Paperbound $1.75

SONGS OF CHILDHOOD, Walter De La Mare. Published (under the pseudonym Walter Ramal) when De La Mare was only 29, this charming collection has long been a favorite children's book. A facsimile of the first edition in paper, the 47 poems capture the simplicity of the nursery rhyme and the ballad, including such lyrics as I Met Eve, Tartary, The Silver Penny. vii + 106pp. (USO) 21972-0 Paperbound
$1.25

THE COMPLETE NONSENSE OF EDWARD LEAR, Edward Lear. The finest 19th-century humorist-cartoonist in full: all nonsense limericks, zany alphabets, Owl and Pussycat, songs, nonsense botany, and more than 500 illustrations by Lear himself. Edited by Holbrook Jackson. xxix + 287pp. (USO) 20167-8 Paperbound $2.00

BILLY WHISKERS: THE AUTOBIOGRAPHY OF A GOAT, Frances Trego Montgomery. A favorite of children since the early 20th century, here are the escapades of that rambunctious, irresistible and mischievous goat—Billy Whiskers. Much in the spirit of *Peck's Bad Boy,* this is a book that children never tire of reading or hearing. All the original familiar illustrations by W. H. Fry are included: 6 color plates, 18 black and white drawings. 159pp. 22345-0 Paperbound $2.00

MOTHER GOOSE MELODIES. Faithful republication of the fabulously rare Munroe and Francis "copyright 1833" Boston edition—the most important Mother Goose collection, usually referred to as the "original." Familiar rhymes plus many rare ones, with wonderful old woodcut illustrations. Edited by E. F. Bleiler. 128pp. 4½ x 6⅜. 22577-1 Paperbound $1.00

Two Little Savages; Being the Adventures of Two Boys Who Lived as Indians and What They Learned, Ernest Thompson Seton. Great classic of nature and boyhood provides a vast range of woodlore in most palatable form, a genuinely entertaining story. Two farm boys build a teepee in woods and live in it for a month, working out Indian solutions to living problems, star lore, birds and animals, plants, etc. 293 illustrations. vii + 286pp.

20985-7 Paperbound $2.50

Peter Piper's Practical Principles of Plain & Perfect Pronunciation. Alliterative jingles and tongue-twisters of surprising charm, that made their first appearance in America about 1830. Republished in full with the spirited woodcut illustrations from this earliest American edition. 32pp. 4½ x 6⅜.

22560-7 Paperbound $1.00

Science Experiments and Amusements for Children, Charles Vivian. 73 easy experiments, requiring only materials found at home or easily available, such as candles, coins, steel wool, etc.; illustrate basic phenomena like vacuum, simple chemical reaction, etc. All safe. Modern, well-planned. Formerly *Science Games for Children.* 102 photos, numerous drawings. 96pp. 6⅛ x 9¼.

21856-2 Paperbound $1.25

An Introduction to Chess Moves and Tactics Simply Explained, Leonard Barden. Informal intermediate introduction, quite strong in explaining reasons for moves. Covers basic material, tactics, important openings, traps, positional play in middle game, end game. Attempts to isolate patterns and recurrent configurations. Formerly *Chess.* 58 figures. 102pp. (USO) 21210-6 Paperbound $1.25

Lasker's Manual of Chess, Dr. Emanuel Lasker. Lasker was not only one of the five great World Champions, he was also one of the ablest expositors, theorists, and analysts. In many ways, his Manual, permeated with his philosophy of battle, filled with keen insights, is one of the greatest works ever written on chess. Filled with analyzed games by the great players. A single-volume library that will profit almost any chess player, beginner or master. 308 diagrams. xli x 349pp.

20640-8 Paperbound $2.75

The Master Book of Mathematical Recreations, Fred Schuh. In opinion of many the finest work ever prepared on mathematical puzzles, stunts, recreations; exhaustively thorough explanations of mathematics involved, analysis of effects, citation of puzzles and games. Mathematics involved is elementary. Translated by F. Göbel. 194 figures. xxiv + 430pp. 22134-2 Paperbound $3.00

Mathematics, Magic and Mystery, Martin Gardner. Puzzle editor for Scientific American explains mathematics behind various mystifying tricks: card tricks, stage "mind reading," coin and match tricks, counting out games, geometric dissections, etc. Probability sets, theory of numbers clearly explained. Also provides more than 400 tricks, guaranteed to work, that you can do. 135 illustrations. xii + 176pp.

20335-2 Paperbound $1.50

MATHEMATICAL PUZZLES FOR BEGINNERS AND ENTHUSIASTS, Geoffrey Mott-Smith. 189 puzzles from easy to difficult—involving arithmetic, logic, algebra, properties of digits, probability, etc.—for enjoyment and mental stimulus. Explanation of mathematical principles behind the puzzles. 135 illustrations. viii + 248pp.
20198-8 Paperbound $1.75

PAPER FOLDING FOR BEGINNERS, William D. Murray and Francis J. Rigney. Easiest book on the market, clearest instructions on making interesting, beautiful origami. Sail boats, cups, roosters, frogs that move legs, bonbon boxes, standing birds, etc. 40 projects; more than 275 diagrams and photographs. 94pp.
20713-7 Paperbound $1.00

TRICKS AND GAMES ON THE POOL TABLE, Fred Herrmann. 79 tricks and games—some solitaires, some for two or more players, some competitive games—to entertain you between formal games. Mystifying shots and throws, unusual caroms, tricks involving such props as cork, coins, a hat, etc. Formerly *Fun on the Pool Table*. 77 figures. 95pp.
21814-7 Paperbound $1.00

HAND SHADOWS TO BE THROWN UPON THE WALL: A SERIES OF NOVEL AND AMUSING FIGURES FORMED BY THE HAND, Henry Bursill. Delightful picturebook from great-grandfather's day shows how to make 18 different hand shadows: a bird that flies, duck that quacks, dog that wags his tail, camel, goose, deer, boy, turtle, etc. Only book of its sort. vi + 33pp. 6½ x 9¼. 21779-5 Paperbound $1.00

WHITTLING AND WOODCARVING, E. J. Tangerman. 18th printing of best book on market. "If you can cut a potato you can carve" toys and puzzles, chains, chessmen, caricatures, masks, frames, woodcut blocks, surface patterns, much more. Information on tools, woods, techniques. Also goes into serious wood sculpture from Middle Ages to present, East and West. 464 photos, figures. x + 293pp.
20965-2 Paperbound $2.00

HISTORY OF PHILOSOPHY, Julián Marias. Possibly the clearest, most easily followed, best planned, most useful one-volume history of philosophy on the market; neither skimpy nor overfull. Full details on system of every major philosopher and dozens of less important thinkers from pre-Socratics up to Existentialism and later. Strong on many European figures usually omitted. Has gone through dozens of editions in Europe. 1966 edition, translated by Stanley Appelbaum and Clarence Strowbridge. xviii + 505pp. 21739-6 Paperbound $3.50

YOGA: A SCIENTIFIC EVALUATION, Kovoor T. Behanan. Scientific but non-technical study of physiological results of yoga exercises; done under auspices of Yale U. Relations to Indian thought, to psychoanalysis, etc. 16 photos. xxiii + 270pp.
20505-3 Paperbound $2.50

Prices subject to change without notice.
Available at your book dealer or write for free catalogue to Dept. GI, Dover Publications, Inc., 180 Varick St., N. Y., N. Y. 10014. Dover publishes more than 150 books each year on science, elementary and advanced mathematics, biology, music, art, literary history, social sciences and other areas.